Encapsulated
and
Powdered Foods

T0174820

FOOD SCIENCE AND TECHNOLOGY

A Series of Monographs, Textbooks, and Reference Books

Editorial Advisory Board

Encapsulated and Powdered Foods

edited by
Charles Onwulata

CRC Press
Taylor & Francis Group
Boca Raton London New York

CRC Press is an imprint of the
Taylor & Francis Group, an **informa** business
A TAYLOR & FRANCIS BOOK

CRC Press
Taylor & Francis Group
6000 Broken Sound Parkway NW, Suite 300
Boca Raton, FL 33487-2742

First issued in paperback 2019

ISBN-13: 978-0-8247-5327-6 (hbk)
ISBN-13: 978-0-367-39276-5 (pbk)

Library of Congress Card Number 2004065512

Library of Congress Cataloging-in-Publication Data

Encapsulated and powdered foods / edited by Charles Onwulata.
    p. cm. -- (Food science and technology ; 142)
  Includes bibliographical references and index.
  ISBN 0-8247-5327-5 (alk. paper)
  1. Food--Preservation. 2. Dried foods. I. Onwulata, Charles. II. Food science and technology (Marcel Dekker, Inc.) ; 142.

TP371.5.E53 2005
664'.02--dc22
                                                   2004065512

Visit the Taylor & Francis Web site at
http://www.taylorandfrancis.com

and the CRC Press Web site at
http://www.crcpress.com

Preface

A major problem facing researchers, students, food manufacturers, and users of food powders is the lack of a central source of information that provides both theoretical and practical knowledge focused on food powders. It is the goal of this book to provide fundamental, practical information that can be used as a reference by scientists and technologists alike.

Foods in their harvesting, processing, and manufacture are often converted from one form to another. Usually the conversion involves particle size reduction of large granules to small easily handled particles or drying wet slurries into powders using many unit operations. Adding to the complexity of food powders are novel processes used to create functional benefits to food powders such as longer storage, preservation of quality, delivery of content, event-specific triggering such as timed release, encapsulated powders. A volume that seeks to cover such a wide subject matter must be introductory in nature. As such, this book broadly touches the following areas:

- Fundamental and practical information on importance of food powders to the food industry. It highlights problems associated with the use of food powder, and addresses the difficulties associated with their use in product development.
- Characteristics of particulate foods. A discussion of the size, shape, distribution of particles, and the effects of processing on the powder's chemical and physical properties.
- Powder manufacturing processes. Topics on drying technologies, phase transitions during storage, crystallization and glass transition of food powders, and the role of moisture sorption in the development of stickiness are covered.
- Blending and segregation of powders. An in-depth discussion of the basic mechanisms of powder blending, segregation, and practical solutions to powder handling problems.
- Characterization of food powders. The effects of powder density, compressibility, hygroscopicity, and moisture content on flowability of bulk powders are discussed.
- Functional properties of particular food powders such as milk, cocoa, salts, and sugars, are included. The effect of ingredient composition on powder properties and their applications in food manufacturing are presented.
- Creation of specialty ingredients, such as encapsulated powders is discussed. Engineered powders, their measurement, and the impact of form on functionality of spray-dried encapsulated powders containing sensitive constituents are presented. Knowledge of manufacture, handling, and use of encapsulated food powders is provided.

This book is intended as a resource for scientists and technologists who must work with foods as powders, but do not have an in-depth knowledge of particle mechanics, physics, or chemistry that is needed. It is invaluable to those who would like to understand food powders and be able to characterize them and determine the effect of powder form in foods. Moreover, Resources are provided for resolving bulk powder handling issues such as how to break up lumps of powder and prevent bridging in the bin. *Encapsulated and Powdered Foods* is the result of the effort of the scientists and industry practitioners who have worked with food powders, and their contribution is greatly appreciated.

C. I. Onwulata

Contents

List of Contributors

Alan Baldwin
Fonterra Research Center
Palmerston North, New Zealand

Gustavo Barbosa-Cánovas
Biological Systems Engineering
Washington State University
Pullman, Washington

Bhesh R. Bhandari
University of Queensland
St. Lucia, Australia

A. A. Boateng
USDA-ARS
Eastern Regional Research Center
Wyndmoor, Pennsylvania

Scott A. Clement
Jenike & Johanson, Inc.
San Luis Obispo, California

Amanda Coder
Particle Sizing Systems
New Port Richey, Florida

Leanne de Muijnck
Cocoa Division
Archer Daniels Midland Company
Milwaukee, Wisconsin

John J. Fitzpatrick
Department of Process Engineering
University College
Cork, Ireland

Richard W. Hartel
University of Wisconsin
Madison, Wisconsin

Kerry Hasapidis
Particle Sizing Systems
New Port Richey, Florida

Heather Helsing
Particle Sizing Systems
Langhorne, Pennsylvania

Kerry Johanson
University of Florida
Gainesville, Florida

Pablo Juliano
Washington State University
Biological Systems Engineering
Pullman, Washington

Kievan Keogh
Dairy Products Research Center
Cork, Ireland

Marinelli, Joseph
Solids Handling Technologies
Fort Mill, South Carolina

Patrick O'Hagan
Particle Sizing Systems
New Port Richey, Florida

David Oldfield
Institute of Food,
   Nutrition and Human Health
   and Riddet Centre
Massey University
Palmerston North, New Zealand

C. I. Onwulata
U.S. Department of Agriculture
ARS, Eastern Regional Research Center
Wyndmoor, Pennsylvania

Enrique Ortega-Rivas
Food and Chemical
    Engineering Program
University of Chihuahua
Chihuahua, Mexico

David Pearce
Fonterra Research Center
Palmerston North, New Zealand

Micha Peleg
Department of Food Science
University of Massachusetts
Amherst, Massachusetts

Greg Pokrajac
Particle Sizing Systems
Langhorne, Pennsylvania

D. Poncelet
Département de Génie des Procédés
    Alimentaires
École Nationale d'Ingénieurs des
    Techniques des Industries
Agricoles et Alimentaires
Nantes, France

James K. Prescott
Jenike & Johanson, Inc.
Westford, Massachusetts

Harjinder Singh
Institute of Food, Nutrition and Human
    Health and Riddet Centre
Massey University
Palmerston North, New Zealand

E. Teunou
Département de Génie des Procédés
    Alimentaires
École Nationale d'Ingénieurs des
    Techniques des Industries Agricoles et
    Alimentaires
Nantes, France

Peggy M. Tomasula
Dairy Processing and Products Research
USDA-ARS ERRC
Wyndmoor, Pennsylvania

Food Powder Properties

1

The Role of Food Powders

Joseph Marinelli
Solids Handling Technologies
Caille Court
Fort Mill, SC

CONTENTS

This chapter discusses flowability issues, particularly with regard to bulk solids and food powders. It also covers current knowledge concerning the storage and flow of food powders in bins, feeders, and conveyors. Additionally, it discusses the future of this topic.

I.   NOT ALL FOOD POWDERS FLOW IN THE SAME WAY

Handling powders is required in all facets of food production. Quantities of ingredients are stored and handled in containers from 25 lb bags to 30 ft diameter silos. These ingredients are used in countless food products, and reliable powder flow is thus critical to the success of any production. Reliable handling of these products is not only critical for individual ingredients, but also for the final product. This chapter provides insight into typical food powder handling applications, problems, and solutions.

A.   Solids Handling Myths (Solids Are Similar to Fluids)

A common myth surrounding bulk solids flow is that a solid is comparable to a fluid. Solids are not fluids. A solid can maintain shear stresses, for example, by forming a pile. Fluids cannot form piles. Solids have cohesive strength whereas fluids do not. That is why a solid can form arches and ratholes.

   Another common misconception is that materials are alike, such as, flour is flour! Why should one worry about handling a particular flour? This is a common misconception not only with food products such as flour, but with all types of bulk solids. Moisture content, particle size, temperature, time of storage at rest, segregation of particles, etc. affect solids flowability.

B.   There is a Science to Bulk Solids Flow

Fortunately, there is a science to solids' flowability and with today's technology the guess-work is taken out of the bin and feeder design. The field of bulk solids handling was developed as a result of the work of Dr. Andrew W. Jenike, who pioneered the theory of bulk solids flow. Jenike developed a scientific approach to the storage and flow of bulk solids in the 1950s that is still relevant today. In fact, the Jenike approach to shear testing is now the ASTM standard [1] in the United States for determining the cohesive and wall friction properties of bulk solids and has been the standard in Europe for years.

C.   Problems Are Not Insurmountable

1.   Flow Problems

No discussion concerning solids flowability would be appropriate without first discussing flow problems that can occur during storage and handling of bulk solids. Predictable flow is often impeded by the formation of an arch or rathole, which can lead to erratic flow conditions.

   a. No Flow Due to Arching.   This condition usually occurs when flow is initiated by opening a gate or starting a feeder in order to discharge a product to a process, packaging operation, truck, etc. One of two problems occurs immediately: The first problem is the formation of an arch over the outlet preventing discharge of material. An arch (bridge, dome) is a blockage that is capable of supporting the entire contents of the bin or silo above (see Figure 1). Extreme methods may be required to initiate flow. Forces greater than that of

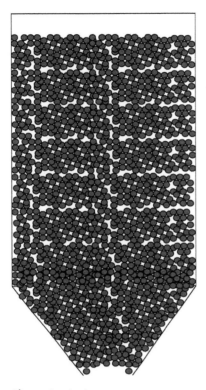

Figure 1 Arch.

gravity, such as sledgehammers, vibrators, and air blasters, are commonly used to overcome the arch and force material flow.

   b. No Flow due to Ratholing.   The second no-flow condition occurs when a preferential flow channel develops in a bin or silo. If the material being handled has some cohesive strength, the flow channel will empty and a stable pipe, commonly referred to as a rathole (pipe, core), will form as shown in Figure 2. In this case, the friction that typically develops between the solid and the hopper wall material is so great that instead of the material flowing along the hopper walls, it flows preferentially on itself. Some material discharges; however, the rest remains stagnant. When this happens, flow stops, usually requiring extreme measures to reinitiate it.

   c. Erratic Flow.   Erratic flow typically consists of a combination of a rathole and an arch occurring in the same bin. If flow is initiated and a stable rathole develops, using some flow aid device such as a sledgehammer or vibrator collapses the rathole and the material arches as it impacts the outlet. Flow has to be reinitiated until a rathole forms again and the scenario repeats itself. Erratic flow can affect solids discharge rates, bulk densities, and even the structural integrity of the bin.

   d. Flow Rate Limitation.   A flow rate limitation is affected by a material's ability to aerate or deaerate. Flow rate limitations typically develop because of a vacuum that is created as air is squeezed out of the voids of a solid as it fills and flows into the hopper. To satisfy this vacuum condition (pressure differential), air or gas from the outlet flows counter to the material being discharged, serving to limit its discharge rate. No matter what speed the feeder is set to, the rate may not increase because it is being impeded by countercurrent air or gas flow.

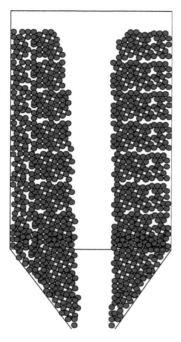

Figure 2   Rathole.

2.   *Handling Issues as a Result of Flow Problems*

The above flow problems have some serious consequences, some of which are listed here.

a. Limited Live Capacity.   The live storage capacity of a bin diminishes if a rathole develops. Only a small portion of the entire contents is actually live, while the rest of the material is left to stagnate in the bin. This results in a bin designed for an acceptable live capacity that is not capable of achieving that capacity.

b. Agglomeration, Spoilage.   Agglomeration can occur as a result of stagnated material that remains in the bin for days, weeks, months, or even years. Because this product cannot move in the bin it remains stagnant. Most materials gain cohesive strength after storage at rest and can pack or agglomerate, creating undesirable and troublesome products and handling problems for customers. Other products can spoil, become infested, or chemically react as they remain in the bin for long periods.

c. Shaking.   Shaking (vibration) occurs as ratholes and arches collapse in a bin. Imagine that a product has ratholed and that this rathole collapses either on its own or owing to some external force. The volume of material that impacts within the bin can cause significant vibrations. Eventually, this fatiguing may affect the structural integrity of the bin.

d. Structural Failure.   Structural failure unfortunately occurs in bins and silos much too often. Failure owing to vibrations is just one area of concern. A preferential flow channel can in fact expose the bin or silo to asymmetric loads. These loads can easily be great enough to cause dents in the silo sidewalls or even collapse the vessel. Figure 3 shows a picture of a silo that has failed because it was not designed to handle the loads applied to its walls. The hopper section dropped and the fast discharging material caused the cylinder to be sucked in.

e. Excessive Power.   Excessive power is required to operate the feeding device that controls discharge from the bin. If a material flows in a channel directly over the outlet, the

Figure 3 Silo failure.

feeder below is supporting this entire column of product. This can lead to high torque and large motor sizes being necessary to achieve the required discharge rates.

3. Flow Patterns in Bulk Storage

To avoid the above flow problems it is critical to understand the types of flow patterns that can develop in a bin or silo, and determine the material's flow properties (discussed later in the chapter). There are essentially two types of flow patterns that can develop: funnel flow and mass flow.

a. Funnel Flow.   Funnel flow occurs when some of the material in a bin moves while the rest remains stationary. The walls of the hopper section are not sufficiently steep or smooth enough to force the material to flow along them. The friction that develops between the hopper walls and material inhibits sliding, which results in the formation of a narrow flow channel, usually directly over the outlet. The first material that enters the bin is usually the last material out (first-in-last-out type of flow sequence).

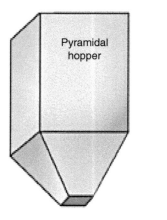 

Figure 4   Funnel flow hoppers.

Some examples of funnel flow bins are shown in Figure 4. Funnel flow bins are suitable for coarse, free-flowing materials that do not degrade, spoil, agglomerate, etc. They can also be used when segregation is not important. The major benefits of a funnel flow bin are reduced headroom requirements and lower fabrication costs. This usually means that shallow cones (60° or less), pyramidal hoppers, and flat-bottomed bins are used. This is why most funnel flow bins are designed: to save height and cost. Additionally, most railcars and bulk trucks will experience a funnel flow pattern. Railcars and hopper trucks have hoppers with very shallow slopes, which leads to funnel flow. The major benefit of funnel flow is low headroom; however, material flow problems usually far outweigh the benefits of low headroom.

b. Mass Flow.   Mass flow occurs when all the material in a bin is in motion whenever any of the material is withdrawn. The material slides along the hopper walls because they are steep and smooth enough to overcome the friction that develops between the wall surface and the bulk solid. The hopper outlet must be large enough to prevent arching (note: mass flow itself does not prevent arching). It is also important to note that stable ratholes will not form in mass flow bins. For these reasons, mass flow bins are suitable for cohesive solids, fine powders, materials that degrade or spoil, and solids that segregate.

Some examples of mass flow bins are shown in Figure 5. The flow sequence is first-in-first-out, which means that mass-flow bins can be used to store solids that develop problematic flow properties if they are allowed to stagnate. Particle segregation is minimized as the fines and coarse particles that have separated are reunited at the outlet. Typically, steep cones and wedge-shaped hoppers (chisels, transition hoppers, etc.) are used to ensure mass flow.[2]

II.   STATE OF THE ART IN FOOD POWDER HANDLING

A.   Measurement of Bulk Solids Flow Properties[4]

Solids flow property tests identify how a solid will flow and allow the design of a bin or hopper to provide reliable product flow. The material's cohesive and wall friction properties, as well as its compressibility, are determined. This scientific approach provides

Figure 5 Mass flow hoppers.

Figure 6 Jenike shear tester.

a complete understanding of how a material will flow in a new design, or why existing systems experience problems.

1. Cohesive Strength

Cohesive strength is measured using a bench scale laboratory testing device known as a direct shear tester (a Jenike shear tester is shown in Figure 6). This device is used to develop the material's "Flow Function" (strength/pressure relationship). Cohesive strength is measured as a function of applied consolidation pressure. In a laboratory, a sample of the material is placed in the shear cell (shown in Figure 7) on a direct shear tester, and both compressive and shear loads are applied to simulate flow conditions in a bin.

This test procedure allows the simulation of several other conditions that affect material flowability. The sample's moisture content and particle size are controlled, while the direct shear tester simulates the effects of temperature and duration of storage. It is very important to simulate the above conditions accurately.

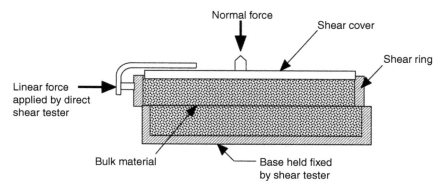

Figure 7 Jenike shear cell.

Once the material in the shear cell is consolidated, its strength is measured by shearing it to failure. By repeating this procedure under different conditions the resulting value of strength versus consolidating pressure can be determined. The process is fairly straightforward; however, it takes some time to simulate a range of pressures acting in a bin.

Several test points (between three and five test points) are developed for each level of pressure. These test points form a locus of points that indicates the strength of the material at a particular pressure level. Typically, three pressure levels are used to simulate the range of pressures that that are representative of the pressures that the material will experience in storage. This results in the generation of a material's flow function.

Once the material's flow function has been developed, the size of the opening that is required to prevent arching and/or ratholing can be determined. Jenike developed "flow factors" that are used to describe a bin's ability to break arches and ratholes. The relationship between the flow function and the flow factor predicts the point at which the strength of the material is overcome by the stresses applied to it by the bin. Simple equations are used to calculate the opening size required to fail the arch and allow the product to flow.

2.   Wall Friction

It is critical to know the required size of the opening to prevent arching and ratholing based on a material's cohesiveness. It is equally important to characterize the wall friction that develops between a solid and the wall surface. Once wall friction values are known, the hopper angles required to ensure mass flow (flow along the walls) are determined.

Wall friction values are expressed as a wall friction angle or coefficient of sliding friction. The lower the coefficient of sliding friction, the less steep the hopper walls need to be to ensure mass flow. The coefficient of sliding friction can be measured using the Jenike shear tester by determining the force required to slide a sample of the bulk solid across a stationary wall surface. The friction that develops between the wall surface and the bulk solid resists this force. A typical test setup is shown in Figure 8. A ring is placed on a sample of wall material and filled with the solid to be tested. As a normal force is applied to the material, the force needed to move the ring and solid is recorded.

Jenike developed design charts (given in Bulletin 123 of The University of Utah Engineering Station [3]) that allow conversion of a material's wall friction angle to the hopper angle required to ensure mass flow. For a given bulk material and wall surface, the wall friction angle is not necessarily a constant but often varies with normal pressure, usually decreasing as normal pressure increases. Oftentimes the tests indicate that at low

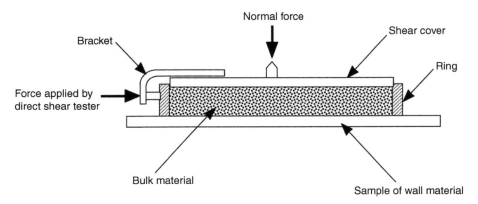

Figure 8   Wall friction test.

consolidation pressures, the wall friction angle is higher than at high pressures. This means that hopper angles typically need to be steeper at low pressures (like those that occur at the outlet of a bin).

3.   *Compressibility*

A material's compressibility is its bulk density/pressure relationship, in other words, the bulk density of the product as it is exposed to the range of pressures or solids' loading experienced due to flow in a bin. Sometimes, bulk density is expressed as loose density or packed density; however, there are more than two values of bulk density.

The information that is gathered during the compressibility test is used to determine the following:

Wall friction angle.   Bulk density values are used in the calculation of hopper wall pressures (the pressure exerted by the material normal to the hopper wall). These pressures are subsequently used to predict the wall friction angle that develops between the material and a sample of wall material.

Loads on walls.   Bulk density directly affects the pressures acting on the walls of a bin. The bulk density/pressure relationship is used in structural calculations to calculate forces acting within the bin.

Loads on feeders.   Knowing a material's range of density helps to predict the solids' loads that will be applied to any feeding device or gate attached to the outlet of the hopper.

Remember that material's moisture content, particle size, duration of storage, and temperature can affect cohesion, wall friction, and compressibility.

B.   Proven Design Technique

1.   *Solids Storage and Handling*

A gravity flow approach to solids handling is always best. Let gravity pull the material from the bin or silo. However, when gravity alone is not successful, flow aids are required; but, keep in mind that flow aid devices are expensive, noisy, require constant maintenance, and do not always work well whereas gravity powered discharge is efficient, cost effective, and reliable. Flow aid devices will be discussed later in this chapter.

2. Silos, Bins, and Hoppers

Many seem to think that 70° is the "magic angle" for mass flow. Remember that the hopper angle for mass flow is dependent on both the smoothness and steepness of the hopper walls and the properties of the bulk solid. Geometries such as cones and wedges will be discussed shortly.

a. Silo and Bin Loading.   When designing bins and hoppers for reliable flow, not only is knowledge of the material's flow properties and flow patterns important, but that of loads on the bin walls and dynamic effects is also critical. Silo failures can range from catastrophic structural collapse to denting of a steel shell. If a bulk solid other than the one for which the silo was designed is deposited in the silo, the structural integrity of the silo may be in jeopardy. Side discharge outlets put in a center discharge silo can impose asymmetric loading on the silo, which can cause failure. Another common problem is the development of mass flow in silos designed structurally for funnel flow. Mass flow loads are greater than those applied by the material in funnel flow. The structural integrity of the silo will be in jeopardy.

b. Cones versus Wedges.   Most hoppers are conical or wedge-shaped. Some hoppers are pyramidal and would be considered as conical from a flow standpoint. Wedge-shaped hoppers are more accommodating than conical hoppers and can handle materials with a wider range of flowability. There are several distinct advantages (and some disadvantages) of wedge-shaped hoppers over conical shaped hoppers; they are as follows:

Advantages of wedge versus cone geometry. Wedge hoppers require less steep hopper walls to ensure mass flow. Cones converge material in all 360° and as such, are required to be steep and smooth. Wedges, however, converge on only one dimension and can typically be about 11° less steep than cones and still promote flow along the walls — mass flow. As a matter of fact, the recommended angle can at times be exceeded without jeopardizing mass flow. This approach will obviously provide a hopper, which is not as tall as a cone, with significant headroom and potentially significant cost savings.

Wedge-shaped hoppers require a smaller outlet width than conical diameters to prevent bridging. For example, say let us assume the material is cohesive enough to require a 12 in. circular opening to prevent arching or bridging. If a wedge configuration is used (with a slotted opening), the slot opening has to be about 6 in. wide by about 18 in. long, to minimize end wall effects to prevent bridging with the same material. This will allow use of a smaller feeder.

Additionally, wedge-shaped hoppers will allow material to flow at a higher discharge rate, simply due to the increased cross-sectional area afforded by the slotted opening.

Disadvantages of wedge versus cone geometry. Wedge-shaped hoppers may cost more to fabricate than conical geometries.

The feeder used to discharge material must be capable of discharging it over the entire cross-sectional area of the slotted outlet. This requires a specially designed screw or belt for use with the hopper (feeders are discussed in the Feeder section, later in the chapter). Keep in mind that this feeder may be more expensive than the one used for a circular opening.

Consider also that this feeder will discharge at its end. If the material has to be discharged along the centerline of the bin, a conical configuration may be more advantageous.

3. Chutes

Oftentimes, it is necessary to transfer solids from the outlet of a bin to a process, truck, or another bin. Conveying material using equipment designed to mechanically transfer solids works quite well; however, for short distances, this can be an expensive approach.

Chutes can be used instead of expensive conveyors to transfer solids for short distances. A chute is simply a pipe or trough that is sized properly and at the correct slope angle to ensure smooth sliding of the material to be transferred from one device or process to another. A chute must be steep enough and smooth enough to ensure sliding along its entire length.

The impact of material on the chute is extremely important. Whether the material drops in free fall from a bin or is being transferred from another chute, impact pressures and the velocities necessary to keep the material moving along a chute are critical. Impact pressure can be calculated as follows:

$$\cong \gamma v^2 \sin^2 \theta / g$$

where σ is the impact pressure, psf, v the velocity before impact, ft/sec, γ the bulk density, pcf, θ impact angle of incoming stream, deg, and g is the acceleration due to gravity, ft/sec^2.

A test can be run in the laboratory to measure the critical chute angle resulting from impact pressure. This information can then be used to develop the minimum chute angle at the point at which it impacts the chute. The test involves placing a sample of solid on a surface representative of the surface to be used as the chute. A load is applied to the sample (using weights) and removed after a short time (say 20 sec) to simulate impact pressure. The surface is then tilted until the sample begins to slide. The angle of the slide is then recorded. This test is repeated several times to ensure accuracy and under a range of loads to simulate a range of impact pressures. A safety factor of about 5° is added to the results.

Other considerations when designing chutes are the chute cross-sectional area and chute wear. Do not converge chutes, because this change in the cross-sectional area will most likely cause the material to stop sliding. Wear can be overcome by using replacement wear liners or abrasion resistant steel. Keep in mind, however, that the rougher the surface, the steeper the angle required to keep material moving in the chute.

4. Feeders

There are several major advantages for using mass flow to provide uniform discharge of product. It is critical that a feeder be designed properly so as to maintain a mass flow pattern. A feeder is required to meter solids at a controlled rate from a bin to a process, truck, etc. It must be designed to:

- maintain uniform flow across the entire outlet cross-sectional area in order to ensure mass flow
- minimize solids loading to reduce feeder torque and motor requirements
- control discharge rate accurately.[5]

These requirements are accomplished by knowing the bulk solids' flow properties, and proper feeder design techniques.

Volumetric feeders discharge a volume of material as a function of time. Volumetric feeding is adequate for many solids' feeding applications. Feed accuracy in the range of 2 to 5% can be achieved with most volumetric designs. Volumetric feeding can become inaccurate if the bulk density of the solid that is being handled varies significantly. The feeder cannot recognize a density change because it simply discharges a certain volume of material, per unit time. Examples of the type of feeders to be discussed in this section are: screws, belts, and rotary valves.

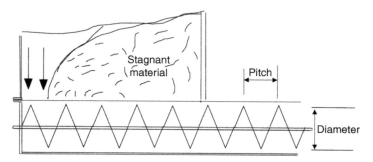

Figure 9 Constant pitch screw feeder.

a. Screw Feeders.   A screw feeder is composed of a series of flights that are welded onto a common shaft. The flights have a particular diameter and pitch (pitch is the distance between the flights). Some screw pitches are constant while others vary. If the screw has significant length, it is important to size the screw shaft to prevent deflection.

The common approach is to use a screw with a constant pitch. The configuration in Figure 9 shows a constant pitch and a constant diameter screw application. This approach typically causes a preferential flow channel to form at the back (over the first flight) of the screw as shown in the figure. This occurs because the first flight fills with material and there is no capacity within the remaining flights to take any more material, because the pitch is constant. This type of flow would obviously destroy a mass flow pattern and create a funnel flow pattern with the problems of ratholing, flooding, segregation, etc.

Screws are commonly used when feed from a slotted outlet is required. Screws are a good choice when: (1) an enclosed device is required, perhaps to handle dusty or toxic materials, (2) when space is restricted and, (3) when attrition (particle breakage) is not a problem.

A properly designed feeder, screw, and even belt (which will be discussed later), must ensure that material is withdrawn over the entire cross-sectional area of the outlet. The constant pitch/constant diameter screw shown in Figure 9 cannot do this.

For the material to be withdrawn over the entire cross-sectional area of the outlet, the flights of the screw are required to increase the capacity in the discharge direction. The approach shown in Figure 10, uses a combination of tapered shaft diameter section (with half pitch screw flights) and an increasing pitch section to accomplish this. This approach results in a uniform, mass flow pattern because the entire outlet cross-sectional area will remain alive.

b. Belt Feeders.   These are also used to feed over long and/or wide slotted openings. Typically belts are used to handle friable, coarse, fibrous, elastic, sticky, or very cohesive solids. They must withdraw material uniformly over the entire outlet.

A belt placed directly below a slotted outlet without any means of providing an increase in capacity can develop a preferential flow channel causing solids' compaction, premature belt wear, and power in excess of what is required to move the belt. A properly designed interface between the bin and belt is recommended to achieve uniform withdrawal over the entire outlet length. An approach that is commonly used is to design an interface as shown in Figure 11.

This interface allows the capacity of the belt to increase in the discharge direction. Note that the slot opening increases from back to front in plan view as well as elevation view. Note also that there is a slanted nose to provide stress relief as material is transferred to the discharge end. Flat or troughed idlers can be used to support and train the belt.

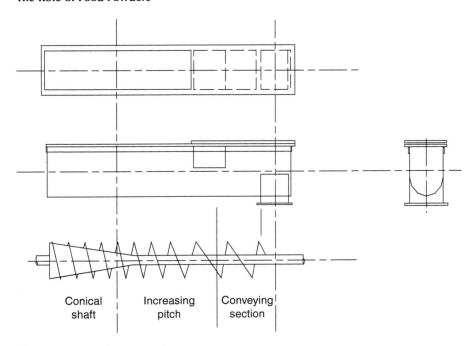

Figure 10   Mass flow screw feeder.

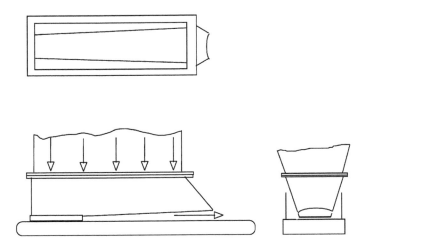

Figure 11   Belt feeder interface.

Troughed idlers will provide a more rigid belt. If flat belts are used, care must be taken to minimize belt sag between the idlers. Significant vibrations can develop as material rides over one idler and sags in between the next. A flat metal or plastic plate can be placed on the idlers for the belt to slide on and not sag. Additionally, the idlers can be closely spaced so as to prevent belt sag. Skirts are also used to prevent spillage and should be mounted in such a way that they expand slightly in the direction of belt travel so as not to interfere with material flow.

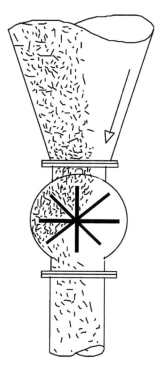 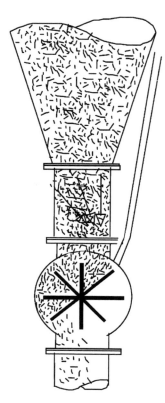

Figure 12   Rotary valve feeder.

c. Rotary Feeders.   These types of devices are commonly used for circular or square outlets that are widely used to discharge solids to pneumatic conveying systems because they are capable of sealing against a pressure gradient usually created by a pneumatic conveying system.

A common problem that exists with rotary valves is that they tend to pull material preferentially from one side of the hopper as the vanes are rotated to pick up solids. This can destroy a mass flow pattern, which is required for most bulk solids. Additionally, once the solid discharges from the vanes, the air that replaces it is effectively pumped back up into the bin. This countercurrent airflow can decrease flow rates and/or cause flooding problems.

A vertical section, as shown in Figure 12 can alleviate the preferential flow problems because the flow channel expands in this area usually opening up to the full outlet.

To be effective, the section's height should be about $1\frac{1}{2}$ to 2 times the diameter of the outlet. To rectify the countercurrent airflow problem, a vent line is required to take the displaced air away to a dust collector or at least back into the top of the bin. Because a rotary valve is a volumetric device, the feed rate is determined by the rotational speed of the vanes.

d. Vibratory Pan Feeders.   Pan feeders can feed material gently and accurately. Unfortunately, they are limited primarily to applications involving round or square outlets. If a long rectangular outlet is used, the feeder must operate across the short dimension of the slot.

e. Louvered Feeders.   A louvered type feeder is shown in Figure 13, and can control discharge from circular, square, or rectangular cross sections using the material's natural

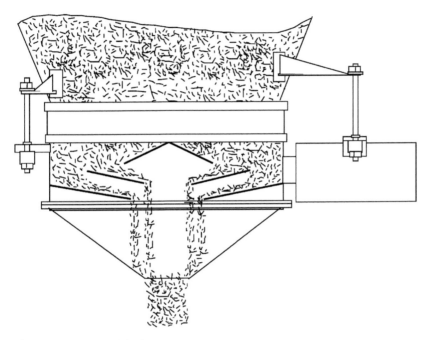

Figure 13 Louvered feeder.

angle of repose. When the drive is energized, the material's angle of repose is overcome and the material discharges usually very evenly and accurately through louvers that are angled and positioned to control material flow. When the drive is stopped, the material stops flowing. These types of devices can also be used to provide a stream or "curtain" of material if required.

5. Conveyors

A distinction must be made between "feeders" and "conveyors." A feeder is used to control discharge rate from a storage hopper or bin and is subjected to varying head loads from the hopper. A conveyor requires a regulated feed rate and must not operate under varying headload conditions.

Conveying systems have one thing in common: their primary purpose is to move product from point "A" to point "B," while completely containing the product within the conveying mechanism through the entire moving process. There should be no leaks or spills associated with the equipment. Conveying systems must be built in such a way that they do not trap product or ingredients and can be easily cleaned.

a. Pneumatic Conveying.   Materials are conveyed pneumatically in essentially two ways: by dilute phase and dense phase conveying. In dilute phase conveying the conveying gas volume and velocity is sufficient to keep the material that is being transported in suspension. This means that the material is being conveyed in a continuous manner, and is not accumulating at the bottom of the conveying line at any point. Dense phase conveying consists of moving plugs of material along the conveying line, more like separate trains on a common track. There is a gap between these plugs, keeping the length of each plug rather uniform and the pressure drop across each plug similar.

Dense phase.   Dense phase conveying is usually accomplished at low velocities (500–1000 ft/min), and high pressures (30–50 psi). Dense phase conveying can be accomplished with the use of many line-charging methods, such as rotary feeders, screw pumps, pressure tanks, etc. Boosters are sometimes required and their number and positioning depends on the physical characteristics of the material being conveyed. Testing is required to ensure reliable dense phase conveying system design.

Dilute phase.   Dilute phase conveying is generally accomplished at high velocities (4000–5000 ft/min) and low air pressure (10–12 psi). Dilute phase conveying can be done either in a vacuum mode or in a pressure mode. Because the bulk solids handled in pneumatic conveying systems are generally dry, the conveying pipes are manufactured from either mild steel or aluminum. Other components such as rotary lock feeders, diverters, etc., are fabricated with metals selected according to need.

b. Mechanical Conveying.

Belt conveyors.   Belt conveyors are built in many sizes and shapes. They move products of various shapes and forms to various points within the manufacturing process. They move bulk raw materials, process ingredients, and packaged finished products.

Belt conveyors used to move bulk solids are primarily trough-type, as opposed to flat-type. Trough-type belt conveyors are built to contain the product on the belt without spilling over the edges. The standard design for troughs is 20, 35, or 45°. Trough angle selection should be associated with the flow characteristics of the product.

Screw conveyors.   Screw conveyors are used to convey solid granular materials and can also be used as mixing conveyors. Screw conveyors are made in a variety of lengths and sizes. In outward appearance, they are all very similar. Nonetheless, screw augers take on different appearances depending on the primary use.

Construction materials will vary with the products being handled. In areas and operations where dry commodities or ingredients are handled, mild steel for all screw conveyor components is adequate. In other instances, where conveyors are operating in a wet processing area of the plant, but handling a dry product, construction might call for stainless steel trough and end plates. In other situations, where the conveyor is handling moist solids or sticky ingredients requiring a wet cleanup, all conveyor components should be made of stainless steel.

Vibratory conveyors.   Vibratory conveyors are an extremely useful means of handling powders and granular products. A vibratory conveyor consists of a carrying trough (deck), supporting legs or springs, and a drive mechanism (exciter). The drive system imparts an oscillating motion to the carrying trough of a specific frequency and amplitude. The bulk material on the carrying trough is moved along the trough by means of the periodic trough motion.

Various combinations of frequencies and amplitudes are used in the industry, depending on the application. If the bulk material has an extremely adhesive quality, then special trough coatings or linings together with large amplitudes may be used.

Bucket elevators.   Bucket elevators are capable of handling bulk materials in a vertical direction, either indoors or outdoors. These units are housed inside a steel casing that forms the support structure. The basic types are: centrifugal discharge with spaced buckets or continuous discharge with a continuous series of buckets. The first type scoops the bulk materials in the boot of the elevator into the buckets. However, this will generally cause severe degradation of the particles. The continuous elevator is fed by a feeder, which allows the bulk material to flow gently into the buckets, minimizing any degradation.

The carrying medium for the buckets can either be a belt or a chain. Stainless steel is used for units that operate in a wet environment or that require wet cleaning.

Hinged pivoted bucket conveyor-elevators. These devices are usually loaded at a horizontal section in the buckets by a feeder and then moved horizontally as well as vertically. The buckets are on hinges that allow them to pivot and remain horizontal.

"En-Masse" conveyor-elevators. The term "en-masse" has been used to differentiate between a totally enclosed type of handling a continuous flow through a casing, pipe, or duct as opposed to the previous types discussed, which for the most part are open, but when enclosed is for reasons such as dust, weather, or personnel safety.

The en masse conveyor-elevator system was first developed to convey bulk solids horizontally in a stationary casing by inducing movement with an inner strand of chain fitted at intervals with transverse bars. This formed a skeletonized moving base that would propel the bulk solids at the same speed en masse through the casing.

This technique was subsequently extended to also provide vertical movement in a stationary duct. In this case, the transverse bars are shaped like horseshoes. For elevating the bulk solids have to be introduced into the conveyor by gravity at the horizontal section. The principle of achieving en masse movement is fundamentally dependent on the material being moved and having a higher internal friction than the material's friction against the casing surface.

6. Flow Aids

Flow aid devices are internal or external devices that are used to assist in discharging materials from a bin. Typically, a flow aid device would be used when flow by gravity alone will not work or when it is impractical to provide the design that is required for reliable flow. However, flow aids are usually redundant and in fact, may actually cause flow problems.

a. Vibratory Bin Dischargers. Mechanical flow aid devices rely on internal devices to force material to flow. Probably the most commonly used device is the vibrating bin discharger (bin activator). A vibrating discharger can accommodate hopper openings from about 3 to 15 ft and is intended to keep material completely live over its entire cross-sectional area. This type of device (Figure 14) is hung from a storage bin by hangers and incorporates a rubber skirt to prevent leakage and isolate the bin from the vibrations. Vibration is transmitted through an outer shell and around an internal dome or cone shaped baffle, by a motor with eccentric weights. A cohesive bulk solid can be broken up and made to flow, depending on the amplitude of vibration applied.

There are several considerations when using or deciding to use a vibrating bin discharger:

1. It must discharge over its entire cross-sectional area and be operated on a regular basis, which usually requires it to be cycled on and off. If not, small preferential flow channels will form, effecting flow and potential thereby causing structural problems.
2. If the bin is flowing in a funnel flow pattern, the diameter of the discharger must be larger than the ratholing capability of the material.
3. A discharger cannot control flow rate, *it is not a feeder*, and as such requires a feeder to control discharge rate to the process, system, etc.

b. Vibrators. Vibrators can be mounted on the side of a bin or chute in an attempt to initiate flow. These vibrators can be air or electrically operated and come in all shapes and sizes. There are rotary, piston, turbine, linear, electromagnetic, eccentric, and several other more specific types of vibrators. Some are designed to provide high frequency, low amplitude vibration to a surface. Others are required to provide high amplitude vibrations, and still others are required to provide a "thump."

Figure 14   Vibrating bin discharger.

Vibrators should be used with caution, as follows:

1.   The material in the bin should not be pressure sensitive. If the material can be squeezed and forms a "snowball," it may pack in the bin due to vibration.
2.   Do not operate a vibrator unless the material has somewhere to go. The feeder must be operating or the gate must be open, or else packing will occur.
3.   Chutes are a good place to mount vibrators, as they will enhance flow down a shallow chute.
4.   Be aware of the vibrator's effect on the structural integrity of your bin.

c. Agitators.   Agitators can be used to break up packed material. These devices consist of horizontal or vertical shafted augers, which move material simply by brute force. Other types of agitation devices consist of vibrating screens, which are intended to shake material loose from the walls of a hopper. Some considerations to be aware of are:

1.   The resisting forces may be large enough to render these devices useless.
2.   The motors must be large enough to overcome these forces.
3.   If the device fails, it may be more difficult to empty the bin than without the device in place.

d. Air Blasters.   This type of flow aid device is intended to introduce a "blast" of high-pressure air, strategically aimed at a material arch, rathole, etc. in an attempt to break it up. The high-pressure air or gas is supplied by a cylinder that is mounted to the outside of a bin and typically refilled by plant air at 80 to 100 psi pressure. A quick acting valve is opened when the unit is fired, releasing the air directly at the arch to break it up. The unit is then quickly refilled and ready for the next firing. Several air blasters can be placed and timed to act throughout a large hopper to maintain flow.

Some points to be aware of when using air blasters:

1. These devices are probably the most successful flow aid devices marketed, as they are quite efficient in breaking arches.
2. If material is severely caked, the initial firing will likely break it up; however, the material may only break into large chunks that one may or may not be able to handle or that subsequent firings may not break up.
3. Use multiple units if a single unit does not have the arch-breaking capability that is needed.
4. Be aware that the introduction of high-pressure air into a bin adds to the loads applied to that bin. Be aware especially of the localized area of the bin wall around the blaster nozzle, as reinforcement may be required.

e. Air Slides, Sweeps, and Fluidizers.   Instead of letting material deaerate and pack in a bin, it is sometimes practical to fluidize it. There are several ways to do this such as: (1) air slides, which use an air plenum and porous fabric, (2) air sweeps that provide a pulse of air in a sweeping motion along a bin wall and, (3) fluidizing cones that mount at the outlet and provide fluidizing air through a perforated metal dish hat.

Fluidization of the entire contents of a bin will prevent arching and ratholing and even provide higher discharge rates than by gravity alone. Some points to consider are:

1. The material must not be cohesive or else channels will form.
2. The material's bulk density will be low and nonuniform.
3. Should the material be fluidized at all times or is it allowed to settle when the unit is not in use? Can it be easily refluidized?
4. The cost of fluidizing air may be prohibitive especially if the product is hygroscopic and the fluidizing air needs to be dried.

f. Flow Aid Additives   Typically, a flow aid additive is a solid or liquid that can be mixed with a difficult flowing product to produce an easier flowing product. One of the most popular flow aid additives is silica. Silica can be added in very small quantities and is extremely effective in breaking cohesive bonds. Other materials used to promote flow in cohesive materials are quicklime, clay, zeolite, freeze conditioning agents, fly ash, flour, and starch. This is just a partial list of possible additives that can be used to aid in material flowability.[6]

It is recommended that all means of achieving gravity flow be explored before deciding to use an additive. Selecting the correct flow aid additive is just the beginning in finding a solution to material flow problems. Consider the following: (1) What effect does the addition of a chemical have on product quality? (2) Is your product or process going to be negatively affected by use of a flow aid additive? (3) Perhaps the rheology of the product changes with the addition of, for instance, talc. This is probably the greatest concern when using flow aid additives. (4) How much additive should be used and will it be effective in reducing cohesion? How much additive is required to accomplish the task of converting a nonflowing, cohesive product to one that is easier to handle? This has to be determined through product testing at various concentrations of flow aid additive. (5) Is it going to be too costly? Some of the additives available today can be fairly expensive. Quantities and vendors play a major role in deciding whether you can use a flow aid additive. The cost of applying the additive also has to be considered if it is sprayed on. A gravimetric system may be required to control additive application accuracy. This system can be expensive. How much mixing is required for an additive to be effective? Some products accept an additive such as silica easily. Tumble blending for a short period of time is adequate for mixing the silica and cohesive product. Consider also that conveying the mixed product may in fact remove some additive, such as through a pneumatic conveying system.

7. Segregation Issues

Segregation of particles can occur with many bulk solids as they are being handled, and the results can be quite costly. The pharmaceutical industry is susceptible to problems of particle segregation. If fines and coarse particles segregate, tablet or capsule quality is affected potentially causing the discarding of valuable drugs.

There are several mechanisms that cause segregation. Among them are: sifting, particles' sliding on a surface, and air entrainment.

> *Sifting* occurs when small particles move through large particles. This is the most common means for particles to separate. For sifting to occur, the particles must be free flowing, of different particle size, fairly large (>100 mesh), and have some means of interparticle motion (such as forming a pile). The particles segregate in a horizontal or side-to-side pattern.
>
> *Particles sliding on a surface* can segregate because fine particles tend to be more frictional than the coarse ones. If a chute is used, the fine particles usually settle at the bottom of the chute due to sifting. Increased friction of the finer particles causes drag and velocity differences between particles as they slide on the chute surface. As the particles discharge from the chute, the fines usually concentrate at the end, while the coarse particles have a trajectory that carries them further away.
>
> *Air entrainment* affects fine particles, as they tend to remain airborne longer because they are less permeable to air than coarse or heavier particles. As a bin is being filled, the fine, light material tends to settle on top, while the coarse particles fall rapidly, creating a vertical or top-to-bottom segregation pattern.

Here are some ways to correct the above segregation problems:

- Minimize sifting by ensuring a mass (first-in-first-out) flow pattern. Even though the material has segregated side-to-side, the coarse and fine particles will be reunited at the outlet because of mass flow.
- Keep a head of material above the hopper section of a mass flow bin to minimize the velocity gradient that can occur as material reaches the hopper section of mass flow bins.
- Make the material more cohesive by adding water or oil. One of the prerequisites for segregation is interparticle motion, so increased cohesiveness causes the particles to stick together. Note: do not make the material too cohesive or it will not flow.
- Use a tangential entry into bins when handling a fine solid that segregates by air entrainment. Instead of filling the bin from the top right into the center of the bin, bring the material in from the top in a sideways motion (tangential to the bin). This will actually minimize the chances of a material segregating vertically by causing side-to-side segregation that can be overcome by the use of a mass flow pattern.

III. WHAT DOES THE FUTURE HOLD?

A. Need for Topic Recognition at the University Level

Presently, there are only a handful of colleges and universities in the United States that actually provide solids' handling courses as part of their undergraduate curriculum. There needs to be at least an introduction to the topic so that graduating engineers have a basic understanding of solids' flow principles that can be applied to applications that they will be exposed to in the field. There are industry shows and some universities offer continuing education courses that provide the tools necessary for proper material handling system design. One must first recognize that there is a problem and develop resources to evaluate and solve these problems.

B. Agricultural Baseline Projections: Summary of Projections, 2004–2013

The USDA Baseline consists of ten-year projections for agriculture, assuming continuation of current farm law as well as specific conditions for the economy, the weather, and the global situation. The baseline covers commodities, trade, and aggregate indicators such as farm income and food prices. The projections were prepared from October through December 2003 prior to the diagnosis of a case of bovine spongiform encephalopathy (BSE) in an adult Holstein cow in Washington State in December 2003.

Stronger domestic and international growth, following the economic slowdown of 2001 through early 2003, provides a favorable demand setting for the U.S. agricultural sector. Declines from a recent peak, a relatively strong U.S. dollar (by historical standards), and trade competition from countries such as Brazil, Argentina, and the Black Sea region are constraining factors on U.S. exports for some agricultural commodities. Nevertheless, improving economic growth, particularly in developing countries, will provide a foundation for gains in global consumption and trade, U.S. agricultural exports, and farm commodity prices. Domestic demand will also increase for meat, feeds, horticultural products, corn used in ethanol production, and food use of rice. As a result, market prices and cash receipts will rise, which will help to improve the financial condition of the U.S. agricultural sector. Consumer food prices are projected to continue a long-term trend of rising less than the general inflation rate. The trend in consumer food expenditure toward a larger share for meals eaten away from home is expected to continue.

Agricultural export markets are important for sustaining prices and farm revenues. Exports account for a growing share of U.S. farm cash receipts, and are a key factor in determining gains in gross farm income.

Agricultural trade depends on the economic prosperity of consumers throughout the world. Economic gains and population growth in developing countries will generate most of the increase in global food demand over the next decade. Economic growth in developing countries is important for global agricultural demand because many developing countries have incomes at levels where consumers diversify their diets to include more meat and other higher valued food products, and where consumption and imports of food and feed are particularly responsive to income changes.

Projected growth in the transition economies (countries of the former Soviet Union and Central and Eastern Europe) of over 4% in 2004–2013 is significant in comparison with the economic contraction of the 1990s. Economic reforms undertaken to shift to market economies and European Union (EU) enlargement contribute to the improved growth prospects. This growth will increase consumer income and thereby raise demand for agricultural goods, such as livestock products, for which demand is relatively responsive to income changes. Although declining somewhat in the short term, the U.S. dollar is assumed to stay at historically strong levels throughout the projections as financial market returns attract financial flows into the United States. The strength of the dollar is a constraining factor for U.S. agricultural competitiveness and export growth.

Competition in global agricultural markets will continue to be strong, with expanding production in a number of foreign countries. For example, increasing exports of soybeans and soybean meal from South America reflect a continuing conversion of land to crop production uses, particularly in Brazil (Figure 15). Competition in global wheat trade continues with traditional exporters (Australia, Argentina, Canada, and the EU) as well as with more recent exporters from the Black Sea region. Brazil and Canada provide competition to U.S. pork exports, while U.S. exports of broilers face strong competition from Brazil and Thailand.

The value of U.S. agricultural exports, which fell from a record of almost $60 billion in fiscal year 1996 to $49.1 billion in 1999, is projected to exceed the earlier record from 2005 through 2013 (Figure 16). U.S. agricultural exports face continued strong trade competition

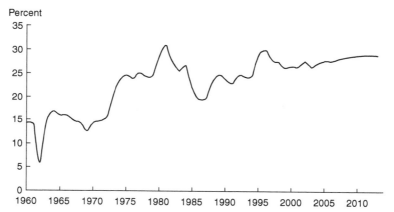

Source: USDA Agricultural Baseline Projection to 2013, February 2004. Economic Research
　　　　Service, USDA.

Figure 15　U.S. agricultural export value relative to total market cash receipts.

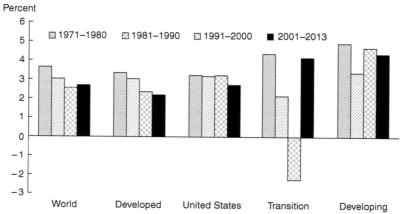

Source: USDA Agricultural Baseline Projection to 2013, February 2004. Economic Research
　　　　Service, USDA.

Figure 16　World gross domestic product (GDP) growth rates, decade averages.

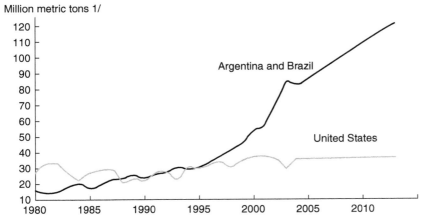

Million metric tons 1/

1/Soybeans soybean meal converted to soybean-equivalent weight.

Source: USDA Agricultural Baseline Projection to 2013, February 2004. Economic Research
　　　　Service, USDA.

Figure 17　Soybean and soybean meal exports: United States, compared to Argentina and Brazil.

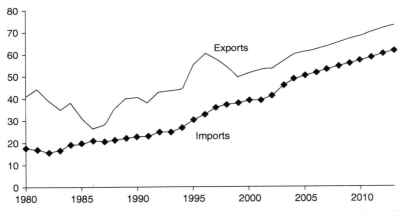

Source: USDA Agricultural Baseline Projection to 2013, February 2004. Economic Research
　　　　Service, USDA.

Figure 18　U.S. agricultural trade value.

throughout the baseline period. A relatively strong U.S. dollar by historical standards, despite declines from a recent peak, also is a constraining factor on U.S. agricultural exports.
Nonetheless, strengthening world economic growth in the long run, particularly in developing countries, provides a foundation for gains in U.S. agricultural exports, which are likely
to increase to about $72 billion by the end of the projections (Figure 17). U.S. agricultural
imports are likely to rise by about the same amount as exports, with the agricultural trade
surplus relatively stable at $10 to $12 billion during the projected period (Figure 18).

　　　Strengthening market conditions lead to rising prices, increases in gross farm income,
and improvement in the financial condition of the U.S. agricultural sector. Net farm income
is likely to decline through much of the projected period from a high 2003 level, reflecting
lower government payments and adjustments in the cattle sector (Figure 19).

　　　Gross cash income is projected to increase gradually as crop and livestock receipts
increase due to growing domestic and export demands. Production expenses are projected

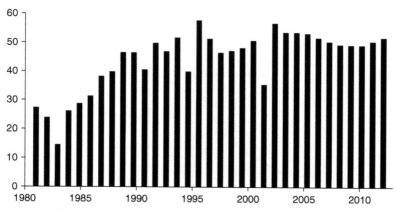

Source: USDA Agricultural Baseline Projection to 2013, February 2004. Economic Research
    Service, USDA.

Figure 19   Net farm income.

to increase at slightly less than the general inflation rate. Cash operating margins tighten
somewhat, with cash expenses increasing from 75% of gross cash income in 2004 to about
78.5% at the end of the projections. Government payments are likely to become relatively
less important over time as a greater share of gross cash income comes from the marketplace.

Net farm income projections for the next decade average about $51 billion, compared
to $47.6 billion in the 1990s. Income increases toward the end of the projections and reaches
$51.5 billion in 2013. Increasing gross cash income and relatively low interest rates through
the baseline assist in asset accumulation and debt management. Debt-to-asset ratios decline
to about 14% in the last several years of the projections, compared with over 20% in the
mid-1980s and 14.7% in 2003.

REFERENCES

1.   American Society for Testing Materials standard D 6128–97.
2.   Bin and hopper design Presented at the 104th Annual Technical Conference and Trade Show
     of the Association of Operative Millers, May 8, 2000.
3.   A. W. Jenike, Storage and Flow of Solids, Bulletin No. 123, University of Utah Engineering
     Experiment Station, salt Lake City, Nov. 1964.
4.   J. A. Marinelli and J. W. Carson, Characterize bulk solids to ensure smooth flow, *Chemical
     Engineering*, April 1994.
5.   Choosing a feeder that works in unison with your bin, *Powder and Bulk Engineering*,
     December 1996.
6.   Additives aid solids flow, *Chemical Processing*, June 2000.

2

Mixtures of Food Powders and Particulates

Micha Peleg
Department of Food Science
Chenoweth Laboratory
University of Massachusetts
Amherst, Massachusetts

CONTENTS

I. INTRODUCTION

A mixture, according to most dictionaries, is the result of combining or putting together two or more components or ingredients. According to this definition, granular foods like rice, lentils, and cornflakes, or food powders like wheat flour, instant milk, or coffee are all mixtures since they contain a variety of compounds belonging to different chemical species. For the following discussion, however, let us define a food particulate mixture as an assembly of particles distinguished by size, shape, over all composition, and any other physical or chemical property. Hence, the smallest elements of the mixture will be the individual particles, irrespective of their nonhomogeneity and size, rather than their chemical, physical, and microstructural components. It will also be assumed that the chemical components, unless otherwise stated, are in thermal and chemical "practical equilibrium," that is, they do not react or exchange moisture on the pertinent timescale.

A deliberate mechanical mixing process is a common way to create powders or granular mixtures. But, they can also be formed spontaneously. Breakage, where the daughter particles have a different composition and size, is the most obvious example. When dealing with the properties and uniformity of powder mixtures, one needs to specify, or at least be aware of the "level of scrutiny" — that is, the pertinent size or volume scale. Thus although all boxes of a given cereal, for example, may have the same amount of raisins or nuts, it would be highly undesirable if all of them were present in a single bowl. Similarly, a cake mix will be a total commercial failure if one cupful contained most of the baking powder while another only flour or sugar. This "level of scrutiny" dictates not only the processes designed to produce any given mixture and assure its uniformity, but also the sampling method to determine its "degree of mixedness," or "level of segregation" wherever a problem. What follows will describe the main factors that determine the physical properties and stability of particulated food mixtures. The focus is on general trends and patterns and not on any particular group of particulated foods. The discussion will draw heavily on work done at the Physical Properties of Foods Laboratory at the University of Massachusetts. No effort has been made to provide a comprehensive and updated literature survey on the subject. The interested reader will find ample references in the publications, which are cited, and in many current textbooks and journals that deal with powders and particulates technology.

II. TYPES OF MIXTURES

As far as structure is concerned, mixtures can be of two major types (Figure 1):

1. "Noninteractive" or "random" mixtures. These are mostly noncohesive, "free flowing," powders or granular materials with more or less uniform particle size. Because of the absence of significant attractive interparticle forces, the particles

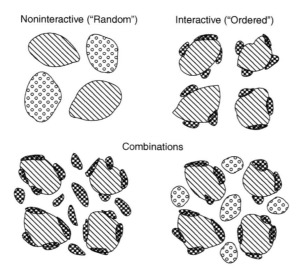

Figure 1 A schematic view of various types of particulates mixtures. For actual micrographs see Barbosa-Canovas et al. (1985).

are free to move one past another with little interruption. Consequently, each particle has an equal probability to be found anywhere in the bulk. Moreover, the particles of any given species are most likely to be uniformly and randomly distributed among the other particles of the mixture. If the size differences between the components are small or nonexistent, such mixtures are inherently stable and they will not segregate during motion (see below).

2. "Interactive" or "ordered" mixtures. Such mixtures are usually formed when there is a "carrier," relatively large particles, with an active surface, and fines, which tend to be attracted to it. The mixture can be created deliberately as in the case when an anticaking agent or a "conditioner" is added to a host powder (Hollenbach et al., 1982, 1983), or when a food colorant is added to a sugar/acid mix to form a dry fruit beverage base (Barbosa-Canovas and Peleg, 1985). It can also be created spontaneously as in the case where fines are deposited on a newly created surface during or immediately after grinding. Either way, the adhering fines are not free to move. Once "caught" they are most likely to stay attached to the surface of the larger "carrier" particles and hence cannot occupy a "random location" in the bulk. The term 'interactive' refers to the cause of this mixture type formation — the attractive forces between the surfaces of particles belonging to different chemical species or size class. The older term 'ordered' refers to the structured character of the particles' spacial distribution in the bed, which results from the formation of mixed aggregates or agglomerates. (The term 'aggregate' commonly refers to a spontaneously formed and usually mechanically unstable lump of adhered particles. The term 'agglomerate' is often used to describe a stable "aggregate" which is deliberately produced by a special process.) As long as all the fines adhere to their host's surface with a sufficient strength, such mixtures tend to be stable, that is, they will not segregate during motion despite a considerable difference between the components' particle size.

The interactive and noninteractive forms of mixture can be the basis of a variety of other mixture types. For example, if an excess of fines is added to a host powder with an active surface, the result will be a mixture of mixed aggregates and the excess fines spread individually or aggregate with their own kind (Hollenbach et al., 1982, 1983; Peleg and Hollenbach, 1984). Or, if a conditioner, for example, is added to a binary mixture and its particles adhere to both components, the result can be a "noninteractive," or "random" mixture of the newly formed mixed aggregates. There are obviously numerous other possible combinations, depending on the number of components and the modes of their interactions as shown schematically in Figure 1. Micrographs of such mixtures can be found in a paper by Barbosa-Canovas et al. (1985b). Suffice to say at this point that the density and mechanical properties of any such mixture will depend on the size and surface properties of the aggregates, which should be treated as if they were a new species of particles.

III. THE BULK DENSITY AND COMPRESSIBILITY OF BINARY MIXTURES

The loose bulk density of a free-flowing noninteractive or random mixture is primarily determined by the particle density of its components. (A particle's density can be lower than its solid density if it contains internal pores, open or closed.) For a noninteractive binary mixture, a plot of the bulk density versus the ingredients ratio composition would

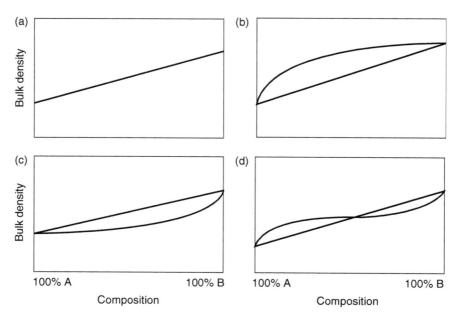

Figure 2 A schematic view of some possible ways in which the type of mixture created can affect the relationship between composition and bulk density. (From Barbosa-Canovas, G.V., Malave-Lopez, J., and Peleg, M. 1987. *J. Food Proc. Eng.* 10: 1–19. With permission.)

look like a straight line connecting the bulk density of the pure components (Figure 2[a]). The situation will be slightly different if one of the components, "A," for example, is small enough to occupy the interparticle space, in which case the plot, theoretically at least, would be above the connecting straight line (Figure 2[b]). When the two ingredients form an interactive mixture the outcome will largely depend, at least in principle, on the nature of the newly formed external surface. If the new surface is inert, as in the case of an added anticaking agent, the result can be a considerable *increase* in the loose bulk density (Figure 2[b]). This is because a much reduced strength of attractive interparticle forces will no more be able to support the open bed structure, which is characteristic of cohesive powders (Figure 3). Consequently, the particles, or rather aggregates, will occupy most of the available space in the containers and hence the higher bulk density.

Theoretically at least, it is also possible that the aggregates are more cohesive than the original particles. In such a hypothetical scenario, the mixture will have an open structure supported by the interaction between the particles' new surfaces, which will lower the mixture's bulk density (Figure 2[c]). In principle, as well as in practice, each of the above mentioned mechanisms can operate at a certain range of the ingredients ratios only, and their role can be reversed in others. This would result in plots of the kind shown schematically in Figure 2(d). The conclusion is that with very few exceptions, it is very hard to estimate a mixture's bulk density from the bulk densities of its components using stoichiometric considerations. Certain trends, though, can be predicted, qualitatively at least, if the nature or mode of the interaction between the components is known. The same can be said about the mixture's compressibility. In general, like in homogeneous powders, a looser structure, that is, low bulk density, would most probably be accompanied by increased compressibility, and vice versa, as shown in Figure 3 (Peleg, 1983). But, since the degree of coating in interactive mixtures can vary considerably experimental determination of

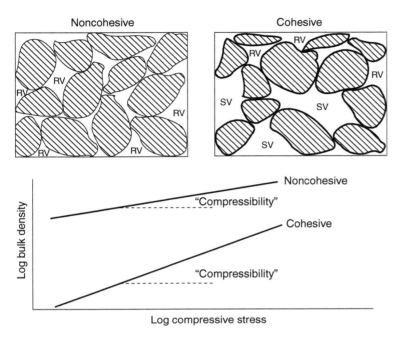

Figure 3 A schematic close up view of the particles' arrangement in noncohesive (free-flowing) and cohesive powder beds (top). (RV are voids created by the random arrival pattern and shape of the particles. SV are "structural voids," created by interparticle attractive forces such as those created through liquid bridges.) The difference in the particles' arrangement is manifested in the powder's bulk density and compressibility as shown at the bottom.

the mixture's density seems to be unavoidable. This would be almost always true when multicomponent mixtures are concerned, in which case the number of possible interactions and structural combinations increases considerably.

IV. MIXING AND SEGREGATION INDICES

A. Mixing Indices

Traditionally the quality of a mixture or its "degree of mixedness," has been determined and quantified in terms of a variety of statistical indices. Examples are (Schofeld, 1976):

$$M = \frac{\sigma_r}{s} \tag{1}$$

$$M = \frac{\sigma_{0-s}}{\sigma_{0-\sigma_r}} \tag{2}$$

$$M = \frac{\sigma_0^2 - s^2}{\sigma_0^2 - \sigma_r^2} \tag{3}$$

where M is the mixing index, σ_0 the standard deviation in the unmixed state, σ_r the theoretical standard deviation of a corresponding "random" mixture, and s the standard deviation of the concentration of the ingredient in question in the examined mixture. The reader will notice that the indices are not identical and that their practicality will depend on whether the component of interest is present in the mixture at a high or low concentration. This will also determine the size of the sample examined. To demonstrate this point, let us assume that the component of interest's concentration is 30%, in which case it is most likely to be found in almost any sample that contains more than a few individual particles. If, however, its expected concentration is only 0.3%, for example, then the sample needs to be considerably larger for its presence to be detected at all. (In some cases in order to guarantee a uniform distribution of a low-concentration component, for example a vitamin or a nutrient, the mixing is done when the components are dissolved or suspended in a liquid and then dried.) In segregating powders, the problem is not only the number and size of the samples that need to be examined, but also *from where* they have to be taken. For the samples to be representative of the lot, or shipment, they cannot be taken solely from the bottom, middle, or top of the bulk (see below). One ought to bear in mind that the magnitude of any mixing index is only a measure of the mixture's *quality at the time when it was determined*. Moving a powder mixture, or merely subjecting it to vibration, can cause segregation and/or breakage, which can affect the spacial distribution of its components. (A familiar food example is the accumulated fines at the bottom of a cereal box. They would rarely, if ever, contain raisins or any other "large" particles like pieces of nuts, which do not break during a normal handling of the product.)

B. Segregation

Segregation almost without exception occurs only in free-flowing powders and particulates. Contrary to intuition, the main factor that affects the rate and extent of particulate segregations is not a difference in the solid or particle density, but a difference in particle size. The larger the difference, the stronger is the segregation tendency (Williams, 1976). Consequently, a most effective way to reduce segregation, or avoid it altogether, is to bring all the components of the mixture to the same particle size. An alternative option would be to create an interactive mixture where the fines are strongly bound to the surface of a carrier large in size or reducing the mixture's mobility by increasing its cohesion. The latter, however, has its own obvious limitations and it can cause other physical, chemical, and biological stability problems, especially in food powders containing soluble compounds. (The most effective way to avoid segregation, as already mentioned, is to spray dry a liquid mixture of the ingredients [a solution, suspension, or emulsion] thus making sure that after dehydration each individual particle contains all the ingredients, and in the right proportions. This however, may not be a viable option economically or technologically, particularly if any of the ingredients is heat labile. Freeze drying is obviously a technological alternative, but for most food powders it is not a feasible option for economical reasons.) A more practical alternative, in certain products, is to fill the ingredients separately in the same individual package. Thus as in the case of dry cocktail base mixes, even if physical segregation of the product does occur, it will not be felt in the reconstituted beverage as long as all the bag's contents are used at once. Individual packaging also helps to avoid the inevitable moisture sorption by the hygroscopic powders. If packed in a large container, which would be repeatedly opened and closed (see below), a hygroscopic product would most probably cake long before it is fully consumed. All the above solutions to a segregation problem considerably increase the production cost, and in some cases, make the product less convenient to use. Still, they are sometimes the only practical solution to the problem that certain mixed products present. The most notable example is noodles containing instant soup mixes where individual packaging is the only practical way to guarantee the same amount of noodles in every cup of soup.

C. Segregation Indices

The tendency of a mixture to segregate, or demix, has been quantified in terms of segregation indices. In principle, any mixing index can also serve as a segregation index, but not vice versa. The simplest test to determine the degree of segregation is to expose a powder specimen in a split cell to vibrations for a selected period of time and then determine the mass fraction of the coarse component in the top and bottom halves of the cell. The segregation index, C_s proposed by Williams (1976), is:

$$C_s = \frac{W_{ct} - W_{cb}}{W_{ct} + W_{cb}} \tag{4}$$

where W_{ct} and W_{cb} are the masses of the coarse component at the top and bottom respectively. The theoretical range of this index is from zero ("perfect" mixing) to one ("total" segregation), but its sensitivity is considerably curtailed if the mixture has only a small fraction of fines. A more sensitive index is the one proposed by Popplewell et al. (1989) where the split cell has only a narrow ring at the bottom. The modified segregation index, I_s, is:

$$I_s = \frac{A - C_0}{1 - C_0} \tag{5}$$

where A is the mass fraction of the fines at the bottom and C_0 the mass fraction of the fine in the whole cell. This index too has the range of zero ($A = C_0$ and hence a "perfect" mix) and one ($A = 1$ and hence "total" segregation), but it is sensitive over a wider range of fines concentration. It is also more sensitive when the imposed vibrations cause not only segregation but compaction of the specimen as well (Popplewell and Peleg, 1991). Both indices can be used to monitor the progress of the segregation process by plotting their magnitude versus time. They can also be used to quantify the effects of factors such as the size difference, the vibrations' amplitude and frequency and the fines concentration on a given mixture's segregation tendency. More recently, the segregation phenomenon in granular materials has been studied and modeled in terms of "convection currents." When these occur the material is being continuously recycled through vertical streams within the vibrated bulk.

V. ATTRITION AND SEGREGATION

Segregation, as already mentioned, is primarily caused by size differences and the particles' chemical composition, shape, and density usually play only a minor role in the process. The source of the fines is therefore unimportant. They can be an admixed ingredient or the result of mechanical breakage or attrition (Malave-Lopez and Peleg, 1986a, 1986b; Popplewell and Peleg, 1989). Familiar examples of the second type are the broken particles at the bottom of a cereal box or the fines at the bottom of the jar of agglomerated and freeze-dried instant coffee (Malave-Lopez, 1986a, 1986b). These "fines" are produced by friction and impact of the parent particles with themselves, the container walls and, during bulk transport, with the hard parts of the equipment. If the daughter particles are small enough, they will migrate and be deposited at the bottom of the container. If very fine, they can create a dust problem instead. A familiar example is the grain dust. When suspended in air at certain concentrations, it creates an explosive mixture, that can be ignited by an electric spark or a lighted cigarette. Complete elimination of the attrition is usually impossible to accomplish. Its extent can be reduced by minimizing vibration and impact during transport and handling and by making sure that the particles, especially if they are fragile agglomerates, are not too dry. Reducing brittleness inhibits failure propagation and with it the tendency to disintegrate upon impact or under stress. Obviously, there must be a very fine balance here. Any excessive amount of moisture, for example, which would plasticize the particles and hence reduce their fragility, can also create biological and chemical stability problems or cause caking.

VI.   MOISTURE MIGRATION BETWEEN PARTICLES OF DIFFERENT SPECIES

Mixing usually involves particulates having different moisture contents. If sealed together in an impermeable container, the relatively "wet" particles will lose moisture to the relatively drier ones. Theoretically, the process will cease when all the particles reach an equilibrium "water activity." Since many foods are usually not at a true thermodynamic equilibrium, the very concept of water activity as a measure of the state of water in them has been challenged (Frank, 2000). Nevertheless, one can consider "practical water activity" as a rough stability criterion, provided that it is only used for processes that take place on a specific and limited timescale. At a constant temperature, the moisture exchange in a sealed container, will in most cases lead to an apparent equilibrium water activity, a_w^*, determined by the individual ingredient's mass, initial moisture content, and moisture sorption isotherm (Figure 4). Neglecting the moisture content in the interparticle space within the container, and assuming that it is indeed hermetically sealed, this apparent a_w^* can be estimated from the expression

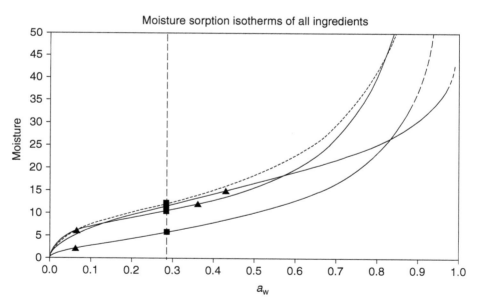

Figure 4   A schematic view of the moisture exchange between the ingredients of a four-component dry food powder mixture in an impermeable package. The filled triangles mark the ingredients' initial moisture and a_w on their respective moisture sorption isotherms. The vertical dashed line marks the mixture's equilibrium water activity, a^*. (From http://www-unix.oit.umass.edu/~aew2000/excelwateract.html   and   http://www-unix.oit.umass.edu/~aew2000/mathwateract.html.   With permission.)

(Peleg and Normand, 1992, Peleg, 1993):

$$a_w^* = \text{Root}\left[m_T = \sum x_i m_i(a_w), a_w\right] \qquad (6)$$

The operation "Root" symbolizes that a_w^* is the solution of the equation inside the brackets with respect to a_w, where m_T is the mixture's total moisture content (on a dry basis), the x_i's are the mixture component's mass fraction (on a dry basis) and the $m_i(a_w)$'s the algebraic equations describing the ingredient's moisture sorption isotherms at the particular temperature. A program to calculate a_w^* numerically using Excel[R] and Mathematica[R] is now available as freeware on the web; see http://www-unix.oit.umass.edu/~aew2000/excelwateract.html and http://www-unix.oit.umass.edu/~aew2000/mathwateract.html. For the user, the entries, of up to 34 different ingredients, are in the form of the actual weights of the ingredients in the formulation and their moisture contents on a wet basis. The level of a_w^* must be low enough so that even the most sensitive ingredient of the mixture will be at a stable state. In other words, for the mixture as a whole to be stable and free flowing it is not just enough if all the ingredients are stable and free flowing before and during the mixing; each must also be in the appropriate state *after* the moisture exchange with the other ingredients.

VII. PARTICLES MIGRATION IN INTERACTIVE ("ORDERED") MIXTURES

When new "naked" particles are admixed with an interactive mixture they can "steal" fines from the surface of already coated carrier particles (Barbosa-Canovas and Peleg, 1985; Sapru and Peleg, 1988). Schematic demonstration of the phenomenon is given in Figure 5. It appears that among crystalline food powders there is a hierarchy of surface affinities. Thus, citric acid crystals will remove absorbed food colorants from sugar crystals but not vice versa. This exchange has not been thoroughly investigated. It may affect the results of mixing powders in different orders. The appearance of fruit flavored dry beverage mixes to which a colorant is added in the form of a fine powder is a case in point. The exchange only occurs when the carrier particles are in motion (as during mixing) and mutual contact is possible. It ceases completely if moisture is absorbed and the mixture becomes cohesive. In such a case the fines are tightly bound to the host particle's surface and cannot be removed by a casual contact with an exposed surface of another particle.

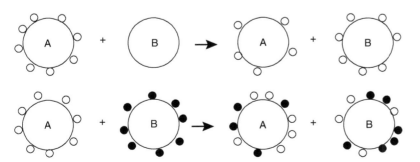

Figure 5   A schematic view of how fines can be exchanged between carrier particles.

VIII.   CONCLUDING REMARKS

Mixtures of food particulates can have very different structures and properties. But because dry food materials are frequently fragile and hygroscopic the individual particles as well as their mixtures can be unstable mechanically as well as chemically or biologically. Since there are numerous possible ways in which particles of different species or sizes can interact, it is usually difficult to predict a mixture's physical properties from mere knowledge of the particles' ingredients and bulk properties. Nevertheless, there are certain idiosyncrasies which once recognized can serve as general guidelines. Although they are neither accurate nor specific, they can still be useful in designing experiments to determine the properties of particulated food mixtures and to assess their physical stability. Mixtures' properties, like those of homogeneous powders, are not static. They evolve in response to the bulk and individual particles' mechanical and thermal histories, and moisture adsorption and migration. These in turn involve a variety of momentum, heat, and mass transfer mechanisms each operating at a different rate. Notable examples are moisture migration within the bulk and in an individual particle, and the selective breakage patterns of particles colliding with equipment walls vis à vis those which only rub one another. Consequently, in any attempt to predict a mixture's properties and performance, one must take into account that there is an unavoidable element of uncertainty. Understanding the factors that affect a mixture's properties and stability, though, can help the food manufacturer to identify the potential causes of problems when they occur and in setting experiments to gather relevant information to achieve the desired mixture properties.

ACKNOWLEDGMENT

Contribution of the Massachusetts Experiment Station at Amherst.

REFERENCES

Barbosa-Canovas, G.V. and Peleg, M. 1985. Migration of absorbed food colorants between powder components. *J. Food Sci.* 50: 1517–1518.

Barbosa-Canovas, G.V., Rufner, R., and Peleg, M. 1985. Microstructure of selected binary food powder mixtures. *J. Food Sci.* 50: 473–477, 481.

Barbosa-Canovas, G.V., Malave-Lopez, J., and Peleg, M. 1987. Density and compressibility of selected food powders mixtures. *J. Food Proc. Eng.* 10: 1–19.

Franks, F. 2000. Personal communication.

Hollenbach, A.M., Peleg, M., and Rufner, R. 1982. Effect of four anticaking agents on the bulk properties of ground sugar. *J. Food Sci.* 47: 538–544.

Hollenbach, A.M., Peleg, M., and Rufner, R. 1983. Interparticle surface affinity and the bulk properties of conditioned powders. *Powder Technol.* 35: 51–62.

Malave-Lopez, J. and Peleg, M. 1986a. Patterns of size distribution changes during the attrition of instant coffee. *J. Food Sci.* 51: 691–694, 702.

Malave-Lopez, J. and Peleg, M. 1986b. Mechanical attrition rate measurements in agglomerated instant coffee. *J. Food Sci.* 51: 687–690, 697.

Peleg, M. 1983. Physical characteristics of food powders. In: Peleg, M. and Bagley, E.B. (Eds.). *Physical Properties of Foods*. AVI Publishing Co., Westport, CT, pp. 293–324.

Peleg, M. 1993. Assessment of a semi-empirical four parameter general model for sigmoid moisture sorption isotherms. *J. Food Proc. Eng.* 16: 21–37.

Peleg, M. and Hollenbach, A.M. 1984. Flow conditioners and anticaking agents. *Food Technol.* 38: 93–102.

Peleg, M. and Normand, M.D. 1992. Estimation of the water activity of multicomponent dry mixtures. *Trends Food Sci. Technol.* 3: 157–160.

Popplewell, L.M., Campanella, O.H., Sapru, V., and Peleg, M. 1989. Theoretical comparison of two segregation indices for binary powder mixture. *Powder Technol.* 58: 55–61.

Popplewell, L.M. and Peleg, M. 1989. An "Erosion Index" to characterize fines production in size reduction processes. *Powder Technol.* 58: 145–148.

Popplewell, L.M. and Peleg, M. 1991. On the segregation of compressible binary powder mixtures subjected to tapping. *Powder Technol.* 67: 21–26.

Sapru, V. and Peleg, M. 1988. Hierarchy in food colorant absorption on selected crystalline powders. *J. Food Sci.* 53: 555–557.

Schofeld, C. 1976. The definition and assessment of mixture quality in mixtures of particulate solids. *Powder Technol.* 15: 969–980.

Williams, J.C. 1976. The segregation of particulate material — a review. *Powder Technol.* 15: 245–251.

3

Physical and Chemical Properties of Food Powders

Gustavo V. Barbosa-Cánovas and Pablo Juliano
Food Engineering Program
Washington State University
Pullman, Washington

Contents

I. INTRODUCTION

The knowledge and understanding of properties is essential to optimize processes, functionality, and reduce costs. Chemical and physical properties are always receiving great attention in all disciplines dealing with powders such as pharmaceuticals, foods, ceramics, metallurgy, detergents, civil constructions, etc., and researchers use them to study the mechanisms of particle interactions and to evaluate particles at a bulk level. Food powders can be observed in several food processing operations, from raw materials' utilization, handling, and processing operations (e.g., blending, solubilization, fluidization, conveying), to storage operations (bulk disposal and packaging).

A powder is a solid material of complex form, made up of a large number of individual particles, each different from its neighbor in many cases. In this chapter, the term "powders" will refer to food particulates in the size range of roughly 50 to 1000 μm (e.g., flour, spices, sugar, soup mixes, and instant beverages). Food powder properties can be classified as either primary, that is, individual or inherent properties (particle density, particle porosity, shape, diameter, surface properties, hardness, or stickiness) or as secondary or bulk properties (bulk density, bulk porosity, particle size distribution, moisture content). Secondary properties provide quantitative knowledge of a powder as bulk.

Food powder properties can also be classified as physical or chemical properties. Physical properties include the particle's shape, density and porosity, surface characteristics, hardness, diameter, and size. On the other hand, the chemical properties of a food material are related to the food's composition and its interaction with other substances like solvents or components within the food structure. In particular, instant properties and stickiness are composition-related and characterize certain processes like dissolution and caking of powders, which are involved in different processing operations such as agglomeration, drying, mixing, and storing. Stickiness can be characterized as having cohesion and properties of cohesion such as tensile strength, angle of repose, and angle of internal friction.

This chapter covers the definitions and determination methods concerning the physical and chemical properties of food powders. Further, different applications of these properties and processing parameters will be discussed, as well as recent developments in the field.

II. PHYSICAL PROPERTIES

The physical properties of powders in general, including food powders, are interdependent. A modification of particle size distribution or moisture content can result in a simultaneous change in bulk density, flowability, and appearance. Physical properties provide a means to quantify the output particulate product in certain drying, grinding, or size enlargement processes, and to compare the effects of composition, structural conformation, material origin, type of equipment, or handling.

A. Density and Porosity

Density (ρ) is defined as the unit mass per unit volume measured in kg/m^3 in SI units, and is of fundamental use for material properties' studies and industrial processes in adjusting storage, processing, packaging, and distribution conditions. Density is given by the particle mass over its volume, and mass can be accurately determined to less than a microgram with precise analytical balances. However, the way that particle volume is measured will depend on the determination method and its application, and will define the type of density considered. There are three main types of densities: true density, apparent or particle density, and bulk density. Bulk density in particular is one of the properties used in specifications of a final product obtained from grinding or drying. Since other authors have used different names for the same measuring condition, it is recommended that the definition of density be verified before using density data.

1. True Particle Density

True particle (or substance) density (ρ_s) represents the mass of the particle divided by its volume, excluding all open and closed pores (Figure 1), that is, the density of the solid material composing the particle. Some metallic powders can present true densities at around $7000 \, kg/m^3$, while most food particles have considerably lower true densities in the range 1000 to $1500 \, kg/m^3$. Most food particles have similar true density (Table 1) due to similarities in their main biological, organic, and inorganic components.

2. Apparent Particle Density

Apparent particle density (ρ_p), also called envelope density, is the mass over the volume of a sample that has not been structurally modified. Volume includes internal pores not externally connected to the surrounding atmosphere, and excludes only the open pores. It is generally measured by gas or liquid displacement methods such as liquid or air pycnometry.

3. Bulk Density

The *bulk density* (ρ_b) of powders is determined by particle density, which in turn is determined by solid density and particle internal porosity, and also by special arrangement of the particles in the container. Bulk density includes the volume of the solid and liquid materials,

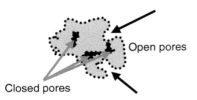

Open pores

Closed pores

Figure 1 Volume definition depends on whether open and closed pores are accounted for.

Table 1 Densities of Common Food Powders

Powder	Particle/true density, ρ_s (kg/m^3)	Bulk density, ρ_b (kg/m^3)
Baby formula		400
Cellulose*	1270–1610	
Citric acid*	1540	
Cocoa powder 10% fat	1450	350–400
Cocoa powder 22% fat	1420	400–550
Coffee (ground and roasted)		310–400
Coffee (instant)		200–470
Coffee creamer		660
Corn meal		560
Egg (whole)		680
Fat*	900–950	
Flour (corn)	1540	500–700
Flour (rye)	1450	450–700
Gelatin (ground)		680
Glucose*	1560	
Microcrystalline cellulose		610
Instant dried whole milk	1300–1450	430–550
Instant dried skim milk	1200–1400	250–550
Oatmeal		510
Onion (powdered)		960
Protein (globular)*	1400	
Rice (polished)	1370–1390	700–800
Salt (granulated)	2160	950
Salt (powdered)	2160	280
Soy protein (precipitated)		800
Starch (corn)*	1500–1620	340–550
Starch (potato)*	1500–1650	650
Sugar (granulated)*	1590–1600	850–1050
Sugar (powdered)*	1590–1600	480
Wheat flour (white/refined)	1450–1490	550–650
Wheat flour (whole grain)		560
Whey		520
Yeast (active dry, baker's)		820

Note: *Main components forming a particle, together with water (1000 kg/m^3), in some cases acting as a binding agent.

Source: From Schubert, H. 1987a. *J. Food Eng.* 6: 22–26; Schubert, H. 1987b. *J. Food Eng.* 6: 83–102; Peleg, M. 1993. *Physical Properties of Foods*. Peleg, M. and Bagley, E.B. (Eds.). Van Nostrand Reinhold/AVI, New York, pp. 293–324. With permission.

and all pores closed or open to the surrounding atmosphere. Four different types of bulk density can be distinguished depending on the method of volume determination:

- *Compact density* is determined after compressing the powder's bulk mass by mechanical pressure, vibration, and impact(s).
- *Tap density* results after a volume of powder has been tapped or vibrated under specific conditions; it is most useful to describe powder behavior during compaction.
- *Loose bulk density* is measured after a powder is freely poured into a container.

- *Aerated bulk density* is used for testing under fluidized conditions or during pneumatic conveying applications when particles are separated from each other by a film of air. In practice, the most loosely packed bulk density is achieved after the powder has been aerated.

Table 1 lists the typical densities of some food powders. It can be observed that inorganic salt presents a notably higher particle density, whereas fat-rich powders present lower densities. In general, bulk densities range from 300 to 800 kg/m^3.

4. Particle Porosity

The volume fraction of air (or void space) over the total bed volume is indicated by porosity (ε) or voidage of the powder. Considering air density as ρ_a, bulk density can be expressed as a function of the solid density and porosity.

$$\rho_b = \rho_s(1 - \varepsilon) + \rho_a\varepsilon \tag{1}$$

Air density is negligible with respect to powder density. Thus, the porosity expression can be calculated directly from the bulk and particle density of a given volume of powder mass.

$$\varepsilon = \frac{(\rho_s - \rho_b)}{\rho_s} \tag{2}$$

In general, the internal, external, and interparticle pores of food powders provide a high porosity rate of 40 to 80%. Bulk porosity can vary significantly due to mechanical compaction, different particle size (concentration of fines), moisture, and temperature. Furthermore, the porosity can be affected by the chemical nature of each constituent powder as well as from the process from which the particles originate. Changes in environmental and time-dependent parameters, such as moisture and temperature during storage, can vary the interactions between particles, thereby affecting their density and porosity as well as the number of interparticle contact points. These changes are mostly due to volume reduction for increased adhesiveness and, to a lesser extent, due to mass variations from water sorption or evaporation, or even due to phase changes with temperature (e.g., in fatty components).

5. Density Determination

A wide range of methodologies have been developed with varied degrees of accuracy for particle and bulk density determination. Apparent particle density can be determined by fluid displacement methods or "pycnometry," using either a liquid or gas (i.e., liquid pycnometry and air pycnometry).

    a. Liquid Displacement Methods.   For fine powders, a pycnometer bottle of 50 ml volume is normally employed (Figure 2), while coarse materials may require larger calibrated containers. The liquid should be a solvent that does not dissolve, react, or penetrate the particulate food solid. The particle density ρ_s can be calculated from the net weight of dry powder divided by the net volume of the powder.

$$\rho_s = \frac{(m_s - m_0)\rho}{(m_l - m_0) - (m_{sl} - m_s)} \tag{3}$$

where m_s is the weight of the bottle filled with the powder, m_0 is the weight of the empty bottle, ρ is the density of the liquid, m_l is the weight of the bottle filled with the liquid, and m_{sl} is the weight of the bottle filled with both the solid and the liquid.

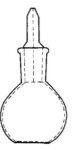

Figure 2   Conventional liquid pycnometer.

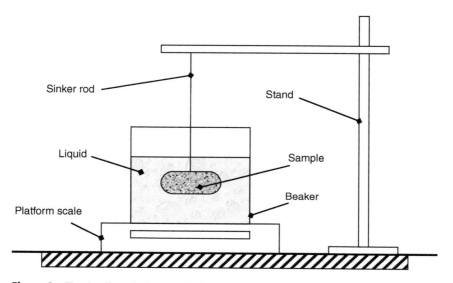

Figure 3   Top-loading platform scale for density determination of irregularly shaped objects.

A top-loading platform scale (Figure 3) is another liquid displacement method, which is lower in accuracy, and can be used for compressed or aggregated bulk powders of considerable size. This test is based on immersing a solid mass of powder and suspending it at the same time in a beaker containing liquid that is placed on top of a scale. The volume of the solid V_s is equivalent to the mass of liquid displaced by its surface (Ma et al., 1998) and can be calculated as:

$$V_s = \frac{m_{LCS} - m_{LC}}{\rho_L} \tag{4}$$

where m_{LCS} is the weight of the container with liquid and submerged solid, m_{LC} is the weight of the container partially filled with liquid, and ρ_L is the density of the liquid.

b. Air Displacement Methods.   Automatic pycnometers can calculate the apparent density of particulate materials, either by using fixed-volume sample chambers of different sizes, or by placing volume-filling inserts inside the chamber. The system consists of two chambers, a pressure measuring transducer, and regulating valves for atmospheric gas removal and replacement with helium gas. Sample volume is calculated from the observed pressure change the gas undergoes when expanding from one chamber containing the sample into another chamber without the sample. Air displacement methods can provide a

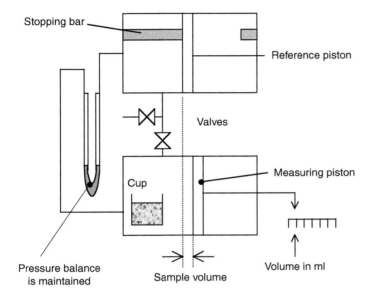

Stopping bar

Reference piston

Valves

Measuring piston

Cup

Volume in ml

Pressure balance
is maintained

Sample volume

Figure 4   An air pycnometer. The reference piston reaches the stopping bar and the measuring piston is displaced at a distance equivalent to the volume occupied by the sample.

reproducibility of apparent density within ±0.01%. This equipment is applicable to several food powders, ranging from coffee creamer to black pepper (Webb and Orr, 1997).

For instance, a special instrument based on two cylinders with one piston each (Figure 4), in which one cylinder is used as reference and the other to contain the powder sample, measures the sample volume accurately. A differential pressure indicator is used to verify that both cylinders maintain the same pressure before and after the test. The extra volume occupied by the solid sample is indicated by the displacement of the piston in the cylinder, which is attached to a scale. The equipment can be calibrated accordingly in cubic centimeters, usually with a digital counter. After the test, the sample volume is divided into sample weight to give the apparent density. Volume occupied by open pores can be calculated, depending upon the sample's hydrophobic properties, by filling pores either through wax impregnation or water addition, and then subtracting the envelope particle volume. By neglecting the closed pores of particles within the sample, this volume difference can give a measure of particle porosity.

The bed pressure method is based on passing gas through the powder bed in a laminar flow regime and measuring two different pressure drops. The Carman–Kozeny equation can be used to determine the particle density; ρ_p is the only unknown variable, which can be readily found by trial and error.

$$\frac{s_1}{s_2} = \left(\frac{\rho_{b1}}{\rho_{b2}}\right) \left[\frac{(\rho_p - \rho_{b2})}{(\rho_p - \rho_{b1})}\right]^3 \tag{5}$$

where conditions 1 and 2 correspond to each set of measurements, s is the gradient of pressure drop, and ρ_b is the powder bulk density.

c. Porosity Measurements.   The volume of the open pores could be determined with a mercury porosimeter. However, this is only suitable for coarse solids and requires costly equipment. The bed voidage method and the sand displacement method can be used alternatively. The bed voidage method (Abrahamsen and Geldart, 1980) consists of filling voids in a powder bed with a much finer powder. The fine powder is of known particle size and used as control. The drawback to this method is in finding control powders that are the same shape and yield the same voidage as the tested powder. The sand displacement method uses a fine sand into which a known amount of coarse particles are mixed. The density of the sample is determined by the difference of the bulk density of the sand alone and that of the mixture. This method is sometimes used to determine the density of coarse bone particles, giving a lower density than that of solid bone as measured by pycnometry.

d. Measurements of Bulk Density.   Aerated bulk density is measured by pouring powder through a vibrating sieve into a cup (Figure 5). A chute is attached to a vibrator of variable amplitude and a stationary chute aligned with the center of a preweighed 100 ml cup (ASTM D6683-01, 2003). Vibration amplitude and sieve aperture size should be set so that the time taken for the powder to fill the 100 ml cup is not less than 30 sec. The Hosakawa powder tester incorporates automated tasks with increased reproducibility for quality control purposes. This instrument can also be used for tap bulk density determination by packing the powder through tapping, jolting, or vibrating the measuring vessel. After tapping, excess powder is scraped from the rim of the cup and the bulk density is determined by weighing the cup.

Another method for tap density determination is the Tap density tester (or the Copley volumeter), which consists of a graduated cylinder and a tapping mechanism (Figure 6). The advantage of the Copley over the Hosakawa tester is that a fixed mass of powder is used and there is no need to remove the container for weighing as the number of taps is progressively increased (Abdullah and Geldart, 1999). The tapping action is provided by a rotating cam driven by a 60 W motor with an output speed of 250 rpm. The tap density tester is designed in compliance with norms established by ASTM, and gives standardized repeatable results for measuring tapped or packed volumes of powders and granulated or flaked materials. These can have digital LED displays and user-selectable counter or timer

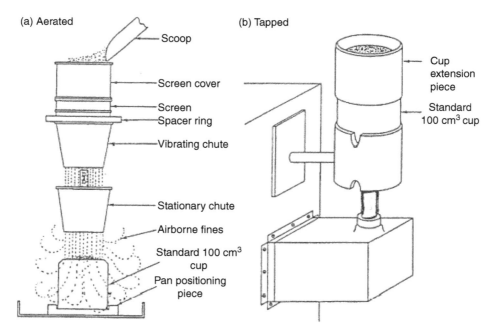

Figure 5   The Hosakawa powder tester is used to measure the aerated and tapped bulk densities. (From Abdullah, E.C. and Geldart, D. 1999. *Powder Technol.* 102: 151–165. With permission.)

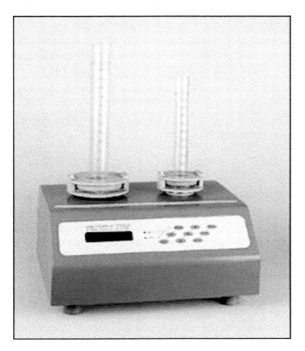

Figure 6   Tap density tester. (From ASTM. 1986. *Compilation of ASTM Standard Definition*. 6th ed. ASTM, Baltimore, MD. With permission.)

operations, including dual nonrotating platform drive units and two graduated funnel top cylinders, which generally are of 100 and 250 ml (Figure 6).

6.   Ultimate Bulk Density

Yan et al. (2001) proposed the concept of "ultimate bulk density" (Barbosa-Cánovas density) while studying density changes in agglomerated food powders under high hydrostatic pressure. When the hydrostatic pressure was higher than a critical value (around 200 MPa), the agglomerated food powders were compressed so densely that all the agglomerates and primary particles were crushed and compressed together, leaving almost no open or closed pores. Because the final compressed bulk density is usually higher than the commonly used apparent "solid density," the bulk porosity (due to the remaining pores being under high pressure) will be a negative value without physical meaning. It is believed that the Barbosa-Cánovas density could be dependent on product formulation, physical properties of product ingredients, and production conditions.

7.   Hausner Ratio

The Hausner ratio can be defined as the ratio of tap bulk density to loose bulk density. The ratio can be used for fluidization studies (Geldart et al., 1984) to evaluate density changes with particle shape (Kostelnik and Beddow, 1970). It can be applied to understand the influence of relative humidity upon process operations. Malavé-Lopez et al. (1985) defined the ratio of asymptotic over initial bulk density by the relationship:

$$\text{HR} = \rho_\infty / \rho_0 \tag{6}$$

where HR is the Hausner ratio, ρ_∞ is the asymptotic constant density after certain amounts of taps, and ρ_0 is the initial bulk density. A more practical equation widely used to evaluate flow properties can be given by the following equation, which calculates powder volume changes in a graduated cylinder after a certain period of time or number of taps (Hayes, 1987):

$$HR = \rho_n/\rho_0 = V_0/V_n \tag{7}$$

where n is the number of taps, ρ_n and ρ_0 are the tapped and loose bulk densities, and V_0 and V_n are the loose and tapped volumes, respectively.

B. Particle Shape

Particle shape influences factors such as flowability of powder, packing, interaction with fluids, and coating of powders such as pigments. However, very little quantitative work has been carried out on these relationships. There are two points of view regarding the assessment of particle shape. One is that the actual shape is unimportant and all that is required is a number for comparison purposes. The other is that it should be possible to regenerate the original particle shape from measurement data.

1. Shape Terms and Coefficients

Different porous structures yield different particle shapes as listed in Table 2. Qualitative terms may be used to give some indication of particle shape and a particle's classification. The earliest method for describing the shape of particle outlines used Heywood's dimensions, that is, length L, breadth or width B, and thickness T in expressions such as the *elongation* ratio (L/B) and the *flakiness* ratio (B/T) (Figure 7). However, one-number shape measurements could be ambiguous and the same single number could be obtained from more than one shape. A better means of definition is the term sphericity, Φ_s, defined by the relation:

$$\Phi_s = \frac{6V_p}{x_p s_p} = \frac{6\alpha_V}{\alpha_s} \tag{8}$$

where x_p is the equivalent particle diameter, s_p is the surface area of the particle, V_p is the particle volume, and α_V and α_s are the volume and surface factors, respectively. The numerical values of the surface factors are all dependent on the particle shape and the

Table 2 General Classification of Particle Shape

Shape name	Shape description
Acicular	Needle shape
Angular	Roughly polyhedral shape
Crystalline	Freely developed geometric shape in a fluid medium
Dentritic	Branched crystalline shape
Fibrous	Regular or irregular thread-like
Flaky	Plate-like
Granular	Approximately equidimensional irregular shape
Irregular	Lacking any symmetry
Modular	Rounded irregular shape
Spherical	Global shape

Source: From Beddow, J.K. 1997. *Image Analysis Sourcebook*. American Universities Science and Technology Press, Santa Barbara, CA. With permission.

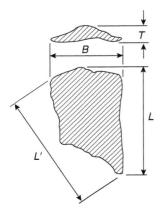

Figure 7 Heywood dimensions for flakiness and elongation ratio determination.

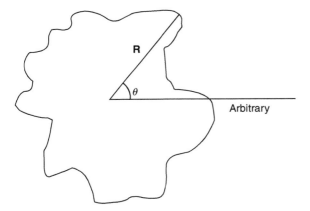

Figure 8 Representation of polar coordinates in an irregular silhouette.

precise definition of the diameter (Parfitt and Sing, 1976). For spherical particles, Φ_s equals unity, while for many crushed materials, its value lies between 0.6 and 0.7.

2. Morphology Studies

Characterization of particle silhouettes for shape characterization can be done using polar coordinates of their peripheries with the center of gravity of the figure as origin (Figure 8). A truncated harmonic series represents the value of the radius vector $\mathbf{R}$ as it is rotated about the origin as a function of the angle of rotation θ.

$$\mathbf{R}(\theta) = A_0 + \sum_{n=1}^{M} A_n \cos(n\theta - \varphi_n) \tag{9}$$

where φ_n is the phase angle of the nth harmonic and A_n is the amplitude. However, a large number of terms must be used for finer details and protuberances. An alternative approach is to represent any closed curve as a function of arc length by the accumulated change in direction of the curve. The outline is described by drawing a tangent around the shape and noting the change in the contour as a function of the angle θ. Computational programs have been designed to iterate and reproduce an equivalent harmonic series closest to its shape (Jones, 1983). The surface of a particle, in terms of roughness, can be completed in detail with optical instruments used in stereomicroscopy, scanning electron microscopy, or confocal scanning optical microscopy.

A mathematical description for modeling the silhouette of a projected particle surface is based on boundary-line analysis by *fractal mathematics* (Mandelbrot, 1977, 1982). The fractal approach analyzes the two-dimensional projections of particles by redefining their contour. A given step length can provide a close polygonal contour at different points of the particle projection that geometrically recreates a self-similar particle silhouette. The procedure is based on estimating the silhouette perimeter by adding the length of all the steps plus the exact length of the last side of the polygon (if different from the step length). Measurement of the perimeter of the silhouette is repeated by using several different step lengths, giving different perimeters in each case. When measuring the length of irregular contours by stepping along each with a pair of dividers, as the step length s decreases, the estimated length L_s will keep increasing without limit. In this case, L_s is related to s by:

$$L_s = K(s)^{1-D} \tag{10}$$

where D is the fractal dimension, which is different for jagged and smoothly rounded contours. Fractal dimension provides useful information about particle shape, openness, and ruggedness in the form of single numerical numbers (Simons, 1996). Fractal characterization can be used as a tool to describe the eroding process in particulate materials during compaction (Barletta and Barbosa-Cánovas, 1993a). In fact, fractal dimension has been used as an attrition index, since attrition causes changes in particle shape and surface on the same scale as the fractal approach (Peleg and Normand, 1985; Olivares-Francisco and Barbosa-Cánovas, 1990).

C.   Strength Properties

There are a number of properties in particulate materials that determine particle breakage and resistance strength during mixing and handling operations, and great attention has been paid to the general behavior of powders under compression stress. Compression and compaction tests are widely used in pharmaceutics, ceramics, metallic powders, and soil mechanics, as well as in the food powder field. They are used by researchers to study the mechanisms of particle interactions and to evaluate particles at bulk levels. Many solid food materials, especially when dry, are brittle and fragile, showing a tendency to break down or disintegrate. Mechanical attrition of food powders usually occurs during handling or processing, when the particles are subjected to impact and frictional forces. Attrition represents a serious problem in most food processes where dry handling is involved, since it may cause undesirable results such as dust formation, health hazard, equipment damage, and material loss. Dust formation may be considered the worst of these aspects, as it may develop into a dust explosion hazard.

1.   Hardness and Abrasiveness

The hardness of powders or granules is the degree of resistance of the particle surface to penetration by another body. It is related to the yield stress in view of the characteristics of the uniaxial stress–strain curve in several types of material failure (i.e., transition between elastic and plastic strains). Hardness can be determined as a qualitative property by using the Mohs hardness scale (Carr, 1976). Ten selected minerals are listed in order of increasing hardness by indicating qualitative resistance to plastic flow. In this way, a material with a given Mohs number cannot scratch a substance with a higher number, but can scratch one with a lower number.

A particle property related to hardness is the crushing strength (ASTM, 1986), which refers to the force required to crush a mass of dry powder, or conversely, the resistance of a mass of dry powder to withstand collapse from external compressive load. Bulk crushing strength can be evaluated by either measuring the amount of fines produced after compression of a fixed volume of particles at a predetermined pressure or measuring the pressure required for producing a predetermined amount of fines.

Abrasiveness of bulk solids, that is, their ability to abrade or wear down the surfaces with which they are in contact, is a property closely related to the hardness of the material. The abrasiveness of food powders can be rated based on the relative hardness of the particles and the contact surfaces, using Mohs hardness scale. From Mohs hardness scale, materials can be generally rated as soft, medium hard, or hard, when values are between 1–3, 3.5–5, and 5–10, respectively. The best way to assess abrasiveness is to use the actual bulk material and the contact surfaces in question. Abrasiveness and hardness are two major factors that govern the choice and design of different types of equipment such as size reduction machines, air classifiers, mixers, dryers, etc. Many food materials are normally soft according to this criterion and, thus, the problems related to strength of materials normally faced in the food industry more often deal with attrition and friability, rather than hardness and abrasion.

2. *Attrition and Friability*

Attrition is deleterious to particle breakdown, which increases the number of smaller particles by reducing particle size and affecting particle size distribution (Figure 9). Except for particle size reduction during comminution or grinding processes, attrition is undesirable in most processes. In fact, it is one of the most common problems experienced by many processing industries dealing with particulate solids. In food powders, it occurs more frequently in agglomerates, mainly because of their multiparticulate structure. Many food agglomerates possess brittle characteristics that make the product susceptible to the vibrational, compressive, shear, or even convective forces applied to the particles during processing. Methods to determine the extent of attrition are, namely, the shear, vibration, and tumbler tests.

Shear cells can be used to study attrition effects in particles under compression (Ghadiri and Ning, 1997). The direct shear cell usually consists of two compartments: one placed on top of the other compartment, and one with a lid that covers the powder and acts like a piston. When a sample is put in the shear cell and compressed by a normal force

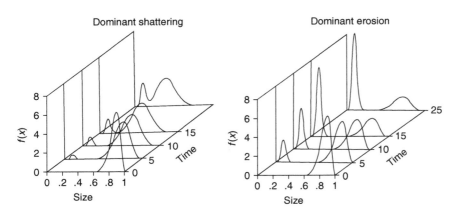

Figure 9   Changes in particle size distribution of disintegrating particulates at different tapping times. (From Popplewell, L.M. and Peleg, M. 1989. *Powder Technol.* 58: 145–148. With permission.)

from the lid, the base compartment can be placed in motion by a horizontal shearing force. Particle size distribution is compared before and after the experiment, and fines generated due to interlocking, frictional, and compaction forces are a measure of attrition.

Attrition of food agglomerates can be determined by using any form of resonance or a simple mechanical motion transmitted from a container to the particles within. The tap density tester (Figure 6) is the one most commonly used in research; it provides vertical vibration and helps to determine the extent of fine formation during handling and transportation. Research has been conducted for instant coffee, milk powders, and other agglomerated food powders (Malave-López et al., 1985; Barletta and Barbosa-Cánovas, 1993a, 1993b; Yan and Barbosa-Cánovas, 2001a). The tumbler test is another standard method (ASTM D441-86, 2002) used for agglomerates. It determines the amount of fines formed from a presieved sample with homogeneous particle size after tumbling for a specified number of times. Fine formation is determined by separating the fines with a sieve having a specific aperture.

Friability is the ability of particles to break down under impact or compressive forces. Impact tests can be used either on single particles or on a quantity of bulk solids. Impact methods measure the powder's ability to resist high-rate loading. Different impact methods can be used in order to characterize powder strength, which include impacting powders with a falling mass, impact tests by ram, and pneumatic dropping of powders on a surface (Mohsenin, 1986; Hollman, 2001). Modern impact test equipment (e.g., Universal testing machines [UTMs]) record the load on the specimen as a function of time and/or specimen deflection prior to fracture or particle breakage, using electronic sensing instrumentation connected to a computer. Friability can be measured in terms of impact energy absorbed by the sample relative to the initial potential energy (gravitational or elastic) of the plunger. Impact tests can be used to determine food powder coating resistance.

A pneumatic dropping impact test was used to study the influence of processing conditions (belt and pneumatic conveying, among other operations) in NaCl crystals (Ghadiri et al., 1991). This test can also be used for powder attrition testing by measuring the crushing strength necessary to produce a certain number of fines, or conversely, to evaluate the state of breakage (Couroyer et al., 2000).

3. *Compression Properties of Food Powders*

The deformation mechanisms occurring during compression of fines and agglomerated foods depend on elastic and viscous flow, in addition to ductile yielding and brittle behavior, and are common in pharmaceutical and food compaction processes (Barletta et al., 1993b). Some particulate materials have plastic and ductile deformations when stress is applied slowly; however, they show elasticity or brittleness under impact stresses. Bulk density and particle size can vary with the rate of the compressive stress applied to a powder. Compressive stress and tensile stress are used to characterize cohesiveness between particles or within a certain powder cake, as well as coating resistance in an encapsulated powder. A parameter that measures the change in bulk density with consolidating pressure acting on a powder bed is termed compressibility. Bulk density (apparent, compact, or tap density) and normal stress have been associated in empirical logarithmic or semilogarithmic relationships. Equation (11) is one of the most studied expressions in food powders, which relates the density fraction with the normal stress applied to a confined powder sample. It has been found valid up to the pressure range of 4.9 kPa with no expectation of particle yield or breakage (Peleg, 1978; Barbosa-Cánovas et al., 1987).

$$\frac{\rho_b - \rho_{b0}}{\rho_{b0}} = C_1 + C_2 \log \sigma \tag{11}$$

where ρ_b is the bulk density under compression stress σ, ρ_{b0} is the powder's bulk density before compression, and C_1 and C_2 are characteristic constants of the powder; C_2 is known as the *compressibility index* representing the change in relative density with applied stress. Results from different confined uniaxial compression tests have evaluated compressibility of the following food powders: fine salt, fine sucrose, cornstarch, baby formula, coffee creamer, soup mix, and active baker's yeast, instant agglomerated coffee, instant agglomerated low-fat (2%) milk, instant agglomerated nonfat milk, instant skim milk, ground coffee, ground corn, cornmeal, salt, sucrose, lactose, and flour (Kumar, 1973; Peleg and Mannheim, 1973; Moreyra and Peleg, 1980; Peleg, 1983; Konstance et al., 1995; Yan and Barbosa-Cánovas, 1997, 2000).

Particle size effect was evaluated in selected instant and noninstant materials (Molina et al., 1990; Yan and Barbosa-Cánovas, 1997, 2000; Onwulata et al., 1998), but no general tendency could be depicted due to differences in porosity and composition, which provided different ductile or brittle behaviors and particle arrangement. On the other hand, compressibility has been shown to increase with water activity in selected powders (Moreyra and Peleg, 1980) and binary mixtures (Barbosa-Cánovas et al., 1987), but tends to decrease in agglomerated powders (Yan and Barbosa-Cánovas, 2000). Other works present studies of compressibility in different cell geometries (Yan and Barbosa-Cánovas, 2000), using different anticaking agents (Hollenbach et al., 1982; Molina et al., 1990; Konstance et al., 1995; Onwulata et al., 1996) and characterizing flowability (Carr, 1965; Peleg, 1978; Ehlermann and Schubert, 1987; Schubert 1987a). In a storage bin, the bulk solids' normal stress increases linearly with depth due to the particles' own weight. Stress on the wall quickly reaches a maximum value as depth increases because part of the bulk solids' weight is transmitted to the walls via friction forces. Loads acting on the feeder or gate depend on the powder's compressibility, which can be used as a parameter for hopper designs. Furthermore, compressibility can be used for quality control in order to determine the material's resistance to breakage throughout production processing up to consumer handling.

The unconfined yield stress, fc, indicates the maximum compressive stress that a cohesive particle array is capable of sustaining at a particular porosity (Peleg, 1978; Mohsenin, 1986; Schubert, 1987a). It also represents the strength of the cohesive material at the surface of an arch (Figure 10), which is opposed by a lower stress induced by its own weight (Teunou et al., 1999). In flowability characterization, the unconfined yield stress refers to a situation in which the physical configuration of the system allows the powder to flow before massive comminution of the particles occur.

4.   Compression Methods

Methods to determine compression behavior can be static (dead load) or based on constantly increasing compression. Compression mechanisms can be approached using different tests

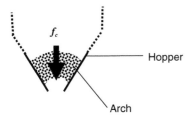

Figure 10   Unconfined yield stress (f_c) represented as the strength necessary to maintain a cohesive arch structure in a hopper.

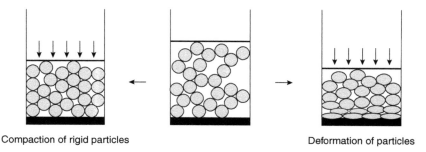

Compaction of rigid particles   Deformation of particles

Figure 11 Confined uniaxial compression test. Bed compression of rigid and deformable particles. (Adapted from Lu, W.-M., Tung, K.-L., Hung, S.-M., Shiau, J.-S., and Hwang, K.-J. 2001. *Powder Technol.* 116: 1–12. With permission.)

or testers, such as the uniaxial confined compression test, cubical triaxial tester, and unconfined yield stress test.

Universal Testing Machines (UTMs) allow the reading of maximum normal and shear forces, creep, and stress relaxation force-deformation graphs. Mechanical compressibility and breaking load under tension can be determined from stress–strain data by also using UTMs. The confined uniaxial compression test using a UTM (Figure 11) involves confining a bed of powder in a cylindrical cell and measuring the force applied to the flat-based piston as a function of the piston's displacement; this is continuously monitored on an electronic recorder chart using a specific computer software package. Different vertical loads can be applied to a bulk solid sample of known mass and compression (Thomson, 1997).

This method can successfully evaluate particle attrition (Bemrose and Bridgwater, 1987), flowability (Peleg, 1978; Schubert, 1987a), compressibility (Malavé-Lopez et al., 1985; Barbosa-Cánovas et al., 1987; Yan and Barbosa-Cánovas, 1997), and agglomerate strength (Adams et al., 1994). In particular, any agglomerate measurement will be affected by both the breakage properties of individual particles and the deformability of their assembly as a whole (Nuebel and Peleg, 1994).

A flexible boundary cubical triaxial test is another commonly used test for compression studies (Kamath 1996; Li and Puri, 1996). It not only allows independent application of the three principal stresses, but also volumetric deformation in three principal directions, which are monitored constantly so that the pressure in these directions is the same. The cubical triaxial tester is useful to investigate anisotropy of cohesive and noncohesive powders and the effect of particle shape and sample deposition methods.

The unconfined yield test (Buma, 1971; Head, 1982) determines the unconfined yield stress of cohesive powder cakes. The load necessary to make the plug fail is defined as the unconfined yield stress. Unconfined yield stress values were obtained as an index of cohesion for whole milk and skim milk powders (Rennie et al., 1999). Dry whole milk was found to be more cohesive than dry skim milk with increasing temperature, indicating the influence of fat in the cohesive mechanism in whole milk. Rumpf (1961) and Pietsch (1969) have discussed the strength of agglomerates and related compression mechanisms (Peleg and Hollenbach, 1984).

5. Compression Mechanisms During Uniaxial Compression Tests

It has been proven from uniaxial compression tests that the compressive mechanisms in fine food powders are different from those in food agglomerates. In fine powders, the first stage of the compression process involves the movement of particles toward filling voids similar to or larger in size than the particles themselves. The packing characteristics of particles or a high interparticulate friction between them will prevent any further interparticulate movement (Nyström and Karehill, 1996). The second stage involves filling of smaller voids by particles that are deformed either elastically (reversible deformation) and/or plastically (irreversible deformation), and eventually broken down (Kurup and Pilpel, 1978; Carstensen and Hou, 1985; Duberg and Nyström, 1986). Most of the organic compounds exhibit consolidation properties, undergoing particle fragmentation during the initial loading, followed by elastic and/or plastic deformation at higher loads.

In agglomerates, bulk compression takes place in three distinct stages: (1) agglomerate particle rearrangement, to fill the voids similar to or larger in size than the agglomerates; (2) agglomerate deformation or brittle breakdown; (3) primary particle rearrangement, elastic and plastic deformation, and fracture (Mort et al., 1984; Nuebel and Peleg, 1994). These steps can be overlapped depending on the type of agglomerate being used. An instant agglomerate obtained from spray- or freeze drying (e.g., coffee, milk, instant juices) will provide a brittle breakdown whereas a granulated material (granulated or encapsulated powders) will show plastic or elastic components during compression. The main difference between bulk compression and individual compression of agglomerates is the bulk's cushioning effect among the particles, which reduces the amount of fracture. Cushioning phenomena is common for brittle cellular solid foams such as coffee and milk agglomerates or powder cheese.

Both compression mechanisms in fine and agglomerated powders are influenced by particle size and size distribution, particle shape, and surface properties. Potato starch and powdered milk powders have been demonstrated to crackle during compression by changing in volume discontinuously (Gerritsen and Stemerding, 1980). If the material is packed in a loose state it shows considerable compressibility. Furthermore, when compressive forces are applied, these forces are transmitted at the contact points.

A decrease in compact porosity with increasing compression load is normally attributed to particle rearrangement with elastic deformation, plastic deformation, and particle fragmentation. Scanning electron microscopy for the qualitative study of volume-reduction mechanisms (Figure 12) has been presented in the literature.

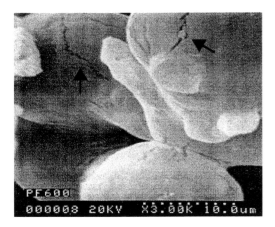

Figure 12 Fracture lines in a milk powder agglomerate after compression. (From Yan, H. and Barbosa-Cánovas, G.V. 2000. Doctoral thesis. Biological Systems Engineering Department, Washington State University, pp. 82–113. With permission.)

D. Surface Area

Another physical characteristic of food powders is the surface area of solids, which can also be a measure of a powder's porosity. The specific surface area of a powder is generally represented by the total particles contained in a unit mass of powder, that is, the internal and external surfaces that can be measured using gases and liquids (Chickaizawa and Takei, 1997) through gas permeametry and gas adsorption methods. If distributions of particle size and shape in a powder are known, the surface area of a powder S_{specific}, can be calculated from the relationship of the particle surface and its volume.

$$S_{\mathrm{specific}} = \frac{S}{V\rho} = \frac{3 \sum_{i=1} R_i^2 N_i}{\rho \sum_{i=1} R_i^3 N_i} \tag{12}$$

where R_i is the average particle radii, N_i is the number of particles, and i is the size range. If particles were perfect spheres, S_{specific} would have a simpler expression (Beddow, 1997):

$$S_{\mathrm{specific}} = \frac{3}{\rho R} \tag{13}$$

Specific surface is important in applications such as powder coating and agglomeration, and for flow and heat transfer studies, where the process is surface dependent. Surface dependent phenomena such as permeametry and gas adsorption (chemisorption, physisorption) can be used for surface area measurement.

Permeametry is based on measuring the permeability of a packed bed of powder in a laminar gas flow, through pressure drop measurements. Evaluation of the resistance of fluid flow through a compact bed of powder provides an idea of the surface area of the powder solids. The fluid mainly used is air since liquids can give erroneous results due to adsorption and aggregation of fine particles. Permeametry is generally suitable for powders of average particle size between 0.2 and 50 μm but it can also be used with coarser powders (up to 1000 μm) using suitably scaled-up test equipment. With highly irregular particles, such as platelets or fibers, errors are mostly particle shape dependent.

Commercial instruments such as the Lea and Nurse apparatus compress the sample to a known porosity in the permeable cells at constant flow or constant pressure drop (Figure 13). Dry air, drawn by an aspirator or a pump, flows through the bed at a constant rate and then passes through a capillary that serves as a flow meter. Static pressure drop across the powder bed is measured with a manometer as static head h_1, while the flow rate is measured by means

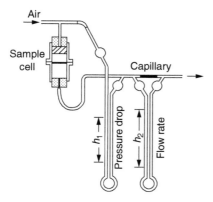

Figure 13   Lea and Nurse apparatus for permeametry measurements.

of the capillary flow meter, giving a reading h_2 on the second manometer. Both pressure drops are small compared with atmospheric pressure and, thus, the compressibility of the gas can be neglected. Permeability types of equipment are applicable for viscous flow using the Carman–Kozeny (1956) equation, which can be simplified to include the static head measurements h_1 and h_2 as follows:

$$S_w = \frac{14}{\rho_s(1 - \varepsilon)} \sqrt{\frac{\varepsilon^3 A h_1}{cLh_2}} \tag{14}$$

where A is the cross-sectional area of the bed, ρ_s is the solid's density, and c is the flow meter conductance. Other constant flow instruments used for constant air flow are the Fisher subsieve sizer, and those by Blaine and Arakawa-Shimadzu (Chikazawa and Takei, 1997). There are also instruments for measuring surface area by permeametry that operate on a variable flow mode (constant volume), such as the Griffin and George permeameter (oil suction), as well as the Reynolds and Branson apparatus (Parfitt and Sing, 1976).

Gas adsorption methods measure the surface area of powders from the amount of gas adsorbed on the powder surface by determining the monolayer capacity and cross-sectional area of an adsorbate molecule. Nitrogen is most commonly used as operating gas, as well as organic adsorptives such as benzene and carbon tetrachloride. The sample adsorbs the adsorptive gas molecules and the vapor pressure gradually decreases until equilibrium is attained. The absorbed amount is determined from the difference between the amount of introduced gas and the residual amount, and measured using helium gas, which is not adsorbed.

III. CHEMICAL AND PHYSICOCHEMICAL PROPERTIES

Chemical reactions in food powders mainly occur due to the presence of free water within the particle or bulk. Temperature gradients can provide opportunities for water diffusion and interactions between particle proteinated and hydrophobic compounds to form new products. However, the most relevant chemistry-related properties in food powder technology are linked to its solubility during processing and consumption, and its caking capacity during storage. For instance, the rate at which dried foods pick up and absorb water has received attention from food drying technologists (Masters, 1985).

A. Instant Properties

The term "instant" is commonly used in the food industry to describe dispersion and dissolution properties of powders. Instant milk powders, coffee, cocoa, baby foods, soups, sauces, soft drinks, sugar mixtures are among the different products available commercially. The instantization process provides food powders with the "instant" attributes so that they can dissolve or disperse more readily in aqueous liquids than when they are in their original powder forms (Schubert, 1980). The instantization process is related to the formation of agglomerates, which have better instant properties than fine particles. Some definitions related to powder dispersion better describe the behavior of agglomerated powders in contact with water.

Instant properties of agglomerates are the most desirable properties in the agglomeration process and they can be measured by the four dissolution properties (Schubert, 1987a) corresponding to the four phases of dissolution, that is, wetting, sinking, dispersing, and solution. *Wettability* refers to liquid penetration into a porous agglomerate system due to capillary action or the ability of agglomerates to be penetrated by the liquid, and describes the capacity of particles to absorb water on their surface (Pietsch, 1999). This property largely depends on particle size. Increasing particle size and/or agglomerating particles can reduce the incidence of clumping. The composition of the particle surface such as the presence of free fat on the surface can affect wettability by reducing it. Surface-active agents, such as lecithin, improve wettability and give instant characteristics to fat-rich powders like cocoa.

Sinkability describes the ability of agglomerates to sink below the water surface quickly, which is controlled by the mass of agglomerates (Pietsch, 1999). It mainly depends on the particle size and density, since larger and denser particles usually sink faster than finer and lighter ones. Particles with a lot of included air may exhibit poor sinkability because of their low bulk density. *Dispersibility* is the ease with which the powder can be distributed as single particles over the surface and throughout the bulk of the reconstituting water. Dispersability is reduced by clump formation and is improved when the sinkability is high. *Solubility* refers to the rate and extent to which the components of the powder particles dissolve in the water. It mainly depends on the chemical composition of the powder and its physical state.

1. Evaluation of Instant Properties

Wetting time is the most important parameter in measuring instant properties. Evaluation procedures are set to a maximum allowable dissolution time when evaluating instant properties for quality control in different food powder products. Standard procedures for measuring instant properties are to define the specific solvent temperature, liquid surface area, amount of material to dissolve, method of depositing a certain amount of material onto the liquid surface, unassisted or predetermined mixing steps, and timing procedure (Pietsch, 1999).

Different protocols have been standardized that portray food powder instant properties. Among them are the penetration speed test, standard dynamic wetting test, and standard methods for dispersibility and wettability. The *penetration speed test* uses a test cell with a screen carrying a layer of agglomerates that is retained with a plexiglass cylinder. The test cell is put into water and penetration time is measured until the entire material bed is submerged (Pietsch, 1999). The *dynamic wetting test* uses a measuring cell attached to a weighing cell that is placed in contact with the liquid by tilting the cell onto the liquid surface (Figure 14). The force measured by the weighing cell is proportional to the liquid volume absorbed due to capillary pressure, and is plotted against time (Schubert, 1980; Pietsch, 1999).

The International Dairy Federation (IDF) established the *dispersibility test* (Schubert, 1987a; Piestch, 1999), which applies light transmission to dispersed agglomerated food powders. Another instrument, using the same light transmission mechanism, measures wettability, dispersibility, and solubility in the same test providing more information on the progress of wetting, dispersion, and dissolution. The *IDF standard method* (IDF, 1979) is specifically designed to determine the dispersibility of instant dried milk and is a rapid routine method to determine wettability (i.e., wetting time).

B. Stickiness in Food Powders

Stickiness is the tendency to adhere to a contact surface. It can refer to particle–particle adhesion or bulk caking (Aguilera et al., 1995), adhesion to packages and container walls, or handling and processing equipment. In fact, stickiness is a prevalent problem that can cause lower product yield, operational problems, equipment wear, and fire hazards (Adhikari

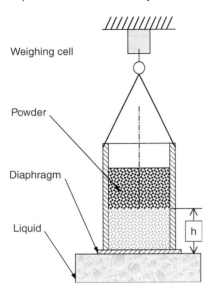

Figure 14 Dynamic wetting test — determination of the amount of liquid volume absorbed by capillary forces. (From Pietsch, W. 1999. *Chem. Eng. Prog.* 8: 67–81. With permission.)

et al., 2001). Interaction of water with solids is the prime cause of stickiness and caking in low-moisture food powders.

Particularly, caking can be an unwanted phenomenon by which amorphous food powders are transformed into a sticky undesirable material, resulting in loss of functionality and lowered quality (Aguilera et al., 1995). Water provides plasticity to food polymeric systems, reducing viscosity and enhancing the molecular mobility of the system, which is linked to liquid and solid bridges formation and caking (Papadakis and Bahu, 1992; Peleg, 1993). Chemical caking is the most common type of caking mechanism, and is caused by chemical reactions in which a compound has been generated or modified, as in decomposition, hydration, dehydration, recrystallization, or sublimation. Plastic-flow caking occurs when the particles' yield values are exceeded and they stick together or merge into a single particulate form as in amorphous materials like tars, gels, lipids, or waxes.

Glass transition temperature can characterize the stickiness during powder storage (Roos and Karel, 1991). The presence of glassy low-molecular-weight materials, such as glucose, fructose, and sucrose in spray- and freeze-dried fruit powders, makes these powders particularly susceptible to stickiness. This is due to their elevated hygroscopicity as well as chemical and conformational changes at increased temperatures. Table 3 describes different types of attractive particle–particle phenomena contributing to powder cohesion and stickiness, which will be described next.

1. Bridging

In general, interaction between particles is regulated by the relationship between the strength of the attractive (or repulsive) forces and gravitational forces. For all particles in the amorphous rubbery state (or above glass transition temperature), forces causing primary particles to stick together are: interparticle attraction forces (van der Waals or molecular forces; electrostatic forces), liquid bridges, and solid bridges (Schubert, 1981; Hartley et al., 1985). Interparticle forces are inversely related to the particle size (Buma, 1971; Rennie et al.,

Table 3 Cohesive Phenomena Contributing to Particle–Particle Attraction and Stickiness

Phenomena	Dissociation energy	Characteristic
Solid bridges	50–200 (kcal/mol)	Solidified from liquid bridges. The magnitude of the adhesion force depends on the diameter of the contact area and the strength of the bridge material
Liquid bridges	Dependent on the composition of the powder	Due to movable liquids (capillary and surface tension properties) and nonfreely movable binder bridges (viscous binders and adsorption layers). Attraction is caused by a potential difference from the imbalance due to the presence of these charges. Becomes important in powders with small particle size (under $100 \, \mu$m)
Intermolecular forces	1–10 (kcal/mol)	Divided into hydrogen bonding and van der Waals forces (from electron motion among dipoles). They act over very short distances within the material structure (prevalent in particles under $1 \, \mu$m)
Electrostatic forces	Dependent on the particle surface, shape, external electric field	Electrostatic forces in the surface from foreign ions or when electron transference occurs by contact charging
Mechanical interlocking	Dependent on the particle surface and shape	Particles of irregular or fibrous shapes under vibration or pressure. Fibrous, bulky, and flaky particles interlock or fold about each other when consolidating stress is applied

Source: From Rumpf, H. 1962. *Agglomeration*. Knepper, W. (Ed.), pp. 379–418; Peleg, M. 1978. *J. Food Process. Eng.* 1: 303–328; Peleg, M. 1983. *Physical Properties of Foods*. Peleg, M. and Bagley, E.B. (Eds.). Van Nostrand Reinhold/AVI, New York, pp. 293–324; Schubert, H. 1987a. *J. Food Eng.* 6: 22–26; Aguilera, J.M., Valle, J.M., and Karel, M. 1995. *Trends Food Sci. Technol.* 6: 149–154. With permission.

1999; Adhikari et al., 2001). van der Waals and electrostatic attraction are not as high as the interparticle connecting force coming from liquid bridges (Schubert, 1987a). A diagram showing the strength of the agglomerate bond as a function of particle size is given in Figure 15.

Van der Waals forces arise from electron motion among dipoles and act over very short distances within the material structure, becoming prevalent when the particle size is less than $1 \, \mu$m (Hartley et al., 1985). Electrostatic forces are longer ranging forces that arise through surface differences on particles and are present when the material does not dissipate electrostatic charge.

Liquid bridges are related to chemical interactions between material particle components and result from the presence of bulk liquid (generally unbound water or melted lipids) between the individual particles. The forces of particle adhesion arise either from surface tension of the liquid/air system (as in the case of a liquid droplet) or from capillary pressure. According to Rumpf (1962), liquid bridges offer strong bonding, which has practical significance in particulate aggregates ranging 1 mm or less in size. The strength of a liquid bridge depends on factors affecting the contact angles (e.g., composition of the solid and nature of the liquid solution) and factors influencing the radius of curvature (particle size, shape,

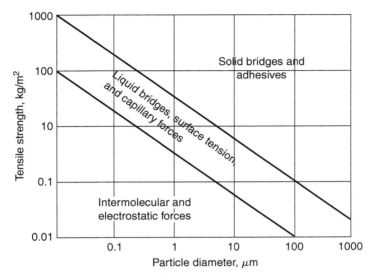

Figure 15 Strength of agglomerates and fine particle bonding as a function of particle size. (From Adhikari, B., Howes, T. Bhandari, B.R., and Truong, V. 2001. *Int. J. Food Prop.* 4: 1–33. With permission.)

interparticle distance, particle roughness, and ratio of liquid amount to solid amount in the agglomerate). Composition of the liquid in the bridge varies in different food materials. The "bridging potential" or "stickiness" is related to factors such as powder moisture, fat or low-molecular-weight sugar content and shape of particles.

Solid bridges form as a result of sintering, solid diffusion, condensation, or chemical reaction, and generally result from solidification at room temperature. The magnitude of the adhesion force depends on the diameter of the contact area and the strength of the bridge material (Loncin and Merson, 1979). Solid bridges are the structural bonds that hold individual particles in an agglomerate together as well as in a powder cake (e.g., bridges formed during fruit or coffee powder storage).

2. Thermodynamic Adsorption

If a type of particle is damped with a spreadable fluid or semisolid food, it will absorb the fluid at different extents depending on the particle surface. Complete wetting of the surface will occur when the surface energy of the adherend is greater than that of the adhesive (Saunders et al., 1992). Low-energy materials absorb strongly on high-energy surfaces, lowering the surface energy of the system.

The mechanism of thermodynamic adsorption is based on Dupre's energy equation (Michalski et al., 1997; Adhikari et al., 2001), which relates the adhesive and adherend surface tensions of the solid and liquid to the work of the adhesion, Ws (J/m^2):

$$Ws = \gamma s + \gamma l \cdot \gamma sl \tag{15}$$

where γs (N/m) and γl (N/m) are the solid and liquid surface tensions and γsl (N/m) is the solid–liquid interfacial tension. Adhesion is attributable to electrodynamic inter-molecular forces acting at the liquid–liquid, liquid–solid, and solid–solid interfaces, and to the interfacial attraction between the adhesive and adherend in terms of the reversible work of adhesion corresponding to material surface tensions. The work of adhesion is now recognized as a function of the reversible work of adhesion and the irreversible deformation of the substrate (Shanahan and Carre, 1995; Michalski et al., 1997).

3. Cohesion and Cohesion Properties

There are two types of powders when referring to powder flowability: noncohesive powders and cohesive powders. Noncohesive (or "free flowing") powders are those in which interparticulate forces are negligible. Most powders are considered noncohesive only when dry or when particle size is above 100 to 200 μm (Peleg, 1978; Teunou et al., 1999). The finer powders are more susceptible to cohesion and their flowability is more difficult (Adhikari et al., 2001).

Powder cohesion occurs when interparticle forces play a significant role in the mechanical behavior of a powder bed. The dynamic behavior of powder seems also to be determined by interparticle forces and packing structure. Compressive and compaction behavior of powders is very important in evaluating flowability, since methods to determine flow properties (e.g., angle of repose and angle of internal friction) account for compressive and compacting mechanisms.

A standard for cohesive strength determination is the ASTM D6128 (or Jenike method), where consolidating conditions found in the depth of a bin are reproduced in a shear cell. The shear cell is used to determine the maximum shear force (or yield point) that a powder can undergo under a predetermined compression load. A plot (*yield locus*) of maximum shear stress versus the corresponding normal consolidating stress gives a curve where two parameters can be obtained (Peleg, 1978): *cohesion C* and *angle of internal friction*.

Cohesion C is a measure of the attraction between particles and is due to the effect of internal forces within the bulk, which tend to prevent planar sliding of one internal surface of the particles on another. Cohesion has been proven proportional to *tensile stress* (Peleg, 1978), which is necessary in separating two bulk portions of a food powder sample. The angle of internal friction is a measure of the interaction between particles and is calculated from the slope of the yield locus. In free-flowing powders, it represents the friction between particles when flowing against each other. Therefore, it depends on their size, shape, roughness, and hardness.

The *static angle of repose* is defined as the angle at which a material will rest on a stationary heap; the angle θ is formed by the heap slope and the horizontal when the powder is dropped on a platform. A higher angle will indicate a greater degree of cohesion, which is also the actual flowability measurement applied by some laboratories in the food industry for quality control.

4. Test Methods

Various test methods have been adapted for characterization of food stickiness. Among them, the shear cell and glass transition methods have received the widest usage and provide a good degree of automation. Other methods such as the optical probe and the sticky point have received less attention. In particular, the optical probe is a novel method that provides a good quantification of stickiness, and may prove very useful for online monitoring of stickiness in powders as a function of moisture and temperature.

Measurement of cohesive and adhesive forces, based on a *shear cell*, has been widely used by powder industries to characterize the degree of powder flowability (Peleg, 1978; Teunou et al., 1999). A shear cell is a box-like structure split in half horizontally, equipped with a provision for applying various normal stresses (σ) and shear stresses (τ) on top. The powder being tested is preconsolidated and presheared with the maximum applicable consolidation load. If a sample is tested at a different normal load, one can obtain a particular shear stress, causing failure (initiation of flow) and disruption of the cohesive structure generally formed in bins and hoppers.

Glass transition temperature of a system predicts the bulk temperature and moisture matrix that renders the product sticky. The differential scanning calorimeter (DSC) measures changes in specific heat capacity (C_p). The Nuclear Magnetic Resonance method tracks the phase change by providing information on the mobility of the entire molecule and that of individual groups or regions within a food structure (Blanshard, 1993). Mechanical–thermal methods measure the changes in loss modulus, elastic modulus, and their ratio as a function of temperature. A sharp change in these properties is indicative of glass transition temperature (Rahman, 1995). Up to now, glass transition temperature has been used to characterize low-molecular-weight carbohydrate systems (Adhikari et al., 2001).

In the *optical probe* method (Lockemann, 1999) a rotating sample of a free-flowing solid with known moisture content, is subjected to a temperature program in an oil bath. Temperature is slowly raised while the test tube is rotated at a slow speed. The motion of the product is observed with an optical sensor and recorded. The sticking point is the temperature at which the product assumes a different flow behavior, thus causing a sharp rise in the reflectance measured by the sensor.

The *sticky point* tester (Wallack and King, 1988) consists of a test tube immersed in a controlled temperature bath. Temperature is slowly raised and the sample is stirred with a machine-driven impeller using double and curved blades imbedded in the sample. When the force experienced to drive the sample stirrer is at maximum, the sticky point temperature is reached. Generally, this test fails to provide information regarding the cohesiveness status of a powder below the sticky point temperature (Peleg, 1993). However, it is based on the softening of powders above their glass transition temperature when the glassy phase yields to a rubbery state, because the plasticization of the particle surface accelerates once the glass transition temperature is exceeded (Peleg, 1993).

C. Water Activity and Glass Transition Temperature

The interaction of water with solids is the prime cause of stickiness and caking in low-moisture particulates. Particle surfaces include adsorbed mono/multilayers of water or free water from capillary condensation. As a plasticizer, water can reduce the surface microroughness of the particles, which allows closer particle–particle contact, thereby increasing forces of attraction between particles. This may be an important cause of caking of particles in the presence of water (Adhikari et al., 2001). In fact, an increase in water content will depress the glass transition temperature (T_g) of particulates.

Generally, food powder surfaces are amorphous in nature and undergo a phase change when they reach T_g. If an amorphous food is stored at a temperature higher than T_g, they are turned into its rubbery state (Figure 16) associated with stickiness and caking (Roos and Karel, 1993; Adhikari et al., 2001). Caking of food powders in storage can be avoided simply by storing them below glass transition temperature (Figure 16), provided moisture adsorption from the environment is avoided (Slade et al., 1993). Since sticky point temperature is 10 to 20°C higher than glass transition temperature (Roos and Karel, 1993; Adhikari et al., 2001), stickiness and adhesion phenomena will not take place below glass transition temperature.

Composition plays a fundamental role in stickiness in combination with factors such as surface relative humidity, temperature, and viscosity. The degree of contribution of food compositional factors to stickiness is presented in Table 4. The T_g in food materials such as sugars, starch, gluten, gelatin, hemicellulose, and elastin decreases rapidly

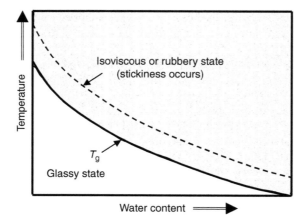

Figure 16   Schematic diagram of the relationship between glass transition temperature T_g, viscosity, water content, and occurrence of stickiness in food powders. (Adapted from Roos, Y.H. 1995. *Phase Transitions in Foods*. Academic Press, New York, pp. 193–245. With permission.)

Table 4   Relative Contribution to Food Powder Stickiness

Factors	Relative contribution to stickiness	
Protein	0	
Polysaccharides	0	
Fats	+	
Low molecular sugars	++	
Water/Relative humidity	+++	
Particle size distribution		+
Compression pressure		++
Temperature	+++	
Viscosity	+++	

Note: 0 base point (negligible), + high, ++ higher, +++ highest.
Source: From Adhikari, B., Howes, T., Bhandari, B.R., and Truong, V.
         2001. *Int. J. Food Prop*. 4: 1–33. With permission.

to about 10°C when dry basis moisture content increases to 30% (Atkins, 1987). Glass transition temperatures of these powders in their anhydrous state range between 5 to 100°C. Low-molecular-weight sugars play important roles in stickiness and caking in food-based particulate systems due to their hygroscopicity and solubility. Low-molecular-weight protein hydrolyzates behave similarly; for example, some amino acids were found to exhibit hygroscopicity and contribute to caking during storage of spray-dried fermented soy sauce powders (Hamano and Aoyama, 1974).

Conversely, food powders with high-fat content could also induce powder caking if stored above the fat melting point. Above room temperature, liquefaction of fat provokes powder softening, thereby increasing the contact area and leading to the formation of liquid bridges (Adhikari et al., 2001).

IV. APPLICATION OF COMPRESSION IN FOODS

The general behavior of powders under compressive stress or compaction due to mechanical motion fits into several processing operations where physical and chemical properties of food powders tend to change under preset temperature and relative humidity conditions. Food powder compression is involved in different industrial applications such as bulk size reduction, grinding, particle size enlargement, food tablet production, encapsulated material resistance, hopper and silo storage, mixing, and packaging. As mentioned before, handling operations are related to the mechanical properties of materials (shear stress, tensile and compressive stress), during which brittle agglomerated food powders can suffer the nondesirable effects of particle breakage or attrition.

A. Particle Size Reduction

Particle size reduction is a comminution process that includes operations such as crushing, grinding, and milling (e.g., milling of cereals, grinding of spices or coffee), a process in which a food is deformed until breakage or failure occurs. During compression, breaking of hard materials can occur along cracks or defects in the structure. Forces commonly used in food processes for particle size reduction are called compressive, impact, attrition (or shear), and cutting forces. For example, crushing rolls type of mills require compressive forces, hammer mills rely on impact forces, disc mills use shear forces for particle attrition, and rotary knife cutter mills use cutting forces. For particle shapes, a large piece of food with many defects can be broken down with low stress and very little deformation, while smaller pieces with fewer defects will require a higher breaking strength. For food materials with fibrous structures, shredding or cutting should be considered for the desired size reduction. Hard brittle materials like sugar crystals can be crushed, broken by impact, or ground by abrasion, while ductile cocoa is better reduced by impact (e.g., in hammer mill).

B. Particle Size Enlargement

Size enlargement operations include agglomeration (by compaction, granulation, tableting, briquetting, sintering) and powder encapsulation. These processes depend on material resistance properties as well as chemical properties like adhesiveness and stickiness. Agglomeration is a process aimed at controlling particle porosity and density and involves the aggregation of dispersed materials into those with larger units held together by adhesive and/or cohesive forces (Bika et al., 2001).

Rewetting agglomeration (or instantization process) is used in food processes partly to improve properties related to handling, while pressure agglomeration involves particulate confinement by compression into a mass that is then shaped and densified. Agglomerates formed by rewetting agglomeration develop lower strength levels primarily because they feature higher porosity from coalescence, while pressure agglomeration causes porosity to decrease as density and strength increase. By agglomerating fine powders to a size of about 100 μm into particles of several millimeters, the wetting behavior of the particles is improved and lump formation can be avoided (Schubert, 1987a). Some instantization processes are classified in Table 5. In rehydration operations, water aided by capillary forces penetrates into the narrow spaces between fine lumps of particles (APV, 1989).

Agglomerate shaping takes place during pressure agglomeration by forcing the material through the holes. Chemical caking plays an important role in briquetting and tableting where compaction of food ingredients such as dextrose, gelatin, glucose, sucrose, lactose, or starch takes place using food gums as binders.

C. Encapsulation

Encapsulation is a process where a continuous thin coating is formed around solid particles (e.g., powdered sweeteners, vitamins, minerals, preservatives, antioxidants, cross-linking

Table 5   Some Examples of Instantization Processes

Agglomeration	Nonagglomeration
Spray drying and agglomeration	Freeze drying
Rewetting agglomeration	Drum drying
Spray-bed dryer agglomeration	Additives (e.g., lecithin)
Press agglomeration	Fat removal through hexane extraction
Thermal treatment using a stream of moist air or a hot surface	

Source: From Pintauro, N. 1972. *Food Process. Rev.* 25. With permission.

agents, leavening agents, colorants, and nutrients) in order to create a capsule wall (King, 1995; Risch, 1995). Encapsulation promotes easier handling of the core or interior material by: (1) preventing lumping, (2) improving flowability, compression, and mixing properties, (3) reducing core particle dustiness, and (4) modifying particle density (Shahidi and Han, 1993). In particular, multiwall structured capsules contain different concentric layers of the same or different composition and earn greater resistance to attrition during handling. Microencapsulation through spray drying of fats reduces the adhesiveness (reducing clumping and caking), and enhances handling properties during storage transport and blending with nonfat ingredients.

D.   Food Powder Handling

Compaction, conveying, mixing, and metering among other types of food powders' handling can provoke attrition (Schubert, 1987a), causing problems such as changes in bulk properties, segregation and, in some agglomerates, loss of instantization ability. Mixing mechanisms are sometimes affected by compaction properties such as mechanical interlocking, surface attraction, plastic welding (from high pressures between small contact areas), and electrostatic attraction, as well as environmental factors such as ambient moisture and temperature fluctuations.

E.   Storage Operations

Evaluation of compression and cohesion properties plays a key role in the study of stresses due to storage in high bins, hoppers, or silos. Mechanical compressibility can give an idea of bulk density changes due to compacting pressure in stored powders (Thomson, 1997). During storage, arches and ratholes are two of the most common flow obstruction problems in powdered materials stored in hoppers and bins, where cohesive strength can cause stagnant material to bind together or interlock, forming a narrow channel above the outlet. Since rathole material remains under storage, it can cake or degrade. Unconfined yield strength of the bulk solid is the main property associated with arching and rathole formation.

V.   RESEARCH UPDATE ON FOOD POWDER PROPERTIES

The latest research on food powder bulk properties includes the study of electrostatic coating, the use of padding materials, and ultimate bulk density.

Electrostatic coating properties. Electrostatic coating of food has been demonstrated as a method to improve the powder coating efficiency of crackers, chips, and shredded cheese, thus reducing powder waste and dust and improving coating evenness (Ricks et al., 2002). The electrostatic coating method is based on charging particles electrically so that they repel each other and seek a grounded surface (to be coated). Flow properties (e.g., cohesion, angle of repose, particle density; including Hausner ratio and flow index) were studied for selected food powders to predict transfer efficiency and dustiness. In this case, particle chargeability played an important role in electrostatic transfer efficiency.

Padding materials. Cushioning materials can be placed in a powder bed of agglomerates or attached to the container wall to absorb or reduce the mechanical or static impact, thereby reducing the degree of attrition. Yan and Barbosa-Cánovas (2001b) studied the padding effect on agglomerated coffee and nonfat milk using pure polyurethane foam as padding material, which was compressed by confined uniaxial compression with a UTM. The padding effects were evaluated according to the padding index, which indicated whether lower deformation occurred when padding material was located on the container wall or in the powder bed itself. The index showed the padding efficiency related to powder attrition, indicating the amount of powders retaining their original particle size when padding material was used. Compression force was lower for instant coffee when padding was added to the compression cell while force-deformation curves for instant milk had an exponential characteristic. The padding efficiency was higher for lower strength milk agglomerates than for more resistant agglomerates (such as coffee) when padding was used.

Ultimate bulk density. High hydrostatic pressure (HHP) is a new possibility among the more traditional powder compression methods. In this case, higher pressure provides higher bulk density up to a critical pressure in which no volume changes occur (ultimate bulk density). Yan et al. (2001) recently studied the effects of HHP processing times and particle size on the bulk density of instant nonfat milk, spray-dried coffee, and freeze-dried coffee. Ultimate bulk density depends on product formulation, physical properties of the ingredients, and production conditions. For similar kinds of agglomerates, even though the initial particle size, bulk density, or water activity may be different, the ultimate bulk density is not significantly different under the same pressure. The ultimate bulk density concept could be a promising tool for use in detecting composition variations due to changes in product formulations or manufacturing conditions.

REFERENCES

Abdullah, E.C. and Geldart, D. 1999. The use of bulk density measurements as flowability indicators. *Powder Technol.* 102: 151–165.

Abrahamsen, A.R. and Geldart, D. 1980. Behavior of gas-fluidized beds of fine powders, Part I. Homogeneous expansion. *Powder Technol.* 26: 35–46.

Adams, M.J., Mullier, M.A., and Seville, J.P.K. 1994. Agglomerate strength measurement using a uniaxial confined compression test. *Powder Technol.* 78: 5–13.

Adhikari, B., Howes, T., Bhandari, B.R., and Truong, V. 2001. Stickiness in foods: a review of mechanisms and test methods. *Int. J. Food Prop.* 4: 1–33.

Aguilera, J.M., Valle, J.M., and Karel, M. 1995. Review: caking phenomena in food powders. *Trends Food Sci. Technol.* 6: 149–154.

ASTM. 1986. *Compilation of ASTM Standard Definitions.* 6th ed. ASTM, Baltimore, MD.

ASTM D441-86. 2002. *Standard Test Method of Tumbler Test for Coal*. ASTM International.

ASTM D6683-01. 2003. *Standard Test Method for Measuring Bulk Density Values of Powders and Other Bulk Solids*. ASTM International.

APV. 1989. *Dryer Handbook (DRH-889)*. APV Crepaco Inc., Rosemont, IL.

Atkins, A.G. 1987. The basic principles of mechanical failure in biological systems. In: *Food Structure and Behavior*. Blanshard, J.M.V. and Lillford, P.J. (Eds.). Nottingham University Press, England, pp. 149–176.

Barbosa-Cánovas, G.V., Malavé-López, J., and Peleg, M. 1987. Density and compressibility of selected food mixtures. *J. Food Eng.* 10: 1–19.

Barbosa-Cánovas, G.V., Ortega-Rivas, E., Juliano, P. and Yan, H. 2005 *Food Powders. Physical Properties, processing and functionality*. Springer-Verlalg, New York, NY.

Barletta, B.J. and Barbosa-Cánovas, G.V. 1993a. Fractal analysis to characterize ruggedness changes in tapped agglomerated food powders. *J. Food Sci.* 58: 1030–1046.

Barletta, B.J. and Barbosa-Cánovas, G.V. 1993b. An attrition index to assess fines formation and particle size reduction in tapped agglomerated food powders. *Powder Technol.* 77: 89–93.

Barletta, B.J., Knight, K.M., and Barbosa-Cánovas, G.V. 1993a. Review: attrition in agglomerated coffee. *Rev. Esp. Cienc. Tecnol. Aliment.* 33: 43–58.

Barletta, B.J., Knight, K.M., and Barbosa-Cánovas, G.V. 1993b. Compaction characteristics of agglomerated coffee during tapping. *J. Texture Stud.* 24: 253–268.

Beddow, J.K. 1997. *Image Analysis Sourcebook*. American Universities Science and Technology Press, Santa Barbara, CA.

Bemrose, C.R. and Bridgwater, J. 1987. A review of attrition and attrition test methods. *Powder Technol.* 49: 97–126.

Bika, D.F., Gentzle, M., and Michaels, J.N. 2001. Mechanical properties of agglomerates. *Powder Technol.* 117: 98–112.

Blanshard, J.M.V. 1993. The glass transition, its nature and significance in food processing. In: *Glassy State in Foods*. Blanshard, J.M.V. and Lillford, P.J. (Eds.). Nottingham University Press, England, pp. 18–48.

Buma, T.J. 1971. Free fat in spray-dried whole milk 5. Cohesion: determination, influence of particle size, moisture content and free-fat content. *Neth. Milk Dairy J.* 25: 107–122.

Carr, E. 1965. Evaluating flow properties of solids. *Chem. Eng.* 72(2): 163–168.

Carr, R.L. 1976. Powder and granule properties and mechanics. In: *Gas–Solids Handling in the Processing Industries*. Marchello, J.M. and Gomezplata, A. (Eds.). Marcel Dekker, New York.

Carstensen, J.T. and Hou, X.P. 1985. The Athy–Heckel equation applied to granular agglomerates of basic Tricalcium Phosphate $[3Ca_3PO_4.Ca(OH)_2]$. *Powder Technol.* 42: 153–157.

Chikazawa, M. and Takei, T. 1997. III. Fundamental Properties of powder beds adsorption characteristics. In: *Powder Technology Handbook*. Gotoh, K., Masuda, H., and Higashitani, K. (Eds.). Marcel Dekker, New York, pp. 245–264.

Couroyer, C., Ning, Z., and Ghadiri, M. 2000. Distinct element analysis of bulk crushing: effect of properties and loading rate. *Powder Technol.* 109: 241–254.

Duberg, M. and Nyström, C. 1986. Studies of direct compression of tablets. XVII. Porosity–pressure curves for characterization of volume reduction mechanisms in powder compression. *Powder Technol.* 46: 67–75.

Ehlerrmann, D.A.E. and Schubert, H. 1987. Compressibility characteristics of food powders: characterizing the flowability of food powders by compression tests. In: *Physical Properties of foods-2*. Jowitt, R., Escher, F., Kent, M., McKenna, B., and Roques, M. (Eds.). *Cost 90bis Final Seminar Proceedings*. ECSC, EEC, EAEC, Brussels and Luxembourg.

Geldart, D., Harnby, N., and Wong, A.C.Y. 1984. Fluidization of cohesive powders. *Powder Technol.* 37: 25–37.

Gerritsen, A.H. and Stemerding, S. 1980. Crackling of powdered materials during moderate compression. *Powder Technol.* 27: 183–188.

Ghadiri, M. and Ning, Z. 1997. Effect of shear strain rate on attrition of particulate solids in a shear cell. Powders & Grains 97. *Proceedings of the 3rd International Conference on Powders & Grains*, Durham, NC.

Ghadiri, M., Yuregir, K.R., Pollock, H.M., Ross, J.D.J., and Rolfe, N. 1991. Influence of processing conditions on attrition of NaCl crystals. *Powder Technol.* 65: 311–320.

Hamano, M. and Aoyama, Y. 1974. Caking phenomena in amorphous food powders. *Trends Food Sci. Technol.* 6: 149–155.

Hartley, P.A., Parfitt, G.D., and Pollack, L.B. 1985. The role of Van der Waals force in agglomeration of food powders containing submicron particles. *Powder Technol.* 42: 35–46.

Hayes, G.D. 1987. *Food Engineering Data Handbook.* John Wiley and Sons Inc., New York, p. 83.

Head, K.H. 1982. *Manual of Soil Laboratory Testing*, Vol. 2. Pentech Press, London, pp. 581–585.

Hollenbach, A., Peleg, M., and Rufner, R. 1982. Effect of four anticaking agents on the bulk characteristics of ground sugar. *J. Food Sci.* 47(2): 538–544.

Hollman, J.P. 2001. *Experimental Methods for Engineers.* McGraw Hill, New York, p. 340.

IDF. 1979. *International IDF Standard 87: 1979.* International Dairy Federation, Brussels, Belgium.

Jones, S. 1983. The problem of closure in the Zahn–Roskies method of shape description. *Powder Technol.* 34: 93–94.

Kamath, S. 1996. Constitutive parameter determination for food powders using triaxial and finite element analysis of incipient flow from hopper bins. Doctoral thesis. Pennsylvania State University.

King, A.H. 1995. Encapsulation of food ingredients. In: *Encapsulation and Controlled Release of Food Ingredients.* Risch, S.J. and Reineccius, G.A. (Eds.). American Chemical Society, Washington, DC, pp. 26–39.

Konstance, R.P., Onwulata, C.I., and Holsinger, V.H. 1995. Flow properties of spray-dried encapsulated butteroil. *J. Food Sci.* 60: 841–844.

Kostelnik, M.C. and Beddow, J.K. 1970. New techniques for tap density. In: *Modern Developments in Powder Metallurgy.* Hausner, H.H. (Ed.). Plenum Press, New York.

Kumar, M. 1973. Compaction behavior of ground corn. *J. Food Sci.* 38: 877–878.

Kurup, T.R.R. and Pilpel, N. 1978. Compression characteristics of pharmaceutical powder mixtures. *Powder Technol.* 19: 147–155.

Li, F. and Puri, V.M. 1996. Measurement of anisotropic behavior of dry cohesive and cohesionless powders using a cubical triaxial tester. *Powder Technol.* 89: 197–207.

Lockemann, C.A. 1999. A new laboratory method to characterize the sticking properties of free flowing solids. *Chem. Eng. Process.* 38: 301–306.

Loncin, M. and Merson, R.L. 1979. *Food Engineering: Principles and Selected Applications.* Academic Press, New York, pp. 229–271.

Lu, W.-M., Tung, K-L., Hung, S.-M., Shiau, J.-S., and Hwang, K.-J. 2001. Compression of deformable gel particles. *Powder Technol.* 116: 1–12.

Ma, L., Davis, D.C., Obaldo, L.G., and Barbosa-Cánovas, G.V. 1998. Mass and spatial characterization of biological materials. In: *Engineering Properties of Foods and Other Biological Materials.* ASAE publication 14–98 Imprint: Washington State University, Pullman, WA.

Malavé-Lopez, J., Barbosa-Cánovas, G.V., and Peleg, M. 1985. Comparison of the compaction characteristics of selected food powders by vibration, tapping and mechanical compression. *J. Food Sci.* 50: 1473–1476.

Mandelbrot, B.P. 1977. *Fractals, Form, Chance and Dimension.* Freeman, San Francisco, CA.

Mandelbrot, B.P. 1982. *The Fractal Geometry of Nature.* Freeman, San Francisco, CA.

Masters, K. 1985. *Spray Drying Handbook.* 4th ed. George Godwin, London.

Michalski, M.C., Desobry, S., and Hardy, J. 1997. Food material adhesion: a review. *Crit. Rev. Food Sci. Nutr.* 37: 591–619.

Mohsenin, N.N. 1986. *Physical Properties of Plant and Animal Materials.* 2nd ed. Gordonand Breach Science Publishers, Inc. Amsterdam, The Netherlands.

Molina, M., Nussinovitch, A., Normand, M.D., and Peleg, M. 1990. Selected physical characteristics of ground roasted coffees. *J. Food Process. Preservation* 14: 325–333.

Moreyra, R. and Peleg, M. 1980. Compressive deformation patterns of selected food powders. *J. Food Sci.* 45: 864–868.

Mort, P.R., Sabia, R., Niesz, D.E., and Rimon, R.E. 1984. Automated generation and analysis of powder compaction diagram. *Powder Technol.* 46: 67–75.

Nyström, C. and Karehill, P.-G. 1996. The importance of intermolecular bonding forces and the concept of bonding surface area. In: *Pharmaceutical Powder Compaction Technology.* Alderborn, G. and Nytröm, G. (Eds.). Marcel Dekker, New York, p. 17.

Nuebel, C. and Peleg, M. 1994. A research note: compressive stress–strain relationships of agglomerated instant coffee. *J. Food Process Eng.* 17: 383–400.

Olivares-Francisco, C. and Barbosa-Cánovas, G.V. 1990. Characterization of the attrition process in agglomerated coffee by natural fractals. Presented at the IFT annual meeting, Anaheim, CA.

Onwulata, C.I., Konstance, R.P., and Holsinger, V.H. 1996. Flow properties of encapsulated milkfat powders as affected by flow agent. *J. Food Sci.* 1: 1211–1215.

Onwulata, C.I., Smith, P.W., and Holsinger, V.H. 1998. Properties of single- and double-encapsulated butteroil powders. *J. Food Sci.* 63: 100–103.

Papadakis, S.E. and Bahu, R.E. 1992. The sticky issue of drying. *Drying Technol.* 10: 817–837.

Parfitt, G.D. and Sing, K.S.W. 1976. *Characterization of Powder Surfaces.* Academic Press, New York.

Peleg, M. 1978. Flowability of food powders and methods for its evaluation: a review. *J. Food Process. Eng.* 1: 303–328.

Peleg, M. 1983. Physical characteristics of powders. In: *Physical Properties of Foods.* Peleg, M. and Bagley, E.B. (Eds.). Van Nostrand Reinhold/AVI, New York, pp. 293–324.

Peleg, M. 1993. Glass transition and physical stability of food powders. In: *Glassy State in Foods.* Blanshard, J.M.V. and Lillford, P.J. (Eds.). Nottingham University Press, England, pp. 18–48.

Peleg, M., and Hollenbach, A. 1984. Flow conditioners and anticaking agents. *Food Technol.* 38(3): 93–102.

Peleg, M. and Mannheim, H. 1973. Effect of conditioners on the flow properties of powdered sucrose. *Powder Technol.* 7: 45–50.

Peleg, M. and Normand, M.D. 1985. Mechanical stability as the limit to the fractal dimension of solid particle silhouettes. *Powder Technol.* 43: 187–188.

Pietsch, W. 1999. Readily engineer agglomerates with special properties from micro- and nanosized particles. *Chem. Eng. Prog.* 8: 67–81.

Pintauro, N. 1972. Agglomeration processes in food manufacture. *Food Process. Rev.* 25.

Popplewell, L.M. and Peleg, M. 1989. An "Erosion Index" to characterize fines production in size reduction processes. *Powder Technol.* 58: 145–148.

Rahman, S. 1995. *Food Properties Handbook.* CRC Press, Boca Raton, FL.

Rennie, P.R., Chen, X.D., Hargreaves, C., and Mackereth, A.R. 1999. A study of the cohesion of dairy powders. *J. Food Eng.* 39: 277–284.

Ricks, N.P., Barringer, S.A., and Fitzpatrick, J.J. 2002. Food powder characteristics important to nonelectrostatic and electrostatic coating and dustiness. *J. Food Sci.* 67: 2256–2263.

Risch, S.J. 1995. Review of patents for encapsulation and controlled release of food ingredients. In: *Encapsulation and Controlled Release of Food Ingredients.* Risch, S.J. and Reineccius, G.A. (Eds.). American Chemical Society, Washington, DC, pp. 197–203.

Roos, Y. and Karel, M. 1991. Plasticizing effect of water on thermal behavior and crystallization of amorphous food model. *J. Food Sci.* 56: 38–43.

Roos, Y. and Karel, M. 1993. Effects of glass transitions on dynamic phenomena in sugar con- taining food systems. In: *Glassy States in Foods.* Blanshard, J.M.V. and Lillford, P.J. (Eds.). Nottingham University Press, England, pp. 207–222.

Roos, Y.H. 1995. Time-dependent phenomena. In: *Phase Transitions in Foods.* Academic Press, New York, pp. 193–245.

Rumpf, H. 1962. The strength of granules and agglomerates. In: *Agglomeration.* Knepper, W. (Ed.). Interscience Publishers, New York. pp. 379–418.

Saunders, S.R., Hamann, D.D., and Lineback, D.R. 1992. A systems approach to food material adhesion. *Lebensam-Wiss-u-Technol.* 25: 309–315.

Schubert, H. 1980. Processing and properties of instant powdered food. In: *Food Process Engineering.* Vol. 1. Linko, P., Mälkki, Y., Olkku, J., and Larinkari, J. (Eds.). Applied Science Publishers, London, pp. 675–684.

Schubert, H. 1981. Principles of agglomeration. *Int. Chem. Eng.* 6: 1–32.

Schubert, H. 1987a. Food particle technology. Part I: properties of particles and particulate food systems. *J. Food Eng.* 6: 22–26.

Schubert, H. 1987b. Food particle technology. Part II: some specific cases. *J. Food Eng.* 6: 83–102.

Shahidi, F. and Han, X.-Q. 1993. Encapsulation of food ingredients. *Crit. Rev. Food Sci. Human Nutr.* 33: 501–547.

Shanahan, M.E.R.; Carre, A. Viscoelastic Dissipation in Wetting and Adhesion Phenomena. *Langmuir.* 1995, 11: 1396–1402.

Simons, S.J.R. 1996. Modeling of agglomerating systems: from spheres to fractals. *Powder Technol.* 87: 29–41.

Slade, L., Levine, H., Ievolella, J., and Wang, M. 1993. The glassy state phenomenon in applications for food industry: application of food polymer science approach to structure–function relationships of sucrose in cookie and cracker systems. *J. Sci. Food Agric.* 63: 133–176.

Teunou, E., Fitzpatrick, J.J., and Synnott, E.C. 1999. Characterization of food powder flowability. *J. Food Eng.* 39: 31–37.

Thomson, F.M. 1997. Storage and flow of particulate solids. In: *Handbook of Powder Science & Technology.* Fayed, M.E. and Otten, L. (Eds.). Chapman and Hall, New York, pp. 389–436.

Wallack, D.A. and King, C.J. 1988. Sticking and agglomeration of hygroscopic, amorphous carbohydrate and food powders. *Biotechnol. Prog.* 4: 31–35.

Webb, A.P. and Orr, C. 1997. *Analytical Methods in Fine Particle Technology.* Micrometrics Instrument Corp., Norcross, GA, USA.

Yan, H. and Barbosa-Cánovas, G.V. 1997. Compression characteristics of agglomerated food powders: effect of agglomerated size and water activity. *Food Sci. Technol. Int.* 3: 351–359.

Yan, H. and Barbosa-Cánovas, G.V. 2000. Compression characteristics of selected food powders: the effect of particle size, mixture composition, and compression cell geometry. Doctoral thesis. Biological Systems Engineering Department, Washington State University, pp. 82–113.

Yan, H. and Barbosa-Cánovas, G.V. 2001a. Attrition evaluation for selected agglomerated food powders: the effect of agglomerate size and water activity. *J. Food Process Eng.* 24: 37–49.

Yan, H. and Barbosa-Cánovas, G.V. 2001b. The effect of padding foam on the compression characteristics of some agglomerated food powders. *Food Sci. Technol. Int.* 7: 417–423.

Yan, H., Barbosa-Cánovas, G.V., and Swanson, B.G. 2001. Density changes in selected agglomerated food powders due to high hydrostatic pressure. *Lebensmittel-Wiss-u-Technol.* 34: 495–501.

Food Powder Processing

4

Handling and Processing of Food Powders and Particulates

Enrique Ortega-Rivas
Food and Chemical Engineering Program
University of Chihuahua
Chihuahua, Mexico

Contents

I. INTRODUCTION

The food processing industry is one of the largest manufacturing industries worldwide. Undoubtedly, it possesses global strategic importance, so it has a critical need for growth based on future research directions detected by an integrated interdisciplinary approach to problems in food process engineering. This industry, like many other processing industries, handles and processes numerous raw materials and finished products in powdered and particulate form. In this sense, future competitiveness may be critically dependent on knowledge originated by research activities in the field known as powder technology or particle technology, which deals with the systematic study of particulate systems in a broad sense. For the case of food products and materials some important applications of powder technology can be mentioned. For example, particle size in wheat flour is an important factor in the functionality of food products, attrition of instant powdered foods reduces their reconstitutability, uneven powder flow in extrusion hoppers may affect the rheology of the paste, and an appropriate characterization of fluid–particle interactions could optimize clarification of juices. The optimum operation of many food processes rely heavily on a sound knowledge of the behavior of particles and particle assemblies, either in dry form or as suspensions.

Research efforts in particle technology have shown a tremendous growth recently. European and Asian professional associations and societies have recognized the importance of powder technology for some time. Some of these associations, such as the Institution of Chemical Engineers (IChemE) and the Society of Chemical Industry (SCI) in the United Kingdom include established research groups on the topic, and organize meetings and conferences on a regular basis. In the United States the importance of powder technology was recognized only in 1992 when a division, known as the Particle Technology Forum, was formed within the American Institute of Chemical Engineers (AIChE). In a more global context, diverse international associations organize conferences and congresses on particle technology with delegates from around the world attending to present the most advanced developments in the area. With reference to powder and particle technology applied specifically to food and biological materials, there has not been as much activity as in the field of inert materials. The main conferences and meetings of professional associations and societies related to food processing, such as the Institute of Food Technologists (IFT) in the United States, the Institute of Food Science and Technology (IFST) in the United Kingdom and the global International Union of Food Science and Technology (IUFoST), do not normally include sessions about food powders. Some isolated efforts have been made to promote exchange of ideas among powder technologists with an interest in food and biological materials and some publications have resulted from such efforts (Schubert, 1993; Ortega-Rivas, 1997).

This chapter deals with the process and design aspects of unit operations involving particulate solids, within the context of food and biological materials. Theoretical considerations, operating principles, and applications of different techniques used to handle and process powders and particulates in the food industry, are reviewed. It attempts to provide criteria and information for students, academics, and industrialists, who may perceive that future growth of this strategic industry may be dependent on a deep knowledge and understanding of this focused discipline.

II.   RELEVANT PROPERTIES OF POWDERED FOOD MATERIALS

Particle characterization, that is, description of the primary properties of food powders in a particulate system, underlies all work in particle technology. Primary properties of particles such as particle shape and particle density, together with the primary properties of a fluid (viscosity and density), together with the concentration and state of dispersion, govern the secondary properties such as settling velocity of particles, rehydration rate of powders, resistance of filter cakes, etc. It could be argued that it is simpler, and more reliable, to measure the secondary properties directly without reference to the primary ones. Direct measurement of secondary properties can be done in practice, but the ultimate aim is to predict them from the primary ones, as when determining pipe resistance to flow from known relationships, feeding in data from primary properties of a given liquid (viscosity and density), as well as properties of a pipeline (roughness). As many relationships in powder technology are rather complex and often not available in many areas, such as food powder processing, particle properties are mainly used for qualitative assessment of the behavior of suspensions and powders, for example, as an equipment selection guide. Since a powder is considered to be a dispersed two-phase system consisting of a dispersed phase of solid particles of different sizes and a gas as the continuous phase, complete characterization of powdered materials is dependent on the properties of the particle as an individual entity, the properties of the assembly of particles, and the interactions between those assemblies and a fluid.

A.   Size, Shape, and Distribution of Particles

There are several single particle characteristics that are very important to product properties (Davies, 1984). They include particle size, particle shape, surface density, hardness, adsorption properties, etc. Of all the mentioned features, particle size is the most essential and important one. The term "size" of a powder or particulate material is very relative. It is often used to classify, categorize, or characterize a powder, but even the term powder is

Table 1 Terms Recommended by the British
Pharmacopoeia for Use with Powdered Materials

Powder type	B.S. meshes	
	All passes	Not more than 40% passes
Coarse	10	44
Moderately coarse	22	60
Moderately fine	44	85
Fine	85	—
Very fine	120	—

Table 2 Approximate Ranges of the Median
Sizes of Some Common Food Powders

Commodity	B.S. mesh	Microns
Rice and barley grains	6–8	2800–2000
Granulated sugar	30–34	500–355
Table salt	52–72	300–210
Cocoa	200–300	75–53
Icing sugar	350	45

not clearly defined and common convention considers that for a particulate material to be considered powder its approximate median size (50% of the material is smaller than the median size and 50% is larger) should be less than 1 mm. It is also common practice to talk about "fine" and "coarse" powders; several attempts have been made at standardizing particle nomenclature in certain fields. For example, Table 1 shows the terms recommended by the *British Pharmacopoeia* with reference to standard sieve apertures. Also, by convention, particle sizes may be expressed in different units depending on the size range involved. Coarse particles may be measured in centimeters or millimeters, fine particles in terms of screen size, and very fine particles in micrometers or nanometers. However, due to the recommendations of the International Organization for Standardization (ISO), SI units have been adopted in many countries and, thus, particle size may be expressed in meters when doing engineering calculations, or in micrometers in virtue of the small range normally covered or when plotting graphs. A wide variety of food powders may be considered in the fine size range. Some median sizes of common food commodities are presented in Table 2.

1. Equivalent Diameters

The selection of a relevant characteristic particle size to start any sort of analysis or measurement often poses a problem. In practice, the particles forming a powder will rarely have a spherical shape. Many industrial powders are of mineral (metallic or nonmetallic) origin and have been derived from hard materials by any sort of size reduction process. In such a case, the comminuted particles resemble polyhedrons with nearly plane faces, in a number from 4 to 7, and sharp edges and corners. The particles may be compact, with length, breadth, and thickness nearly equal but, sometimes, they may be plate-like or needle-like. As the particles get smaller, and by the influence of attrition due to handling, their edges may become smoother; thus, they can be considered spherical. The term "diameter" is therefore, often used to refer to the characteristic linear dimension. All these geometrical features of important industrial powders, such as cement, clay, and chalk, are related to the intimate structure of their forming elements, whose arrangements are normally symmetrical with definite shapes like cubes, octahedrons, etc. On the other hand, particulate food materials are mostly organic in origin, and their individual grain shapes could have a great diversity of structures, since their chemical compositions are more complex than those of inorganic industrial powders. Shape variations in food powders are enormous ranging from extreme degrees of irregularity (ground materials like spices and sugar), to an approximate sphericity (starch and dry yeast) or well-defined crystalline shapes (granulated sugar and salt).

Considering the aspects mentioned above, expressing a single particle size is not simple when its shape is irregular. This is often the case with many applications, mostly when dealing with food powders of truly organic origin. Irregular particles can be described by a number of sizes. There are three groups of definitions, as listed in Tables 3, 4, and 5: equivalent sphere diameters, equivalent circle diameters, and statistical diameters. In the first group, the diameters of a sphere, which would have the same property of the particle itself, are found (e.g., the same volume, the same settling velocity, etc.). In the second group, the diameters of a circle, which would have the same property of the projected outline of the particle, are considered (e.g., projected area or perimeter). The third group is obtained when a linear dimension is measured (usually by microscopy) parallel to a fixed direction. The most relative measurements of the diameters mentioned above would probably be the statistical diameters because they are practically determined by direct microscopy observations. Thus, for any given particle Martin's and Feret's diameters could be radically different both from a circle of equal perimeter or equal area (see Figure 1). In practice, most of the equivalent diameters will be measured indirectly for a given number of particles taken from a representative sample and, therefore, it would be most practical to use a quick, less accurate measure on a large number of particles than a very accurate measure on very few particles. Also, it would be rather difficult to perceive the equivalence of the actual particles with an ideal sphericity. Furthermore, such equivalence would depend on the method employed to determine the size. For example, Figure 2 shows an approximate equivalence of an irregular particle depending on different equivalent properties of spheres.

Taking into account the concepts presented above, it is obvious that the measurement of particle size results depend upon the conventions involved in the particle size definition and also the physical principles employed in the determination process (Herdan, 1960). When different physical principles are used in particle size determination, it can hardly be assumed that they would give identical results. For this reason it is recommended to select

Table 3 A List of Definitions of "Equivalent Sphere Diameters"

Symbol	Name	Equivalent property of a sphere
x_v	Volume diameter	Volume
x_s	Surface diameter	Surface
x_{sv}	Surface–volume diameter	Surface-to-volume ratio
x_d	Drag diameter	Resistance to motion in the same fluid at the same velocity
x_f	Free-falling diameter	Free-falling speed in the same liquid, same particle density
x_{st}	Stokes' diameter	Free-falling speed if Stokes' law is used ($Re_p < 0.2$)
x_A	Sieve diameter	Passing through the same square aperture

Table 4   A List of Definitions of "Equivalent Circle Diameters"

Symbol	Name	Equivalent property of a circle
x_a	Projected area diameter	Projected area if particle is resting in a stable position
x_p	Projected area diameter	Projected area if particle is randomly orientated
x_c	Perimeter diameter	Perimeter of the outline

Table 5   A List of Definitions of "Statistical Diameters"

Symbol	Name	Dimension measured
x_F	Feret's diameter	Distance between two tangents on opposite sides of particle
x_M	Martin's diameter	Length of the line which bisects the image of particle
x_{SH}	Shear diameter	Particle width obtained with an image shearing eyepiece
x_{CH}	Maximum cord diameter	Maximum length of a line limited by the contour of the particle

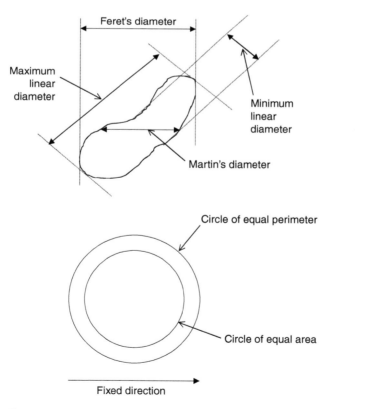

Figure 1   Methods used to measure diameter of nonspherical particles.

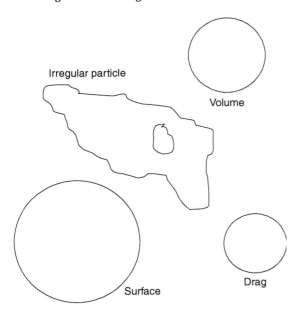

Irregular particle

Volume

Surface

Drag

Figure 2   Equivalent spheres.

Table 6   General Definitions of Particle Shape

Shape name	Shape description
Acicular	Needle shape
Angular	Roughly polyhedral shape
Crystalline	Freely developed geometric shape in a fluid medium
Dentritic	Branched crystalline shape
Fibrous	Regularly or irregular thread-like
Flaky	Plate-like
Granular	Approximately equidimensional irregular shape
Irregular	Lacking any symmetry
Modular	Rounded irregular shape
Spherical	Global shape

a characteristic particle size that can be measured according to the property or the process, which is under study. Thus, for example, in pneumatic conveying or gas cleaning, it is more relevant to choose to determine the Stokes' diameter, as it represents the diameter of a sphere of the same density as the particle itself, which would fall into the gas at the same velocity as the real particle. In flow through packed or fluidized beds, on the other hand, it is the surface–volume diameter, that is, the diameter of a sphere having the same surface-to-volume ratio as the particle, which is more relevant to the aerodynamic process.

2.   Shape of Particles

General definitions of particle shapes are listed in Table 6. It is obvious that such simple definitions are not enough to compare particle size measured by different methods or to incorporate it as a parameter into equations where particle shapes are not the same (Herdan, 1960; Allen, 1981). Shape, in its broadest meaning, is very important in particle behavior and just looking at the particle shapes, with no attempt at quantification, can

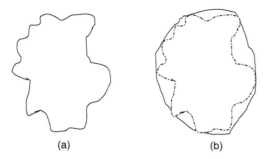

(a)   (b)

Figure 3   Relation between: (a) perimeter, and (b) convex perimeter of a particle.

be beneficial. Shape can be used as a filter before size classification is performed. For example, as shown in Figure 3, all rough outlines could be eliminated, by using the ratio: (perimeter)/(convex perimeter), or all particles with an extreme elongation ratio. The earliest methods of describing the shape of particle outlines used length L, breadth B, and thickness T, in expressions such as the elongation ratio (L/B) and the flakiness ratio (B/T). The drawback with simple, one-number shape measurements is the possibility of ambiguity; the same single number may be obtained from more than one shape. Nevertheless, a measurement of this type, which has been successfully employed for many years, is the so-called sphericity, Φ_s, defined by the relation:

$$\Phi_s = \frac{6V_p}{x_p s_p} \tag{1}$$

where x_p is the equivalent diameter of particle, s_p is the surface area of one particle, and V_p is the volume of one particle. For spherical particles Φ_s equals unity, while for many crushed materials its value lies between 0.6 and 0.7. Since direct measurement of particle volume and surface is not possible, to evaluate such variables a specific equivalent diameter should be used to perform the task indirectly. For example, when using the mean projected diameter x_a, as defined in Table 4, the volume and surface of particles may be calculated using:

$$V_p = \alpha_v x_p^3 \tag{2}$$

and

$$s_p = \alpha_s x_p^2 \tag{3}$$

where α_v and α_s are the volume and surface factors, respectively, and their numerical values are all dependent on the particle shape and the precise definition of the diameter (Parfitt and Sing, 1976). The projected diameter x_p is usually transferred into the volume diameter x_v of a sphere particle, as defined in Table 3, which is used as a comparison standard for the irregular particle size description, thus the sphere with the equivalent diameter has the same volume as the particle. The relationship between the projected and the equivalent diameters in terms of volume is expressed as follows:

$$x_v = x_p \left[\frac{6\alpha_v}{\pi} \right]^{1/3} \tag{4}$$

where x_v is the equivalent diameter of the sphere of the same volume as the particle. When the mean particle surface area is known, the relationship between those two diameters is:

$$x_v = x_p \left[\frac{\alpha_s}{\pi} \right]^{1/2} \tag{5}$$

where all the variables have been previously defined.

3. Particle Size Measurement

Particle size distribution measurement is a common method in any physical, mechanical, or chemical process because it is directly related to material behavior and/or physical properties of products. Foods are frequently in the form of fine particles during processing and marketing (Schubert, 1987), and their bulk density, compressibility, and flowability are highly dependent on the particle size and its distribution (Barbosa-Cánovas et al., 1987). Segregation will take place in a free-flowing powder mixture because of the differences in particle size (Barbosa-Cánovas et al., 1985). Size distribution is also one of the factors affecting the flowability of food powders (Peleg, 1977). For quality control or system property description, the need to represent the particle size distribution of food powders becomes paramount and proper descriptors in the analysis of the handling, processing, and functionality of each food powder.

In measuring particle size two most important decisions have to be made before a technique can be selected for the analysis; these are concerned with the two variables measured, the type of particle size and the occurrence of such size. Particle size was previously discussed and, emphasizing what has been already presented, it is important to bear in mind that great care must be taken when selecting the particle size, as an equivalent diameter, in order to choose the one most relevant to the property or process that is to be controlled. The occurrence of amount of particle matter belonging to specified size classes may be classified or arranged by diverse criteria as to obtain tables or graphs. In powder technology the use of graphs is convenient and customary for a number of reasons. For example, a particular size that is to be used as the main reference of a given material is easily read from a specific type of plot.

There are four different particle size distributions for a given particulate material, depending on the quantity measured: by number $f_N(x)$, by length $f_L(x)$, by surface $f_S(x)$, and by mass (or volume) $f_M(x)$. Of the above, the second mentioned is not used in practice as the length of a particle by itself is not a complete definition of its dimensions. These distributions are related but conversions from one to another are possible only in cases when the shape factor is constant, that is, when the particle shape is independent of the particle size. The following relationships show the basis of such conversions:

$$f_L(x) = k_1 \cdot x \cdot f_N(x) \tag{6}$$

$$f_S(x) = k_2 \cdot x^2 \cdot f_N(x) \tag{7}$$

$$f_M(x) = k_3 \cdot x^3 \cdot f_N(x) \tag{8}$$

where constants k_1, k_2, and k_3 contain a shape factor that may often be particle size dependent making an accurate conversion impossible without full quantitative knowledge of its dependence on particle size. If the shape of the particles does not vary with size, the constants

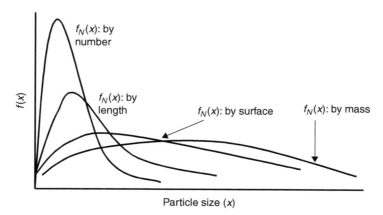

Figure 4   Four particle size distributions of a given particle population.

mentioned above can be easily found by definition of the distribution frequency:

$$\int_0^\infty f(x)\, dx = 1 \tag{9}$$

therefore, the areas under the curve should be equal to 1.

Different methods give different types of distributions and the selection of a method should be based on both the particle size and the type of distribution required. In food processes, many types of specific distributions would be most relevant. For example, when clarifying fruit juices, for primary removal of suspended solids the size distribution by mass should be the one of interest because this particular stage would be defined by gravimetric efficiency. Final clarification, however, would be better described by surface, or even number, distribution because of the low concentration of solids, which cause turbidity. Ortega-Rivas et al. (1997) successfully described suspended solids' removal in apple juice using particle size distributions by mass. Figure 4 shows the four types of distribution.

There is an abundance of methods available for measurement of particle size distributions and several textbooks, such as some referred to in this chapter (Allen, 1981; Kaye, 1981), are available, and review the field in great depth. Table 7 gives a schematic review of the methods available, size ranges covered, and the types of particle size and size distribution that are measured. Only a preliminary selection can be attempted using Table 7, because it is impossible to list all the important factors influencing the choice, such as type of equivalent diameter required, quantity to be measured, size range, quantity of sample available, degree of automation required, etc.

B.   Types of Densities

The density of a particle is defined as its total mass divided by its total volume. It is considered quite relevant for determining other particle properties such as bulk powder structure and particle size, therefore it requires careful definition (Okuyama and Kousaka, 1991). Depending on how the total volume is measured, different definitions of particle density can be given: the true particle density, the apparent particle density, and the effective (or aerodynamic) particle density. Since particles usually contain cracks, flaws, hollows, and closed pores, it follows that all these definitions may be clearly different. The true

Table 7 Analytical Techniques of Particle Size Measurement

Technique	Approximate size range (μm)	Type of particle size	Type of size distribution
Sieving			
Woven wire	37–4000	x_A	By mass
Electro formed	5–120	x_A	By mass
Microscopy			
Optical	0.8–150	x_a, x_F, x_M	By number
Electron	0.001–5	x_{SH}, x_{CH}	
Gravity sedimentation			
Incremental	2–100	x_{st}, x_f	By mass
Cumulative	2–100	x_{st}, x_f	By mass
Centrifugal sedimentation			
Two layer-incremental	0.01–10	x_{st}, x_f	By mass
Cumulative			
Homogeneous-incremental			
Flow classification		x_{st}, x_f	
Gravity elutriation (dry)	5–100	x_{st}, x_f	By mass
Centrifugal elutriation (dry)	2–50	x_{st}, x_f	By mass
Impact separation (dry)	0.3–50	x_{st}, x_f	By mass or number
Cyclonic separation (wet or dry)	5–50	x_{st}, x_f	By mass
Particle counters			
Coulter principle (wet)	0.8–200	x_v	By number

Table 8 Densities of Common Food Powders

Powder	Density (kg/m^3)
Glucose	1560
Sucrose	1590
Starch	1500
Cellulose	1270–1610
Protein (globular)	~1400
Fat	900–950
Salt	2160
Citric acid	1540

particle density represents the mass of the particle divided by its volume excluding open and closed pores, and is the density of the solid material of which the particle is made. For pure chemical substances, organic or inorganic, this is the density quoted in reference books of physical/chemical data. Since most inorganic materials consist of rigid particles, whereas most organic substances are normally soft, porous particles, the true density of many food powders would be considerably lower than those of mineral and metallic powders. Typical nonmetallic minerals, as some previously mentioned, would have true particle densities well over 2000 kg/m^3, while some metallic powders can present true densities of the order of 7000 kg/m^3. By contrast, most food particles have considerably lower densities of about 1000 to 1500 kg/m^3. Table 8 lists typical densities for some food powders. As can be observed, salt (which is of inorganic origin) presents a notably higher density than the other substances that are listed. The apparent particle density is defined as the mass of a particle divided by its volume excluding only the open pores, and is measured by gas or liquid displacement methods such as liquid or air pycnometry. The effective particle density is referred to as the mass of a particle divided by its volume including both open and closed pores. In this case, the volume is within an aerodynamic envelope as "seen" by a gas flowing past the particle and, as such, this density is of primary importance in applications involving flow round particles, for instance, in fluidization, sedimentation, or flow through packed beds.

None of the three particle densities defined above should be confused with bulk density of materials, which includes the voids between the particles in the volume measured. The different values of particle density can also be expressed in a dimensionless form, as relative density or specific gravity, which is simply the ratio of the density of the particle to the density of water. It is easy to determine the mass of particles accurately but difficult to evaluate their volume because they have irregular shapes and voids between them.

1. Particle Density

The apparent particle density, or if the particles have no closed pores also the true density, can be measured by fluid displacement methods, that is, pycnometry, which are in common use in the industry. The displacement can be carried out using either a liquid or a gas, with the gas employed normally being air. Thus, the two known techniques to determine true or apparent density, when applicable, are liquid pycnometry and air pycnometry.

Liquid pycnometry can be used to determine particle density of fine and coarse materials depending on the volume of the pycnometer bottle that is used. For fine powders a pycnometer bottle of 50 ml volume is normally employed, while coarse materials may require larger calibrated containers. Figure 5 shows a schematic diagram of the sequence of events involved in measuring particle density using a liquid pycnometer. The particle density ρ_s is clearly the net weight of dry powder divided by the net volume of the powder,

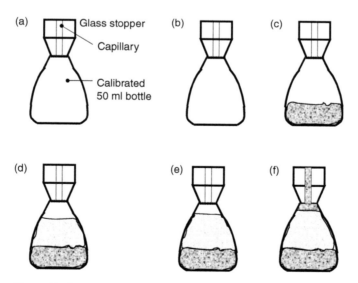

Figure 5 Descriptive diagram of density determination by liquid pycnometry: (a) description of pycnometer, (b) weighing, (c) filling to about 1/2 with powder, (d) adding liquid to almost full, (e) eliminating bubbles, (f) topping and final weighing.

calculated from the volume of the bottle, subtracting the volume of the added liquid, that is:

$$\rho_s = \frac{(m_s - m_o)\rho}{(m_l - m_o) - (m_{sl} - m_s)} \tag{10}$$

where m_s is the weight of the bottle filled with the powder, m_o is the weight of the empty bottle, ρ is the density of the liquid, m_l is the weight of the bottle filled with the liquid, and m_{sl} is the weight of the bottle filled with both the solid and the liquid. Air bubbles adhering to particles and/or liquid absorbed by the particles can cause errors in density measurement. Therefore, a liquid that absorbs particles slowly and that has low surface tension should be selected. Sometimes, when heating or boiling procedures are needed to do the gas evacuation, the liquid that has a high boiling point and does not dissolve the particle should be used (Okuyama and Kousaka, 1991). When the density of larger, irregularly shaped solid objects, such as compressed or aggregated bulk powders is needed, a method available to evaluate fruit or vegetable volumes may be used. A schematic diagram of a top-loading platform scale for volume and density measurement is shown in Figure 6. A beaker big enough to host the solid is partially filled with some kind of liquid that will not dissolve the solid. The weight of the beaker with the liquid in it is recorded and the solid object is completely immersed and suspended at the same time, using a string, so that it does not touch either the sides or the bottom of the beaker. The total weight of this arrangement is recorded again, and the volume of the solid V_s can be calculated (Ma et al., 1997) by:

$$V_s = \frac{m_{LCS} - m_{LC}}{\rho_L} \tag{11}$$

where m_{LCS} is the weight of the container with liquid and submerged solid, m_{LC} is the weight of the container partially filled with liquid, and ρ_L is the density of the liquid.

Air pycnometry can be performed in an instrument, which usually consists of two cylinders and two pistons, as shown in Figure 7. One is a reference cylinder, which is always

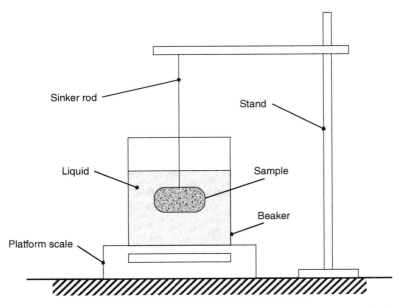

Figure 6 Top-loading platform scale for density determination of irregularly shaped objects.

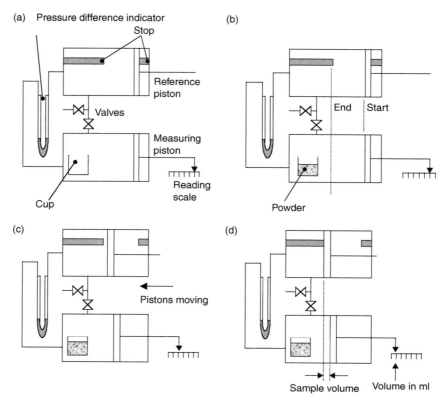

Figure 7 Descriptive diagram of density determination by air pycnometry: (a) description of instrument, (b) filling of cup, (c) pistons displacement, (d) reading.

empty, and the other has a facility for inserting a cup with the sample of the powder. With no sample present, the volume in each cylinder is the same so that, if the connecting valve is closed and one of the pistons is moved, the change must be duplicated by an identical stroke in the other so as to maintain the same pressure on each side of the differential pressure indicator. But, if a sample is introduced in the measuring cylinder (Figure 7), and the piston in the reference cylinder is advanced all the way to the stop, to equalize the pressures, the measuring piston will have to be moved by a smaller distance because of the extra volume occupied by the sample. The difference in the distance covered by the two pistons, which is proportional to the sample volume, can be calibrated to read directly in cubic centimeters, usually with a digital counter. The method will measure the true particle density if the particles have no closed pores, or the apparent particle density if there are any closed pores, because the volume measured normally excludes any open pores. If, however, the open pores are filled either by wax impregnation or by adding water, the method will also measure the envelope volume. By measuring the difference between the two volumes the open pore volume will be obtained, and can be used as a measure of porosity.

2. Bulk Density and Porosity

The bulk density of food powders is so fundamental to their storage, processing, and distribution that it merits particular consideration. When a powder fills a vessel of known volume V and the mass of the powder is m then the bulk density of the powder is m/V. However, if the vessel is tapped, it will be found in most cases that the powder will settle and more powder has to be added to fill the vessel completely. If the mass now filling the vessel is m', the bulk density is $m'/V > m/V$. Clearly, this change in density just described has been caused by the influence of the fraction of volume not occupied by a particle, known as porosity. The bulk density is, therefore, the mass of particles that occupies a unit volume of a bed, while porosity or voidage is defined as the volume of the void within the bed divided by the total volume of the bed. These two properties are in fact related via the particle density in that, for a unit volume of the bulk powder, there must be the following mass balance:

$$\rho_b = \rho_s(1 - \varepsilon) + \rho_a\varepsilon \tag{12}$$

where ρ_b is the powder bulk density, ρ_s is the particle density, ε is the porosity, and ρ_a is the air density. As the air density is small relative to the powder density, it can be neglected and the porosity can thus be calculated as:

$$\varepsilon = \frac{(\rho_s - \rho_b)}{\rho_s} \tag{13}$$

Equation (13) gives the porosity or voidage of the powder, and whether or not this includes the pores within the particles depends on the definition of particle density used in such evaluation.

Over the years, in order of increasing values, three classes of bulk density have become conventional: aerated, poured, and tap. Each of these classes depends on the treatment to which the sample is subjected and, although there is a move toward standard procedures, these are far from being universally adopted. There is still some confusion in the open literature with regard to the interpretation of these terms. Some people consider the poured bulk density as loose bulk density while others refer to it as apparent density. The actual meaning of the term aerated density can also be considered quite confusing. Strictly speaking, it means that individual particles are separated by a film of air and are not in direct contact with each other. Some authors, however, interpret it to mean the bulk density after the powder has been aerated. Such interpretation yields, in fact, the most loosely packed bulk density when, for cohesive materials, the strong interparticle forces prevent the particles from rolling over each other. Considering this second interpretation, aerated and bulk densities could both be simply regarded as loose bulk density, and this approach is implied in many investigations when dealing with cohesive powders. For many food powders, which are more cohesive in behavior, the terms more commonly used to express bulk density are loose bulk density, and as poured and tapped bulk density after vibration. Another way to express bulk density is in the form of a fraction of its particles' solid density, which is sometimes referred to as the "theoretical density." This expression, as well as the use of porosity instead of density, enables and facilitates the unified treatment and meaningful comparison of powders having considerably different particle densities.

The aerated bulk density is, in practical terms, the density when the powder is in its most loosely packed form. Such a form can be achieved by dropping a well-dispersed "cloud" of individual particles down into a measuring vessel. Aerated bulk density can be determined using an apparatus like the one illustrated in Figure 8. As shown, an assembly of screen cover, screen, a spacer ring, and a chute is attached to a mains-operated vibrator of variable amplitude. A stationary chute is aligned with the center of a preweighed 100 ml cup. The powder is poured through a vibrating sieve and allowed to fall from a fixed height

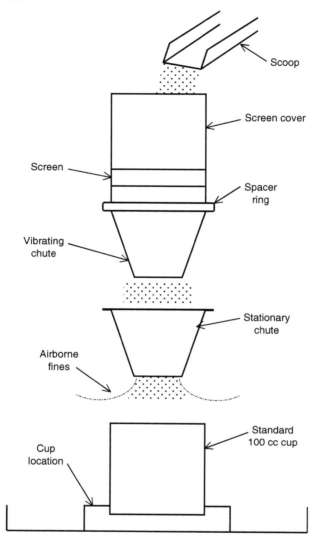

Figure 8 Determination of aerated bulk density.

of 25 cm approximately through the stationary chute into the cylindrical cup. The amplitude of the vibration is set so that the powder will fill the cup in 20 to 30 sec. The excess powder is skimmed from the top of the cup using the sharp edge of a knife or ruler, without disturbing or compacting the loosely settled powder.

Poured density is widely used, but the measurement is often performed in a manner found suitable for the requirements of the individual company or industry. In some cases the volume occupied by a particular mass of powder is measured, but the elimination of operation judgment, and thus possible error, in any measurement is advisable. To achieve this, the use of a standard volume and the measurement of the mass of powder to fill it are needed. Certain precautions are to be taken, for example, the measuring vessel should be fat rather than slim, the powder should always be poured from the same height, and the possibility of bias in the filling should be made as small as possible. Although measurement of poured bulk density is far from being standardized, many industries use a sawn-off funnel with a trap door or stop, to pour the powder through into the measuring container.

The tap bulk density, as its name implies, is the bulk density of a powder that has settled into a packing closer than that which existed in the poured state, by tapping, jolting, or vibrating the measuring vessel. As with poured bulk density, the volume of a particular mass of powder may be observed, but it is generally better to measure the mass of powder in a fixed volume. Although many people in industry measure tap density by tapping the sample manually, it is best to use a mechanical tapping device so that the conditions of sample preparation are more reproducible. An instrument that is useful in achieving such reproducibility is the Hosakawa powder characteristic tester, which has a standard cup (100 ml) and a cam-operated tapping device that moves the cup upward and drops it periodically (once in every 1.2 sec). A cup extension piece has to be fitted and powder added during the sample preparation to ensure that the powder never packs below the rim of the cup. After the tapping, excess powder is scraped from the rim of the cup and the bulk density is determined by weighing the cup.

Approximate values of loose bulk density of different food powders are given in Table 9. As can be seen, with very few exceptions, food powders have apparent densities in the range of 300 to 800 kg/m^3. As previously mentioned, the solid density of most food powders is about 1400 kg/m^3, so these values are an indication that food powders have high porosity, which can be internal, external, or both. There are many published theoretical and experimental studies of porosity as a function of the particle size, distribution, and shape. Most of them pertain to free-flowing powders or models (e.g., steel shots and metal powders), where porosity can be treated as primarily due to geometrical and

Table 9 Approximate Bulk Density and Moisture of Different Food Powders

Powder	Bulk density (kg/m^3)	Moisture content (%)
Baby formula	400	2.5
Cocoa	480	3–5
Coffee (ground and roasted)	330	7
Coffee (instant)	470	2.5
Coffee creamer	660	3
Corn meal	560	12
Corn starch	340	12
Egg (whole)	680	2–4
Gelatin (ground)	680	12
Microcrystalline cellulose	610	6
Milk	430	2–4
Oatmeal	510	8
Onion (powdered)	960	1–4
Salt (granulated)	950	0.2
Salt (powdered)	280	0.2
Soy protein (precipitated)	800	2–3
Sugar (granulated)	480	0.5
Sugar (powdered)	480	0.5
Wheat flour	800	12
Wheat (whole)	560	12
Whey	520	4.5
Yeast (active dry baker's)	820	8
Yeast (active dry wine)		8

statistical factors (Gray, 1968; McGeary, 1967). Even though in these cases porosity can vary considerably, depending on factors such as the concentration of fines, it is still evident that the exceedingly low density of food powders cannot be explained by geometrical considerations alone. Most food powders are known to be cohesive and, therefore, an open bed structure supported by interparticle forces is very likely to exist (Scoville and Peleg, 1980; Moreyra and Peleg, 1981; Dobbs et al., 1982). Since the bulk density of food powders depends on the combined effect of interrelated factors, such as the intensity of attractive interparticle forces, the particle size, and the number of contact points (Rumpf, 1961), it is clear that a change in any of the powder characteristics may result in a significant change in the powder bulk density. Furthermore, the magnitude of such change cannot always be anticipated. There is an intricate relationship between the factors affecting food powder bulk density, as well as surface activity and cohesion.

C. Failure properties

To make powders flow, their strength should be less than the load put on them, that is, they must fail. The basic properties describing this condition are known as "failure properties" and they are: the angle of wall friction, the effective angle of internal friction, the failure function, the cohesion, and the ultimate tensile strength. The failure properties take into account the state of compaction of the powder as this strongly affects its flowability unless the powder is cohesionless, like dry sand, and it gains no strength on compression. These properties may also be strongly affected by humidity and, especially in the case of food and biological materials, by temperature. The time of consolidation can also have an effect on failure properties of powders. It is therefore important, to test such properties under controlled conditions using sealed powder samples or air-conditioned rooms or enclosures. Also, time consolidating samples must be tested to simulate storage conditions.

The angle of wall friction ϕ is equivalent to the angle of friction between two solid surfaces except that one of the two surfaces is a powder. It describes the friction between the powder and the material of construction used to confine the powder, for example, a hopper wall. The wall friction causes some of the weight to be supported by the walls of a hopper. The effective angle of internal friction δ is a measure of the friction between particles and depends on their size, shape, roughness, and hardness. The failure function FF is a graph showing the relationship between unconfined yield stress (or the strength of a free surface of the powder) and the consolidating stress, and gives the strength of the cohesive material in the surface of an arch as a function of the stress under which the arch was formed. The cohesion is, as mentioned earlier, a function of interparticle attraction and is due to the effect of internal forces within the bulk, which tend to prevent planar sliding of one internal surface of particles upon another. The tensile strength of a powder compact is the most fundamental strength mechanism, representing the minimum force required to cause separation of the bulk structure without major complications of particle disturbances within the plane of failure.

1. Determinations Using Shear Cells

There are several ways, direct or indirect, of testing the five failure properties defined above. All of them can be determined using a shear cell, but simplified or alternative procedures can be adopted when the aim is to monitor the flowability of the output from a process or to compare a number of materials. There are basically two types of shear cells available for powder testing: the Jenike shear cell, also known as the translational shear box, and the annular or ring shear cell, also called the rotational shear box. The Jenike shear cell is circular in the cross section, with an internal diameter of 95 mm. A vertical cross section through the cell is shown in Figure 9. It consists of a base and a ring, which can slide horizontally over the base. The ring and base are filled with the powder and a lid is placed in position. By means of a weight carrier, which hangs from a point at the center of the lid, a vertical compacting load can be applied to the powder sample. The lid carries a bracket with a projecting pin and a measured horizontal force is applied to the bracket, causing the ring and its contents, as well as the lid, to move forward at a constant speed. The shear force needed to make the powder flow can thus be obtained. Five or six different vertical loads are applied to a set of identical samples and the shear force needed to initiate flow is found in each case. Every determined force is divided by the cross-sectional area of the cell, in order to calculate the stress. The derived shear stresses are, then, plotted against normal stresses. The resulting graph is a yield locus (Figure 10), and it is a line that gives the stress conditions needed to produce flow for the powder when compacted to a fixed bulk density.

If the material being tested is cohesive, the yield locus is not a straight line and does not pass through the origin. It can be shown that the graph when extrapolated downward cuts the horizontal axis normally. As shown in Figure 10, the intercept T is the tensile strength of the powder compacts tested and the intercept C is called the cohesion of the powder; the yield locus ends at the point A. A yield locus represents the results of a series of tests on samples that have the initial bulk density. More yield loci can be obtained by changing the sample preparation procedure and a family of yield loci can be obtained. This family of yield loci contains all the information needed to characterize the flowability of a particular material; it is not, however, in a convenient form. For many powders, yield locus curves

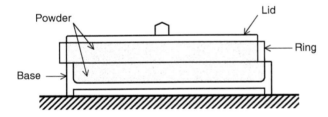

Figure 9 Diagram of Jenike shear cell.

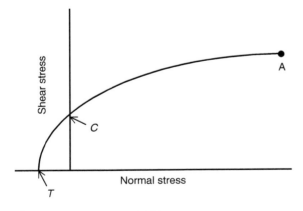

Figure 10 The Jenike yield locus.

can be described by the empirical Warren–Spring equation (Chasseray, 1994):

$$\left(\frac{\tau}{C}\right)^n = \frac{\sigma}{T} + 1 \tag{14}$$

where τ is the shear stress, C the material's cohesion, σ the normal stress, T the tensile stress, and n the shear index ($1 < n < 2$). As mentioned before, cohesion is a very important property in food powders. Table 10 lists cohesion values for several food powders.

Two important properties defined above can be obtained from the yield loci: the first is the effective angle of internal friction while the second is the failure function of the powder. The angle of wall friction can be measured by replacing the base of a Jenike shear cell by a plate of the material of which the hopper (or any sort of container) is to be made. The ring from the shear cell is placed on the plate and filled with powder and the lid is put in position. The shear force needed to maintain uniform displacement of the ring is found for different vertical loads on the lid. The slope of the graph of shear force against normal force gives the angle of friction between the particles and the wall, or angle of wall friction. This measure would complete the testing of a particulate material using only a Jenike shear cell.

In annular shear cells, the shear stress is applied by rotating the top portion of an annular shear, as represented in Figure 11. These devices allow much larger shear distances to be covered both in sample preparation and its testing allowing a study of flow properties after testing, but their geometry creates some problems. The distribution of stress is not

Table 10   Cohesion for Some Food Powders

Material	Moisture content (%)	Cohesion (g/cm^2)
Corn starch	<11.0	4–6
Corn starch	18.5	13
Gelatin	10.0	1
Grapefruit juice	1.8	8
Grapefruit juice	2.6	10–11
Milk	1.0	7
Milk	4.4	10
Onion	<3.0	<7
Onion	3.6	8–15
Soy flour	8.0	1

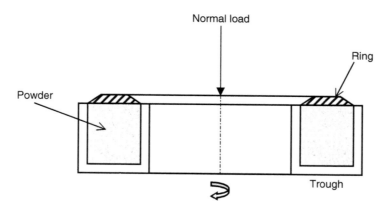

Figure 11   Diagram of an annular shear cell.

uniform in the radial direction but, for the ratio of the inner and outer radii of the annuli greater than 0.8, the geometrical effects are often considered negligible. The annular shear cells tend to give lower values for yield strength than the Jenike shear cell tester. There is another type of shear cell known as the ring shear or ring shear tester. In this device the cell is in the form of a full ring and is rotated like the annular shear cell. It has been reported to have the advantages of nearly unlimited shear deformation, possibility of measurements at very low consolidation stresses, ease of operation, and possibility of time consolidation measurements using a consolidation bench (Schulze, 1996). In contrast to the annular shear cell, the results obtained using a ring shear tester are in reasonable agreement with those obtained with the Jenike shear cell.

2. Direct Measurements

The angle of internal friction can be measured directly by the "grooved plate" method. The base of the Jenike shear cell is replaced by a metal plate, in which a number of saw-toothed grooves are cut (Figure 12). These grooves are filled with the powder to be tested. The ring from the Jenike cell is then placed on the plate and filled with the powder and the lid is placed into position. A load is placed on the lid and the ring is pushed across the grooves until the shear force settles out at a constant value, which is measured, and this action is repeated for different vertical loads. The graph of shear force against normal force will be a straight line with its slope being the angle of internal friction of the powder.

For direct measurement of the failure function, a split cylindrical die as shown in Figure 13 is used. The bore of the cylinder may conveniently be about 50 mm and its height should be a little more than twice the bore. The cylinder is clamped so that the two halves cannot separate and it is filled with the powder to be tested, which is then scraped off, level with the top face. By means of a plunger the specimen is subjected to a known consolidating stress. The plunger is then removed and the two halves of the split die are separated, leaving a self-standing column of the compacted powder. A plate is then placed on top of the specimen and an increasing vertical load is applied to it until the column collapses. The stress at which this occurs is the unconfined yield stress, that is, the stress that has to be applied to the free vertical surface on the column to cause failure. If this is repeated for different compacting loads and the unconfined yield stress is plotted against the compacting stress, the failure function of the powder will be obtained. Although the results of this method can be used for monitoring or for comparison, the failure function obtained will not be the same as that given by shear cell tests, due to the effect of die wall friction when forming the compact. A method of correcting for friction has been described elsewhere (Williams et al., 1971).

Two methods can be used for direct measurement of tensile strength. In the first, a mold of the same diameter as the Jenike cell is split across a diameter. The base of

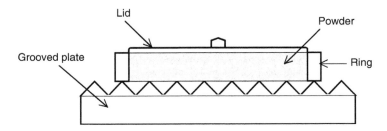

Figure 12   The grooved plate method for measuring the angle of internal friction.

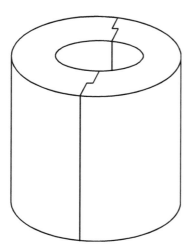

Figure 13   Device for direct measurement of failure function.

the cell is roughened, by sticking sandpaper to it. The two halves are clamped together, the cell is filled with the powder, and then a lid is placed in position. The specimen is compacted by the application of a known vertical force to the lid and this, along with the clamp, are removed. The two halves of the cell, containing the specimen, rest on a base plate in which slots have been cut to form an air bearing. Air is introduced so that the cell can move horizontally without friction and the force needed to pull the two halves of the specimen are determined. Knowing the cross-sectional area of the specimen allows the tensile strength to be found. Measurements are made for a number of compacting loads and the tensile strength is plotted against compacting stress. This method is quite difficult to perform properly, requiring careful attention to details. The second method is easier to use and gives results with less scatter. In this case a mold of the same diameter as the Jenike shear cell, and a lid which just fits inside it, are used. The base of the cell and the lower face of the lid are covered with sticking tape on which glue is spread. The cell is filled with the powder, which is scraped level with the top of the cell and the lid is placed in position. A compacting load is applied to the lid by means of a weight hanger and left in position until the glue has hardened. The lid is then attached through a tensile load cell to an electric motor, by which the lid is slowly lifted. The stress required to break the specimen is thus obtained. After failure, the lid and the base of the cell are examined and the result is accepted only if both are completely covered with powder, showing that tensile failure has occurred within the powder specimen and not at the surface. Figure 14 presents a diagram of these two methods for direct measurement of tensile strength.

D.   Reconstitution Properties

In the context of food drying, reconstitutability is the term that is used to describe the rate at which dried foods pick up and absorb water reverting to a condition that resembles the undried material, when put in contact with an excessive amount of this liquid (Masters, 1976). In the case of powdered, dried biological materials, a number of properties may influence the overall reconstitution characteristics. For instance, wettability describes

(a)

(b)

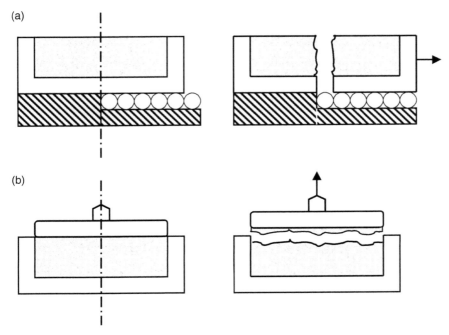

Figure 14 Methods for measuring tensile strength: (a) horizontal split cell, (b) lifting lid cell.

the capacity of the powder particles to absorb water on their surface, thus initiating reconstitution. Such a property depends largely on particle size. Since small particles have a large surface area : mass ratio, they may not be wetted individually. In fact, they may clump together sharing a wetted surface layer. This layer reduces the rate at which water penetrates into the particle clump. Increasing particle size and agglomerating particles can reduce the incidence of clumping. The nature of the particle surface can also affect wettability. For example, the presence of free fat on the surface reduces wettability. The selective use of surface-active agents, such as lecithin, can sometimes improve wettability in dried powders containing fat.

Another important property is sinkability, which describes the ability of the powder particles to sink quickly into the water. This depends mainly on the size and density of the particles. Larger, denser particles sink more rapidly than finer, lighter ones. Particles with a high content of occluded air may be relatively large but exhibit poor sinkability because of their low density. Finally, dispersability describes the ease with which the powder may be distributed as single particles over the surface and throughout the bulk of the reconstituting water, while solubility refers to the rate and extent to which the components of the powder particles dissolve in the water. Dispersability is reduced by clump formation and is improved when the sinkability is high, whereas solubility depends mainly on the chemical composition of the powder and its physical state.

Food dried powders and particulates are normally reconstituted for consumption. For a dried product to exhibit good reconstitution characteristics there has to be a correct balance between the individual properties discussed above. In many cases, alteration of one or two of these properties can markedly change the rehydrating behavior. Several measures can be taken in order to improve reconstitutability of dried food products. The selected drying method and adjustment of drying conditions can result in a product with good rehydration properties. For example, it is well known that freeze drying is a process by which ice crystals are produced and sublimated at very low pressures (Heldman and Singh, 1981). This procedure results in food particles with an open pore structure, which absorb water easily when they are reconstituted.

Another alternative is the use of the so-called combined methods, such as osmotic dehydration followed by conventional drying. In osmotic dehydration, food particles are immersed in a concentrated solution. By osmotic pressure, the water inside the particles tends to migrate to the solution in order to equate water activities on both sides of the cellular wall (Monsalve-Gonzalez et al., 1993). This partial dehydration will aid in the final stage of drying, and textural damage of the biological materials will be minimized. In this sense, biological materials dehydrated by combined methods will also have an open pore structure and, similar to freeze-dried materials, will present good reconstitution properties. Beltran-Reyes et al. (1996) developed an apple-powdered ingredient by grinding dried apples obtained by osmotic dehydration followed by conventional heated air drying. They determined that the firmness of the rehydrated mash, measured as an extrusion force in a texture analyzer, was a direct function of the particle size. For the same ingredient, Ortega-Rivas and Beltran-Reyes (1997), reported that rehydration improved as particle size decreased.

The most efficient method to improve the rehydration characteristics of dried food powders is probably the use of agglomeration (Barletta and Barbosa-Cánovas, 1993). In order to agglomerate particles, the powder is treated with steam or warm, humid air such that condensation occurs on the particle surface. Interparticle contact is promoted, often by swirling the wetted powder in a vortex. Agglomerates are formed and then dried in a chamber and cooled on a vibrating fluidized bed. Agglomeration, as a unit operation, will be discussed further in a subsequent section.

III. HANDLING OF FOOD PARTICULATE SYSTEMS

The food and related industries handle considerable quantities of powders and particulate materials every year. Materials handling mainly implies storage and conveying and is concerned with open and confined storing, as well as with movement of materials in different cases, for instance, from supply point to store or process, between stages during processes, or to packing and distribution.

The handling of materials is a crucial activity, which adds nothing to the value of the product, but can represent an added cost if not managed properly. For this reason, responsibility for materials' handling is normally vested in specialist handling engineers, and many food manufacturers adopt this procedure. When a specific materials' handling department has not been provided, the responsibility for efficient handling of materials falls on the production manager and his/her staff. It is important, therefore, for production executives to have a sound knowledge of the fundamentals of good handling practice.

Silos, bins, and hoppers that are used to store materials in the food industry vary in capacity from a few kilos to multiton-capacity vessels. Startup delays and ongoing inefficiencies are common in solids processing plants. An important cause for these problems is the improper design of bulk solids handling equipment. For bulk particulate or powdered food materials, which fall within the scope of this chapter, a convenient classification of conveyors would comprise those using direct mechanical power to move the conveying elements, such as belt, chain, and screw conveyors, and those relying on a stream to carry particulates in it, such as hydraulic and pneumatic conveyors.

A. Storage

The design of bins, hoppers, and silos has never been given the attention it deserves. Approaches using properties such as angle of repose or angle of spatula in design considerations are ineffective, because the resulting values bear no relation to the design parameters needed to ensure reliable flow, mainly because particulate solids tend to compact or consolidate when stored. Attempts to try and model bulk solids as fluids also leads to a bottleneck, due to the fact that flowing bulk solids generate shear stresses and are able to maintain these stresses even when their flow rate is changed dramatically. It is also improper to consider bulk solids as having viscosity since almost all bulk solids exhibit flow properties that are flow-rate independent. The systematic approach for designing powder handling and processing plants started in the mid-1950s with the pioneering work of Andrew W. Jenike. His concept was to model bulk solids using the principles of continuum mechanics. The resulting comprehensive theory (Jenike, 1964) describing the flow of bulk solids has been applied and perfected over the years, but is generally recognized worldwide as the only scientific guide to bulk solids' flow.

The procedures for the design of a bulk solids' handling plant are well established and follow four basic steps: (1) determination of the strength and flow properties of the bulk solids for the worst likely flow conditions expected to occur in practice; (2) calculation of the bin, stockpile, feeder, or chute geometry to give the desired capacity to provide a flow pattern with acceptable characteristics, in order to ensure that discharge is reliable and predictable; (3) estimation of the loadings on the bin and hopper walls and on the feeders and chutes under operating conditions; (4) design and detailing of the handling plant including the structure and equipment.

1. Elements of Bulk Solids' Gravity Flow

Only fluids can flow; bulk solids under gravity forces can fall, slide, or roll, but against gravity, they must be lifted by mechanical means. Solids cannot be pumped by centrifugal or reciprocating pumps, so they have to be suspended in liquids or gases. There is no satisfactory term for "flow" of bulk solids, as they do not follow strict definitions of fluid behavior, since a fluid is considered to be a continuum in which there are no voids. For a fluid, when the shear rate is linearly proportional to the shear stress, it is said to be Newtonian and the coefficient of proportionality is called absolute viscosity. Any deviation from this definition makes the fluid non-Newtonian. Solids in suspension can be referred to as non-Newtonian mixtures to differentiate them from a number of non-Newtonian fluids, which are a continuum or are perfectly homogeneous liquids. For all these reasons, bulk solids in suspension are occasionally referred to as "imperfect fluids."

The gravity flow of bulk solids occurs under the pressure corresponding to the equivalent of a "static head" of the material. Such head would be caused by the height of a solid column in a bin, but in practice is often not available to produce the flow due to phenomena known as "arching" or "bridging." The velocity head at discharge from the bin is usually a small fraction of the head, with the major part being consumed by the friction of the moving solids against the walls of the bin, as well as against like solids. Friction is the resistance that one body offers to the motion of a second body when the latter slides over the former. The friction force is tangent to the surfaces of contact of the two bodies and always opposes motion. The coefficient of static friction μ for any two surfaces is the ratio of the limiting friction to the corresponding normal pressure, that is:

$$\mu = \frac{F}{N} \tag{15}$$

where F is the maximum friction of impeding motion and N is the normal pressure.

If a body rests on an inclined plane and if the angle of inclination of the plane to the horizontal, α, is such that the motion of the body impends, this angle α is defined as the angle of repose, so it follows that:

$$\mu = \tan \alpha \tag{16}$$

When two surfaces move relative to each other, the ratio of the friction created to the normal pressure is called the coefficient of kinetic friction and is independent of the normal pressure. The coefficient of kinetic friction is also less than the coefficient of static friction and independent of the relative velocity of the rubbing surfaces. There is experimental evidence, which supports the theory that the value of the kinetic friction coefficient increases as the velocity decreases, and passes without discontinuity into that of static friction. All these principles would hold under conditions of a particular test, but must be modified in order to be applied to different conditions. Herein lies the main difficulty in applying existing test data on series of new tests because of the great variety in flowing conditions of bulk solids. The problem is particularly complicated as the properties of the flowing material depend on time and method of storage.

When granular solids are stored in an enclosed container, the lateral pressure exerted on the walls at any point is less than that predicted from the head of the material above such a point. There is usually friction between the wall and the solid particles, and the interlocking of these particles causes a frictional effect throughout the bulk solid mass. The frictional force at the wall tends to offset the weight of the solid and reduces the effect of the head of solids on the floor of the container. In an extreme case, such frictional force causes the mass of bulk solids to arch, or bridge, so that it would not fall even if the material below is discharged. For many granular solids, when the height of the solid bed reaches about three times the diameter of the bin, additional head of material shows virtually no effect on the pressure at the bin floor.

Solids tend to flow out of any opening near the bottom of a bin but are best discharged through an opening in the floor. The pressure at a side outlet is smaller than the vertical pressure at the same level and removal of solids from one side of a bin considerably increases the lateral pressure on the opposite side while the solids flow. When an outlet at the bottom of a bin containing free-flowing solids is opened, the material immediately above such an opening begins to flow. A central column of solids moves downward without disturbing the material at the sides. Eventually lateral flow begins, starting from the top layer of solids and a conical depression forms in the surface of the mass of bulk solids being discharged. The material slides laterally into the central column moving at an angle approaching the angle of internal friction of the solids and the solids at the bin floor are the last to move. If additional material is added at the top of the bin, at the same rate as the material leaving through the bottom outlet, the solids near the bin walls remain stagnant and do not discharge as long as flow persists. The rate of flow of granular solids by gravity through a circular opening in the bottom of a bin is dependent on the diameter of the opening as well as on the properties of the solid and is independent, within wide limits, on the head or height of the solids.

2. Flow Patterns in Storage Bins

The general theory pertaining to gravity flow of bulk solids has been documented over the years (Arnold et al., 1982; Roberts, 1988) and from a standpoint of flow patterns, there are basically three types of flow in symmetrical geometry: mass flow, funnel flow and expanded flow (Figure 15). In mass-flow bins, the flow is uniform and the bulk density of

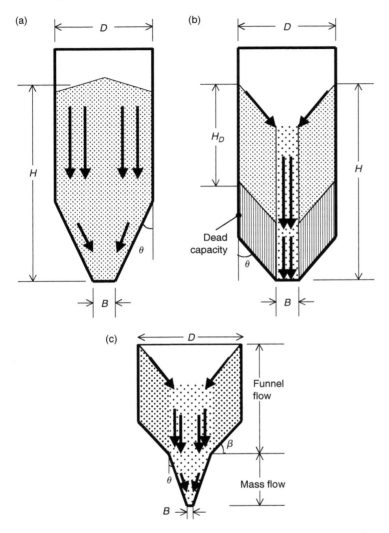

Figure 15 Types of flow patterns in hoppers: (a) mass flow, (b) funnel flow, and (c) expanded flow.

the feed is practically independent of the head of solids in the bin. Mass flow guarantees complete discharge of the bin's contents at predictable flow rates. When properly designed, a mass-flow bin can remix the bulk of the solid during discharge even if segregation is promoted during filling. Mass-flow bins are generally recommended for cohesive materials, for materials that degrade with time, for fine powders, and for particulate systems that have to be prevented from segregating. Normally food powders are highly cohesive and, therefore, the use of mass-flow bins would represent a preferred alternative for their storage.

Funnel flow occurs when the hopper is not sufficiently steep and smooth to force the bulk solid to slide along the walls. It also occurs when the outlet of the bin is not fully effective, due to poor feeder or gate design. In a funnel-flow bin the stored material flows toward the outlet through a vertical channel forming within the stagnant solids. The diameter of such a channel approximates the largest dimension of the effective outlet. Flow out of this type of bin is generally erratic and gives rise to segregation problems. However, flow will continue until the level of the bulk solid in the bin drops an amount equal to the drawdown. At this level the bulk strength of the contained material is sufficient to sustain a stable rathole as illustrated in Figure 15(b). Once the level defined by H_D in Figure 15(b) is reached, there is no further flow and the material below this level represents dead storage. For complete discharge, the bin opening has to be at least equal to the critical rathole dimension, determined at the bottom of the bin corresponding to the bulk strength at this level. For many cohesive bulk solids, and for normal consolidation heads occurring in practice, ratholes measuring several meters high are often observed. This makes control of the product discharge rate quite difficult and funnel flow somewhat impractical. Funnel flow has the advantage of providing wear protection of the bin walls as the material flows against stationary material. Funnel-flow bins are suitable for coarse, free-flowing, or slightly cohesive, nondegrading solids when segregation is unimportant. With reference to food systems, funnel-flow bins may be used for grains, pulses, oilseeds, and so on, mainly for the application of feeding such materials to processing directly, as in cereal extrusion or cereal milling.

The expanded-flow bin combines characteristics of mass flow and funnel flow. The higher part of the hopper operates in funnel flow while the lower operates in mass flow. The mass-flow outlet usually requires a smaller feeder than would be the case for funnel flow. The mass-flow hopper should expand the flow channel to a diagonal or diameter equal to or greater than the critical rathole diameter, thus eliminating the likelihood of ratholing. Funnel-flow bins provide the wall protection of funnel flow, together with the reliable discharge of mass flow. Expanded flow is ideal where large tons of bulk solids need to be stored and is particularly suitable for storing large quantities of bulk solids while maintaining acceptable head heights. The concept of expanded flow may be used to advantage in the case of bins or bunkers with multiple outlets. Expanded flow bins are recommended for the storage of large quantities of nondegrading solids. This design is also useful as a modification of existing funnel-flow bins to correct erratic flow caused by arching, ratholing, or flushing.

3. Wall Stresses in Bins and Silos

The prediction of wall loads in bins is an important piece of information for their design. It is necessary to estimate the pressures at the wall, which are generated when the bin is operated, in order to design the bin structure efficiently and economically. The approaches to the study of bin wall loads are varied and involve analytical and numerical techniques, such as finite element analysis. Despite these varied approaches, it is clear that the loads are directly related to the flow patterns that are created in the bin. The flow pattern in mass-flow bins is reasonably easy to predict but in funnel-flow bins such prediction becomes quite a difficult task. For this reason, unless there are compelling reasons to do otherwise, bin shapes should be kept simple and symmetric.

Research relating to wall stresses dates back to the 1800s when Janssen (1895) published his now famous theory. More recent investigations include those reported by Walker (1966), Walker and Blanchard (1967), Jenike and Johanson (1968, 1969), Walters (1973), Clague (1973), Arnold et al. (1982), Roberts (1988), and Thompson et al. (1997). Examination of these papers shows that the solution of the problem of stress distributions in bins is extremely complex. However, most researchers agree that the loads acting on a bin wall are different during the initial stage of filling and during the stage of flowing in discharge. When bulk solids are charged into an empty bin, with the gate closed or with the feeder at rest, the bulk solids settle as the solids' head rises. In this settlement, the solids contract vertically in the cylindrical section and partially vertically in the hopper section. The principal stress tends to align with the direction of contraction of the solids, forming what is termed as an active or peaked stress field. It is assumed that the solids are charged into the bin without significant impact to cause packing, and that powders are charged at a sufficiently low rate so that they deaerate. It is also assumed that the bin and feeder have been designed correctly for the solids to flow without obstruction. When the gate is fully opened or the feeder operates so that the solids start flowing out of the outlet, there is vertical expansion of the solids within the forming flow channel and the flowing mass of solids contract laterally. The principal stresses within the flow channel tend to align with the lateral contractions and the stress field is said to be passive or arched.

The region of switch from an active to a passive stress field originates at the outlet of the bin, when the gate is opened, or the feeder is started, and rapidly moves upward into the bin as the solids are withdrawn. At the switch level a fairly large overpressure may be present, and it is assumed to travel upward with the switch at least to the level at which the channel intersects the vertical section of the bin. For a typical bin consisting of a hopper plus a cylindrical section above it, five stress fields have been recognized during the fill and discharge sequences: (1) in the cylindrical section during initial filling, where the state of stress is peaked or active; (2) in the cylindrical section during emptying, where the state of stress is either peaked or changes to arched, depending on whether the switch level is assumed to be caught at the transition; (3) in the converging hopper section during filling, where the state of stress is assumed to be peaked; (4) in the converging hopper section during emptying, where the state of stress is assumed to be arched; and (5) the switch field, that is, the region in the bin where the peaked stress field established during initial filling is transformed into the arched stress field. This switch starts at the outlet of the hopper, if newly filled from being completely empty, and then travels up very quickly as emptying continues, generally to become caught in the transition. Most of the researchers mentioned agree upon a wall pressure or stress distribution as shown in Figure 16.

The Janssen theory mentioned above (Janssen, 1895) includes, possibly, the oldest reported attempt to calculate pressures in silos. Janssen derived an equation for the calculation of vertical and horizontal pressures and wall shear stresses. He assumed a vertical force balance at a slice element spanning the full cross section of a silo being filled with bulk solids (see Figure 17), and determined the wall friction coefficient with a shear tester as well as the horizontal pressure ratio from pressure measurements in a model bin. He also assumed a constant vertical pressure across the cross section of the slice element and restricted his evaluation to vertical silo walls. The Janssen equation for the vertical pressure p_v on dependence of the depth z below the bulk solids' top level reads as follows for a cylindrical silo:

$$p_v = \frac{g \rho_b D}{4 \mu' K'} \left[1 - e^{-\left(\frac{4 \mu' K' z}{D} \right)} \right] \tag{17}$$

where g is the acceleration due to gravity, ρ_b is the bulk density of solids, D is the silo diameter, μ' is the sliding friction coefficient along the wall, and K' is the ratio of the horizontal to the vertical pressure, which can be expressed as:

$$K' = \frac{1 - \sin \delta}{1 + \sin \delta} \tag{18}$$

where δ is the angle of internal friction of solids.

The advantage of the Janssen equation is its simplicity, that is, of an analytical equation and its general good agreement with pressure measurements in silos for the state of filling. The disadvantages are its nonvalidity for the hopper section, its assumption of a constant vertical stress across the cross section, and its assumption of plastic equilibrium throughout

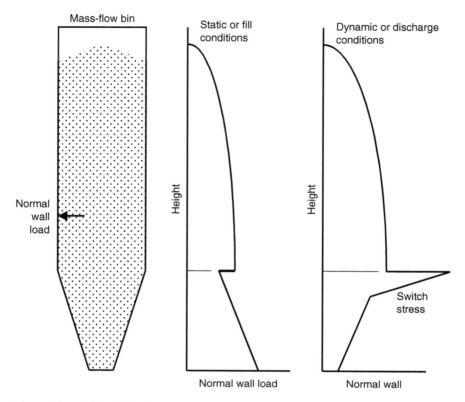

Figure 16   Wall load distribution in silo.

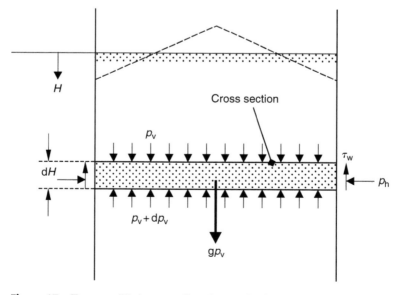

Figure 17   Force equilibrium at a slice element for full cross section of silo.

the stress field in the silo. But the Janssen equation cannot explain the increase in horizontal pressure when discharge is initiated. This disadvantage can only be overcome by using finite element methods, more sophisticated yield criteria, and a very high degree of computational effort (Häußler and Eibl, 1984).

Simplicity and analytical solution made the Janssen equation the basis for the first standard for the calculation of loads in silos more than 35 years ago. This equation is still the most widely used analytical solution for the calculation of pressures in silos. However, the structural design of silos requires the incorporation of experience, measurement results in model bins, and full-scale silos, as well as accepted safety margins for uncertainties. Major factors contributing to the loads in silos are the flow profile, the flow behavior, the interaction between wall material and bulk solids, and the performance of feeders and discharge aids (Jenike, 1964). The loads in silos are influenced by many factors. Some of them are related to the bin structure, its construction material and size. Many other factors, however, depend on the bulk solids' flow properties, the design of outlet size, the type of feeder, the discharge aids, and the operating conditions. These factors are especially important for nonfree-flowing, cohesive bulk solids.

4.   *Solids Discharge*

The amount of solids discharged through an opening at the bottom of a bin, and failure to restart the flow after intermission, depend on the bin design, shape, and the location of the opening, apart from the flow properties of the solid. The flow properties of the granular material include grain nature, size, moisture content, temperature, adhesion, cohesion and, above all, time of consolidation at rest. There are very few solids that are free-flowing and that will restart flow after an extended period at rest. Examples include inert materials such as graded gravel and dry sand. In contrast, as has been stated, food powders are mostly cohesive and their flow is very difficult, even without consolidation time. Pressure distribution within a bin affects its design for strength, but does not enter into calculation of the solids flow from the hopper. It has been mentioned that the volume of bulk solids discharged is independent of the head above the orifice, due to the arching effect. Therefore, the design of bins or silos in terms of their ability to initiate flow without any aid is based on solid mechanics theories taking into consideration only the hopper of the container.

Food powders are complex because of their composition (Schubert, 1987), their large distribution in particle size and the presence of solid–liquid–gas phases in the particle. Moisture has a great influence on flowability and its presence and proportion within the food powder depend on the relative humidity of the surrounding atmosphere. The remaining factors that affect most powders' flowability, that is, failure properties, particle and bulk density, and so on, also affect food powder flowability and, therefore, have a direct influence on the design of bin geometry for mass flow. Teunou et al. (1999) characterized representative food powders for their flowability and design of hoppers for mass flow, and their main findings are summarized in Table 11.

In order to ensure flow from bins, even after the hopper geometry has been determined following careful calculations, flow promotion may be necessary. Classification of flow promotion may be termed as passive and active, involving energy. A third class of flow promotion may also be considered, that is, the use of feeders, which are useful not only to promote flow, but also to control the flow rate. Passive devices normally include inserts usually placed within the hopper section of a bin, with the purpose of expanding the size of the active flow channel in a funnel-flow bin so as to approach mass flow. Another aim of an insert is to relieve pressure at the outlet region. Inverted cones and pyramids have been used for years in this regard, but with limited success. Vibrating hoppers by the use of electrical

Table 11   Physical Properties and Hopper Dimensions of Food
Powders*

Powder	X (%)	ff	$\delta(°)$	$\phi(°)$	$\theta(°)$	B (mm)
Flour	12.6	2.71	32	12.6	37	110
Skim milk	4.6	11.04	50	13.0	32	270
Tea	6.6	4.22	43	15.0	31	130
Whey-permeate	3.8	5.85	49	15.0	30	180

Note: *X is the water content in wet basis while the remaining variables are
        as defined in this text.
Source: Adapted from Teunou, E., Fitzpatrick, J.J., and Synnott, E.C. (1999).
        J. Food Eng. **39**: 31. With permission.

Table 12   Guide for Selection of Feeders

Bulk solid characteristics	Type of feeder
Fine, free-flowing solids	Apron, vibratory, screw, star
Nonabrasive, granular materials	Apron, vibratory, screw
Difficult-to-handle (abrasive, hot, etc.) materials	Apron, vibratory
Heavy, lumpy, or highly abrasive materials	Apron, vibratory

motors, pneumatic knockers, eccentric drives, or electromagnetic units, are one of the most
important and versatile flow assisters or active device types. Air cannons or air blasters are
also commonly used to promote gravity flow in bins.

In designing hoppers for silos, the procedure described so far would consist of making
calculations to determine optimum hopper slope and outlet opening in order to ensure flow.
In case flow does not occur, a flow promotion device can be selected after careful study
of conditions and factors. Once flow out of a bin is guaranteed, the next step, to complete
the proper design of the bulk storage plant, would be to control the flow rate, so as to
provide adequate feed to any given food powder process. In order to do so, the use of
a feeder will become necessary. A feeder is a device used to control the flow of bulk
solids from a bin. A feeder must be selected to suit a particular bulk solid and the range
of feed rates required. It is particularly important to design the hopper and feeder as an
integral unit, to ensure that the flow from the hopper is fully developed with uniform draw
of material from the entire hopper outlet. There are several types of feeders but the most
common are the belt or apron feeder, the screw feeder, the vibratory feeder, and the star
feeder. Careful considerations, such as those described above, should be taken in selecting
a feeder for a particular application. Table 12 provides a preliminary guide in choosing
a feeder.

B.   Mechanical Conveying

The main types of mechanical conveyors are belt, chain, and screw conveyors. In the food
industry belt conveyors are not widely used. The main application in food systems has
been in conveying grains. In fact, it has been mentioned (Wright et al., 1997) that the belt
conveyor drive power calculation has its origin in grain handling in the late 1700s in the
United States. Possibly, the most common devices for transportation of food granules and
particulates, using mechanical forces, are the bucket elevator and the screw conveyor.

1. Bucket Elevators

Bucket elevator systems comprise high capacity units primarily intended for bulk elevation of relatively free-flowing materials and may be considered a special adaptation of chain conveying. Bucket elevators are the simplest and most dependable equipment units for vertical lifting of different types of granular materials. They are available in a wide range of capacities and may operate either entirely in the open or be totally enclosed. High efficiency in bucket elevators results from the absence of frictional loss from sliding the material on the housing, and this feature distinguishes it from the vertical, or nearly vertical, scraper conveyor. The material carrying element of this sort of conveying is the bucket, which may be enclosed in a single housing called a leg, or two legs may be used. The return leg may be located some distance from the elevator leg. A single or double chain is used to attach the buckets. The most important considerations affecting the design and operation of bucket elevators are: (1) the physical properties of the conveyed material, (2) the shape and spacing of the buckets, (3) the speed at which the elevator is driven, (4) the method of loading the elevator, and (5) the method of discharging the elevator.

Important physical properties of the material being elevated are particle size, lump size, moisture content, angle of repose, flowability, abrasiveness, friability, etc. The design of the buckets has to do, principally, with capacity and ease of discharge. They may be constructed out of malleable iron or steel and can be shaped with either sharp or round bottoms. Mounting and spacing of the buckets will conform to a specific elevator design. Some typical bucket elevators are shown schematically in Figure 18. They may be fastened to the chain at the back (Figure 18[a]) or at the side if mounted on two chains (Figure 18[b]). Guides are sometimes used for two-chain installations, particularly in the upward leg. Single-chain installations have no guides or supports between the head and foot wheels except, possibly, an idler or two placed at strategic points to eliminate whip. The center spacing of buckets varies with their size, shape, speed, as well as head and foot wheel diameter. The buckets must be placed so that the centrifugal discharging grain does not hit the bucket ahead of the one discharging. For general purposes, the spacing will be 2 to 3 times the projected width. The speed of the drive in bucket elevators, although much

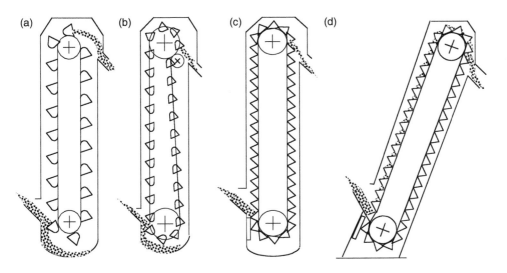

Figure 18 Bucket elevators: (a) centrifugal-discharge spaced buckets, (b) positive-discharge spaced buckets, (c) continuous bucket, (d) supercapacity continuous bucket.

depending on the type of material, is mainly controlled by the rate and method of discharge. Bucket elevators can be mainly loaded in three different ways. Spaced buckets receive part of the charge directly from a chute and part of the change by scooping; continuous buckets are filled as they pass through a loading leg with a feed spout above the tail wheel and, they can also be loaded in a bottomless boot with a cleanout door. Three main types of discharge are generally recognized: centrifugal, positive, and continuous. A fourth type of discharge may sometimes be considered: gravity discharge, in which buckets are carried pivoted on two chains and are tipped mechanically to facilitate discharge.

Except for overlapping buckets, which are not extensively used in processing, discharge depends upon centrifugal force in part or in full, or the ability of the material to be thrown into a chute as the buckets go over the head pulley. The characteristics of this feature and, in particular, the trajectory of the material after it leaves the bucket, are important to properly design and operate bucket elevators. Centrifugal discharge requires the speed of the chain to be held within close limits so that the trajectory will fall within a specified region. Figure 19 shows a head wheel and a bucket in a series of positions. A unit mass of grain is subjected to two forces at the point the bucket starts to turn around the pulley. These forces are the weight of the unit volume W and the centrifugal force F_c acting radially which is:

$$F_c = \frac{Wv_t^2}{3600gr} \tag{19}$$

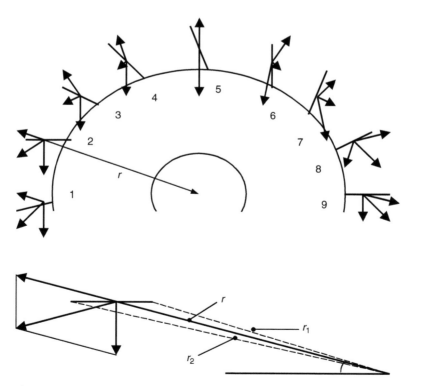

Figure 19 Force diagram of the loads in a head-wheel bucket in a number of different positions. The effective radius of the head-wheel bucket varies from r_1 to r_2.

where W is the weight of elemental mass, v_t is the tangential velocity, g is the acceleration due to gravity, and r is the effective radius.

The resultant of these forces R, shown in Figure 19, determines the point at which discharge takes place and its characteristics. The resultant for positions 1 to 4 in Figure 19 is of such a direction that the material is held in the bucket, at position 5, F_c and W are opposing and R is zero, so there is no force on the material. Discharge begins at this point, the initial velocity and trajectory being that of the projected speed of the wheel at that point. Note that R in positions 6 to 8 is nearly in the same direction of motion of the bucket forcing, discharge. In order to produce this condition, F_c and W must be equal at a point near the top of the travel, that is:

$$F_c = W = \frac{Wv_t^2}{3600gr} \tag{20}$$

so that

$$v^2 = 3600gr \tag{21}$$

since

$$v = 2\pi rN \tag{22}$$

where N is r/min, then:

$$N = 54.19 \left[\frac{1}{\sqrt{r}} \right] \tag{23}$$

Equation (23) shows the relationship between the effective head-wheel radius and its revolutions per minute for the most satisfactory discharge conditions. Discharge is not uniform or instantaneous because the effective radius varies from r_1 to r_2 as shown in Figure 19. Thus, the material at the outer edge of the bucket discharges first.

Bucket elevator horsepower can be calculated quite easily using the following equations:

1. Horsepower, HP, for spaced buckets and digging boots:

$$HP = \frac{TH}{152} \tag{24}$$

2. Horsepower, HP, for continuous buckets with loading leg:

$$HP = \frac{TH}{167} \tag{25}$$

In Equations (24) and (25), T is the bucket capacity in tons/h and H is lift in meters. Both equations include normal drive losses, as well as loading pickup losses, and are applicable for vertical or slightly inclined lifts.

As previously mentioned, bucket elevators are by far the most efficient way of elevating granular and particulate materials in a number of processing industries. In the food industry they are employed extensively for elevating a variety of commodities such as sugar, beans, oilseeds, salt, and cereals.

2. Screw Conveyors

These conveyors are used to handle finely divided powders, damp materials, hot substances that may be chemically active, and granular materials of all types. They operate on the principle of a rotating helical screw, moving material in a trough or casing. Flights are made out of stainless steel, copper, brass, aluminum, or cast iron, principally. They may be hard-surfaced with Stellite or similar materials to resist highly abrasive materials. Although screw conveyors are simple and relatively inexpensive, power requirements are high and single sections are limited in length. The standard pitch screw has a pitch approximately equal to the diameter and is used on most horizontal installations and on inclines up to 20°. Half-standard pitch screws may be used for inclines greater than 20°. Double-flight and triple-flight, variable-pitch and steeped-diameter screws are available for moving difficult materials and controlling feed rate. Ribbon screws are used for wet or sticky substances, while specially cut flight and ribbon screws are used for mixing. Figure 20 shows the main components of screw conveyors.

Screw conveyors are usually made up of standard sections coupled together, so special attention should be given to bending stresses in the couplings. Hanger bearings supporting the flights can obstruct the flow of material when the trough is loaded above its level. Thus, with difficult materials, the load in the trough must be kept below this level. Alternatively, special hanger bearings, which minimize obstruction, should be selected. Since screw conveyors operate at relatively low rotational speeds, the fact that the outer edge of the flight may be moving at a relatively high linear speed is often neglected. This may create a wear problem, and if wear is too severe it can be reduced by the use of hard-surfaced edges, detachable, hardened flight segments, rubber covering, or high-carbon steels.

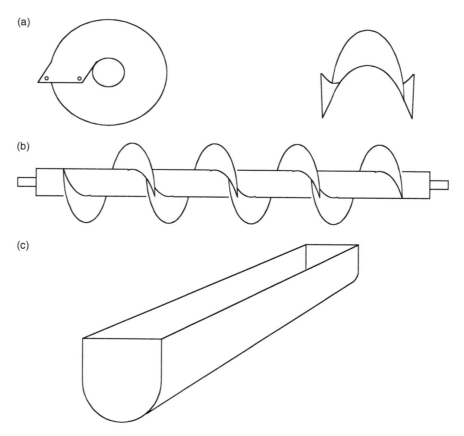

Figure 20   Screw conveyor components: (a) flight, (b) screw, formed by mounting flights on an axle, (c) trough.

Concise data and formulae are normally not available for individual screw conveyor design and it is best to consult specialized engineers when designing and installing large screw conveying systems. Data that could be available to assist in selection and design are normally empirical in nature. Roberts (1999, 2000), presents analytical data to predict the performance of screw conveyors. The power requirement of a screw conveyor is a function of its length, elevation, type of hanger brackets, type of flights, the viscosity or internal resistance of the material, the coefficient of friction of the material on the flights and housing, and the weight of the material. Consideration must also be given to additional power needed to start a full screw, to free a jammed screw, or to operate with material that has a tendency to stick to the trough sides. The HP required to drive a screw conveyor depends upon the dimensions of the system and the characteristics of the material. A rough approximation for normal horizontal operation can be determined from the following relation:

$$HP = \frac{CL\rho_b F}{4500} \tag{26}$$

where C is the capacity in m^3/min, L is the conveyor length in meters, ρ_b is the apparent density of material in kg/m^3, and F is a factor depending on the type of material, as appearing in Table 13.

In Equation (26) if HP is less than 1, it should be doubled; if it ranges from 1 to 2, it should be multiplied by 1.5; if it ranges from 2 to 4, it should be multiplied by 1.25; and if it ranges from 4 to 5, it should be multiplied by 1.1. No correction is necessary for values above 5 hp.

Screw conveyors are very versatile devices for handling a wide variety of materials horizontally, at an inclination, and even vertically. They are suited for both dry bulk materials as well as semiliquid, nonabrasive materials. In the food industry the applications are numerous and they have been used (1) for conveying different grains and oilseeds such as barley, corn, rice, rye, wheat, cottonseed, and soy beans; (2) for moving fine food powders such as flour, icing sugar, starch, and powdered milk; and (3) for handling viscous food materials such as sugar-beet pulp, peanut butter, and comminuted meat.

Table 13 Material Factors for Horizontal Screw Conveyors

Type a ($F = 1.2$)[a]	Type b ($F = 1.4$–1.8)[b]	Type c ($F = 2.0$–2.5)[c]	Type d ($F = 3.0$–4.0)[d]
Barley	Soy meal	Granular moist malt	Raw sugar
Granular dried malt	Cacao seeds	Cocoa	Bone meal
Corn flour	Coffee seeds	Dehydrated milk	
Cotton seed flour	Corn	Starch	
Wheat flour	Corn meal	Icing sugar	
Malt	Jelly granules		
Rice			
Wheat			

Notes:
[a] Light, fine, nonabrasive, free-flowing materials ρ_b:480–640 kg/m^3.
[b] Nonabrasive, granular or fines mixed with lumps ρ_b: up to 830 kg/m^3.
[c] Non and mildly abrasive, granular or fines mixed with lumps ρ_b: 640–1200 kg/m^3.
[d] Mildly abrasive or abrasive, fine, granular or fines with lumps ρ_b: 830–1600 kg/m^3.

C. Pneumatic Conveying

One of the most important bulk solids handling techniques in food processing is the movement of material suspended in a stream of air over horizontal, inclined, or vertical surfaces, ranging from a few to several hundred meters. This type of conveying is one of the most versatile, handling materials ranging from fine powders through 6.35 mm pellets and bulk densities of 16 to more than 3200 kg/m³. As compared with previously discussed methods, pneumatic conveying offers the containment and flexibility of pipeline transport for bulk solids that, otherwise, will be exposed to direct contact with moving mechanical parts. Most of the food powders and particulates handled in the food processing industries would present hygiene and contamination problems when conveyed through the opening; in such an instance, pneumatic conveying represents an obvious choice for duties in which integrity of handled products is paramount.

Pneumatic conveying has been used extensively for many years in many food process operations. In fact, as reported by Reed and Bradley (1991), one of the earliest recorded uses was for unloading wheat from barges to flour mills at the end of the 19th century in London. Some other grains, as well as different cargo such as alumina, cement, and plastic resins are still unloaded using the same basic methods. Other common applications include unloading trucks, railcars, and barges, transferring materials to and from storage vessels, injecting solids into reactors and combustion chambers, and collecting fugitive dust by vacuum. The limitations on what can be conveyed depend more upon the physical nature of the material than on its generic classification. Particle size, hardness, resistance to damage, and cohesive properties are key factors in determining whether a material is suitable for this sort of conveying. Cohesive or sticky materials are often difficult to handle in a pneumatic conveyor. Moist substances that are wet enough to stick to the pipeline walls usually cannot be conveyed successfully. Materials with high oil or fat contents can also cause severe buildup in pipelines making conveying impractical.

1. Theoretical Aspects

In contrast with the conveying methods previously discussed, pneumatic conveying can be perfectly identified as a case of two-phase flow, which is a topic well covered by fluid mechanics. Flow of gas in a pipeline is well understood with the conveying gas obeying the ideal gas law, and its density ρ_g being a function of pressure and temperature, as given by:

$$\rho_g = \frac{P}{RT} \tag{27}$$

where P is the absolute pressure, R is the universal gas constant, and T is the absolute temperature.

Mean gas velocity v in a pipeline is a function of mass flow rate of the gas and the density of the flow area:

$$v = \frac{\dot{m}}{\rho_g A} \tag{28}$$

where $\dot{m}$ is the mass flow rate of gas and A is the flow area.

By combining Equations (27) and (28), it follows that the mean gas velocity is a function of the gas pressure, that is:

$$v = \frac{\dot{m}RT}{PA} \tag{29}$$

Assuming that the mass flow rate of the gas and the flow area are constant, as well as the gas temperature, the velocity at any two points in the line becomes proportional to the absolute gas pressure:

$$\frac{v_2}{v_1} = \frac{P_1}{P_2} \tag{30}$$

where P_1 and P_2 are absolute pressures.

The relationship between gas velocity and pressure drop, in a straight pipe, is found by the following simple equation:

$$\Delta P = f\left(\frac{L}{D}\right)\left[\frac{(\rho_g u^2)}{2}\right] \tag{31}$$

where f is the fanning friction factor, L is the pipe length, D is the pipe diameter, and u is the local gas velocity.

As Equation (31) indicates, the pressure drop in a pipe is approximately proportional to the square of the gas velocity. The increase in velocity from one end of the pipe to the other, results in a difference in pressure drop per unit length of more than 2. This illustrates the significance of density changes in the gas, as flow proceeds through the pipeline. Changes in the gas velocity also affect the suspension of solids in the gas stream. At low velocities particles may be sliding on the bottom of the pipe, while at higher velocities particles will be fully suspended by the gas.

The moving gas stream applies drag and lift to the particles. For particles to be conveyed in such a gas stream, the velocity of the gas must be sufficiently high to stop particles from settling out. In flow through horizontal pipes, the minimum air velocity to stop particles settling to the bottom of the pipe is called the saltation velocity. The equivalent velocity for flow through vertical pipes is known as the choking velocity. The saltation velocity is a function of the density of the gas and the solids, as well as particle and pipeline diameter (Cabrejos and Klinzing, 1994). There is also a direct relationship between the saltation velocity and the solids loading ratio. Generally, saltation occurs at higher velocities when the solids loading ratio is also high. In terms of designing equipment for pneumatic conveying, there is another type of velocity, that is, the minimum conveying velocity, that is used to describe the correlation of gas velocity to the behavior of solid particles inside a pipeline. This velocity is the lowest one necessary to prevent plugging the line in a given system for a given material. Some researchers have suggested using the saltation velocity with a safety factor, while others have developed empirical correlations. Some of these correlations, however, often predict widely differing velocities for the same set of conditions (Wypych, 1999).

2. Classification of Conveying Systems

Pneumatic conveying systems can be categorized in a number of ways depending on their function as well as type and magnitude of operating pressure. Solids loading is a useful criterion to classify pneumatic conveyors, which can run over a wide range of conditions, bounded on one end by gas alone with no entrained solids, and at the other end by a completely full pipe where the solids are plugging the line. Most industrial conveying systems operate somewhere in between these two extremes, being ranked broadly as either dense-phase or dilute-phase systems, depending upon the relative solids loading and velocity of the system. This is best illustrated graphically in a general state diagram, which is a plot of pressure per unit length of pipe as a function of conveying gas velocity, with constant solids flow rate. As shown in Figure 21, at higher velocities particles are generally suspended in the

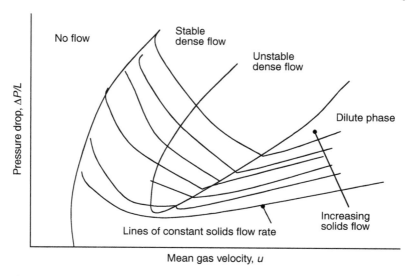

Figure 21   General state diagram for flow of solids in a pipe.

gas with low solids loading ratio, typically below 15, and termed as dilute-phase conveying. If the gas velocity is slowly decreased, the pressure required to convey a constant amount of solids also drops. After reaching a minimum, a further reduction in gas velocity results in an increase in pressure, as particles begin to fall out of suspension and interparticle collision increases. This region, with a solids loading ratio typically higher than 15 and gas velocity below the saltation velocity, is that of dense-phase conveying. With many materials it is difficult to establish a definite boundary separating dense-phase and dilute-phase regions, and conveying can occur over a continuous range from a fully suspended to a slow moving bed. With other materials very distinct regions are observed and the conveying progresses in either a very stable or unstable way.

Dense-phase conveying, also termed "nonsuspension" conveying, is normally used to discharge particulate solids or to move materials over short distances. There are several types of equipment such as plug-phase conveyors, fluidized systems, blow tanks, and, more innovative, long-distance systems. Dilute-phase, or dispersed-phase conveyors, are more versatile in use and can be considered the typical pneumatic conveying systems as described in the literature. The most accepted classification of dilute-phase conveyors comprises: pressure, vacuum, combined, and closed-loop systems.

3. Dilute-Phase Conveyors

Dilute-phase conveying is the commonly employed method for transporting a wide variety of suspended solids using air flowing axially along a pipeline. The method is mainly characterized by the low solids-to-air ratio and by the fact that air and solids flow as a two-phase system inside a pipeline. Figure 22 shows the four main types of dilute-phase conveying systems. The pressure system, also called positive-pressure, or push system, operates at superatmospheric pressure and is used for delivery to several outlets from one inlet (Figure 22[a]). Although most applications of these systems lie within the scope of dilute-phase conveying, under certain arrangements they can also operate as high-pressure, dense-phase conveyors. In general, pressure systems can hold higher capacities and longer conveying distances than negative pressure systems. The vacuum, negative-pressure or pull

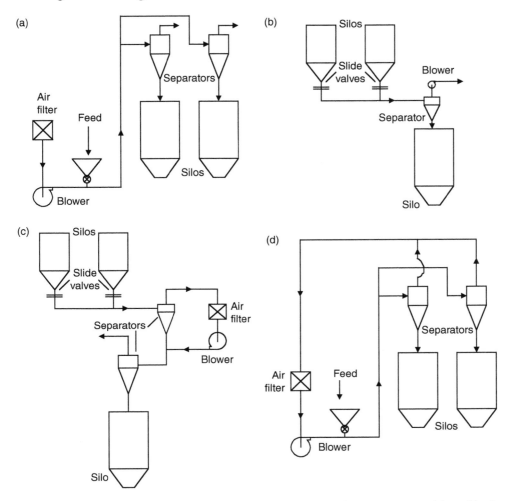

Figure 22 Dilute-phase pneumatic conveyors: (a) pressure system, (b) vacuum system, (c) combined system, (d) closed-loop system.

system, works at subatmospheric pressure and is used for delivery to one outlet from several inlets (Figure 22[b]). Vacuum systems are usually limited to shorter distances than positive ones and are more restrained to operate with dilute, low solids loading than pressure systems. When both features of pressure and vacuum systems are combined in a unit, the advantages of each can be exploited (Figure 22[c]). This arrangement consists of two sections: a pull/push system with a negative-pressure front end, followed by a positive-pressure loop. The benefit is that they capitalize on the ease of feeding into a vacuum and combine this with the higher capacity and longer conveying distance when using positive pressure. Recirculation of the conveying air, as in the closed-loop system (Figure 22[d]), reduces contamination of product by air and limits product dehydration. However, such systems are often difficult to control and an intercooler may be required to prevent the pump from overheating the recirculated air.

Most food materials may be conveyed satisfactorily at air speeds within the range of 15 to 25 m/sec. Above this, abrasion of tube bends and product damage may represent difficulties. At extremely low speeds, solids tend to settle out and block horizontal pipe runs. In terms of pressure drop, if air at high pressure is used, its correspondingly high initial energy will enable more conveying to be accomplished, per kilogram of air, than if low-pressure air is used. High-pressure systems are, however, proportionately more expensive than low-pressure ones. The maximum pressure recommended for general purpose, dilute-phase conveying of food materials is about 170 kPa. With regard to solid–air ratio, for maximum efficiency this should be as high as possible, but without invading the range of dense-phase conveying. For flour and salt such ratio may be up to 80 kg of solid/m^3 of air, while for wheat it may be limited to 30 kg of solid/m^3 of air. There is an upper limit for this ratio for specific materials; exceeding it will cause blockage of the system due to saltation. Finally, material properties such as size, shape, density, and surface properties, need to be carefully considered in the operation and selection of dilute-phase conveying systems. Other important properties are friability, hygroscopicity, as well as susceptibility to impact, abrasion damage, or oxidation.

4. Applications

Pneumatic conveying is, possibly, the most applicable type of technology for diverse conveying tasks in the food industry. All the many advantages previously mentioned pertaining to the various types of pneumatic conveyors described, suit food materials perfectly. The predominant features of many food powders and particulates, in the sense of being susceptible to damage when handled, makes pneumatic conveying an obvious alternative for food processing. It is worth pointing out, however, that a number of examples already mentioned such as sugar, spray-dried milk powder, as well as powdered and granulated coffee, have found in this type of conveying the most appropriate way of being transported with minimum amount of damage. A final reference of the advantages of pneumatic conveying systems related to handling of food materials, has to do with their self-cleaning capacity, virtual dustless operation, and general sanitary conditions

IV. PROCESSING OF FOOD POWDERS

Relevant recent developments in instrumentation and measuring techniques, has made the deeper understanding of foods microstructure and properties possible. The detailed study of physical properties of foods has revealed an important impact on how these properties may affect food processes. Among these properties, those related to bulk particulate systems such as particle size distribution and particle shape, are directly involved in an important number of unit operations such as size reduction, mixing, agglomeration, dehydration, and filtration. The optimum operation of many food processes relies heavily on a good knowledge of the behavior of particles and particle assemblies, either in dry form or as suspensions. Some relevant unit operations involving food particles and particulates will be reviewed.

A. Size Reduction

In many food processes it is frequently required to reduce the size of solid materials for different purposes. For example, size reduction may aid other processes such as expression and extraction, or may shorten heat treatments such as blanching and cooking. Comminution is the generic term used for size reduction and includes different operations such as crushing, grinding, milling, mincing, and dicing. Most of these terms are related to a particular

Table 14 Types of Force Used in Size Reduction
Equipment

Force	Principle	Examples of equipment
Compressive	Nutcracker	Crushing rolls
Impact	Hammer	Hammer mill
Attrition	File	Disk attrition mill
Cut	Scissors	Rotary knife cutter

application, for example, milling of cereals, mincing of beef, dicing of tubers, or grinding of spices. The reduction mechanism consists of deforming the food piece until it breaks or tears. Breaking of hard materials along cracks or defects in their structures is achieved by applying diverse forces. The types of forces commonly used in food processes are compressive, impact, attrition or shear, and cutting. In a comminution operation more than one type of force is usually acting. Table 14 summarizes these types of forces commonly used in some of the mills in the food industry.

1. Comminution Laws

In the breakdown of hard and brittle food solid materials, two stages of breakage are recognized: (1) initial fracture along existing fissures within the structure of the material, and, (2) formation of new fissures or crack tips followed by fracture along these fissures. It is also accepted that only a small percentage of the energy supplied to the grinding equipment is actually used in the breakdown operation. Figures of less than 2% efficiency have been quoted (Coulson and Richardson, 1992) and, thus, grinding is a very inefficient process, perhaps the more inefficient of the traditional unit operations. Much of the input energy is lost in deforming the particles within their elastic limits and through interparticle friction. A large amount of this wasted energy is released as heat, which, in turn, may be responsible for heat damage of biological materials. Theoretical considerations suggest that the energy required to produce a small change in the size of unit mass of the material can be expressed as a power function of the size of the material, that is:

$$\frac{dE}{dx} = -\frac{K}{x^n} \tag{32}$$

where dE is the change in energy, dx is the change in size, K is a constant, and x is the particle size. Equation (32) is often referred to as the general law of comminution and has been used by a number of workers to derive more specific laws depending on the application.

Rittinger considered that for the grinding of solids, the energy required should be proportional to the new surface produced and gave to the power n the value of 2, obtaining thus the so-called Rittinger's law by integration of Equation (32):

$$E = K\left[\frac{1}{x_2} - \frac{1}{x_1}\right] \tag{33}$$

where E is the energy per unit mass required for the production of a new surface by reduction, K is called Rittinger's constant and is determined for a particular equipment and material, x_1 is the average initial feed size, and x_2 is the average final product size. Rittinger's law has been found to hold better for fine grinding, where a large increase in surface results.

Kick reckoned that the energy required for a given size reduction was proportional to the size reduction ratio and took the value of the power n as 1. In such a way, by integration of Equation (32), the following relation, known as Kick's law is obtained:

$$E = K \left[\ln \frac{x_1}{x_2} \right] \tag{34}$$

where x_1/x_2 is the size reduction ratio. Kick's law has been found to hold more accurately for coarser crushing where most of the energy is used in causing fracture along existing cracks.

A third version of the comminution law is the one attributed to Bond, who considered that the work necessary for reduction was inversely proportional to the square root of the size produced. In Bond's consideration n takes the value of $\frac{3}{2}$, giving the following version (Bond's law), also by integrating Equation (32):

$$E = 2K \left[\frac{1}{\sqrt{x_2}} - \frac{1}{\sqrt{x_1}} \right] \tag{35}$$

When x_1 and x_2 are measured in micrometers and E in kWh/ton, $K = 5E_i$, where E_i is the Bond Work Index, defined as the energy required to reduce a unit mass of material from an infinite particle size to a size such that 80% passes a 100 μm sieve.

2. Size Reduction Equipment

Size reduction is a unit operation widely used in a number of processing industries. Many types of equipment are used in size reduction operations. In a broad sense, size reduction machines may be classified as crushers used mainly for coarse reduction, grinders employed principally in intermediate and fine reduction, ultrafine grinders utilized in ultrafine reduction, and cutting machines used for exact reduction (McCabe et al., 1992). Table 15 lists the principal size reduction machines for applications in food processing, while a general guide to equipment selection, as a function of food material and reduction range, is presented in Table 16.

In crushing rolls, two or more heavy steel cylinders revolve toward each other (Figure 23) so particles of feed are nipped and pulled through. The nipped particles are subjected to compressive force causing reduction in size. In some designs differential speed is maintained so as to exert shearing forces also on the particles. The roller surface can be smooth or can carry corrugations, breaker bars, or teeth, as a manner of increased friction and facilitate trapping of particles between the rolls. Tooth-roll crushers can be mounted in pairs, like the smooth-roll crushers, or with only one roll working against a stationary curved breaker plate. Tooth-roll crushers are much more versatile than smooth-roll crushers but they cannot handle very hard solids. They operate by compression, impact, and shear and not by compression alone, as do smooth-roll crushers.

Figure 24 shows a hammer mill equipment that contains a high-speed rotor turning inside a cylindrical case. The rotor carries a collar bearing a number of hammers around

Table 15 Size Reduction Machines Used in Food Process Engineering

Range of reduction	Generic name of equipment	Type of equipment
Coarse and intermediate	Crushers	Crushing rolls
Intermediate and fine	Grinders	Hammer mills
		Disk attrition mills
		Tumbling mills (rod mills)
Fine and ultrafine	Ultrafine grinders	Hammer mills
		Tumbling mills (ball mills)

Table 16 Application Examples of Size Reduction Machines

	Crushing rolls	Hammer mills	Attrition mills	Tumbling mills
Fineness range				
Coarse	•			
Intermediate	•	•	•	•
Fine and ultrafine		•	•	•
Chocolate	•			•
Cocoa			•	•
Corn (wet)			•	
Dried fruits		•		
Dried milk		•		
Dried vegetables		•		
Grains	•		•	
Pepper		•	•	
Pulses			•	
Roasted nuts			•	
Salt		•		•
Spices		•		
Starch (wet)			•	
Sugar		•		•

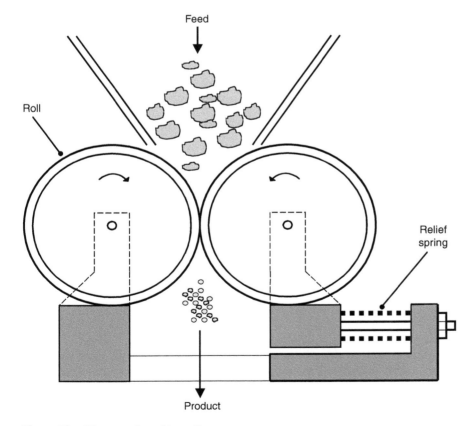

Figure 23 Diagram of crushing rolls.

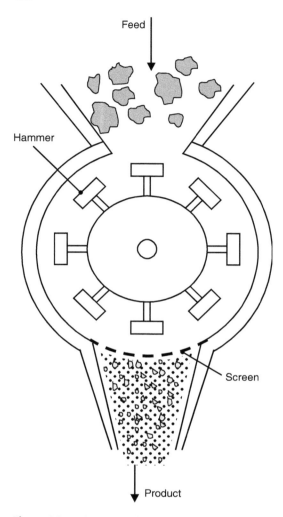

Figure 24 A hammer mill.

its periphery. By a rotatory action, the hammers swing through a circular path inside the casing containing a toughened breaker plate. Feed passes into the action zone with the hammers driving the material against the breaker plate and forcing it to pass through a bottom-mounted screen by gravity when the particles attain a proper size. Reduction is mainly due to impact forces, although under choke-feeding conditions, attrition forces can also play a part in such reduction. By replacing the hammers with knives or other elements, tough, ductile, or fibrous materials can be handled. The hammer mill is a very versatile piece of equipment, which gives high-reduction ratios and may handle a wide variety of materials from hard and abrasive to fibrous and sticky.

Disk attrition mills, as those illustrated in Figure 25, make use of shear forces for size reduction, mainly in the fine size range of particles. There are several basic designs of attrition mills. The single-disk mill (Figure 25[a]) has a high-speed rotating grooved disc leaving a narrow gap with its stationary casing. Intense shearing action results in comminution of the feed. The gap is adjustable, depending on feed size and product requirements. In the double-disk mill (Figure 25[b]) the casing contains two rotating disks that rotate in

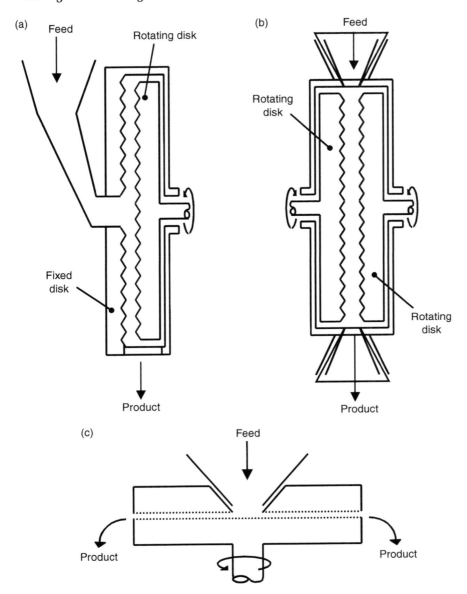

Figure 25 Disk attrition mills: (a) single disk mill, (b) double disk mill, (c) Buhr mill.

opposite directions giving a greater degree of shear compared with the single-disk mill. The pin-disk mill carries pins or pegs on the rotating elements. In this case impact forces also play an important role in particle size reduction. The Buhr mill (Figure 25[c]), which is the older type of attrition mill originally used in flour milling, consists of two circular stones mounted on a vertical axis. The upper stone is normally fixed and has a feed entry port, while the lower stone rotates. The product is discharged over the edge of the lower stone.

Tumbling mills are used in many industries for fine grinding. They basically consist of a horizontal slow-speed rotating cylinder partially filled with either balls or rods. The cylinder shell is usually of steel, lined with carbon-steel plate, porcelain, silica rock, or rubber. The balls are normally made out of steel or flint stones, while the rods are usually manufactured by high carbon steel. The reduction mechanism is carried out as follows: as the cylinder rotates, the grinding medium is lifted up the sides of the cylinder and drops on to the material being comminuted, which fills the void spaces between the medium. The grinding medium components also tumble over each other, exerting a shearing action on the feed material. This combination of impact and shearing forces brings about a very effective size reduction. As a tumbling mill basically operates in a batchwise manner, different designs have been developed to make the process continuous. As illustrated in Figure 26(a), in a trunnion overflow mill, the raw material is fed in through a hollow trunnion at one end of the mill and the ground product overflows at the opposite end. Putting slotted transverse partitions in a tube mill converts it into a compartment mill (Figure 26[b]). One compartment may contain large balls, another small balls, and a third pebbles, thus achieving a segregation of the grinding media with the consequent rationalization of energy. A very efficient way of segregating the grinding medium is the use of the conical ball mill shown in Figure 26(c).

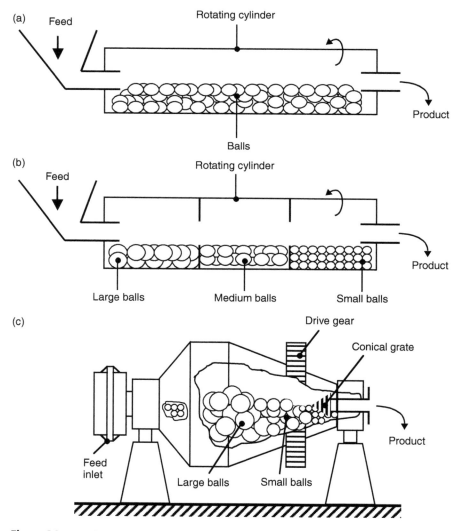

Figure 26   Tumbling mills: (a) trunnion overflow mill, (b) compartment mill, (c) conical mill.

While the solid feed enters from the left into the primary grinding zone where the diameter of the shell is maximum, the comminuted product leaves through the cone at the right end where the diameter of the shell is minimum. As the shell rotates, the large balls move toward the point of maximum diameter, and the small balls migrate toward the discharge outlet. The initial breaking of feed particles is performed, therefore, as the largest balls drop to the greatest distance. On the other hand, final reduction of small particles is carried out because the small balls drop a smaller distance. In such an arrangement, the efficiency of the milling operation is greatly increased.

3. Applications

For the food industry size reduction is, without doubt, one of the most important processing steps. Size reduction is normally applied in an infinite variety of grinding processes. These range from readily grindable (sugar and salt), through tough-fibrous (dried vegetables) and very tough (gelatin), to those materials, which tend to deposit (full-fat soy, full-fat milk powder). The fineness requirements may vary immensely from case to case. Some examples of applications of size reduction in food processes are the milling of wheat, the refining of chocolate, the grinding of spices and dried vegetables, the breaking of cocoa kernels, the preparation of cocoa powder, the degermination of corn, the production of fish-meal, the manufacture of chocolate, etc.

B. Size Enlargement

Size enlargement operations are used in the process industries with different aims such as improving handling and flowability, reducing dusting or material losses, producing structurally useful forms, enhancing appearance, etc. Size enlargement operations are known by many names, including compaction, granulation, tableting, briquetting, pelletizing, encapsulation, sintering, and agglomeration. While some of these operations could be considered rather similar, for example, tableting and pelletizing, some others are relevant to a specific type of industry, for example, sintering in metallurgical processes. In the food industry, the term agglomeration is applied to the process, where the main objective is to control porosity and density of materials in order to influence properties like dispersibility and solubility. In this case the operation is also often referred to as instantizing, because rehydration and reconstitution are important functional properties in food processes. On the other hand, when size enlargement is used with the objective of obtaining definite shapes, the food industry takes advantage of a process known as extrusion, which can shape and cook at the same time. In a more general context, however, instantizing and extrusion of food processes are the two common categories of agglomeration: tumble/growth and pressure agglomeration, and are referred to as such in the literature.

1. Aggregation Fundamentals

Agglomeration can be defined as the process by which particles join or bind with one another in a random way, resulting in an aggregate of porous structure much larger in size than the original material. The term includes varied unit operations and processing techniques aimed at agglomerating particles (Green and Maloney, 1999). As mentioned above, agglomeration is used in food processes mainly to improve properties related to handling and reconstitution. Figure 27 shows some common binding mechanisms of agglomeration with bridges or force fields at the coordination points between particles (Pietsch, 1991). The two-dimensional structure represented in the figure is, in reality, three-dimensional

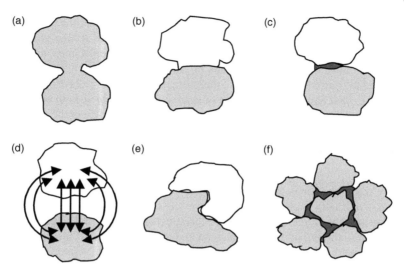

Figure 27 Different binding mechanisms in agglomeration: (a) partial melting sinter bridges, (b) chemical reaction hardening binders, (c) liquid bridges hardening binders, (d) molecular and like-type of forces, (e) interlocking bonds, (f) capillary forces.

containing a large number of particles. Each particle interacts with several others surrounding it and the points of interaction may be characterized by contact, or by a distance small enough for the development of binder bridges. Alternatively, sufficiently high attraction forces can be caused by one of the short-range force fields. The number of all interaction sites of one particle within the agglomerate structure is called the coordination number. Particles in an agglomerate could be quite numerous, making it difficult to estimate the coordination number. Indirect measurement of the coordination number can be made as a function of other properties of the agglomerate. In regular packs of monosized spherical particles the coordination number k and the porosity or void volume ε, are related by:

$$k\varepsilon \approx \pi \tag{36}$$

Equation (36) gives good approximation of the coordination numbers of ideal agglomerate structures. Table 17 lists several values of coordination numbers calculated using Equation (36) and compared with the ideal number for different structures, such as those illustrated in Figure 28.

A general relation describing the tensile strength of agglomerates σ_t held together by binding mechanisms acting at the coordination points is:

$$\sigma_t = \frac{1-\varepsilon}{\pi} k \frac{\sum_{i=1}^{n} A_i(x,\dots)}{x^2} \tag{37}$$

where A_i is the adhesion force caused by a particular binding mechanism and x is the representative size of the particles forming the agglomerate.

Substituting Equation (36) into Equation (37), the following relation is obtained:

$$\sigma_t = \frac{1-\varepsilon}{\varepsilon} \frac{\sum_{i=1}^{n} A_i(x,\dots)}{x^2} \tag{38}$$

Table 17 Geometric Arrangement, Porosity, and Coordination Number of Packings of Monosized Particles

Geometric arrangement	Porosity (ε)	Coordination number (π/ε)	k
Cubic	0.476	6.59	6
Orthorhombic	0.395	7.95	8
Tetragonal–spheroidal	0.302	10.39	10
Rhombohedral (pyramidal)	0.260	12.08	12
Rhombohedral (hexagonal)	0.260	12.08	12

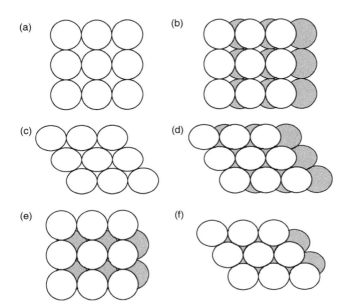

Figure 28 Packings of monosized spherical particles: (a) cubic, (b) and (c) orthorhombic, (d) tetragonal–spheroidal, (e) rhombohedral (pyramidal), (f) rhombohedral (hexagonal).

A further simplification results because many binding mechanisms are a function of the representative particle size x and thus:

$$\sigma_t = \frac{1-\varepsilon}{\varepsilon} \frac{\sum_{i=1}^{n} A_i(x, \ldots)}{x} \tag{39}$$

The three dots in parentheses in Equations (37), (38), and (39) indicate that A_i is also a function of other, unknown, parameters.

When liquid bridges form at the coordination points, A_i depends on the bridge volume and the wetting characteristics represented by the wetting angle. There are models available for predicting adhesion forces of various types (Pietsch, 1991), but A_i might be of different magnitude at each of the many coordination points, due to roughness or microscopic structure of particulates forming the agglomerates.

The representative particle size most appropriate to describe the agglomeration process is the surface equivalent diameter, x_{sv}, because porosity is surface dependant. Such diameter is the size of a spherical particle, which, if the powder consists of only these particles, would have the same specific surface area as the actual sample. When determining the specific surface area, methods must be chosen that only measure the outer particle surface excluding the accessible inner surface due to open particle porosity.

From the previous paragraphs, it can be gathered that the strength of agglomerate structures held together by bonding mechanisms is highly dependent on porosity and particle size, or, more properly, specific surface area. The relationship would be inversely proportional in both cases, that is, higher strength at lower porosities and lower surface areas.

Agglomerates that are completely filled with liquid obtain strength from the negative capillary pressure in the structure. A relationship for this case is:

$$\sigma_t = c \frac{1 - \varepsilon}{\varepsilon} \alpha \frac{1}{x_{sv}} \tag{40}$$

where c is a correction factor, α is the surface tension of the liquid and x_{sv} is the surface equivalent diameter of the particle. In order to apply Equation (40), there must be a complete wetting of the solids by the liquid.

For high-pressure agglomeration and the effect of matrix binders general formulas have not yet been developed. It can be considered, however, that the effects of variables would follow the trend described before, with porosity, particle surface, contact area, and adhesion, all playing an important role. For nonmetallic powders, the following equation can be used to evaluate the needed applied pressure p to agglomerate:

$$\log p = m V_R + b \tag{41}$$

where V_R represents the relationship V/V_s, V being the compacted volume at a given pressure and V_s the volume of the solid material to be compacted; m and b are constants.

2. Agglomeration Methods

With few exceptions, agglomeration methods can be classified into two groups: tumble/growth agglomeration and pressure agglomeration. Also, agglomerates can be obtained using binders or in a binderless manner. The tumble/growth method produces agglomerates of approximate spherical shape by buildup during tumbling of fine particulate solids, the resulting granules are at first weak and require binders to facilitate formation, and posttreatment is needed to reach final and permanent strength. On the other hand, products from pressure agglomeration are made from particulate materials of diverse sizes, and are formed without the need of binders or posttreatment, and acquire immediate strength.

The mechanism of tumble/growth agglomeration is illustrated in Figure 29. As shown, the overall growth process is complex and involves both disintegration of weaker bonds and reagglomeration by abrasion transfer and coalescence of larger units (Cardew and Oliver, 1985). Coalescence occurs at the contact point when, at impact, a binding mechanism that is stronger than the separating forces develops. Additional growth of the agglomerate may proceed by further coalescence, or by layering, or both. The most important and effective separation force counteracting on the bonding mechanism is the weight of the solid particle. For particles below 10 μm approximately, the natural attraction forces, such as molecular, magnetic, and electrostatic, become significantly larger than the separation forces due to particle mass and external influences. In such a way, natural agglomeration occurs.

The conditions needed for tumble/growth agglomeration can be provided by inclined disks, rotating drums, any kind of powder mixer, and fluidized beds (Figure 30). In general terms, any equipment or environment creating random movement is suitable for carrying out tumble/growth agglomeration. In certain applications, very simple tumbling motions, such as on the slope of storage piles or on other inclined surfaces, are sufficient for the

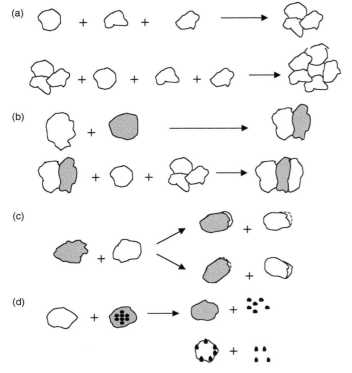

Figure 29 Kinetics of tumble/growth agglomeration: (a) nucleation, (b) random coalescence, (c) abrasion transfer, (d) crushing and layering.

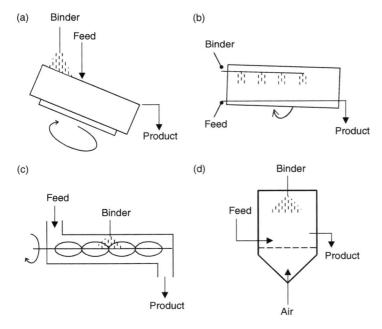

Figure 30 Equipment for tumble/growth agglomeration: (a) inclined rotating disk, (b) inclined rotating drum, (c) ribbon powder blender, (d) fluidized bed.

formation of crude agglomerates. The most difficult task of tumble/growth agglomeration is to form stable nuclei due to the presence of few coordination points in small agglomerates. Also, since the mass of particles and nuclei are small, their kinetic energy is not high enough to cause microscopic deformation at the contact points, which enhances bonding. Recirculation of undersized fines provides nuclei to which feed particles adhere more easily to form agglomerates. In the whole process, tumble/growth agglomeration renders first weak agglomerates known as green products. These wet agglomerates are temporarily bonded by surface tension, and by capillary forces of the liquid binder. This is the reason why, in most cases, tumble/growth agglomeration requires some sort of posttreatment. Drying and heating, cooling, screening, adjustment of product characteristics by crushing, rescreening, conditioning, and recirculation of undersize material, are some of the processes, which have been used as posttreatment in tumble/growth agglomeration. Sometimes, a large percentage of recycle must be rewetted for agglomeration and needs to be processed again, causing economical burden to this technology (Pietsch, 1983).

In contrast to tumble/growth agglomeration where no external forces are applied, in pressure agglomeration pressure forces act on a confined mass of particulate solids, which is then shaped and densified (Engelleitner, 1994). Pressure agglomeration is normally carried out in two stages. The first one comprises a forced rearrangement of particles due to a little applied pressure, while the second step consists of a steep pressure rise during which brittle particles break and malleable particles deform plastically (Pietsch, 1994). The mechanism of pressure agglomeration is illustrated in Figure 31. There are two important phenomena, which may limit the speed of compaction and, therefore, the capacity of the equipment: compressed air in the pores and elastic springback. Both can cause cracking and weakening which, in turn, may lead to destruction of the pressure-agglomerated products. The effect of these two phenomena can be reduced if the maximum pressure is maintained for some time, known as dwell time, prior to its release.

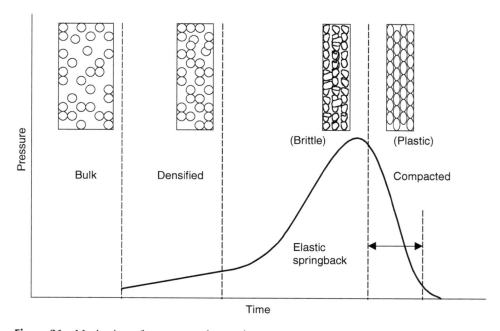

Figure 31   Mechanism of pressure agglomeration.

Pressure agglomeration can be carried out in different types of equipment. Generally, low- and medium-pressure agglomeration is achieved in extruders including the screen extruder, the screw extruder, and the intermeshing gears extruder. On the other hand, high-pressure agglomeration is performed in presses such as the punch-and-die press, the compacting roller press, and the briquetting roller press. Low- and medium-pressure agglomeration yield relatively uniform agglomerates of elongated spaghetti-like or cylindrical shape, whereas high-pressure agglomeration produces pillow or almond-like shapes. Figure 32 presents the equipment used for low- and medium-pressure agglomeration, while Figure 33 illustrates some common machinery for high-pressure agglomeration.

3. Applications

Agglomeration has many applications in food processing. In the context of instantizing, tumble/growth agglomeration is used in the food industry to improve reconstitutability of a number of products including flours, cocoa powder, instant coffee, dried milk, sugar, sweeteners, fruit beverage powders, instant soups, and diverse spices. With regard to

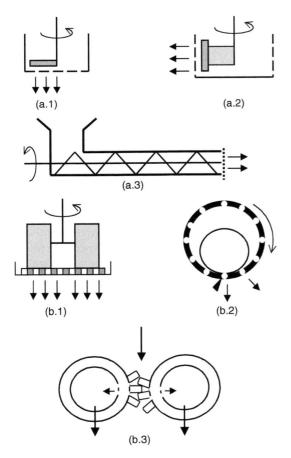

Figure 32 Equipment used for (a) low- and (b) medium-pressure agglomeration: (a.1) screen extruder, (a.2) basket extruder, (a.3) cylindrical-die screw extruder, (b.1) flat-die extruder, (b.2) cylindrical-die extruder, (b.3) intermeshing-gears extruder.

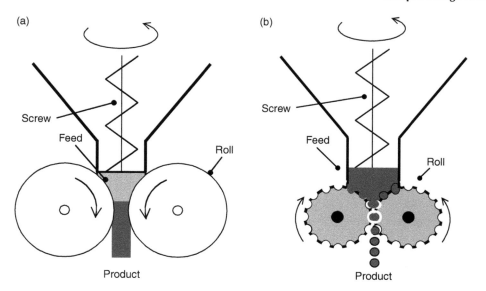

Figure 33 Equipment used for high-pressure agglomeration: (a) compacting roller press, (b) briquetting roller press.

shaping, extrusion has been extensively used in grain process engineering to obtain an array of products from diverse cereals, principally ready-to-eat breakfast cereals.

C. Mixing

The unit operation in which two or more materials are interspersed in space with one another is one of the oldest and yet one of the least understood of the unit operations of process engineering. Mixing is used in the food industry with the main objective of reducing differences in properties such as concentration, color, texture, taste, and so on, between different parts of a system. Since the components being mixed can exist in any of the three states of matter, a number of mixing possibilities arise. The mixing cases involving a fluid, for example, liquid–liquid and solid–liquid, are most frequently encountered, so they have been extensively studied. Despite the importance of the mixing of particulate materials in many processing areas, fundamental work of real value for either designers or users of solids mixing equipment is still relatively sparse. It is through studies in very specific fields, such as powder technology and multiphase flow, that important advances in the understanding of mixing of solids and pastes have been made.

Mixing is more difficult to define and evaluate with powders and particulates than it is with fluids, but some quantitative measures of dry solids mixing may aid in evaluating mixer performance. In actual practice, however, the proof of a mixer is in the properties of the mixed material it produces. A significant proportion of research efforts in the food industry is directed toward the development of new and novel mixing devices for food materials. These devices may be effective for many applications since they deliver a mixed product with the required blending characteristics. Due to the complex properties of food systems, which can themselves vary during mixing, it is extremely difficult to generalize or standardize mixing operation for wider applications of mixing devices, either novel or traditional. Developments in mathematical modeling of food-mixing processes are scarce and there are no established procedures for process design and scale-up. As a result, it is virtually impossible to devise relationships between mixing and quality (Niranjan, 1995), especially for blending of food powders. With reference to solid foods, Niranjan and de Alwis (1993) mentioned as characteristic features of food mixing, the fragile and differently sized nature of food products, as well as the segregating tendency of blended food systems on discharge. These characteristics, along with some others like cohesiveness and stickiness, makes food particulate mixing a complicated operation.

1. Mixing Mechanisms

Three mechanisms have been recognized in solids mixing: convection, diffusion, and shear. In any particular process one or more of these three basic mechanisms may be responsible for the course of the operation. In convective mixing masses or group of particles transfer from one location to another, in diffusion mixing individual particles are distributed over a surface developed within the mixture, whereas in shear mixing groups of particles are mixed through the formation of slipping planes that develop within the mass of the mixture. Shear mixing is sometimes considered as part of a convective mechanism. Pure diffusion, when feasible, is highly effective, producing very intimate mixtures at the level of individual particles but at an exceedingly slow rate. Pure convection, on the other hand, is much more rapid but tends to be less effective, leading to a final mixture that may still exhibit poor mixing characteristics on a fine scale. These features of diffusion and convective mixing mechanisms suggest that an effective operation may be achieved by a combination of both, in order to take advantage of the speed of convection and the effectiveness of diffusion.

Compared with fluid mixing, in which diffusion can be normally regarded as being spontaneous, particulate systems will only mix as a result of mechanical agitation provided by shaking, tumbling, vibration, or any other mechanical mean. Mechanical agitation will provide conditions for the particles to change their relative positions either collectively or individually. The movement of particles during a mixing operation, however, can also result in another mechanism, which may retard, or even reverse, the mixing process and is known as *segregation*. When particles differing in physical properties, particularly size and/or density are mixed, mixing is accompanied by a tendency to unmix. Thus, in any mixing operation, mixing and demixing may occur concurrently and the intimacy of the resulting mix depends on the predominance of the former mechanism over the latter. Apart from the properties already mentioned, surface properties, flow characteristics, friability, moisture content, and the tendency to cluster or agglomerate, may also influence the tendency to segregate. The closer the ingredients are in size, shape, and density, the easier the mixing operation and the more the intimacy of the final mix. Once the mixing and demixing mechanisms reach a state of equilibrium, the condition of the final mix is determined and further mixing will not produce a better result.

A general theory of segregation, regardless of the particular circumstances in which the operation takes place, has not yet been offered to explain the segregation phenomena in particulate systems. In any blending operation the mixing and demixing mechanisms will be acting simultaneously. The participation of each of these two sets of mechanisms will be dictated by the environment and by the tendency of each component to segregate out of the system. Since these two mentioned sets of mechanisms will be acting against each other, an equilibrium level will be obtained as the final state of the mixture.

2. Degree of Mixing

Over the years many workers have attempted to establish criteria for the completeness and degree of mixture. In order to accomplish this, very frequent sampling of the mix is usually required and, tending to be statistical in nature, such an exercise is often of more interest to mathematicians than to process engineers. Thus, in practical mixing applications, an ideal mixture may be regarded as the one produced at minimum cost and which satisfies the product specifications at the point of use.

Food mixing is a complicated task not easily described by mathematical modeling. Mixture quality results from several complex mechanisms operating in parallel, which are hard to follow and fit to a particular model. Dankwertz (1952) has defined the scale and intensity of segregation as the quantities necessary to characterize a mixture. The scale of segregation is a description of unmixed components, while the intensity of segregation is a measure of the standard deviation of composition from the mean, taken over all points in the mixture. In practice it is difficult to determine these parameters, since they require concentration data from a large number of points within the system. They provide, however, a sound theoretical basis for assessing mixture quality. Taking into account the complexity of components and interactions in food solids mixing, it would be rather difficult to define a unique criterion to assess mixture quality. A mixing endpoint or optimum mixing time can also be considered a very relative definition due to the segregating tendency of food powder mixing. The degree of uniformity of a mixed product may be measured by analysis of a number of spot samples. Food powder mixers act on two or more separate materials to intermingle them. Once a material is randomly distributed through another, mixing may be considered to be complete. Based on these concepts, the well-known statistical parameters, mean and standard deviation of component concentration, can be used to characterize the state of a mixture. If spot samples are taken at random from a mixture and analyzed, the standard deviation of the analyses s, about the average value of the fraction of a specific powder $\bar{x}$ is estimated by the following relation:

$$s = \sqrt{\frac{\sum_{i=1}^{N} (x_i - \bar{x})^2}{N - 1}} \tag{42}$$

where x_i is every measured value of fraction of one powder and N is the number of samples.

The standard deviation value on its own may be meaningless, unless it can be checked against limiting values of either complete segregation s_0, or complete randomization s_r. The minimum standard deviation attainable with any mixture is s_r and it represents the best possible mixture. Furthermore, if a mixture is stochastically ordered, s_r would equal zero. Based on these limiting values of standard deviations, Lacey (1954) defined a mixing index M_1 as follows:

$$M_1 = \frac{s_0^2 - s^2}{s_0^2 - s_r^2} \tag{43}$$

The numerator in Equation (43) would be an indicator of how much mixing has occurred, while the denominator would show how much mixing can occur. In practice, however, the values of s, even for a very poor mixture, lie much closer to s_r than to s_0. Poole et al. (1964), suggested an alternative mixing index, that is:

$$M_2 = \frac{s}{s_r} \tag{44}$$

Equation (44) clearly indicates that for efficient mixing or increasing randomization M_2 would approach unity. The values of s_0 and s can be determined theoretically. These values would be dependent on the number of components and their size distributions. Simple expressions can be derived for two-component systems, while for a binary multisized particulate mixture Poole et al. (1964) demonstrated that:

$$s_r^2 = \frac{pq}{[w/(q(\sum f_a w_a)_p + p(\sum f_a w_a)_q)]} \tag{45}$$

where p and q are the proportions by weight of components within a total sample weight w and f_a is the size fraction of one component of average weight w_a in a particle size range. For a given component in a multicomponent and multisized particulate system, Stange (1963) presented an expression for s_r, as follows:

$$s_r^2 = \frac{p^2}{w} \left\{ \left[\frac{1-p}{p} \right]^2 \cdot p \left(\sum f_a w_a \right)_p + q \left(\sum f_a w_a \right)_q + r \left(\sum f_a w_a \right)_r + \cdots \right\} \tag{46}$$

Equations (43) and (46) can be used to calculate mixing indices defined by Equation (42). Another suggestion for the characterization of the degree of homogeneity in mixing of powders, has been reviewed by Boss (1986), with the degree of mixing M_3 defined as:

$$M_3 = 1 - \frac{s}{s_0} \tag{47}$$

Some other mixing indices have been reviewed by Fan and Wang (1975).

McCabe et al. (1992), presented the following relationship to evaluate mixing time t for solids blending:

$$t = \frac{1}{k} \ln \frac{1 - 1/\sqrt{n}}{1 - 1/M_2} \tag{48}$$

where k is a constant and n is the number of particles in a spot sample. Equation (48) can be used to calculate the time required for any required degree of mixing, provided k is known and the segregating forces are not active. Mixing times should not be very long due to the unavoidable segregation nature of most food solids mixtures. Instead of improving efficiency, long mixing times often result in poor blending characteristics. A graph of the degree of mixing versus time is recommended to select the proper mixing time quantitatively. Most cases of mixing of powders will attain maximum degree of homogeneity in less than 15 min when the proper type of machine and working capacity have been chosen.

3. Powder Mixers

In general terms, mixers for dry solids have nothing to do with mixers involving a liquid phase. According to the mixing mechanisms previously discussed, solids mixers can be classified into two groups: segregating mixers and nonsegregating mixers. The former operate mainly by a diffusive mechanism while the latter practically involve a convective mechanism. Segregating mixers are normally nonimpeller type units, such as tumbler mixers, whereas nonsegregating mixers may include screws, blades, and ploughs in their designs, and examples include horizontal trough mixers and vertical screw mixers. Food powders can also be mixed by aeration using a fluidized bed. The resulting turbulence of passing air through a bed of particulate material causes material to blend. Mixing times required in fluidized beds are significantly lower than those required in conventional powder mixers. Van Deemter (1985) discussed different mixing mechanisms prevailing in fluidized beds.

Tumbler mixers operate by tumbling the mass of solids inside a revolving vessel. These vessels take various forms, such as those illustrated in Figure 34, which may be fitted with baffles or stays to improve their performance. The shells rotate at variable speeds having values up to 100 r/min with working capacities around 50 to 60% of the total. They

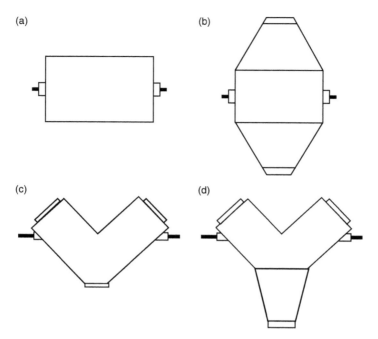

Figure 34 Tumbler mixers used in food powder blending: (a) horizontal cylinder, (b) double cone, (c) V-cone, (d) Y-cone.

are manufactured using a wide variety of materials, including stainless steel. This type of equipment is best suited for gentle blending of powders with similar physical characteristics. Segregation can represent a problem if particles vary, particularly in size and shape.

Horizontal trough mixers consist of a semicylindrical horizontal vessel in which one or more rotating devices are located. For simple operations single or twin screw conveyors are appropriate and one passage through such a system may be good enough. For more demanding duties a ribbon mixer, like the one shown in Figure 35, may be used. A typical design of a ribbon mixer will consist of two counteracting ribbons mounted on the same shaft. One moves the solids slowly in one direction while the other moves it quickly in the opposite direction. There is a resultant movement of solids in one direction, so the equipment can be used as a continuous mixer. Some other types of ribbon mixers operate on a batch basis. In these designs troughs may be closed, as to minimize dust hazard, or may be jacketed to allow temperature control. Due to small clearance between the ribbon and the trough wall, these kinds of mixers can cause particle damage and may consume high amounts of power.

In vertical screw mixers a rotating vertical screw is located in a cylindrical or cone-shaped vessel. The screw may be mounted centrally in the vessel or may rotate or orbit around the central axis of the vessel near the wall. Such mixers are schematically shown in Figure 36(a) and 36(b) respectively. The latter arrangement is more effective and stagnant layers near the wall are eliminated. Vertical screw mixers are quick, efficient, and particularly useful for mixing small quantities of additives into large masses of material.

4. Applications

Applications of powder mixing in food systems are diverse and varied and include blending of grains prior to milling, blending of flours and incorporation of additives to

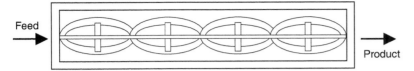

Figure 35 Plain view of an open ribbon mixer.

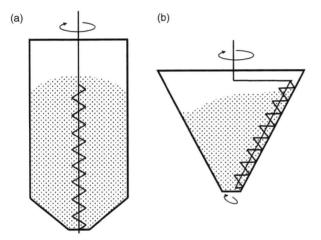

Figure 36 Vertical screw mixers: (a) central screw, (b) orbiting screw.

flours, preparation of custard powders and cake mixes, blending of soup mixes, blending of spice mixes, incorporation of additives in dried products, preparation of baby formula, etc.

D. Cyclonic Separations

Separation techniques are involved in a great number of processing industries and represent, in many cases, the everyday problem of a practicing engineer. In spite of this, the topic is normally not covered efficiently and sufficiently in higher education curricula of some engineering programs, mainly because its theoretical principles deal with a number of subjects ranging from physics principles to applied fluid mechanics. In recent years, separation techniques involving solids have been considered in the general interest of powder and particle technology, as many of these separations involve removal of discrete particles or droplets from a fluid stream.

Separation techniques are defined as those operations, that isolate specific ingredients of a mixture without a chemical reaction being carried out. Several criteria have been used to classify or categorize separation techniques. One such criteria consists in grouping them according to the phases involved, that is, solid with liquid, solid with solid, liquid with liquid, etc. A classification based on this criterion is shown in Table 18. Dry separation techniques would, therefore, constitute all those cases in which the particle to be isolated or segregated from a mixture is not wet, and would include particular examples of the solid mixtures and gas–solid mixture cases listed in Table 18. The most important dry separation techniques in processing industries have been reviewed by Beddow (1981). In food processing, there are important applications of dry separation techniques, such as the removal of particles from dust-laden air in milling operations or the recovery of the dried product in spray dehydration.

Table 18 Classification of Separation Techniques According to
Phases Involved

Type of mixture	Technique
Liquid–Liquid	Distillation
	Extraction
	Decantation
	Dialysis and electrodialysis
	Parametric pumping
Solid–Solid	Screening
	Leaching
	Flotation
	Air classification
Solid–Gas	Cycloning
	Air filtration
	Scrubbing
	Electrostatic precipitation
Solid–Liquid	Sedimentation
	Centrifugation
	Filtration
	Membrane separations

In many processes of food and related industries, separating solids from a gas stream is very important. A typical example is one of the risk of dust explosion in the dry milling industry. It has been found that not only in this industry, but also in many others the atmosphere may become dust-laden with particles from different sources representing even a health risk. In other cases the suspension of particles in a gas stream has been promoted, as in pneumatic conveying or spray drying, but at the end of the process there is the need to separate the phases. Separation of solids from a gas is accomplished using many different devices. Perhaps the devices most commonly used to separate particles from gas streams are cyclones.

1. Operating Principles

Cyclones are by far the most common type of gas–solids separation device used in different industrial processes. They have no moving parts, are inexpensive compared to other separation devices, can be used at high temperatures, produce a dry product, have low energy consumption, and are extremely reliable. Their primary disadvantage is that they have relatively low collection efficiency for particles below about 15 μm. As illustrated in Figure 37, a cyclone consists of a vertical cylinder with a conical bottom, a tangential inlet near the top, and outlets at the top and the bottom respectively. The top outlet pipe protrudes into the conical part of the cyclone in order to produce a vortex when a dust-laden gas (normally air) is pumped tangentially into the cyclone body. Such a vortex develops centrifugal force and, because the particles are much denser than the gas, they are projected outward to the wall flowing downward in a thin layer along this wall in a helical path. They are eventually collected at the bottom of the cyclone and separated. The inlet gas stream flows downward in an annular vortex, reverses itself as it finds a reduction in the rotation space due to the conical shape, creates an upward inner vortex in the center of the cyclone, and then exits through the top of the cyclone. In an ideal operation, in the upward flow, there is only gas whereas the downward flow has all the particles fed with the stream. Cyclone diameters

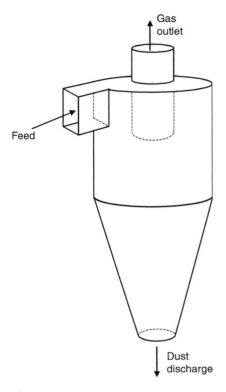

Gas
outlet

Feed

Dust
discharge

Figure 37   Schematic diagram of a cyclone.

range in size from less than 0.05–10 m, feed concentrations cover values from 0.1 to about 50 kg/m³, while gas inlet velocities may be in the order of 15–35 m/sec.

A cyclone is in fact a settling device in which a strong centrifugal force operates, acting radially, instead of the relatively weak gravity force that acts vertically. Due to the small range of particles involved in cyclone separation (the smallest particle that can be separated is about 5 μm), it is considered that Stokes' law primarily governs the settling process. The common form of Stokes' law is:

$$u_t = \frac{x^2(\rho_s - \rho_g)g}{18\mu_g} \tag{49}$$

where u_t is the terminal settling velocity, x is the particle diameter, ρ_s is the solids density ρ_g is the gas density, and μ_g is the gas viscosity.

Cyclones can generate centrifugal forces between 5 and 2500 times the force of gravity, depending on the diameter of the unit. When particles enter into the cyclone body, they quickly reach their terminal velocities corresponding to their sizes and radial position in the cyclone. The radial acceleration in a cyclone depends on the radius of the path being followed by the gas and is given by the equation:

$$g = \omega^2 r \tag{50}$$

where ω is the angular velocity and r the radius. Substituting Equation (49) into Equation (50):

$$v_t = \frac{x^2(\rho_s - \rho_g)\omega^2 r}{18\mu_g} \tag{51}$$

where v_t is the terminal velocity of the particle.

Also, the centrifugal acceleration is a function of the tangential component of the velocity $v_{tan} = \omega r$, and thus, considering this, Equation (51) becomes:

$$v_r = \frac{x^2(\rho_s - \rho_g)v_{tan}^2}{18\mu_g r} \tag{52}$$

Multiplying Equation (52) by g/g, the resultant equation gives:

$$v_t = \left[\frac{x^2(\rho_s - \rho_g)g}{18\mu_g} \right] \frac{v_{tan}^2}{gr} = (u_t)\frac{v_{tan}^2}{gr} \tag{53}$$

where u_t is the terminal settling velocity defined by Equation (49). As can be implied, according to Equation (53), the higher the terminal velocity the easier it is to "settle" a particle within a cyclone.

For a given particle size, the terminal velocity is a maximum in the inner vortex, where r is small, so the finest particles separated from the gas are eliminated in the inner vortex. They migrate through the outer vortex to the wall of the cyclone and drop, passing the bottom outlet. Smaller particles, which do not have time to reach the wall, are retained by the air and carried to the top outlet. Although the chance of a particle to separate decreases with the square of the particle diameter, the fate of a particle also depends on its position in the cross section of the entering stream and on its trajectory in the cyclone. Thus, the separation according to size is not sharp. A specific diameter, called the *cut diameter* or *cut size*, can be defined as that diameter for which one-half of the inlet particles, by mass, are separated while the other half are retained by the gas. The cut size is a very useful variable to determine the separation efficiency of a cyclone. Since a given powder to be separated in a cyclone would have an extremely fine half of its distribution, such half may not be easily separated using conventional pressure drops. Therefore, it is advisable to make it the cut size to coincide with the mean size of a powder particle size distribution to guarantee separation of the coarse part of such distribution, as the fine one may unattainable due to the small range involved.

2. Dimensionless Scale-Up Approach

Experience and theory have shown that there are certain relationships among cyclone dimensions that should be observed for efficient cyclone performance (Geldart, 1986), and which are generally related to the cyclone diameter. There are several different standard cyclone "designs" and a very common one is called the *Stairmand* design, whose dimensions are shown in Figure 38. Using standard geometries of cyclones is much easier to predict effects on variables changes and scale-up calculations are greatly reduced. Such calculations may be carried out by means of dimensionless relationships. Selection and operation of cyclones can be described by the relationship between the pressure drop and the flow rate, and the relationship between separation efficiency and flow rate (Svarovsky, 1981). The pressure drop versus volumetric flow rate relationship is usually expressed as $Eu = f(Re)$, where Eu is the Euler number and Re is the Reynolds number. The Euler number is in fact a pressure loss factor, easily defined as the limit on the maximum characteristic velocity v obtained by

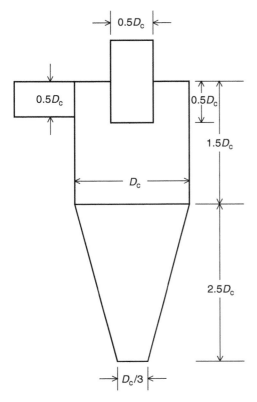

Figure 38 Dimensions of a Stairmand standard cyclone.

a certain pressure drop ΔP across the cyclone. It can be expressed as:

$$Eu = \frac{2\Delta P}{\rho_g v^2} \tag{54}$$

where, as previously defined, ρ_g is the gas density.

The Reynolds number defines flow characteristics of the system and, in the case of cyclones, the characteristic dimension may be taken as the cyclone body diameter D_c. The Reynolds number for this case is, therefore, represented by:

$$Re = \frac{D_c v \rho_g}{\mu_g} \tag{55}$$

where, as already defined, μ_g is the gas viscosity.

The relationship between separation efficiency and flow rate is not significantly influenced by operational variables, so it is commonly expressed in terms of cut size x_{50}. The use of cut size to define efficiency of cyclones is of utmost importance since their performance is highly dependent on particle size. Considering that cut size implies the size of particles to be separated it follows that such particles must be influenced by forces exercised on the suspension. The forces developed in a cyclone can be analyzed by sedimentation theory, and a dimensionless group thus derived, the Stokes number Stk, will include the cut size. The Stokes number is a very useful theoretical tool and, for the case of cyclones, its derivation may be carried out as follows.

The radial settling velocity in a cyclone is due to the centrifugal acceleration, which is proportional to the square of the tangential velocity of the particle and indirectly proportional to the radius of the particle position. As the tangential motion of the particle is unopposed, the tangential particle velocity can be taken as equal to the tangential component of the fluid velocity at the same point. For the same flow regime, the velocities anywhere in the flow in a cyclone are proportional to a characteristic velocity v, function of the cyclone cylindrical geometry, called also the body velocity. The position radii are proportional to the cyclone diameter D_c. Under such assumptions, Equation (52) can be approximated to:

$$v = \frac{x^2(\rho_s - \rho_g)v^2}{18\mu_g D_c} \tag{56}$$

Reexpressing Equation (56) in dimensionless form, the Stokes' number Stk is obtained as:

$$Stk = \frac{x^2(\rho_s - \rho_g)v}{18\mu_g D_c} \tag{57}$$

Since the value of the gas density, usually air, is negligible in comparison with the solids density, Equation (57) can also take the following form:

$$Stk = \frac{x^2 \rho_s v}{18\mu_g D_c} \tag{58}$$

Furthermore, if the dimension x is replaced by the specific cut size x_{50}:

$$Stk_{50} = \frac{x_{50}^2 \rho_s v}{18\mu_g D_c} \tag{59}$$

Equations (54), (55), and (56), defining Euler Eu, Reynolds Re, and Stokes Stk_{50} numbers respectively, are related by specific functions, which can be plotted as shown in Figure 39 and Figure 40, for a given cyclone geometry. The cyclone inside diameter D_c, is the one shown in Figure 38 and, as previously mentioned, all geometrical proportions are related to it. In the case of scale-up procedures, proportions must be maintained. The cyclone

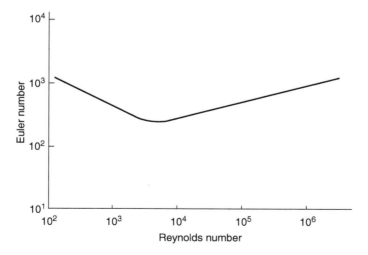

Figure 39 A typical plot of Eu versus Re for cyclones.

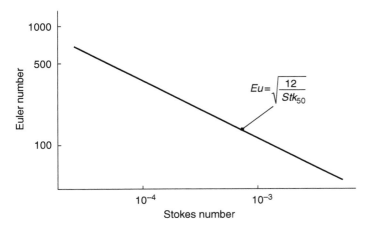

Figure 40   A typical plot of Eu versus Stk_{50} for cyclones.

body velocity v is the characteristic velocity which can be defined in various ways, but the simplest one is based on the cross section of the cylindrical body so that:

$$v = \frac{4Q}{\pi D_c^2} \tag{60}$$

where Q is the gas flow rate.

3.   Applications

As mentioned before, cyclones are extensively used in the food industry to reduce particle load to safe levels in dry milling, as well as in classification of particles in closed circuit grinding operations. They are also employed in recovering fines from spray drying and fluidized bed drying processes. Another important application is in pneumatic conveying of diverse food products, such as grains and flours.

V.   CONCLUSIONS

Powder technology is of dynamic significance to the world economy with a broad range of industries taking advantage of the rapidly growing knowledge in this discipline. Research within universities and similar institutions, coupled with the vested interest from the industrial community, has stimulated relevant results that have been applied to make a number of processes more efficient. Periodical meetings and specialized publications are spreading the most relevant and recent advances in diverse topics such as materials handling, particle formation, mixing, grinding, and separation. As mentioned at the beginning of this chapter, there has been a rapid growth of adapting knowledge of particle technology worldwide in recent years.

In the case of the application of particle technology principles to biological materials, specifically food powders and particulates, there is much to be done as very few research groups can be identified in this particular field around the globe. Since many strategic food industries, such as those based on grain processing, rely heavily on a firm understanding of powder technology, there is a lot of potential for research and development in the years to come.

Food powder processing encompasses subjects of established disciplines such as particle technology, as well as food and chemical engineering. There are a lot of activities in fundamental and applied research, to provide the food industry with theoretical tools to increase competitiveness in a great number of processes, which involve food powders.

REFERENCES

Allen, T. (1981). *Particle Size Measurement*. Chapman & Hall, London.

Arnold, P.C., McLean, A.G., and Roberts, A.W. (1982). *Bulk Solids: Storage, Flow and Handling*. The University of Newcastle Research Associates (TUNRA), Australia.

Barbosa-Cánovas, G.V., Málave-López, J., and Peleg, M. (1985). Segregation in food powders. *Biotechnol. Prog.* **1**: 140–146.

Barbosa-Cánovas, G.V., Málave-López, J., and Peleg, M. (1987). Density and compressibility of selected food powders mixture. *J. Food Process Eng.* **10**: 1–19.

Barletta, B.J. and Barbosa-Cánovas, G.V. (1993). An attrition index to assess fines formation and particle size reduction in tapped agglomerated food powders. *Powder Technol.* **77**: 89–93.

Beddow, J.K. (1981). Dry separation techniques. *Chem. Eng.* **88**: 70–84.

Beltran-Reyes, B., Ortega-Rivas, E., and Anzaldua-Morales, A. (1996). Characterization of reconstituted apple paste in terms of rehydration and firmness. *Food Sci. Technol. Int.* **2**: 307–313.

Boss, J. (1986). Evaluation of the homogeneity degree of a mixture. *Bulk Solids Handl.* **6**: 1207–1210.

Cabrejos, F.J. and Klinzing, G.E. (1994). Minimum conveying velocity in horizontal pneumatic transport and the pickup and saltation mechanisms of solid particles. *Bulk Solids Handl.* **14**: 541–550.

Cardew, P.T. and Oliver, R. (1985). Kinetics and Mechanics in Multiphase Agglomeration Systems. Notes of course on Agglomeration Fundamentals. 4th International Symposium on Agglomeration. Toronto University, Waterloo, Ontario, Canada.

Chasseray, P. (1994). Physical Characteristics of Grains and Their Byproducts. In *Primary Cereal Processing*, Godon, B. and Willm, C. (Eds.). VCH Publishers, New York.

Clague, K. (1973). The Effects of Stresses in Bunkers. Ph.D thesis. University of Nottingham, England.

Coulson, J.M. and Richardson, J.F. (1996). *Chemical Engineering*. Vol. 2. Butterworth-Heinemann, Stoneham, MA, USA.

Dankwertz, P.V. (1952). The definition and measurement of some characteristics of mixing. *Appl. Sci. Res.* **3A**: 279–281.

Davies, R. (1984). Particle size measurement: experimental techniques. In *Handbook of Powder Science and Technology*, Fayen, M.E. and Otten, L. (Eds.). Van Nostrand Reinhold, New York.

Dobbs, A.J., Peleg, M., Mudget, R.E., and Rufner, R. (1982). Some physical characteristics of active dry yeast. *Powder Technol.* **32**: 75–81.

Engelleitner, W.H. (1994). Method Comparison. Notes of course on Briquetting, Pelletizing, Extrusion, and Fluid Bed/Spray Granulation. The Center for Professional Advancement, Chicago, IL.

Fan, L.T. and Wang, R.H. (1975). On mixing indices. *Powder Technol.* **11**: 27–32.

Geldart, D. (1986). *Gas Fluidization Technology*. John Wiley & Sons, London.

Gray, W.A. (1968). *The Packing of Solid Particles*. Chapman & Hall, London.

Green, D.W. and Maloney, J.O. (1999). *Perry's Chemical Engineers' Handbook*. McGraw-Hill, New York.

Häußler, U. and Eibl, J. (1984). Numerical investigations of discharging silos. *J. Eng. Mech.* **100**: 957.

Heldman, D.R. and Singh, R.P. (1981). *Food Process Engineering*. Van Nostrand Reinhold, New York.

Herdan, G. (1960). *Small Particle Statistics*. Butterworths, London.

Janssen, H.A. (1895). Versuche über getreidedruck in silozellen. *Zeitschrift des Vereins Deutscher Ingenieure* **39**: 1045.

Jenike, A.W. (1964). Storage and Flow of Solids. Bulletin No. 123, Utah Engineering Experiment Station. Salt lake City, UT.

Jenike, A.W. and Johanson, J.R. (1968). Bin loads. *J. Struct. Div.* **95**: 1011.

Jenike, A.W. and Johanson, J.R. (1969). On the theory of bin loads. *Trans. ASME Series B* **91**: 339.

Kaye, B.H. (1981). *Small Characterization of Fine Particles*. John Wiley & Sons, New York.

Lacey, P.M.C. (1954). Developments on the theory of particle mixing. *J. Appl. Chem.* **4**: 257–268.

Ma, L., Davis, D.C., Obaldo, L.G., and Barbosa-Cánovas, G.V. (1997). Mass and Spatial Characterization of Biological Materials. In *Engineering Properties of Foods and Other Biological Materials*. Washington State University Publisher, Pullman.

Masters, K. (1976). *Spray Drying Handbook*. John Wiley & Sons, New York.

McCabe, W.L., Smith, J.C., and Harriot, P. (1992). *Unit Operations in Chemical Engineering*. McGraw-Hill, New York.

McGeary, R.K. (1967). Mechanical Packing of Spherical Particles. In *Vibratory Compacting*, Hausner, H.H., Roll, K.H., and Johnson, P.K. (Eds.). Plenum Press, New York.

Monsalve-González, A., Barbosa-Cánovas, G.V., and Cavalieri, R.P. (1993). Mass transfer and textural changes during processing of apples by combined methods. *J. Food Sci.* **58**: 1118–1124.

Moreyra, R. and Peleg, M. (1981). Effect of equilibrium water activity on the bulk properties of selected food powders. *J. Food Sci.* **46**: 1918–1922.

Niranjan, K. (1995). An Appraisal of the Characteristics of Food Mixing. In *Food Process Design and Evaluation*, Singh, R.K. (Ed.). Technomics, Lancaster, PA, USA.

Niranjan, K. and de Alwis, A.A. (1993). Agitation and Agitator Design. In *Encyclopedia of Food Science, Food Technology and Nutrition,* Vol. 1, Macrae, R., Robinson, R.K., and Sadler, M. (Eds.). Academic Press, London.

Okuyama, K. and Kousaka, Y. (1991). Particle Density. In *Powder Technology Handbook*, Linoka, K., Gotoh, K., and Higashitani, K. (Eds.). Marcel Dekker, New York.

Ortega-Rivas, E. (1997). Guest editor of special issue on handling and processing of food powders and suspensions. *Food Sci. Technol. Int.* **3**: 317–390.

Ortega-Rivas, E. and Beltran-Reyes, B. (1997). Rehydration properties of apple powders and particulates. *Powder Handl. Process.* **9**: 245–248.

Ortega-Rivas, E., Meza-Velásquez, F., and Olivas-Vargas, R. (1997). Reduction of solids by liquid cyclones as an aid to clarification in apple juice processing. *Food Sci. Technol. Int.* **3**: 325–331.

Parfitt, G.D. and Sing, K.S.W. (1976). *Characterization of Powder Surfaces*. Academic Press, New York.

Peleg, M. (1977). Flowability of food powders and methods for its evaluation: a review. *J. Food Process Eng.* **1**: 303–328.

Pietsch, W. (1983). Low-Energy Production of Granular NPK Fertilizers by Compaction-Granulation. *Proceedings of Fertilizer'83*, pp. 467–479. British Sulphur Corp., London, UK.

Pietsch, W. (1991). *Size Enlargement by Agglomeration*. John Wiley & Sons, Chichester, UK.

Pietsch, W. (1994). Parameters to be Considered During the Selection, Design, and Operation of Agglomeration Systems. Preprints of 1st International Particle Technology Forum, Part I, pp. 248–257. AIChE, New York.

Poole, K.R., Taylor, R.F., and Wall, G.P. (1964). Mixing powders to fine scale homogeneity: studies of batch mixing. *Trans. Inst. Chem. Eng.* **42**: T305–T315.

Reed, A.R. and Bradley, M.S.A. (1991). Advances in the design of pneumatic conveying systems: a United Kingdom perspective. *Bulk Solids Handl.* **11**: 93–97.

Roberts, A.W. (1988). *Modern Concepts in the Design and Engineering of Bulk Solids Handling Systems*. The University of Newcastle Research Associates (TUNRA), Australia.

Roberts, A.W. (1999). The influence of granular vortex motion in the volumetric performance of enclosed screw conveyors. *Powder Technol.* **104**: 56–67.

Roberts, A.W. (2000). Predicting the Performance of Enclosed Screw Conveyors. *From Powder to Bulk Conference*. IMechE, London, June 2000.

Rumpf, H. (1961). The Strength of Granules and Agglomerates. In *Agglomeration*, Knepper, W.A. (Ed.). Industrial Publishers, New York.

Schubert, H. (1987). Food particle technology part I: properties of particles and particulate food systems. *J. Food Eng.* **6**: 1–32.

Schubert, H. (1993). Powder Technology. *Food Ingredients Europe Conference Proceedings 1993.* Expoconsult Publishers, Maarsen, The Netherlands.

Schulze, D. (1996). Flowability and time consolidation measurements using a ring shear tester. *Powder Handl. Process.* **8**: 221–226.

Scoville, E. and Peleg, M. (1980). Evaluation of the effect of liquid bridges on the bulk properties of model powders. *J. Food Sci.* **46**: 174–177.

Stange, K. (1963). Die mischgute einer Zufallsmischung aus drei und mehr Komponenten. *Chem. Ing. Tech.* **35**: 580–582.

Svarovsky, L. (1981). *Solid-Gas Separation.* Elsevier, Amsterdam, The Netherlands.

Teunou, E., Fitzpatrick, J.J., and Synnott, E.C. (1999). Characterisation of food powder flowability. *J. Food Eng.* **39**: 31.

Thompson, S.A., Galili, N., and Williams, R.A. (1997). Lateral and vertical pressures in two different full-scale grain bins during loading. *Food Sci. Technol. Int.* **3**: 371.

Van Deemter, J.J. (1985). Mixing. In *Fluidization*, 2nd ed. Davidson, J.F., Clift, R., and Harrison, D. (Eds.). Academic Press, London.

Walker, D.M. (1966). An approximate theory for pressure and arching in hoppers. *Chem. Eng. Sci.* **21**: 975.

Walker, D.M. and Blanchard, M.H. (1967). Pressures in experimental coal hoppers. *Chem. Eng. Sci.* **22**: 1713.

Walters, J.K. (1973). A theoretical analysis of stresses in silos with vertical walls. *Chem. Eng. Sci.* **28**: 13.

Williams, J.C., Birks, A.H., and Bhattacharya, D. (1971). The direct measurement of the failure function of a cohesive powder. *Powder Technol.* **4**: 328–337.

Wright, H., McElhinney, I., and Lemmon, L. (1997). Current UK drive power formulae for belt conveyors: fact or friction. *Bulk Solids Handl.* **17**: 201–204.

Wypych, P.W. (1999). The ins and outs of pneumatic conveying. *International Symposium on Reliable Flow of Particulate Solids III.* Porsgrunn, Norway, August 11–13, 1999.

5
Rotary Drum Processing

A. A. Boateng
Research Chemical Engineer
Eastern Regional Research Center
Agricultural Research Service USDA
Wyndmoor, Pennsylvania

CONTENTS

I. INTRODUCTION

The rotary drum is widely used by industries to carry out a wide variety of process applications. This is certainly not different in the food industry where it is employed for several

---

Mention of trade names or commercial products in this article is solely for the purpose of providing specific information and does not imply recommendation or endorsement by the U.S. Department of Agriculture.

food processing operations. Applications such as food powder mixing and demixing, simple heating or cooling of grains, evaporative drying, baking of food and food products all use the rotating drum processor.

The widespread usage of the rotary drum can be attributed to such factors as: (1) the ability to handle varied feedstock, for example, slurries or granular materials including powders with large variations in particle size, (2) the long residence time, of one or two orders of magnitude over other contactors like the fluidized bed, and (3) the ability to maintain distinct environments, that is, the feed bed and the surrounding open space for gas flow called the freeboard. For example, reduced conditions within the bed coexisting with an oxidizing freeboard is a unique feature of the rotary drum processor that is not easily achieved in other reactor contactors. The two most important pieces of information required for the design of a rotary drum contactor to process foods are the dispersion of the solids/powders and heat transfer.

As a mixer, the rotary drum possesses two mixing coefficients, one in the axial direction, that is, the axial mixing coefficient, and the other in the transverse direction, often called the radial mixing coefficient. The former is attributed to the mechanism that results in an overall convection, causing the bulk of the material to move from the inlet of the drum to the outlet with an average velocity equal to the mean velocity profile. The latter, however, involves mechanisms on a smaller scale that cause local constraints on individual particles and result in velocity components both in the axial and the transverse directions. Both axial and transverse mixing coefficients tend to increase with an increase in drum rotational speed. For low rates of rotation one expects a spread in the residence time distribution due to the influence of the velocity profile (Wes et al., 1976). The effect of the drum size, particle rheology, and the drum's internal features are therefore major design considerations. The effect of the drum rotational speed on the transverse flow patterns is illustrated later.

As a thermal processor, the rotary contactor is basically a heat exchanger in which energy from a hot phase is extracted by the bed material. Therefore, during its passage along the cylinder, the bed material will undergo various heat exchange processes, a typical sequence being drying, heating, and chemical or physical reactions, that cover a broad range of temperatures. Rotary contactors allow direct contact between the freeboard gas and the bed material being processed or noncontact, that is, externally heated drum processors. In certain food processing applications where contamination of the process gas is undesirable, such noncontact heat exchange arrangement is the preferred process design route. For contact heat exchange, however, the most common configuration is the countercurrent flow type whereby the bed and the freeboard gas are in opposite directions (Figure 1) although the cocurrent flow type is utilized in some instances such as drying of foods.

The focus of this chapter is to present the features of the rotary drum as a food processor by providing a quantitative description of the transverse bed motion and segregation of bed materials and the resultant effect of these phenomena on the bed heat transfer during food processing applications such as drying, heating or cooling, roasting, bloating, and even cooking.

II. ROTARY DRUM FEATURES

As its name implies, the rotary processor is a partially filled horizontal drum rotating on its horizontal axis. The sizing of the drum depends on the application, the feed rate, and related transport properties such as temperature, gas flow rates, and bed material velocities that ultimately will determine the process duration or the residence time. For example, in food

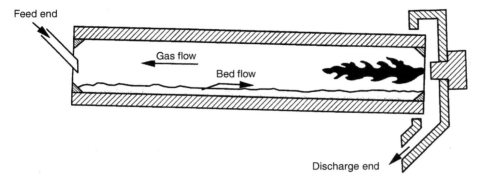

Figure 1   Countercurrent flow rotary drum heat exchange where the charge is the bed and the gas flow area is the freeboard space.

drying the drum diameter-to-length ratios of the order of 5 to 10 are typical depending on whether the heat exchange is contact or noncontact. Such L/D ratios can result in residence times in the 20 to 60 min range depending upon the drum speed, the type of internal flights, if any, and the slope in the longitudinal direction typically of the order of 1 to 3°.

The movement of a charge (feed) in a rotating cylinder can be resolved into two components mentioned earlier, that is, movement in the axial direction, which determines residence time, and movement in the transverse plane, which influences most of the primary bed processes such as charge or particle mixing, heat transfer, and reaction rate (physical or chemical) as well as the axial progress of the charge. Although this link between particle motion in the transverse plane and particle velocity in the axial direction was established several decades ago, the literature generally dealt with these two types of bed motion as independent phenomena until recent advances in the characterization and application of granular flow theories to powder processing in such devices (Boateng, 1998).

Because the drum is usually partially filled, the transport phenomenon within the cross section of the rotary drum is complex. However, it determines the process outcome. The variations in the conditions at any cross section in the drum processor, which are defined by the differential heat and mass balance and the transfer rate mechanisms, determine the state of the dependent variables such as cooling temperature, moisture content, etc. of the processing material or food along the cylinder length.

For any material's processing including foods, the particulate material or slurry that forms the bed or the charge is set in motion by the drum rotation. In the transverse plane, the most extreme class of this motion may be characterized as centrifuging, which occurs at critical and high speeds, in which all the bed material rotates with the drum wall. Cascading, which also occurs at relatively high rates of rotation, is a condition in which the height of the leading edge (shear wedge) of the powder rises above the bed surface and particles cascade or shower down on the free surface as depicted in Figure 2. Although operation of the contactor in either of these conditions is rare because of attrition and dusting issues, certain food drying applications take advantage of the high particle-to-heat transfer fluid exposure of the cascading mode and the separation effect caused by the centrifugal force component. For example, starting at the other extreme, that is, at very low rates of rotation and moving progressively to higher rates, the bed will typically move from slipping, in which the bulk of the bed material, en masse, slips against the wall, to slumping whereby a piece of the bulk material at the shear wedge becomes unstable, yields and empties down the incline, to rolling, which involves a steady discharge onto the bed surface. In the slumping mode, the angle subtended by the bed's free surface "the dynamic angle of repose" varies in a cyclical

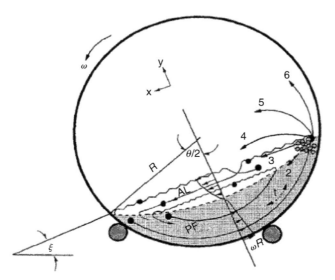

Figure 2 Modes of radial bed motions in a rotary drum: 1. Slipping; 2. Slumping; 3. Rolling; 4. Cascading; 5. Cataracting; 6. Centrifuging. R is the drum radius, θ is the bed angle, ξ is the dynamic angle of repose, and ω is the drum's rotational speed. AL and PF are active and passive regions of the bed respectively. (From Boateng, A.A. [1998]. *J. Chem. Eng. Comm.*, **170**, 51–66. With permission.)

manner while in the "rolling mode" the angle of repose remains constant. It has been established (Rutgers, 1965) that the dynamic similarity of the rotary drum behavior, and hence the type of transverse bed motion that occurs during powder processing, is dependent upon the rotational Froude number, defined by the ratio of the rotational acceleration to the acceleration due to gravity, that is, $Fr = \omega^2 R/g$ where the critical condition implies $Fr = 1$ when material centrifuges. The various regimes of the bed behavior modes for a 12% loading (fraction of cross section occupied by the charge) and powder particle angle of repose of 35% are shown in the table below.

Mode	$Fr \{= \omega^2 R/g\} @ \phi = 35°$
Slipping	$Fr < 1.0 \times 10^{-5}$
Slumping	$1.0 \times 10^{-5} < Fr < 0.3 \times 10^{-3}$
Rolling	$0.5 \times 10^{-3} < Fr < 0.2 \times 10^{-1}$
Cascading	$0.4 \times 10^{-1} < Fr < 0.8 \times 10^{-1}$
Cataracting	$0.9 \times 10^{-1} < Fr < 1$
Centrifuging	$Fr > 1.0$

In the rolling mode, where rotary drum mixing is maximized, two distinct regions can be discerned, the shearing region called active layer formed by particles near the free surface and the passive or plug flow region at the bottom where the shear rate is zero. The particular mode chosen for an operation is dependent upon the intent of the application. A survey of various rotary drum type operations (Rutgers, 1965) has indicated that most operations are in the 0.04 to 0.2 range of N-critical, which is well below the centrifuging mode and probably the cascading mode as well. In food processing, all the above-mentioned modes are possible depending on the type of food, its state, and the intended function.

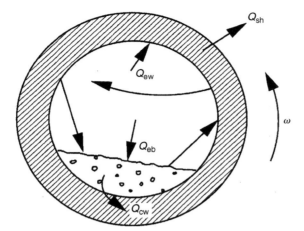

Figure 3   Rotary drum heat transfer paths. Q_{ew} is heat transfer to exposed wall, Q_{eb} is that to the exposed bed, Q_{cw} is heat transfer from bed to covered wall, and Q_{sh} is that lost through the shell. (From Boateng, A.A. and Barr, P.V. [1996]. *Int. J. Heat Mass Transfer*, **39**[10], 2131–2147. With permission.)

As mentioned earlier, for thermal treatment, the differential heat and mass balance and their rates at the transverse plane define the process conditions at any axial location of the rotary drum. The heat transfer between the freeboard and the bed is rather complex and occurs by all the paths and processes shown in Figure 3. This is true for contact or noncontact heat exchange processes except that the heat source will be, internal or external, respectively, thereby influencing the direction of the shell losses. The paths shown involve heat transfer from the freeboard to the charge directly through the exposed bed surface and indirectly through the exposed wall. Within the charge itself, heat conduction is similar to that of packed beds, that is, particle-to-particle conduction and radiation as well as through the interstitial gas depending upon the size distribution of the powder or grains being processed (Figure 4). However, superimposed on this effective heat transfer are convective and granular diffusion components induced by granular motion resulting from the drum rotation. The magnitude and extent of the latter are dependent upon the rheological properties of the material or food powder being processed. The flow behavior of rice grains in a 1 m rotary drum is hereby presented to illustrate the importance of such features.

III.   POWDER/GRANULAR FLOWS IN THE ROTARY DRUM

Experiments carried out on the continuous flow of rice grains in the transverse plane of a rotating drum provide understanding of the rheological behavior of grain products in rotary drums. The main rheological property with the greatest impact on the granular flow behavior is the coefficient of restitution of the grain defined as the ratio of its speed of separation upon impact to the speed of approach during a collision. Figures 5 through 7 show the granular transport properties such as surface velocity, velocity fluctuations, and linear concentrations of long grain rice as a function of the degree of fill and the drum peripheral speed.

In light of the shape and the thickness of the active layer, experiments have shown that the transition of particles from the plug flow region of the transverse plane, where the material moves in rigid motion with the cylinder, to the active layer (the thin layer on

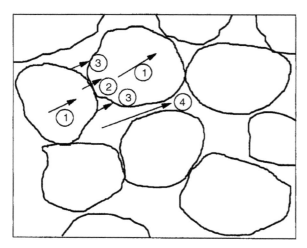

Figure 4   Bed heat conduction: 1. Within particle conduction; 2. Particle-to-particle conduction; 3. Particle-to-particle radiation; 4. Particle-to-particle convection. (From Boateng, A.A. [1993]. Ph.D. Dissertation. The University of British Columbia, Vancouver. With permission.)

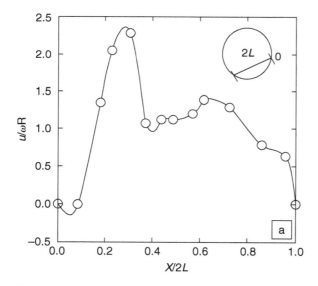

Figure 5   Free surface velocity profile for rice grain processing. $2L$ is the chord length, u is the linear velocity of particles at free surface, R is the drum radius and ω is the rotational rate. (From Boateng, A.A. [1993]. Ph.D. Dissertation. The University of British Columbia, Vancouver. With permission.)

the exposed bed surface), where particles are continuously shearing, does not only depend on the material's angle of repose but also on physical properties such as the coefficient of restitution of the material. Here, because the shear stress comprises mainly collisional elasticity, it is observed that for industrial materials such as polyethylene pellets with a small dynamic angle of repose of 22% the transition from potential energy in the plug flow region into kinetic energy in the active layer is readily established. However, for low coefficient of restitution materials such as grains, for example, rice grains, the friction angle is relatively

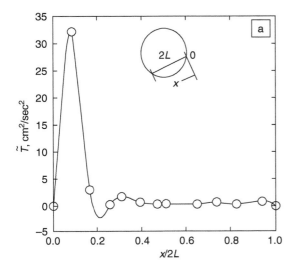

Figure 6 Surface velocity fluctuations for rice grain processing. $2L$ is the free surface length and x is any location beginning from the apex, 0. (From Boateng, A.A. [1993]. Ph.D. Dissertation. The University of British Columbia, Vancouver. With permission.)

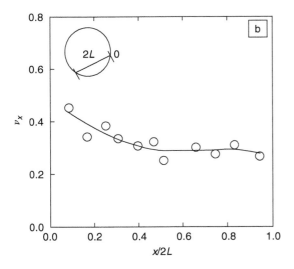

Figure 7 Surface linear concentration profile for rice grain processing. $2L$ is the free surface length. (From Boateng, A.A. [1993]. Ph.D. Dissertation. The University of British Columbia, Vancouver. With permission.)

high ($\sim$40%) and the energy dissipation is low. As a result, high potential energy buildup is encountered during the transition prior to release into the active layer. This manifests itself as an overburden and results in instabilities as is evident by the formation of multiple dynamic angles of repose thereby giving way to unsteady distribution of velocity at the exposed bed surface. Processing such materials might require internal baffles to break the flow and to readily enable the transition.

Generally, the depth of the active layer as well as the rheological characteristics such as dilation and granular temperature within the layer depend on the coefficient of restitution of the material. Granular temperature, which is a measure of kinetic energy per unit mass in random motion of particles, is high where the concentration is low in a region of high mean velocity. This notwithstanding, the shape of the active layer is always parabolic regardless of grain type. However, the depth or bed thickness depends on the physical properties of the grain and the operational conditions of the drum. Moreover, it has been observed that (Boateng, 1993) for a given grain type and a given drum speed, increasing the degree of fill reduces the percent increase in active layer depth.

The parabolic surface velocity profile observed here implies that, after transferring from the plug flow region into the active layer, particles could accelerate rapidly up to around "mid-chord" of the free surface plane before decelerating. Aside from interparticle collision, gravity plays a major role in particulate momentum transfer. Thus, increasing the degree of fill provides a longer chord length for material to traverse and for larger drums the velocity at the exposed bed surface may tend to approach fully developed flow condition by mid-chord. For deep beds surface velocities can reach as high as 4.5 to 7.5 times the tangential drum velocity.

Because of the instabilities encountered during the processing of materials with low coefficient of restitution such as food products, dilation or shearing of the charge to achieve thorough mixing is often an industrial problem. Hence stirrers and mixers are often employed in the form of tumblers to assist the dilation in improving the mixing coefficient.

IV.   MIXING AND SEGREGATION IN DRUM PROCESSING

It is evident both from the literature and operator experience that thorough mixing of particles in the transverse plane of a rotary drum is fundamental to achieving uniform heating, cooling, or drying of the charge and, ultimately, to the production of a homogeneous product. Good mixing assumes that particles are evenly sized, which effectively means that (statistically) exposure to the freeboard will be the same for each primary powder particle. Unfortunately, when there is a significant variation in grain or particle size, superimposed on this mixing will be the tendency for small particles in the active layer to sift downward through the matrix of larger particles. Thus, the bed motion described in Section 3 tends to concentrate the finer grains within the core (see, Figure 8). The material within the core, because it has very little chance of reaching the exposed bed surface for direct heat transfer from the freeboard (for drums heated from within), tends to a lower temperature than the surrounding material. Thus, segregation can counteract the convective transport of energy to the material being heated and thereby promote temperature gradients within the charge. The net effect is not necessarily negative because, for example, when processing grains of variable sizes and shapes, smaller ones may react faster to temperature and segregation of fines to the cooler core may be essential to obtain uniform cooking. This suggests that particle size distribution in the feed material might need some optimization.

A segregation model that considers a binary mixture of small and large particles in the continuously shearing active layer of the rotary drum provides insights to the rearrangement of grains and the mixing patterns. The model result shown in Figure 9 employs continuum equations to describe the mixing and segregation rates in the transverse plane of the bed, which result from both particle percolation and diffusional mixing (Boateng and Barr, 1996b). The diffusion coefficients and the convective terms for material concentration are obtained from the granular flow behavior described earlier.

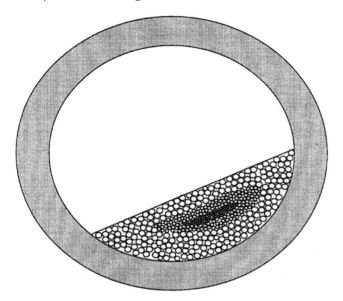

Figure 8 Segregated core in a rotary drum. Note fines form a "kidney" at center core after only a few drum revolutions.

Figure 9 Temperature distribution in a segregated core. The higher temperatures are at the periphery where large particles are situated and the colder regions are the core ("kidney") where finer particles are located. (Boateng and Barr, 1996a.)

V. ROTARY THERMAL PROCESSES

When the rotary drum is employed for high temperature applications (e.g., $>700°F$ [645 K] where radiation heat transfer is important) it is classified as a kiln. In most kiln applications, the objective is to drive specific bed reactions, which for either kinetic or thermodynamic reasons, often require high temperatures. The energy necessary to raise the bed temperature to the level required for the intended process, and in some instances, to drive the reactions themselves, originates in most operations with the combustion of hydrocarbon fuels in the freeboard near a burner that is subsequently transferred by heat exchange between the freeboard and the bed. Heat transfer between the freeboard and the bed is rather complex and occurs by all the paths and processes shown earlier. Because the analytical tools for handling freeboard transport phenomena have been the subject of considerable research, for example, the zone method (Guruz and Bac, 1981) for determining radiative heat transfer, or commercial software for calculating fluid flow (and occasionally combustion processes as well), our ability to simulate the freeboard conditions exceeds our ability to accurately determine conditions experimentally within the bed. Although numerous rotary kiln models have been proposed in the past for industrial calcination processes (e.g., see Wes et al., 1976; Brimacombe and Watkinson, 1979; Tscheng and Watkinson, 1979; Gorog et al., 1981, etc.), models for food processing are virtually nonexistent. Most heat transfer models assume that, at each axial position the bed is well-mixed in the transverse plane; that is, the bed material is isothermal over any transverse section of the kiln. However, many kiln operations suffer considerable difficulty in achieving a uniform product due primarily to instabilities in the flow behavior especially when processing grains of low coefficient of restitution and thereby poor flowability as shown earlier for rice grains. Evidence such as this, as well as operator experience, suggests that substantial transverse temperature nonuniformity is generated within the bed. Thus, the well-mixed assumption, although expedient to the modeling of the rotary drum processor, is clearly deficient because it ignores the motion of the bed in the transverse plane or, more precisely, because it ignores the effect of this motion on the redistribution within the bed material of energy absorbed at the bed–freeboard interfacial surfaces.

VI. SIMPLE HEATING AND COOLING MODELING

In most rotary kiln operations the primary objective is to transfer the maximum amount of energy to (heating) or in some instances from (cooling) the bed in order to drive the various processes such as drying, chemical reaction, or simple heating (or cooling). Virtually all this energy input occurs as heat transfer, since the power required to rotate the drum (and hence drive the bed motion) is small compared with that supplied by fuel combustion. Although the prediction of the net heat transfer to the bed is paramount in the development of any kiln heating process, the task is not an easy one owing to the complex interaction of the various heat transfer paths and processes as shown earlier. The source of energy for heat transfer is usually combustion (although electrical heating is possible in some special applications), and this energy is transferred directly to the walls and bed surface by radiation (either from the emitting products of combustion, that is, H_2O and CO_2, or by particulate emission in the flame region) and freeboard convection. However, the wall interacts radiatively with the bed surface (and vice versa) as well as with other sections of the wall. In addition, only a portion of the energy absorbed by the inside wall surface during exposure to the freeboard is lost through the shell, with the remainder being transferred to the bottom surface of the bed. This regenerative action may enhance heat transfer to the bed or, under some circumstances, may operate in reverse drawing energy from the bed (Barr, 1986). To complicate the situation still further the "receptiveness" of the bed surfaces to heat transfer will depend on the ease with which energy received at the surface is dispersed within the bed. As may be inferred from the previous section, a bed that is characterized by poor mixing (particularly in the transverse plane) will also be a poor receptor for heat transfer since the temperature gradients associated with a large conductive component within the bed will mean high bed surface temperatures and hence reduced driving force for heat transfer from the freeboard. Although virtually all past models for kiln heat transfer have been concerned primarily with directing energy to the bed surface (both the top surface, which is exposed to the freeboard and the bottom surface, which contacts the wall), a truly complete calculation model must also include the subsequent distribution of energy within the bed. The phenomenon described herein sets out the heat transfer problem in rotary kiln processing. Some mathematical models for heating or cooling are hereby illustrated.

A.   One-Dimensional Thermal Model Applied to the Bed and Freeboard of a Rotary Kiln

Several thermal models for the rotary kiln (e.g., see Sass, 1967; Tscheng and Watkinson, 1979; Barr et al., 1989) have the capability of predicting "average" conditions within both the bed and the freeboard as functions of axial position. The thermal component of these one-dimensional models can be derived by considering the transverse slice, which divides the section into separate control volumes of freeboard gas and bed material. Under steady-state conditions energy conservation for any control volume requires that,

$$\dot{Q}_{\text{NET}} = \sum (\dot{n}H)_{\text{out}} - \sum (\dot{n}H)_{\text{in}} \tag{1}$$

where $\dot{Q}$ is the heat transfer, H is the enthalpy, and $\dot{n}$ is the component species. If conditions in the freeboard and bed are each assumed to be uniform in the transverse plane (the plug flow assumption), the application of Equation (1) to the control volume of freeboard gas and bed material, respectively, yields a pair of ordinary differential equations relating axial gradients of temperature and composition to the net rates of heat transfer for each control volume,

$$\dot{Q}_g = \sum_{i=1}^{N_g} \left[\dot{n}_{gi} c_{p_{gi}} \frac{dT_g}{dz} + H_i \frac{d\dot{n}_i}{dz} \right] \tag{2a}$$

$$\dot{Q}_b = \sum_{j=1}^{N_b} \left[\dot{n}_{bj} c_{p_{bj}} \frac{dT_b}{dz} + H_b \frac{d\dot{n}_j}{dz} \right] \tag{2b}$$

where, N represents the total number of species in each region and T is the average or bulk temperature at that axial position. In the absence of any chemical reaction or phase transformations, that is, simple heating/cooling, these equations simplify to:

$$\sum \dot{m}_g c_{p_g} \frac{dT_g}{dz} = Q_{g \to ew} + Q_{g \to eb} \tag{3a}$$

$$\sum \dot{m}_b c_{p_b} \frac{dT_b}{dz} = Q_{g \to eb} + Q_{ew \to eb} + Q_{cw \to cb} \tag{3b}$$

where Q represents the various heat transfer paths. One additional condition that must be met is that no net energy accumulation can occur within the wall, which yields an auxiliary condition:

$$Q_{g \to ew} + Q_{eb \to ew} + Q_{cb \to cw} = Q_{\text{shell}} \tag{4}$$

The system of Equations (3a) and (3b) can therefore be solved for successive axial positions by any of a variety of available techniques (e.g., Runge Kutta) provided that the various heat transfer terms are characterized in terms of the local gas, bed, and wall temperatures. Thus, by starting at either end of the kiln, a complete solution of the thermal problem can be developed. It is chiefly the methodology employed in evaluating the heat transfer terms that distinguishes the various one-dimensional models.

As already mentioned, heat transfer at the interfacial surfaces is complex and involves radiation, convection and, at the covered bed/covered wall interface, conduction as well. Although a heat transfer coefficient can be allocated to each transport path as per Gorog et al. (1983) this should not obscure the difficulty associated with realistic determination of values for these coefficients. In the complete thermal modeling, often the one-dimensional model is only required to produce a framework from which to operate a two- or three-dimensional model for the bed. Existing models used to evaluate heat transfer at the interfaces can be found in the literature. In the freeboard certain models, for example, that developed by Barr (1986), can predict coefficients for radiative heat transfer, that is, $h_{r,g-ew}, h_{r,g-eb}, h_{r,eb-ew}, h_{r,ew-ew}$. Heat convection to the exposed wall and exposed bed may be calculated using convective flow models, for example, Gorog et al. (1983).

$$h_{g \to ew} = 0.036 \frac{k_g}{D_k} Re^{0.8} Pr^{0.33} \left(\frac{D_k}{L_k} \right)^{0.055} \tag{5}$$

$$h_{g \to eb} = 0.4 G_g^{0.62} \tag{6}$$

where k is thermal conductivity, D and L are the diameter and length of the kiln respectively, Re is the Reynolds number for gas flow, and Pr is the Prandtl number. The gas flux, expressed in [kg/m^2/h], is given as a function of the mass flow rate of the freeboard gas and its flow area,

$$G_g = 3600 \dot{m}_g / A_g \tag{7}$$

In employing these expressions, the hydraulic diameter and transverse area of the freeboard must be utilized in evaluating the dimensionless grouping and mass flux terms. At the covered wall/covered bed interface, models exist to calculate the transfer coefficients, for example, Schlunder (1982). Although perhaps some of these are inappropriately complex for most heating/cooling applications, they take into account such factors as single particle heat transfer coefficient, wall-to-bed radiation, solid-to-solid heat conduction, and the continuum heat conduction through the gas gap between the bed and the wall surface.

Since one should not restrict any modeling exercise to nonreactive conditions in the bed and freeboard, Equation (3) can be expanded to include the reactive terms, which were originally present in Equation (2), to yield the system of equations for calculating axial temperature profiles;

$$\sum n_i c_{p_i} \frac{dT_g}{dz} = h_{ew} A_{ew} (T_g - T_w) + h_{eb} A_{eb} (T_g - T_b) + \sum \gamma_i A_g \tag{8a}$$

$$\sum n_j c_{p_j} \frac{dT_b}{dz} = h_{eb} A_{eb} (T_b - T_g) + h_{cw} A_{cw} (T_b - T_w) + \sum \gamma_j A_b \tag{8b}$$

where T_b and T_w are the average temperature over the interfacial surfaces and γ are the production rates for various species involved in either chemical reactions, for example, freeboard combustion, or phase changes (such as evaporation of free or bound moisture) each to be determined by the appropriate kinetic expressions. Because mass must also be conserved for the control volumes of the bed and freeboard, a system of ordinary differential equations representing the mass balance also results from these same kinetic expressions, for example, when drying is the primary process,

$$\frac{d\dot{n}_b}{dz} = \gamma_b A_b = A_o \cdot \exp(E/RT) \tag{9}$$

where the drying rate, γ, is expressed by an Arrhenius type equation activation energy, E, and temperature via the universal gas constant, R.

In developing a global solution for the kiln model, the complete system of ordinary differential equations (i.e., the two energy balance equations, the mass balance equations, and the auxiliary energy condition for the wall) must be solved simultaneously.

B. Two-Dimensional Rotary Kiln Bed Modeling

Although useful results can be obtained from one-dimensional models, the assumption that conditions will be uniform across any transverse section of the bed material will hold only for a well-mixed bed. Since segregation is known to occur within the bed, a two-dimensional model provides an opportunity to examine the effects on kiln performance of "de-mixing" within the bed. As mentioned earlier, segregation in the transverse plane is driven by the bed motion established by the rotation of the kiln but, since no adequate model for this motion has previously appeared, attempts to predict conditions within the bed in two or even three dimensions have been rare.

One can model the rolling bed mode, which is the preferred mode for most kiln operations. This comprises two distinct regions, that is, the thinner active layer at the region near the surface, and the relatively thicker plug flow region near the covered wall. Heat transfer within the active layer occurs by conduction (diffusion) and advection (convection). Energy conservation for a control volume in the active layer can be written as:

$$\frac{\partial}{\partial x}\left(k_{\mathrm{eff}}\frac{\partial T}{\partial x}\right) - \rho c_{\mathrm{p}} u_x \frac{\partial T}{\partial x} + \frac{\partial}{\partial y}\left(k_{\mathrm{eff}}\frac{\partial T}{\partial y}\right) - \rho c_{\mathrm{p}} u_x \frac{\partial T}{\partial y} + \dot{m}_{\mathrm{b}} c_{\mathrm{p_b}}\frac{\mathrm{d}T_{\mathrm{b}}}{\mathrm{d}z} = 0 \qquad (10)$$

If it is further assumed that mixing is sufficient then, within the active layer, the temperature gradient in the axial direction of the kiln (i.e., $\mathrm{d}T_{\mathrm{b}}/\mathrm{d}z$) in this expression is uniform. The last term in Equation (10), the axial gradients of temperature in the active layer, accounts for the removal of energy from the control volume by axial bed flow. Thus, the mass flux ($\dot{m}$) is determined by the kiln feed rate and the transverse area of the active layer, A_{al}, that is, $\dot{m} = \dot{M}/A_{\mathrm{al}}$.

In contrast to the active layer, the plug flow region is relatively deep and, since it rotates as a rigid body about the kiln axis, a cylindrical coordinate system may be more applicable to this region as well as for the kiln wall itself. Energy conservation for the plug flow region and the wall yields two simultaneous equations to be solved such as:

$$\frac{\partial}{\partial r}\left(k_{\mathrm{pf}}\frac{\partial T}{\partial r}\right) + \frac{1}{r}\frac{\partial}{\partial \theta}\left(k_{\mathrm{pf}}\frac{\partial T}{r\partial \theta}\right) - \rho c_{\mathrm{p_{pf}}}\omega r\frac{\partial T}{r\partial \theta} = 0 \qquad (11a)$$

$$\frac{\partial}{\partial r}\left(k_{\mathrm{w}}\frac{\partial T}{\partial r}\right) + \frac{1}{r}\frac{\partial}{\partial \theta}\left(k_{\mathrm{w}}\frac{\partial T}{r\partial \theta}\right) - \rho c_{\mathrm{p_w}}\omega r\frac{\partial T}{r\partial \theta} = 0 \qquad (11b)$$

where ω is the drum rotational rate and θ is the polar coordinate. The first two terms constitute the radial and transverse conduction respectively, and the third term represents the movement of energy through the control volume due to the rotation of the kiln. Since the bed particles are assumed to advance axially only in the active layer, Equation (11a) does not include any term for energy transport in the axial direction, it being implicitly assumed that axial conduction in the plug flow region will be negligible.

The effective thermal conductivity might be calculated by correlations such as that developed by Diessler and Eian (1952) in combination with an interparticle radiation model by Schotte (1960) to achieve:

$$k_b^r = \frac{1 - e_o}{1/k_s + 1/4\sigma\,e d_p T^3} + e_o 4\sigma\,e d_p T^3 \tag{12}$$

which is expressed in terms of solids thermal conductivity, k_s, bed porosity, ε_o, and emissivity, e with σ as the Stefan Boltzmann's constant. Because of the mixing in the active layer, the effective heat conductance there is greatly enhanced by granular diffusion and interparticle collisions. The effective thermal conductivity in the active layer may therefore comprise a factor relating the mass transfer to heat diffusion in the active layer, that is,

$$k'_{eff} = k_{eff} + \rho c_p D_y \tag{13}$$

where D_y is the mass diffusion coefficient calculated in the flow model using the granular temperature. The prime denotes the enhanced conductivity several times greater than the packed bed conduction, k_{eff}, ρ is the bed density.

VII. PERTINENT DISCUSSIONS

A. Synchronous and Asynchronous Solutions

Synchronous solutions of the formulations described can be sought to establish temperature and concentration profiles. With this method the heat transfer problem external to the bed might be determined using current values of the freeboard gas temperature and mean bed temperature. This would be accomplished in the same manner as Barr (1986) in which a transient one-dimensional model of the wall was employed to ensure closure of the energy balance over the cross section, or alternatively, using a two-dimensional, steady-state formulation of the wall problem. Once the net rate of heat transfer to the bed is determined (including the flux distribution to the covered wall surface) the axial gradients of the freeboard gas and mean bed temperatures would be available for use in advancing these temperatures to the next axial position by solving Equation (8) to determine temperature distribution and concentration gradients within the bed at the current axial position.

Alternatively, an asynchronous solution of the one- and two-dimensional problem may be sought to develop the one-dimensional solution over the entire kiln length before returning to the charge end to expand the axial bed temperature profile into the transverse plane. Implicit in doing so is the assumption that, at any axial position, heat transfer in the transverse plane (within the bed) will not significantly alter the freeboard, bed, and wall.

The heat fluxes and their interactions presented in Figure 10 in the form of heat transfer coefficients are from the radiative heat transfer zone model (Barr, 1989). Except for the gas-to-exposed wall component, the heat transfer to the various surfaces increases toward the location of the flame (in this case a countercurrent operation). There is a predominance of radiation heat transfer of up to about 95% with convection playing a minor role. However, this is not the case when the freeboard gas temperatures are low in cases like low-temperature drying.

The calculated bed effective thermal conduction as a function of kiln or dryer length and freeboard temperature is shown in Figure 11 as feed material of two different particle size distributions. The kiln rotation is about 1.5 rpm. Although the gas temperatures for Trial 2 are higher than that for Trial 1, the thermal conductivity of the bed is lower because the particle size of the powder for Trial 2 is smaller. In the larger particle bed, particle-to-particle heat transfer is enhanced by larger void spaces. Notwithstanding, the self-diffusion

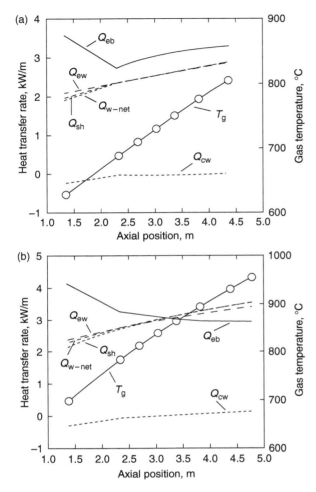

Figure 10 Rotary drum axial heat fluxes profiles. Q is the heat transfer rate and T_g is the freeboard gas temperature. (Boateng and Barr, 1996.)

coefficient component for either case increases with increased rotation and enhances the effective conduction of heat into the bed from the freeboard.

Results of two-dimensional temperature distribution calculations are typical of what is depicted in Figure 12. For a well-mixed bed, enabled by high drum rotation, an isothermal bed can be readily achieved. However, upon bed segregation, it is probable that there will be temperature gradients with the coldest regions coinciding with the segregated core.

B. Rotary Drying

For drying applications temperature gradients are not acceptable since drying fronts showing moisture gradients can be developed. Hence dryers are designed to operate in the cascading or even the cataracting mode and assumed well mixed. As a result the one-dimensional representation presented in Equation (8) coupled with Equation (9) will suffice in modeling the extent of drying in horizontal rotary kilns.

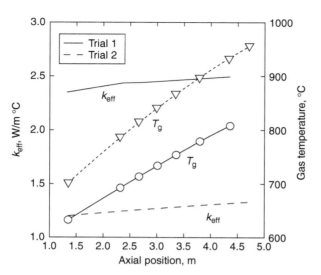

Figure 11 Calculated thermal conduction as function of dryer length. Trial 1 and Trial 2 are runs with two different particle size materials. k_{eff} is the effective thermal conductivity and T_g is the freeboard gas temperature. (Boateng and Barr, 1996.)

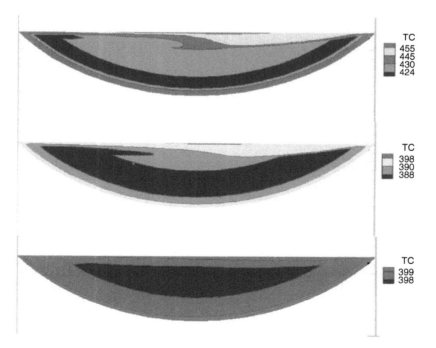

Figure 12 Two-dimensional temperature distribution of a pilot (0.41 m ID × 5.5 m long) kiln bed. The drum rotation rate increases from 1.5 rpm (top), to 3 rpm (middle) and 5 rpm (bottom). (Boateng and Barr, 1996.)

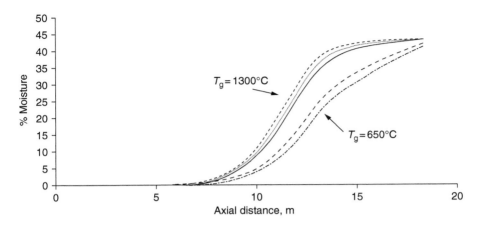

Figure 13 Countercurrent rotary drum drying curves. Wet material is fed from the right and hot gas from air heater enters from the left at the stipulated temperature.

Figure 13 presents results of drying a 40% moisture laden charge as function of axial distance and inlet gas temperature. The dryer used here is 18.3 m (60 ft) long with a 1.83 m (6 ft) ID operating in a countercurrent heat exchange mode. The activation energy and the frequency factor used in Equation (9) for free moisture evaporation are 90 kJ/kg K and 10×10^6 1/sec respectively. As seen here, the gas entrance temperature is an important design criterion since it determines the dryer length and hence the economics of the operation. The higher the gas inlet temperature, the shorter is the dryer but this must be balanced with the material's sensitivity to such temperatures.

C. Rotary Dryer Sizing

Dryers can be sized by experience then modeled for optimization using the equations presented with little data. They are designed to a specific moisture laden in the freeboard usually expressed in for example, kg H_2O/m^3/h and an L/D of about 7 to 8. Hence, after calculating the total moisture to be evaporated based on the feed and the desired product moisture the designed load can be used to estimate the volume of the vessel required to do the job. With that established, the heat requirements can be calculated based on the latent and the sensible heat of the charge material. An example of the calculation for a simple design to dry a 200 kg/h product from 50 to 10% moisture is shown in the Appendix.

NOTATION

A Heat transfer area [m^2], Arrhenius preexponential factor [1/sec]
c_p Specific heat capacity at constant pressure [J/kgC]
d_p Particle diameter [m]
D_y Diffusion coefficient [m^2/sec]
e Emissivity [−]
e_o Void fraction [−]
e_p Coefficient of restitution of particles [−]

E Activation energy [J/mol]
Fr Rotational Froude number [$\omega^2 R/g$]
g Gravitational acceleration [m/sec^2]
g_0 Radial distribution function [$-$]
G Gas flow rate [kg/h]
h Heat transfer coefficient [W/m^{2}C]
H Enthalpy [J/kg]
k Thermal conductivity [W/mC]
k' Enhanced thermal conductivity [W/mC]
L Drum length [m]
m Mass flux [kg/m^{2}sec]
n Species
Pr Prandtl number [$-$]
Q Heat transfer rate [W]
R Drum radius [m], Universal gas constant [J/molK]
Re Reynolds number [$-$]
T Thermodynamic temperature [°C]
$\tilde{T}$ Granular temperature [m^2/sec^2]
u Velocity parallel to bed surface [m/sec]

Greek

α Inclined angle subtended by bed surface (dynamic angle of repose) [rad]
γ Species production rate [kg/sec]
ρ Bulk density [kg/m^3]
ρ_p Particle density [kg/m^3]
σ Stefan–Boltzmann constant [5.67×10^{-8} W/m^{2}K]
ν Solid concentration (solids fraction) [$-$]
ω Angular velocity [1/sec]
θ Fill angle [rad]

SUBSCRIPT/SUPERSCRIPT

al   Active layer
b   Bed
cb   Covered bed
cw   Covered wall
eb   Exposed bed
ew   Exposed wall
eff   Effective
g   gas
p   Particle
pf   Plug flow
r   Radiation
s   Surface, solids
w   Wall

APPENDIX

DRYER DESIGN

	English units		SI units	
Kiln product [kg/h]			200	[kg/h]
Feed moisture content	50	[%]		
Dryer product moisture	10	[%]		
Wet feed into dryer	881.85	[lb/h]	400	[kg/h]
Total moisture evaporated [lb/h]	352.74	[lb/h]	160	[kg/h]
Max dryer product @ 10% H_2O [lb/h]	529.11	[lb/h]	240	[kg/h]
Max dry feed [lb/h]	440.92	[lb/h]	200	[kg/h]
Max water in product [lb/h]	52.91	[lb/h]	24	[kg/h]
Co-current dryer design condition:				
Use 6–7 lb H_2O per ft^3/h 7–8 L/D				
Use 3.5 lb-H_2O/ft^3/h 8 L/D				
Vessel volume [cu-ft]	100.78	[cu-ft]	2.85	[m^3]
Vessel diameter [ft]	2.52	[ft]	0.77	[m]
Vessel length [ft]	20.17	[ft]	6.15	[m]
Dryer dimension selected	2.5 × 20 ft		762 × 6096 mm	

DRYER HEAT REQUIREMENTS

Criteria				
Exit gas temp. [°F]	250	[°F]	121	[°C]
Material temp. [°F]	225	[°F]	107	[°C]
Ambient temp. [°F]	70	[°F]	21	[°C]
Latent heat H_2O	970	[Btu/lb]	2255	[kJ/kg]
Heat load [Btu/lb-H_2O]	1765	[Btu/lb]	4103	[kJ/kg]
Heat requirements				
Evaporation	399,350.2	[Btu/h]	117.01	[kW]
Heat in product	17,085.81	[Btu/h]	5.01	[kW]
Sensible heat in H_2O	4,100.59	[Btu/h]	1.20	[kW]
Subtotal	420,536.6	[Btu/h]	123.22	[kW]
Heat for shell radiation 10%	42,053.66	[Btu/h]	12.32	[kW]
Total net heat	462,590.2	[Btu/h]	135.54	[kW]
Gross heat	622,584.7	[Btu/h]	182.42	[kW]
Choose dryer inlet gas temp.				
Air heater outlet temp	500	[°F]	260	[°C]
Ave. temp inlet/outlet	375	[°F]	190.6	[°C]
CP	0.24			
DEN.	0.067			
Air heater				
**Gas mas flow rate [lb/h]	10,376.41		4,706.67	[kg/h]
Volume flow rate of gas ACFM	2,581.20		4,385.56	[Am^3/h]
Dryer exit gas ACFM	1,909.01		3,243.49	[Am^3/h]
Volume of evap. water ACFM	180.23		306.21	[Am^3/h]
Exit gas flow ACFM	2,089.24		3,549.70	[Am^3/h]
Dryer inside velocity, FPM	531.95		2.70	[m/sec]

HEATER DESIGN

Insert fuel flow rate until exhaust gas mass is equal to that of the air heater for 200% ea.

	Gas	Solid				
Fuel flow rate			240.28	[lb/h]	108.99	[kg/h]
Vol. percent O_2 in exhaust gas, %			14			
Fuel Analysis, wt.%:	Gas	Solid				
C	67.48	75.6	67.48			
H	20	24.4	20			
O	12.36	0	12.36			
N	0.16	0	0.16			
S	0	0	0			
HHV, Btu/lb	17,819.3	23,098	17,819.25		41,423.17	[kJ/kg]
Weight/Molecular wt.			10.2428			
Exhaust gas temp. [F,R,C]			250	710	121	[°C]
Ambient Temp. [F,R,C]			70	530	21	[°C]
Constant dependent on fuel		0.9	1			
Excess air percent			200	%		
Stoichiometric Air/Fuel ratio			14.06131	wt/wt		
Stoic. air flow rate			3,378.65	lb/h	1,532.53	[kg/h]
Actual air flow rate			10,135.96	lb/h	4,597.60	[kg/h]
** Exhaust gas flow rate			10,376.24	lb/h	4,706.59	[kg/h]
2581.153 ACFM		at 500°F				
Total Heat with exhaust gas			4, 59, 904.1	Btu/h	134.75	[kW]

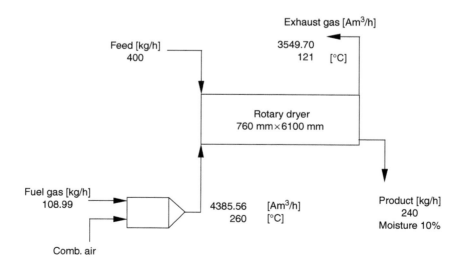

REFERENCES

Barr, P.V. (1986). Heat Transfer Processes in Rotary Kilns, Ph.D. Dissertation. The University of British Columbia, Vancouver.

Barr, P.V., Brimacombe, J.K., and Watkinson, A.P. (1989). A heat-transfer model for the rotary kiln: Part II. Development of a cross-sectional model, *Met. Trans. B*, **20B**, 403–419.

Boateng, A.A. (1993). Rotary Kiln Transport Phenomena — Study of the Bed Motion and Heat Transfer, Ph.D. Dissertation. The University of British Columbia, Vancouver.

Boateng, A.A. (1998). On flow induced kinetic diffusion and rotary kiln bed burden heat transport, *J. Chem. Eng. Comm.*, **170**, 51–66.

Boateng, A.A. and Barr, P.V. (1996a). A thermal model for the rotary kiln including heat transfer within the bed, *Int. J. Heat Mass Transfer*, **39**(10), 2131–2147.

Boateng, A.A. and Barr, P.V. (1996b). Modeling of particle mixing and segregation in the transverse plane of a rotary kiln, *J. Chem. Eng. Sci.*, **51**(17), 4167–4181.

Brimacombe, J.K. and Watkinson, A.P. (1979). Heat transfer in a direct fired rotary kiln: Part I— Pilot plant and experimentation, *Met. Trans. B*, **9B**, 201–208.

Diessler, R.G. and Eian, C.S. (1952). National Advisory Comm., Aeronaut., Rm. ESZCJOS.

Gorog, J.P., Brimacombe, J.K., and Adams, T.N. (1981). Radiative heat transfer in rotary kilns, *Met. Trans. B*, **12B**, 55–70.

Gorog, J.P., Adams, T.N., and Brimacombe, J.K. (1983). Heat transfer from flames in a rotary kiln, *Met. Trans. B*, **14B**, 411–424.

Guruz, H.K. and Bac, N. (1981). Mathematical modeling of rotary cement kilns by the zone method, *Can. J. Chem. Eng.*, **59**, 540–548.

Rutgers, R. (1965). Longitudinal mixing of granular material flowing through a rotary cylinder: Part I — Description and theoretical, *Chem. Eng. Sci.*, **20**, 1079–1087.

Sass, A. (1967). Simulation of heat transfer phenomenon in a rotary kiln, *I&EC Process Res. Dev.*, **6**(4), 532–535.

Schotte, W. (1960). Thermal conductivity of packed beds, *AIChE J.*, **6**(1), 63–67.

Schlunder, E.V. (1982). Particle Heat Transfer, *Proceedings of the 7th International Heat Transfer Conference*, Munchen, FRG.

Tscheng, S.H. and Watkinson, A.P. (1979). Convective heat transfer in rotary kilns, *Can. J. Chem. Eng.*, **57**, 433–443.

Wes, G.W.J., Drinkenburg, A.A.H., and Stemerding, S. (1976). Heat transfer in a horizontal rotary drum reactor, *Powder Technol.*, **13**, 185–192.

6

Supercritical Fluid Processing to Control Particle Size of Food Ingredients

Peggy M. Tomasula
Dairy Processing and Products Research Unit
United States Department of Agriculture
Agricultural Research Service
Eastern Regional Research Center
Wyndmoor, Pennsylvania

CONTENTS

Mention of trade names or commercial products in this chapter is solely for the purpose of providing specific information and does not imply recommendation or endorsement by the U.S. Department of Agriculture.

I.   INTRODUCTION TO SUPERCRITICAL FLUIDS

A.   Supercritical Fluid Extraction

Supercritical fluid extraction (SFE) is the most familiar of the supercritical processing methods currently used by the food processing industries. SFE is carried out using supercritical carbon dioxide (SCO_2), the most common supercritical fluid (SF), as the solvent. SCO_2 is an environmentally benign and nontoxic solvent that is easily handled and easy to obtain. Food grade CO_2 is a by-product of corn wet-milling operations and other fermentation processes. In a SFE process, CO_2 may be recycled and reused — with small amounts of makeup SCO_2 added in each cycle — thereby lowering process operation costs relative to processes that use organic solvents. It is also nonflammable and does not leave solvent residue that can contaminate the food substrate or extract. Supercritical ethane and propane are also suitable solvents for SFE and preferable to SCO_2 in some cases; however, their use in food processing is unlikely because of the difficulty in obtaining these solvents for food use and the great care that must be exercised in handling them because of their flammability.

B.   Applications of Supercritical Fluid Extraction

Supercritical carbon dioxide is an ideal solvent for solubilizing lipophilic substances. The use of SCO_2 as a replacement for hexane and other organic solvents for extraction, fractionation, refining, and deodorization of lipids or essential oils and flavor and aroma isolation was demonstrated by many researchers. Commercially, SCO_2 is used to decaffeinate coffee and tea, for the recovery of extracts from beer brewing, hops, and in limited cases, to extract spices and flavors. In various studies, SCO_2 has been used to extract oils from soybean, rice bran, sunflower, rapeseed, wheat germ, and corn germ. It has also been used to extract evening primrose oil, shark liver oil, lemon oil, bergamot oil, paprika oleoresin, beta carotene, tocopherols, black pepper, red pepper, ginger, turmeric, nutmeg, vanilla, phenols and tannins from grape seeds, coffee aroma extracts, cholesterol from milk fat and egg yolk, and oil from potato chips. This list is not exhaustive. The references for these studies are too numerous to list here.

C.   Other Supercritical Fluids Processes

Other processes based on SCO_2 have been developed but few have been applied to food systems (Tomasula, 2003). SCO_2 was used to modify mixtures of chitosan with glucose or malto-oligosaccharides and phosphorylation of amylose and polyvinyl alcohol carbohydrates (Yalpani, 1993). An SF extrusion technology for production of low-density expanded products, such as bread, has also been developed (Mulvaney and Rizvi, 1993). The use of SF as reaction media for enzymatic catalysis is reviewed in Kamat et al. (1995). Tomasula et al. (1998) developed a process utilizing high pressure and SCO_2 to create enriched fractions of α-lactalbumin and β-lactoglobulin from whey protein concentrate. Additional information on SF is available from many sources (Paulaitis et al., 1983; Cygnarowicz and Seider, 1991; McHugh and Krukonis (1994); Rizvi, 1994; Kaiser et al., 2001).

II.   PROPERTIES OF SUPERCRITICAL FLUIDS

Carbon dioxide may exist as a solid, liquid, or gas. A typical pressure–temperature phase diagram is shown in Figure 1. CO_2 is considered an SF when process temperature and

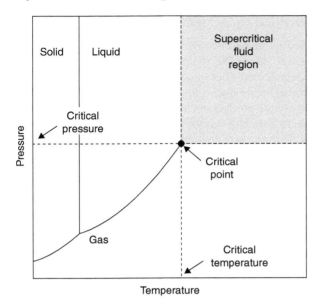

Figure 1 Pressure–temperature diagram showing the critical point and the SF region (McHugh and Krukonis, 1994).

pressure exceed its critical temperature, T_c, of 31°C and critical pressure, P_c, of 7.38 MPa (72.9 atm). In this region, CO_2 has properties of a liquid and a gas and the distinction between the liquid and gas phase is removed.

Changes in the density, viscosity, diffusivity, solubility, and bulk dielectric constant of CO_2 are induced with changes in temperature or pressure and are enhanced with further increases in temperature and pressure, allowing improved solubility of a component. The density of CO_2 at ambient temperature and atmospheric pressure is only 0.002 g/cc but at its critical point, the density of CO_2 is 0.469 g/cc. Because they are gas-like, SF have lower viscosities, ranging from 0.02 to 0.12 cP, and higher diffusion coefficients, on the order of 0.001 cm^2/sec, than conventional liquid solvents. This enhances their ability for extraction. Therefore, the solvent strength of the SF can be tuned as needed through small changes in temperature (T) and pressure (P) to vary solvent properties between those of a liquid and a gas. The low, critical temperature of SCO_2 is also an advantage in processing, not only for energy savings, but because heat-induced damage to the substrate is minimized. Small additions of ethanol or other solvents, added as a cosolvent or entrainer, increase its efficiency in extracting significant amounts of nonvolatile polar substances. Because it is nonpolar, SCO_2 does not solubilize polar substances, proteins or amino acids, sugars, esters, ethers, lactones, or cellulose.

III. PARTICLE SIZING USING SUPERCRITICAL FLUIDS

Supercritical fluid processes for particle sizing and design represent a more recent use for SF. These processes have been proposed for, or currently are used by, the pharmaceutical industries to create micro and nanosized particles with controlled particle size and particle size distribution. In general, particle sizing is accomplished by contacting the product with a SF and then forcing the mixture through a nozzle.

A. Rapid Expansion of the Supercritical Fluids Method

Several different approaches are used to create microparticles. One of the most familiar is the rapid expansion of supercritical solutions (RESS) method, which is used to recover products that are poorly soluble in aqueous media but soluble in the SF (Matson et al., 1987; Debenedetti et al., 1993). In this method, the product is first dissolved in the SF. Cosolvents may be used to increase the solubility of the product in the SF. The solution is then rapidly depressurized from high to low pressure through a capillary or pinhole nozzle to create very fine particles. The product is insoluble in the now low-pressure gas and is supersaturated in the stream. Crystalline growth of the particles occurs. This method is of limited use for food applications because of the low solubility of most food components in SF, the low yields obtained, and it has not been demonstrated for production scale.

B. Gas Antisolvent and Supercritical Antisolvent Methods

In the gas antisolvent (GAS) and supercritical antisolvent (SAS) processes, either a dense gas or an SF is mixed with an organic solution containing the dissolved product. The dense gas or SF should be miscible with the solvent but immiscible with the product. In a typical operation, the gas or SF is introduced to the solution expanding it. This leads to a reduction in the bulk density of the solvent and causes a reduction of the solubility of the solute in the expanded solution. Precipitation of the solute occurs upon sufficiently high antisolvent concentration. In one example, crystals of β-carotene smaller than 1 μm were obtained from solutions of ethyl acetate or dichloromethane with SCO_2 as the antisolvent (Cocero and Ferrero, 2002). The process may also be used to create particles with different morphologies such as threads, sponges, foams, and films and has been demonstrated in the processing of proteins for pharmaceutical applications and proposed as an alternative to spray drying or lyophilization, followed by milling, which may protect proteins from denaturation (Elvassore et al., 2001).

C. Aerosol Solvent Extraction System and Solution Enhanced Dispersion by Supercritical Fluids

Similar to the GAS and SAS processes are the ASES (aerosol solvent extraction system) and SEDS (solution enhanced dispersion by supercritical fluids) processes (Jung and Perrut, 2001). In the ASES process, the solution is aerosolized by spraying it through a nozzle into the dense gas antisolvent, which induces precipitation. In the SEDS process, a coaxial nozzle is used to simultaneously spray the SF and the solution containing the product into a vessel. The SF breaks up the solution into small droplets and extracts the solvent from the solution at the same time. These processes are not useful for creating particles from food components, such as sugars, which are soluble in water only, or proteins. Proteins are soluble in organic solvents but organic solvents add significantly to operating costs, must be generally recognized as safe (GRAS) for food applications, and may denature the protein. Proteins are also soluble in water but contacting water and SCO_2 leads to the formation of carbonic acid, which may lower pH to the isoelectric point of the protein causing it to precipitate. However, as discussed in Jung and Perrut (2001), an improvement in SEDS processing that includes the introduction of a second solvent, ethanol, miscible in the solution and soluble in the SF, led to recovery of the sugars — lactose, maltose, and sucrose — and the model proteins — lysozyme and trypsin with little denaturation. Several other modifications of SEDS processing are described, some leading to production of micro and nanoencapsulated products. These processes are for pharmaceutical applications but may be useful for creating nutraceutical ingredients with high bioavailability. Scale-up issues relative to the GAS and ASES processes are discussed in Thiering et al. (2001).

D. Particles from Gas-Saturated Solutions–Suspensions

Another process for producing particles using SFs is the particles from gas-saturated solutions–suspensions (PGSS) method. In this method, the SF is solubilized into a melt or mixture. The gas saturated melt or mixture is then sprayed through a nozzle, forming small solid particles. This method and variations on it are discussed in Jung and Perrut (2001). This technology may hold promise in the food industries through the creation of novel sugar and protein coatings.

IV. FORMATION OF PARTICLES WITH CO_2 AS A PRECIPITANT

A. Precipitation of Casein from Milk Using CO_2 — Batch Processing

Another process for producing particles uses either high pressure or supercritical precipitation with CO_2 as an alternative to acid precipitation. This process is unlike those discussed above because in this case, CO_2 is dissolved in an aqueous solution comprised of a protein or another water-soluble component and hydrolyzes to form carbonic acid according to the equation

$$CO_2 + H_2O = H^+ + HCO_3^-$$

The protein or other substance then precipitates. Jordan et al. (1985) demonstrated this process on the laboratory scale for isolation of casein from milk. Casein was precipitated from milk at pressures below 7.38 MPa and temperatures in the range from 40 to 70°C.

Tomasula et al. (1995) isolated casein from milk using CO_2 in a large batch reactor to determine the effects of pressure and temperature on yield of casein and casein quality. Initial experiments showed that high yields of casein were obtained only when CO_2 was sparged through milk in the form of droplets, not when it was allowed to flow through a thin tube. A sparger was fabricated from a 2 μm porous metal filter. Stirring of reactor contents did not influence yield, an indication that the reaction was not controlled by the rate of mass transfer. Stirring was not used in subsequent studies.

The effects of CO_2 pressure on pH of milk at 38°C are shown in Figure 2 (Tomasula et al., 1999a). Initial pH of the milk was 6.6. The pH decreases linearly with addition of CO_2 to a value of 5.4, corresponding to a pressure of 2.8 MPa. After this point, pH decreases at a lower rate with further addition of CO_2 to pH of 4.8, corresponding to a pressure of 14 MPa. Upon depressurization of the vessel at this pressure, casein pH was 4.8 and pH of the associated whey was only 6.0. The higher pH for the whey indicates that much of the CO_2 was evolved upon release of pressure while casein was converted to its acid form. Pressures greater than 14 MPa were not investigated.

In experiments, temperature was varied from 32 to 60°C and pressure was varied over the range from 2.76 to 5.52 MPa at each temperature. For a 500 g milk sample, results indicated that yields approached 100% in the temperature range from 38 to 49°C at each pressure studied. The impact of CO_2 pressure on particle morphology was also evident. With the exception of runs conducted at 60°C, discernible particles or granules of casein were observed for runs conducted from 32 to 49°C. At 32°C and constant pressure, the casein had a creamy appearance, but samples produced at 38°C and the same pressure

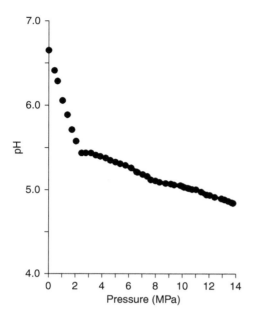

Figure 2   Effect of CO_2 pressure on pH of milk at 38°C.

was drier in appearance, and casein quality was friable. However, at constant temperature and with increase in pressure from 2.76 to 5.52 MPa, a decrease in curd particle size was observed. Casein precipitated at 60°C was dry and stringy with no discernible particles and had a rubbery feel.

The major advantage of using CO_2 as a substitute for mineral acid precipitation is elimination of the precipitant from the product upon release of pressure. Washing steps are not needed to remove the precipitant from the product and neutralization of the acid product stream is not required. Less handling of the product in subsequent washing stages means that the potential for particle breakage and loss is minimized.

1.   Particle Size Distribution Studies

Particle size distribution (PSD) studies were carried out on the wet casein products using the wet sieving method developed by Jablonka and Munro (1985). The results are plotted in Figure 3 to show the effect of system pressure and temperature on casein PSD. For casein made at 38°C, the mean curd particle size (d_{50}) decreased from 3.35 mm at 2.76 MPa to 841 μm at 5.52 MPa. However, d_{50} increased to 2.00 mm at 43°C and 5.52 MPa, but many of the particles appeared to be fused together. In all runs though, particles as large as 6 mm were obtained. Particle size may have been influenced not only by use of a sparger with a 2 μm pore size to introduce CO_2, the lack of an acidulation period — a practice often used when processing proteins to allow the proteins to agglomerate before further processing, but also shearing of the protein induced by depressurization of the vessel.

2.   Microstructure of CO_2 Precipitated Casein

The microstructures of casein precipitated from whole milk at 38°C using CO_2 pressures of 6.2 and 9.0 MPa, respectively, are compared to those of casein precipitated from skim milk at 38°C, CO_2 pressure of 9 MPa, and a control (Figure 4). The control casein was

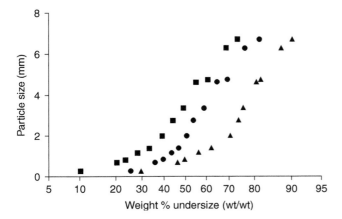

Figure 3   Effect of pressure on casein particle size distribution at 38°C and 2760 kPa (■), at 38°C and 5520 kPa (▲), and at 43°C and 5520 kPa (●).

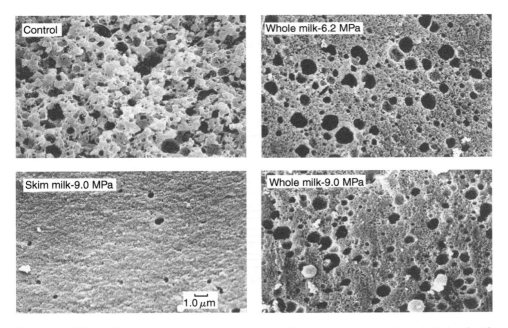

Figure 4   Effect of pressure on the microstructure of casein aggregates formed during batch processing using CO_2 as the precipitant at 38°C compared to a control using 1 M HCl as precipitant.

precipitated from whole milk at 38°C using 1 M HCl. Comparison of all CO_2-casein samples with the control shows that the control is comprised of irregularly shaped particles while CO_2-casein is comprised of much smaller particles, as discussed above, that appear to be compressed. Comparison of the whole milk casein samples obtained at the two pressures shows no difference in microstructure. However, comparison of the skim and whole milk samples at the same pressure shows that the skim milk sample is more cohesive possibly because of the absence of fat globules.

B. Precipitation of Casein from Milk Using CO_2 — Continuous Processing

Because the batch process is not scaleable for commercial application, a continuous process for casein precipitation from milk using CO_2 was developed (Tomasula et al., 1997). The continuous process is composed of steps for contacting high pressure or SCO_2 with a pressurized milk stream, a stage for reaction and precipitation of the milk and CO_2, and a stage for removing the casein product. Two separate reactor/precipitators were designed and tested. In one of the reactors, milk was sprayed into a column filled with pressurized CO_2; in the other, five streams of liquid CO_2 were pumped into a stream of pressurized milk, mixed in a static mixer and then pumped through a heated double-pipe tubular reactor. Hot water flowed in the outer pipe of the reactor to control temperature. A diagram of the process is shown in Figure 5. The diagram of the process using the spray reactor is similar with the spray reactor replacing the tubular reactor. The process is novel in that the product may be removed from the reactor without depressurizing the system as required in batch operation.

Depressurization occurs stepwise, reducing pressure only about 0.48 MPa/stage, so that particle size is maintained throughout processing. Depressurization through a valve or nozzle directly from the reactor/precipitation step was shown to limit particle size to 45 μm. In this study, a 12-stage progressing cavity pump, operated in reverse and fed through its outlet port was used to depressurize the reactor/precipitator, although the method is not limited to use of a pump to accomplish depressurization stagewise. The pump had a 316 SS rotor turning in a food grade nitrile stator.

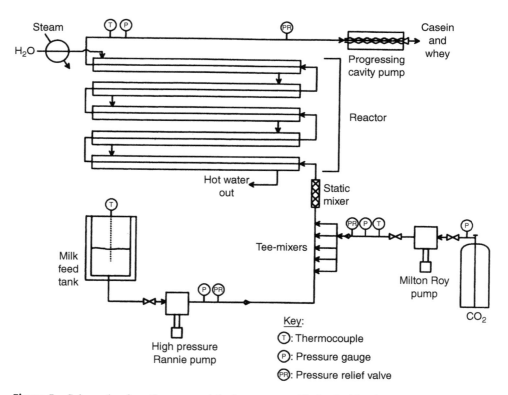

Figure 5   Schematic of continuous precipitation process with the double-pipe reactor/precipitator.

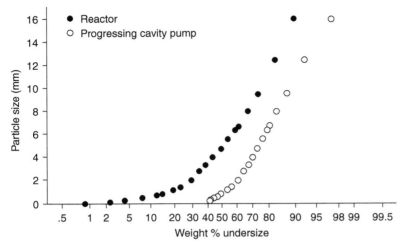

Figure 6 A comparison of particle size distribution of casein precipitated at 38°C and 5520 kPa during continuous processing before and after pumping.

1. Particle Size Distribution Studies

   a. Effects of Pumping.   Studies comparing the PSD of casein removed directly from the tubular reactor/precipitator to that removed from the progressing cavity pump are shown in Figure 6 for a run at 38°C and 5.52 MPa. The particles from the tubular reactor/precipitator were obtained by stopping the pump, opening the line leading to the pump taking care to slowly depressurize the reactor, and then flushing out the reactor with water. As discussed in Tomasula et al. (1997), breakup of casein curd occurred as it moved through the pump during depressurization. Although the cavities of the pump could handle large particles of casein, some casein was broken up in the smallest gaps between the rotor and stator as the casein was moved from stage to stage. This caused a shift in the value of d_{50} from 4.8 mm for particles formed in the reactor/precipitator to 841 μm for the particles as they left the depressurizing pump. However, using the continuous process, casein particles as large as 16 mm were obtained compared to 6 mm for casein particles formed in the batch reactor.

   b. Effects of Contact Pattern.   Figure 7 compares PSD for casein produced at 38°C in either the spray or tubular reactor, after removal from the depressurization stage, with casein produced in the batch reactor. PSD studies were conducted for casein produced at 4.13 MPa in the spray reactor and 5.52 MPa in the tubular and batch reactors. d_{50} of casein produced in the spray reactor was approximately 6 mm and particles slightly larger than 6 mm were obtained, while d_{50} for the tubular reactor was only 850 μm with particle sizes extending in size up to 16 mm. Casein obtained from the spray reactor was watery and highly compressible making it less prone to breakage as it passed through the pump. The lower precipitation pressure corresponds to higher precipitation pH. Jablonka and Munro (1986) showed that casein particle size increases when precipitated at higher pH. The particles also tended to slip through the sieves used in the particle studies and settle at the 6 mm sieve, so larger particles may have been obtained but were difficult to collect. The continuous reactors yield casein with larger d_{50} and maximum particle size than batchwise operation but the stepwise depressurization step using a pump sheared the particles. Casein precipitated at higher pressure (lower pH) was firmer and therefore more susceptible to shear while that precipitated at lower pressure (higher pH) was softer and less susceptible to shear while

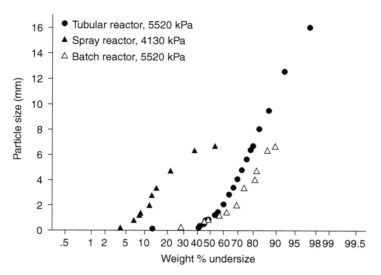

Figure 7   Effect of CO_2–milk contacting pattern on particle size distribution of casein precipitated at 38°C.

passing through the pump. Differences in particle sizes between batch and continuous tubular operation may be attributed to the different methods used for addition of CO_2. Sparging of CO_2 into the batch reactor is a slow process with gradual pressure elevation and pH depression until system pressure is reached compared to the rapid mixing of the CO_2 and milk streams already at system pressure that may impact precipitation rates and particle formation. The impact of residence time on PSD could not be elucidated. The residence time for reaction in the continuous systems was 1 min compared to an approximate residence time of 5 min in the batch reactor system.

C.   Precipitation of Other Proteins Using CO_2

Tomasula et al. (1999b) also demonstrated the use of high pressure and supercritical CO_2 to prepare a soy protein isolate containing more than 90% protein. The process was modified further to prepare isolates for other plant proteins but will not be discussed here. In this method, 10% by weight solutions of soy meal, soy flour, or soy flakes in water were prepared. Other concentrations may also be used. The solution was then adjusted to pH 9 with 1 M NaOH. pH in the range from 7 to 11 precipitate out the insoluble carbohydrates and proteins. Soy proteins were approximately 90% soluble in water adjusted to pH 9. While yield increased slightly with increasing pH, damage to the sulfur-containing amino acids of the protein remaining in solution was limited in the range from 7 to 9. The solution was then centrifuged and the solid precipitate was discarded. The extract solution was then contacted with CO_2 at 38°C and pressures ranging from 4.14 to 7.58 MPa using the pilot scale apparatus described above for preparation of casein. In all cases, a soy protein isolate resulted with protein content >90%. The product was very attractive, containing large, uniformly sized particles of the protein. For comparison, soy extract was precipitated with HCl to prepare a soy isolate. The resulting product was pasty in appearance with few discernible particles. In similar work, Hofland et al. (2000) used electron microscopy to compare the appearance of soy protein isolate prepared using CO_2 to that using H_2SO_4. The soy protein aggregates resulting from CO_2 precipitation were discrete and regular spheres while aggregates precipitated using H_2SO_4 were irregular. The regular shape of the particles formed using CO_2 as precipitant was attributed to the precise control over pH and final particle size was attributed to initial protein concentration. Mean aggregate size increased with initial protein concentration.

V. CONCLUSIONS

New processes based on SF as an alternative to those utilizing hazardous organic solvents or acids are rapidly being developed. Supercritical extraction processes that utilize SCO_2, mostly as a replacement for hexane, either to remove an undesirable component or to isolate a valuable component from a food substrate, are already in commercial operation. Processes based on SF to create microparticles with controlled particle size distribution are still new, limited to small-scale production, and used mostly for production of pharmaceutical ingredients. The processes may be useful in the future for production of microsized nutraceutical ingredients or additives for food products, with their small size ensuring improved distribution in food matrices or coatings. Processes that utilize high pressure CO_2 or supercritical CO_2, as a replacement for mineral acids, to precipitate protein from solutions have been demonstrated for isolation of casein from milk, soy protein isolate from a soy protein extract, and for other plant proteins. This method eliminates acid containing product streams and produces protein products with improved morphology compared to methods utilizing mineral acids. These processes are under commercial development in the US.

REFERENCES

Cocero, M.J. and Ferrero, S. 2002. Crystallization of β-carotene by a GAS process in batch. Effect of operating conditions. *J. Supercritical Fluids* 22: 237–245.

Cygnarowicz, M.L. and Seider, W.D. 1991 Chapter 11. Design and Control of Supercritical Extraction Processes: A Review. In *Supercritical Fluid Technology: Reviews in Modern Theory and Applications*. Bruno, T.J. and Ely, J.F., Eds. CRC Press pp. 383–403.

Debenedetti, P.G., Tom, J.W., Kwauk, X., and Yeo, S.-D. 1993. Rapid expansion of supercritical solutions (RESS): fundamentals and applications. *Fluid Phase Equilib.* 82: 311.

Elvassore, N., Baggio, M., Pallado, P., and Bertucco, A. 2001. Production of different morphologies of biocompatible polymeric materials by supercritical CO_2 antisolvent techniques. *Biotechnol. Bioeng.* 73: 449–457.

Hofland, G.W., de Rijke, A., Thiering, R., van der Wielen, L.A.M., and Witkamp, G.J. 2000. Isoelectric precipitation of soybean protein using carbon dioxide as a volatile acid. *J. Chromatogr. B* 743: 357–368.

Jablonka, M.S. and Munro, P.A. 1985. Particle size distribution and calcium content of batch-precipitated acid casein curd: effect of precipitation temperature and pH. *J. Dairy Res.* 52: 419–428.

Jablonka, M.S. and Munro, P.A. 1986. Effect of precipitation temperature and pH on the continuous pilot-scale precipitation of acid casein curd. *NZ J. Dairy Sci. Technol.* 21: 111–123.

Jordan, P.J., Lay, K., Ngan, N., and Rodley, G.F. 1985. Casein precipitation using high pressure carbon dioxide. *NZJ. Dairy Sci. Technol.* 22: 247–256.

Jung, J. and Perrut, M. 2001. Particle design using supercritical fluids: literature and patent survey. *J. Supercritical Fluids* 20: 179–219.

Kaiser, C.S., Rompp, H., and Schmidt, P.C. 2001. Pharmaceutical applications of supercritical carbon dioxide. *Pharmazie* 56: 907–926.

Kamat, S.V., Beckman, R.J., and Russell, A.J. 1995. Enzyme activity in supercritical fluids. *Crit. Rev. Biotechnol.* 15: 41–71.

Matson, D.W., Fulton, J.L., Petersen, R.C., and Smith, R.D. 1987. Rapid expansion of supercritical fluid solutions: solute formation of powders, thin films, and fibers. *Ind. Eng. Chem. Res.* 26: 2298–2306.

McHugh, M.A. and Krukonis, V.J. 1994. *Supercritical Fluid Extraction: Principles and Practice*, 2nd ed. Butterworth-Heinemann Series in Chemical Engineering: London.

Mulvaney, S.J. and Rizvi, S.S.H. 1993. Extrusion processing with supercritical fluids. *Food Technol.* 12: 74–82.

Paulaitis, M.E., Penninger, J.M.L., Gray, R.D., Jr., and Davidson, P., Eds. 1983. *Chemical Engineering at Supercritical Fluid Conditions*. Ann Arbor Science: Ann Arbor, MI.

Rizvi, S.S.H., Ed. 1994. *Supercritical Fluid Processing of Food and Biomaterials*. Chapman and Hall, Blackie Academic and Professional: London.

Thiering, R., Dehghani, F., and Foster, N.R. 2001. Current issues relating to anti-solvent micronisation techniques and their extension to industrial scales. *J. Supercritical Fluids* 21: 159–177.

Tomasula, P.M., Craig, J.C., Jr., Boswell, R.T., Cook, R.D., Kurantz, M.J., and Maxwell, M. 1995. Preparation of casein using carbon dioxide. *J. Dairy Sci.* 78: 506–514.

Tomasula, P.M., Craig, J.C., and Boswell, R.T. 1997. Continuous process for casein production using high-pressure carbon dioxide. *J. Food Eng.* 33: 405–419.

Tomasula, P.M., Parris, N., Boswell, R.T., and Moten, R. 1998. Preparation of enriched fractions of α-lactalbumin and β-lactoglobulin from cheese whey using carbon dioxide. *J. Food Process. Preservation* 22: 463–476.

Tomasula, P.M., Boswell, R.T., and Dupre, N.C. 1999a. Buffer properties of milk treated with high pressure carbon dioxide. *Milchwissenschaft* 54: 667–670.

Tomasula, P.M. (1999b). Production of high protein concentrates. US Patent and Trademark Office, Application No. 09/247,219, Filed 02/10/1999.

Tomasula, P.M. 2003. Supercritical Fluid Extraction of Foods in Encyclopedia of Agricultural, Food and Biological Engineering: Heldmen, D., Ed. http://www.Dekker.com.

Yalpani, M. 1993. Supercritical fluids: Puissant media for the modification of polymers and biopolymers. *Polymer* 34: 1102–1105.

7

Dry Coating

E. Teunou and D. Poncelet
Département de Génie des Procédés Alimentaires
École Nationale d'Ingénieurs des Techniques des Industries Agricoles et Alimentaires
Nantes, France

CONTENTS

I. SUMMARY

Dry coating is suitable for coating of core materials that must not be wetted during processing, for example, some pharmaceuticals and most of the neutraceuticals. Different dry coating processes are reviewed, particularly dry impact-blending, coating with plasticizer, and hot-melt coating. A specific case is presented in which spherical granules made of crystalline cellulose (750 to 800 mm) are coated by a fine powder (AQOAT) using triethyl citrate as plasticizer. The examined characteristics are the water barrier properties, their water content, and their quick solubility. The results show that the AQOAT film presents fast dissolution in water (2.5 min) and relative protection. Unfortunately this protection is not sufficient in certain cases, for example, the 2% moisture content of the resulting coated granules is not satisfactory for probiotics requirements, even if this value is smaller than the one obtained with the aqueous coating system. Finally, dry coating using a plasticizer appears quite adequate for coating moisture-sensitive food products but this requires some improvement of the processing conditions.

II. INTRODUCTION

Dry coating in the present case must be considered as all coating processes where the core materials are not wetted (by water) during processing. It is a response to the coating of all products that are sensible to water and hence to all traditional aqueous coating methods such as some pharmaceuticals and most of neutraceuticals (specially probiotics). These types of products are of great interest for different industries. This interest is based on production of new, healthy, and natural tablets and innovative ingredients for improvement of human health. One of the important factors, which have an influence on human health, is the quality and security of nutrition. Today, it must be admitted, not almost unanimously, that the inclusion of functional foods or neutraceuticals in foods can help to fulfill these objectives.

Although there is no specific definition of neutraceuticals, they can be defined as ordinary foods that have components or ingredients incorporated into them to give them a specific medical or physiological benefit, besides a purely nutritional effect. This definition can include supplementation of food (cereals, grains, drinks, and yoghurts) with minerals, vitamins, antioxidants, probiotics, prebiotics, herbs, botanicals, and oils [1]. Probiotics may be defined as life mono or mixed cultures of microorganisms which, when applied to animals or humans, have a beneficial effect on the host by improving the properties of the indigenous flora. The presence of probiotics in the gastrointestinal tract, for example, prevents colonization with photogenic microorganisms [2]. Moreover, they can influence the secretion of intestinal mucosa, corresponding with pathogenic attachment in intestines. The beneficial effects of probiotic bacteria are also related to (1) significant suppression of the tumor growth [3,4]; (2) antiallergy effect [5]; (3) inhibition of *Helicobacter pilori*, which is associated with chronic gastritis, peptic ulcers, and risk of gastric cancer [6]; (4) decreasing the cholesterol level [7]; and (5) improving lactose utilization.

Most of these products, which are in granular form, are very sensitive to water. It is quite understandable that the most important thing, for preserving the positive effects of these functional foods, is the quantity of life (probiotic bacteria) or active (minerals or fats) components in the final ready-to-use products (granules, pellets, or tablets). This quantity depends mainly on the moisture content of the product, which in itself is related to the production technique and also to the storage conditions. Spray and freeze drying and different aqueous coating techniques have been successfully used for this purpose but when the active components are too sensitive to moisture, these methods are not adapted. Indeed, during the last twenty years, pharmaceutical coating technology has been based on the aqueous systems, which do not cause environmental pollution [8]. During the aqueous film coating process, the humidity in the coating chamber increases significantly and Clementi and Rossi [9] demonstrated that the viability of dry probiotics decreases rapidly, especially with high moisture content. Hence aqueous coating systems are not applicable for encapsulation of dry probiotics. Dry coating is an alternative to these conventional methods, which can allow maximal saving of the viability and activity of the components by minimizing the effects of atmospheric moisture on their shelf life and by slowing down the rate of viability loss at room temperature. The growing demand for functional foods and additives (e.g., the European market was estimated at £1000 million in 2000) is a consequence of two parallel trends in the food and life science industries: consumers now spend less time cooking per day, down to an average of 20 min from 50 years ago and they also demand more and more safe, hygienic, and healthy foods, this last point being amplified by the aging of the population [10]. There is a real need to find new dry coating techniques, which propose better conditions to preserve the sensitive dried powders against water vapor and oxygen and offer excellent potential for commercial application. These techniques are grouped into four main approaches:

1. The use of nonaqueous solvent in traditional coaters.
2. The hot-melt coating.
3. The dry particles coating where the fine particles are fixed on the surface of the core material by higher agitation using mechanochemical treatment with no wetting.
4. Coating with plasticizer where the fine particles are fixed on the core using a plasticizer.

The aim of this chapter is to describe these different approaches and to analyze a case study where a cellulose granule (as a probiotic model) is dry coated by a fine powder (AQOAT) using a plasticizer. The formed coating shell was tested, keeping in mind its permeability, its water content, and its dispersion in water, and compared to a membrane obtained by aqueous coating.

III. DRY COATING

Dry coating is not really a new technology. Indeed, dry coating was first used in the preparation of pharmaceuticals, cosmetics, dental materials, HPLC packing, ceramics, copy toners, cements, inks, pigments, detergents, and so on [11]. Its application in chemical [12], aeronautical, metallurgical [12–17], wood and paper industries consists, most of the time, in coating or recovering woods, metals, or any desired surface by paints, varnishes, different polymers, or precious metals to protect them against corrosion or to reinforce them [18]. But its application to granular material and specifically in the food area is quite recent and is growing for various purposes (improvement of powder flowability, control release, and fast-soluble powders) but it can be said that it is still in its infancy.

In its strictest sense, dry coating refers to techniques where the core materials are strongly surrounded by fine particles simply by collision or by the use of nonaqueous plasticizers. But, as specified in the introduction, dry coating for food ingredients refers to all these techniques that allow coating without wetting (with water) a core material. So defined, the use of nonaqueous solvents and hot melt are considered as dry coating techniques. Figure 1 summarizes the principles of the four dry coating approaches. These approaches are briefly described in the next paragraph; the solvent approach is equivalent

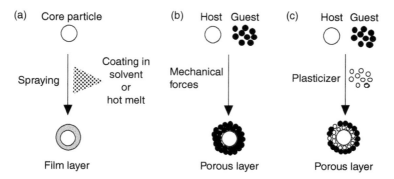

Figure 1 Principles of different dry-coating approaches: (a) solvent and hot-melt, (b) dry particle coating, and (c) use of plasticizer.

to the conventional coating with the difference that the coating material is dissolved in an adequate and suitable nonaqueous solvent. The hot-melt approach requires a coating material that is liquid at relatively higher temperature (hot melt) and solid at handling and processing temperature. The dry particle coating approach requires high-energy throughput to create collision and adhesion between core particles and the fine dry particles while the fourth approach uses a plasticizer to bind them.

IV. DIFFERENT APPROACHES FOR COATING PARTICLES WITHOUT WETTING

A. Use of Nonaqueous Solvents

Vaporizable solvents for coating were first used to enhance the material efficiency of the coating process by increasing the solid content of the sprayed coating material while keeping low viscosity and their application as a dry coating process was limited to nonaqueous solvents. Indeed, the use of the solvents helped to increase the solid content from 4 to 20% and brought some satisfaction to the pharmaceutical industries in the past. The most used solvents are alcohol (methanol, ethanol, n-propanol, and n-butanol), cetones (acetone), ether, and esther (dichloromethane, ethyl acetate, chloroform, etc.).

The process can take place in all types of conventional fluid-bed and pan-coating systems (Figure 2), which are well-described in other chapters. As mentioned above, the coating material is dissolved in a solvent, which is evaporated after spraying on the core. It is quite clear that the main drawbacks of this approach are the emission of solvent into the atmosphere, the solvent residue in the final product, and the risk of explosion. The permissible emission rate for solvents, which is determined by the local authorities, has been further reduced by European regulations. So the application of this approach requires an expensive solvent recovery system (Figure 3) where the solvent can be recycled properly. Nevertheless, in the best cases 5 to 10% of the solvent is not recovered [19].

The use of vacuum installation during this process has been known to reduce the limitations of this technique, specifically the explosion hazard, the environmental strains, and the solvent losses. The system relies on the physical properties of solvent evaporation in vacuum and the process can be optimized to minimize energy consumption. With this improvement the process is sterile and suitable for coating active agents, aromatic substances, easily oxidizable, hydroscopic, and/or heat-sensitive products.

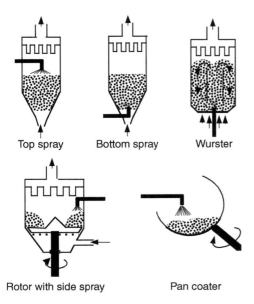

Top spray   Bottom spray   Wurster

Rotor with side spray   Pan coater

Figure 2   Different types of particle coaters.

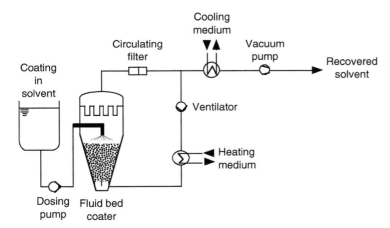

Figure 3   Diagram of a fluid-bed installation with a simplified solvent recovery system.

To summarize, although there have been massive improvements in all aspects of this process, its application to coating in the food area is not easily accepted today in the food industry and even in other industries because the new international regulations are becoming more and more severe about solvent release in the atmosphere and their concentration in food, drugs, and cosmetic products. The explosion hazard and the cost of the installation are also unfavorable factors.

B.   Hot-Melt Coating

This second approach of dry coating was also found indirectly through the research of moisture barrier polymers or membranes. The use of some molten materials such as solid fats, waxes, and some polymers as coating materials provides a shell with very good barrier

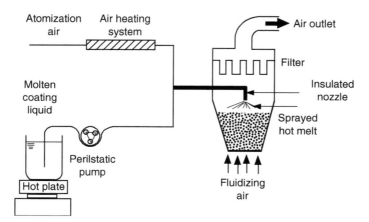

Figure 4   A conventional top spray fluid-bed installation for hot-melt coating.

properties against water vapor and gas [20]. During this process, which can take place in all types of fluid beds (top, bottom, rotor, and Wurster), the coating material, melted by heating (hot melt) is sprayed on the core and solidified directly by cold air. The core material is then never in contact with water.

The process must be conducted in such a way that the molten coating is maintained at a constant temperature (generally on a hot plate), usually 40 to 60°C above its melting point, before any contact with the core. It can be seen from Figure 4, which presents a typical installation for a hot-melt fluid-bed coater, that the atomization air is heated too (generally to the same temperature as the spray liquid) and can serve to keep the molten coating at its application temperature. Adequate insulation is also designed to prevent remelting of coated particles.

This dry coating approach is known for its low cost as it presents many advantages: (1) no solvent used, that is, sprayed liquid = 100% coating agent, (2) short processing time, (3) no drying process required [21]. The main limitations of hot melt are (1) its inability to adapt to heat-sensitive (biological) products, (2) bad flow properties, and (3) the strength and sometimes the odor (due to oxidation) of the coated particles.

C.   Dry Particle Coating

As shown by the principles in Figure 1(b), this approach consists of coating a relatively large particle size (core material or host) mechanically with fine particles (guest). This leads to coated particles that can dissolve quickly in water because of their high porosity. The mechanisms of dry particles coating is described as follows [22]: initially the agglomerates of fines coating (guests) are separated into their primary particles before adhering to the host particles. This occurs rather quickly and is followed by the dispersion and rearrangement of fines spread over the surface due to collisions between coated hosts and noncoated hosts. No liquid of any kind (solvents, binders, or water) is required, that is, compared to other approaches mentioned above, dry particle coating is the true dry coating.

This process is very similar to a dry mixing of a binary mixture of powders where the smaller particles slowly adhere onto the larger particles after separation and rearrangement and cannot be easily removed from them because, as the size of the guest particles is quite small, van der Waals interactions are strong enough to keep them firmly attached to the host particles. This type of mixing is referred to as ordered mixing or structured mixing [23].

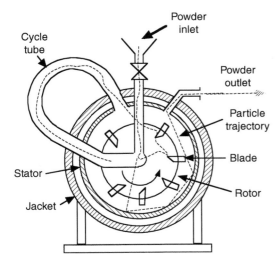

Figure 5   Schematic of a hybridizer.

This similarity between dry coating and dry mixing is clearly described by Pfeffer [24] who underlines the importance of referring to the literature on ordered mixing before conducting any serious study on dry coating since the two types of drying processes are closely related. The difference between the two types of process is the strength of the bonding between the guest and the host particles. Indeed, in ordered mixing, the larger particles are loosely covered with the fine particles while in dry particle coating, the surface covering is more permanent because the physical or chemical bonding is stronger. To achieve this strong mechanical force, higher mechanical energy is required and a number of devices, whose names are generally associated with the process itself, have been designed for this purpose.

1.   Dry Impact Blending (Mechanochemical Treatment or Hybridization)

The hybridizer, shown schematically in Figure 5, was developed by the Japanese company, Nara Machinery Co. and consists of a very high-speed rotating rotor with six blades, a stator, and a powder recirculation circuit. The powder (host and guest particles) placed in the blending chamber is subjected to high impaction and dispersion due to the high rotating speed of the rotor. This agitation breaks down its agglomerates and produces an electric charge by contact and collisions between particles leading to embedding of the fine guest particles onto the surface of the host particles (this phenomenon is called hybridization or mechanochemical treatment).

2.   Mechanofusion

A mechanofusion machine is shown in Figure 6. Its principle is almost the same as a hybridizer with the difference that the outer vessel rotates whereas the inner piece and a scraper are stationary. A measured amount of host and guest particles is placed into the rotating vessel. As the vessel rotates at speeds between 200 and 1600 rpm, the particles, which are forced to pass through the gap between the inner piece and the rotating drum are submitted to intense shearing and compressive forces. These forces generate sufficient heat energy to "fuse" the guest particles onto the surface of the host particles (mechanofusion).

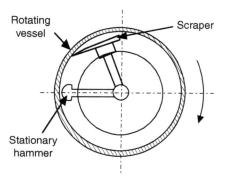

Figure 6   Schematic of the mechanofusion machine.

Mechanofusion produces very strong physical and chemical bonds, which enhance the coating process.

More details about the two above-mentioned devices can be found in [24,25], but it has to be kept in mind that they present short processing time and require such high-energy throughput that the internal temperature of the process can rise up to 60°C, a characteristic that can be useful for mechanofusion or filming of the coating shell by fusion. However, they may not be appropriate for certain applications if there are special constraints such as very low temperature range, material hardness, and cost factors that need to be considered. This is the case for many food and pharmaceutical ingredients, which, being organic and relatively soft, are very sensitive to heat and can quite easily be deformed by severe mechanical forces [24]. For such products, there is real need for a soft dry coating system where the guest particles are attached to the host particles with minimum degradation of particle size, shape, and composition. The literature inventory shows three soft techniques that have been designed for such applications. They are briefly described below from the hardest to the softest.

3.   High-Speed Elliptical-Rotor Type Mixer (HEM or Theta Composer)

A schematic cross-section of the HEM is presented in Figure 7. It consists of a slow rotating elliptical vessel (around 30 rpm) and a faster (500 to 3000 rpm) elliptical rotor. The rotor and the vessel rotate coaxially in opposite directions and the powder mixture inside the vessel, consisting of host and guest particles, is subjected to shear and compressive stresses as it is forced into the small clearance between the vessel and the rotor. Successive and repeated feeding of the minimum clearance region with the powder mixture lead to strong immobilization of the guest particles on the surface of the host particles [24,27].

4.   Rotating Fluidized Bed Coater

This newly developed coating device is described by Pfeffer [24] as a system operating on the principle of a rotating fluidized bed coater (RFBC). The host and guest powder mixture are placed into the rotating bed and is fluidized by the radial flow of gas through the porous wall of the cylindrical distributor, as seen in Figure 8. Due to high rotating speeds, very high centrifugal and shear forces are developed within the fluidized gas-powder system leading to the breakup of the agglomerates of the guest particles and their adhesion on the host particles.

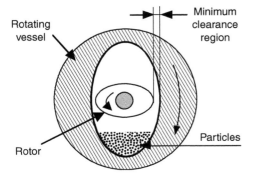

Figure 7 Schematic cross section of a high-speed elliptical-rotor mixer (HEM).

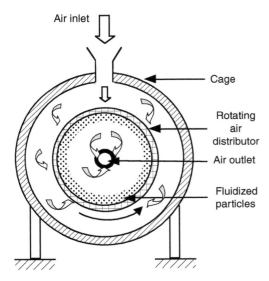

Figure 8 A rotating fluidized bed coater.

5. Magnetically Assisted Impaction Coater

A schematic of the magnetically assisted impaction coater (MAIC) device, proposed by Ramlakhan [26], is shown in Figure 9. A measured amount of magnetic, host and guest particles are placed into a processing vessel (200 ml glass bottle). The magnetic particles are made of barium ferrite and coated with polyurethane to help prevent contamination of the coated particles. An external oscillating magnetic field is created using a series of electromagnets surrounding the processing vessel. When a magnetic field is created, the magnetic particles are excited and move furiously inside the vessel resembling a gas-fluidized bed system, but without the flowing gas. These agitated magnetic particles then impart energy to the host and guest particles, causing collisions and allowing coating to be achieved by means of impaction or peening of the guest particles onto the host particles. Note that this apparatus is very versatile (i.e., can operate both as a batch and continuous system and for various types of materials) and generates no increase in temperature of the material although there is some heat generated on a microscopic level.

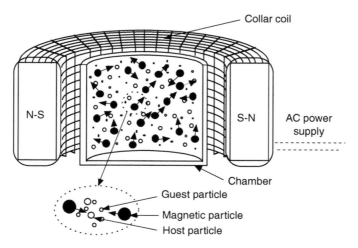

Figure 9   Schematic of a magnetically assisted impaction coating device.

D.   Use of Plasticizer

Despite the improvement brought about by the soft dry particle coating systems described above, their application to certain foods or ingredients, especiallly food, is still critical for various reasons, particularly the investment in a new blending machine. The idea here is to simply replace the required mechanical forces for dry coating by a nonaqueous binder or a plasticizer (Figure 1[c]). The coating process consists of two stages. The first stage involves feeding simultaneously the powder mixture (host and guest) and a spray of plasticizing mixture. At this stage, the layer formed around the host particles is just a deposit of guest particles. This deposit is turned into film in the second stage by curing, which is done by heating for a short time. This novel technique was developed by the Japanese company Shin-Etsu for enteric coating of tablets and pellets with a mixture of guest particles (Shin-Etsu AQOAT and talk) using a mixture of triethyl citrate and acetylated mono glyceride as plasticizer [11–28]. Note that the technique can be extended to other applications providing the formulation of suitable guest particles and plasticizer for a given purpose. The technique was quite welcomed in the food and pharmaceutical industries because of obvious advantages: (1) no deformation of the resulting coated particles and (2) no investment in one of the new blending machines mentioned above as the process can take place in any conventional coating system (Figure 2) providing a powder feeding unit as specified in Figure 10.

V.   COATING OF CRYSTALLINE CELLULOSE GRANULES USING AQOAT — A CASE STUDY

Our research was concentrated on the characterization of coated crystalline cellulose granules using the dry-coating technique developed by Shin-Etsu. As mentioned above, a large and growing number of companies are trying to adapt this technology to various applications (powders or granular materials), so knowledge of the parameters for its optimum application is needed. The objective of this study, which was done partly in collaboration with Glatt Pharmatech-Binzen, was to compare the water content in the beads during coating with aqueous and dry-coating systems with particular reference to coating water-sensitive products. The experiments were carried on successfully with a modified pan coater

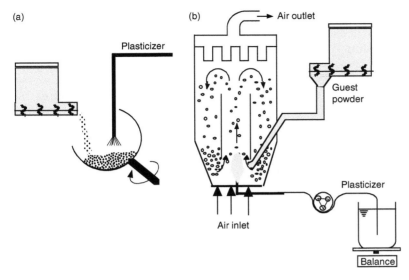

(a)

(b)

Air outlet

Plasticizer

Guest
powder

Plasticizer

Balance

Air inlet

Figure 10 Dry feeder units associated to (a) pan dry coater and (b) Wurster dry coater.

(Figure 10[a]) and a modified Wurster fluid-bed coater (Figure 10[b]) but only the results from the fluid bed will be presented here.

A. Experiment

1. Materials

White spherical beads of crystalline cellulose (IPS – Cellts), with mean particle diameter equal to 835 μm and density 828 kg/m^3, were used as core materials (host particles). Their water content (6%) was too high compared to the recommended 0.5% for probiotics. So, in order to assimilate these core materials to probiotics, the latter were dehydrated in a drying oven for 48 h at 100°C and cooled after using a desiccator before coating.

A 3% solution of SEPIFILM LP 010 (SEPIC, Paris — France) was used as coating solution for aqueous coating. SEPIFILM consists of hydroxypropyl methylcellulose, microcrystalline cellulose, and stearatic acid.

The materials for dry film coating were a combination of coating polymer powders AQOAT (SYNTAPHARM — Mülhein — Germany), consisting of hydroxypropyl methylcellulose acetate succinate and talk (guest particles), and a liquid plasticizer (triethyl citrate — Sigma and acetylated monoglyceride) in the following proportion: 7 ml of plasticizer for 30 g of AQOAT. The particle size of the guest particles was not more than 10 μm.

Adequate dyes were added to the SEPIFILM solution and the plasticizer solution to color the formed coating shell. This precaution was necessary for the dissolving test.

A classical lab-scale Wurster (a bottom-spray coater), UNIGLATT (Glatt-Binzen) was used for aqueous coating. A powder-feeding device was adapted to the system for dry coating as specified in Figure 10(b).

2. Procedure

Different operating parameters of the fluid-bed coater (flow rate of the plasticizer through the nozzle, air pressure for atomization, and powder feed rate) for aqueous and dry coating were optimized to reduce agglomeration and coating material lost during the process. These parameters are reported in the following descriptions.

a. Aqueous-Coating With SEPIFILM LP 010. 500 g of dried core materials were first fluidized, then the 3% aqueous solution of SEPIFILM was sprayed at a rate of 5 g/min through the nozzle in the fluid-bed chamber for coating. The atomization pressure was set to 1.5 bars. The temperature of the inlet and outlet air was respectively 60 and 42°C. The experiment was run for 60 min. The temperature of the atmosphere around the fluid bed was 20°C and the relative humidity was around 45%.

b. Dry-Coating System With AQOAT and Plasticizer. The metal tube ($d = 1.5$ cm) of the powder feeding system was first positioned between the nozzle and the insert to allow homogeneous dispersion of the guest particles in the coating zone. 500 g of dried core materials were then fluidized, followed by a simultaneous atomization of the plasticizer at a rate of 1 ml/min and the dispersion of the guest particles at a rate of 2 g/min through the nozzle in the fluid-bed chamber for coating. The atomization pressure was set to 1.5 bars. The temperature of the inlet and outlet air was respectively 40 and 27°C. The experiment was run for 30 min with no curing to keep in mind the application to sensitive neutraceuticals.

3. Analysis

a. Physical Properties of the Granules. The water content of different granules was determined by the classical gravity method. Samples (approximately 10 g of the encapsulated beads) were maintained at 100°C for 48 h in the oven and were weighed until constant weight.

The diameter was measured using a graduated microscope. The mean diameter was calculated from the diameter measurements of 50 beads. The thickness of the membrane was calculated as the difference between the initial and the final diameter of the granules.

b. Permeability of the Granules. Different methods exist for measuring the permeability of water vapor through a membrane. The principle of these methods is to determine the quantity of water that diffuses through the film or the membrane separating the humid space from the dry space. This principle was not applied here because large films of coating materials could not be elaborated. So it was decided to measure the hydration of granules during storage in a relatively dry medium.

This hydration was measured using a desiccator. It consists of a small box filled with $Mg(NO_3)_2$ and covered hermetically by a stopper (Figure 11). The granules (10 g approximately) were put in a cup, weighed, and placed in the center of the box. They were allowed to stand for 48 and 168 h in the closed box at room temperature ($\sim$25°C). The final weight of the granules was measured for the calculation of the water uptake. Note that at 25°C, the $Mg(NO_3)_2$ provided a humidity of about 50% RH in the box.

c. Dispersion in Water. For this experiment, 1 g of colored granules was poured into 50 ml of water and the solution was mixed using a magnetic stirrer (800 rpm). The experiment was run and stopped when the colored coating shell disappeared or when the white core beads appeared. This lapse of time was recorded as dispersion time.

VI. RESULTS AND DISCUSSION

The results are presented here as a comparison to the aqueous coating and the dry-coating systems. This approach may appear inadequate with regard to some parameters but it finds its justification in the fact that SEPIFILM and AQOAT are closed in composition and were both designed to improve the stability of moisture-sensitive ingredients.

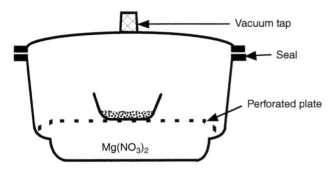

Figure 11   Testing device of granules permeability or hydration.

A.   Coating Processes

During aqueous coating with SEPIFILM LP 010, a dried coating solution was collected on the insert cylinder wall as a thin "paper" sheet, so that the coating material was lost at the end of the experiments. Despite different optimization strategies, this "paper" sheet could not be avoided during coating. At least its quantity could be reduced drastically to approximately 3% of the total coating material when the coating liquid was batch fed, that is, the coating liquid was alternatively sprayed in the system for 5 min and stopped for 5 min allowing some drying. The final mean diameter of the coated granule was 856 μm, that is, a thickness of about 10.5 μm.

It was observed during preliminary experiments of this dry coating with AQOAT that, over a coating flow rate of 2 g/min and a plasticizer flow rate of 1 ml/min, there was agglomeration of granules and a great loss of the guest particles, which were dispersed in the fluid-bed chamber before finally getting stuck on the air filter of the fluid bed. This underlines the difficulty in performing such an experiment that is rather described as simple by Shin-Etsu [28] and the necessity to establish optimum operating conditions for each case. The final mean diameter of the coated granule was 870 μm, that is, a thickness of about 17.5 μm and a coating shell, which consisted of a double layer of guest particles, assuming an ordered adhesion of the guest particles.

Dry coating was a quick process. Its processing time was half of the aqueous coating time and led to a shell thickness that was about two times larger than the SEPIFILM membrane. But its efficiency was much less than that of conventional aqueous coating (96%). Indeed, despite the use of optimized parameters mentioned above, the efficiency of dry coating was 75% — the efficiency was calculated as actual weight gain of coated samples divided by the theoretical weight gain. Note that, it is reported [11] that the efficiency can be increased to more than 95%.

B.   Hydration During Coating Processes

As mentioned above, the core materials were introduced in the fluid-bed chamber at water content optimistically equal to zero. During coating with SEPIFILM, samples of granules were collected for every 100 ml of coating solution and tested to determine their water content. It appears that the granules were always at 4% water content whatever the added quantity of the coating material, while the water content of the granules was 2% during dry coating with AQOAT. This means that granules absorb water during coating and that, with the above process parameters, the air conditions in the fluid-bed chamber were in such a way that the equilibrium water content of particles in the chamber was 4 and 2% respectively during aqueous and dry coating. It was not possible to reduce this equilibrium water content because the initial water content of air could not be controlled on the one hand and, on the other, the water was evaporating continuously during film formation. This second cause is especially valid for aqueous coating. This result, which shows an impact of the fluidizing air properties on the moisture content of granules, has promoted one of the main research topics of our research work: the effect of the fluidizing and atomization air properties on the quality of the coated particles.

C.   Permeability of the Shell (Hydration of Granules During Storage)

The hidden objectives of these experiments was to increase the shelf life of moisture-sensitive product probiotics by reducing the transfer of water vapor in the capsule or the permeability of the coating shell. The permeability of the shell was simply evaluated through the hydration of the granules as mentioned in the sections on materials and procedures. So, the measurement of the final moisture contents after storage showed that the core beads with and without coating membrane absorb the same ratio of water ($\sim$6%) after 48 h. But, after 168 h, this ratio increases to about 8% for beads without membrane and 6% for SEPIFILM- and AQOAT-coated granules. Now, considering the fact that the capsule consists mainly of the core material, it can be said that the SEPIFILM and the AQOAT can protect the core against water vapor, but not sufficiently because the difference of water content between beads with and without the membrane was about 2%. This protection may be improved by an increase of plasticizer concentration as demonstrated by Park and Chinnan [29], who found that the water vapor barrier properties of cellulose films increase as the concentration of plasticizer (Myvacet 7-00 TM) increases, while investigating the gas and water vapor barrier properties of edible films.

Finally, the protection by both SEPIFILM and AQOAT are similar and are not sufficient. So, the only way to retain the advantage brought by about dry coating (low hydration during the coating process) is to provide adequate storage. The relatively high water content of coated particles and their hydration during the dry-coating process and storage underline future challenges to be overcome in order to apply this technology to water-sensitive products.

D.   Dispersion of the Coating Shell in Water

The results are given here as disintegrating time of the coating membrane in water. The final SEPIFILM-coated granules could be dissolved in water in 90 sec (Figure 12), while the AQOAT granules totally disintegrated after 5 min, probably due to the difference in coating thickness (10.5 and 17 μm respectively). Indeed, the time necessary for dissolving the membrane depends on the thickness of the SEPIFILM shell (Figure 12).

Knowing that the thickness is a function of the mass of coating material, it can be said that there is a linear relationship between the thickness of the film and the disintegration time of the coating shell. But the thickness alone cannot explain this difference in disintegration time, and other factors such as the composition of the shell and the difference in moisture content level in the granule after coating (4 and 2% respectively) can be pointed out. Also note that the disintegration time of the AQOAT-coated film may have been longer if the coating was followed by a curing step, which normally reduces the permeability of the shell.

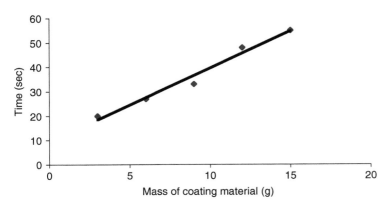

Figure 12   Disintegrating time of granules shell in water versus the thickness of SEPIFILM.

VII.   CONCLUSION AND PERSPECTIVES

Dry-coating processes, where the fine guest particles are spread over the surface of the host particles (core) due to collisions and curing (mechanofusion), without any water, solvent, or binders, are probably the best for coating moisture-sensitive ingredients. A number of works have been reviewed and they reveal a real infatuation for this type of coating processes during the last ten years, owing to the interesting potential that they present for various industrial applications. Unfortunately, their implementation requires an investment in a specific blending machine and higher energy throughput that renders them ineffective for coating soft products such as food ingredients. Some major innovations have been made to develop softer and cheaper dry-coating systems, but the use of plasticizer is the technique that looks promising, from all points of view, for coating food ingredients. This technique has shown low moisture levels compared with aqueous coating systems. However, a very small level of water content was detected in the granules (2%), and this is not acceptable for application to some sensitive products such as probiotics. That is why further experiments must concentrate in designing an air-dry system in order to improve the processing conditions (e.g., drying of the fluidizing and atomization air), leading to very low moisture content (less than 0.5%) of the final granules.

With reference to the properties of the granules, it appears that there is a need for a powerful method that will allow measuring without doubt the percentage of water that can cross a given film on a coating particle. The absence of this method can explain some contradictory results such as the AQOAT shell, which could not be dissolved easily in water and could not protect the core against water vapor. The emphasis must also be on the formulation aspect to find the suitable combination and composition of plasticizers and guest particles that can fulfill the two opposite objectives (fast solubility in water and barrier for water vapor). Hence, beyond all the functional properties of the granules the testing of which is required to check the success of the coating process, the strength of the coated granules is another parameter that must be taken into account when dealing with dry coating. Outstanding research works have been published on this topic [30,31], analyzing the strength of agglomerates and presenting different testing methods; but the selected method should be adequate to report the binding strength of the guest particles to the host particles' surface, for example, an adapted abrasion test was actually developed in our laboratory for this purpose, which was designed in such a way that the mechanisms of constraint during the test correspond to what coated particles can encounter during their handling and processing.

REFERENCES

1. Dowden, (1999). A. Fun and functional foods. DotPharmacy (Accredited distance learning for pharmacists from Chemist & Druggist — http://www.dotpharmacy.co.uk/upneutra.html) June modulus.

2. Pascual, M., Hugas, M., Badiola, J.I., Monfort, J.M., and Garriga, M. (1999). *Lactobacillus salivarius* CTC2197 prevents *Salmonella enterica* colonization in chickens. *Appl. Environ. Microbiol.* 65(11): 4981–4986.

3. Reddy, G.V., Shahanti, K.M., and Banerjee, M.R. (1973). Inhibition effect of yogurt on Ehrlich ascites tumor-cell proliferation. *J. Nat. Cancer Inst.* 50: 815–817.

4. Arai, K., Murata, I., Hayakawa, K., Kataoka, M., and Misuoka, T. (1981). Effect of oral administration of sour milk on the preservation of health. *In intestinal flora and carcinogenesis*, T. Mitsuoka (ed.), Gakkai-Shuppan Center, Tokyo, pp. 105–123.

5. Sanders, E.M. (1999). Probiotics. *Food Technol.* 53(11): 67–77.

6. Aiba, Y., Suzuki, N., Kabir, A.M.A., Ttakagy, A., and Koga, Y. (1998). Lactic acid-mediated suppression of *Helicobacter pylori* by the oral administration of *Lactobacillus salveticus* as in a gnotobiotic murine model. *Amer. J. Gastroenterol.* 93(11): 2097–2101.

7. Gilliland, S.E., Nelson, C.R., and Maxwell, C. (1985). Assimilation of the cholesterol by *Lactobacillus acidophilus*. *App. Environ. Microbiol.* 49(2): 377–381.

8. Nacagami, H., Keshikawa, T., Matsumura, M., and Tsukamoto, H. (1991). Application of aqueous suspension and latex dispersion of water-insoluble polymers for tablet and granule coatings. *Chem. Pharm. Bull.* 39: 1837–1842.

9. Clementi, F. and Rossi, J. (1984). Effect of drying and storage conditions on survival of *Leuconostoc oenos*. *Amer. Enol. Vitic.* 35(3): 18–186.

10. Meesters, G.M.H. (2002). Actual trends in the food and life sciences industry. Proceedings of the 53rd GLATT workshop, Technological Training Center, Binzen, Part 1.

11. Obara, S., Maruyama, N., Nishiyama, Y., and Kokubo, H. (1999). Dry coating: an innovative enteric coating method using a cellulose derivative. *Eur. J. Pharm. Biopharm.* 47: 51–59.

12. Alince, B. and Lepoutre, P. (1983). Viscosity, packing density and optical properties of pigment blends. *Colloids Surf.* 6: 155–165.

13. Arezzo, F., Gimondo, P., Hashimoto, M., Ono, N., and Takahashi, T. (1996). Characterization of TiN films deposited onto stainless steel strips by continuous dry-coating process. *Thin Solid Films* 290–291: 226–231.

14. Babic, R., Metikos-Hukovic, M., and Radovcic, H. (1994). The study of coal tar epoxy protective coatings by impedance spectroscopy. *Prog. Org. Coat.* 23: 275–286.

15. Lin, T.J., Antonelli, J.A., Yang, D.J., Yasuda, H.K., and Wang, F.T. (1997). Plasma treatment of automotive steel for corrosion protection: a dry energetic process for coatings. *Prog. Org. Coat.* 31: 351–361.

16. Miyamoto, Y., Kubo, Y., Hashimoto, M., Ono, N., Takahashi, T., Ito, I., Gimondo, P., and Arezzo, F. (1995). Properties of thin TiN films deposited onto stainless steel by an in-line dry coating process. *Thin Solid Films* 270: 253–259.

17. Nadkarni, S.K. (1997). Corrosion-resistant coating for aluminum. *Met. Finish* 95: 96–97.

18. Perry, A.J. and Treglio, J.R. (1996). Surface processing of sheet metal using metal ion beams. *Surf. Coat. Technol.* 81: 87–91.

19. Glatt International (1999). Solvent recovery. Process information (http://www.glatt.de).

20. Jozwiakowski, M.J., Franz, R.M., and Jones, D.M. (1990). Characterisation of hot-melt fluid bed coating process for fine granules. *Pharmaceutical* 7(11): 3–10.

21. Prasch, A. (1998). Hot-melt coating of biological products. Proceedings of GLATT workshop, Technological Training Center, Binzen, Part 4.

22. Alonso, M. and Alguacil, F.J. (2001). Stochastic modeling of particle coating. *AIChE J.* 47: 1303–1308.

23. Hersey, J.A. (1975). Ordered mixing: a new concept in powder mixing practice. *Powder Technol.* 11: 41–44.

24. Pfeffer, R., Dave, R.N., Wei, D., and Ramlakhan, M. (2001). Synthesis of engineered particulates with tailored properties using dry particle coating. *Powder Technol.* 117: 40–67.

25. Honda, H. and Koishi, M. (1996). Packing structure of monolayer coated powder prepared by dry impact blending process utilizing mechanochemical treatment. *Proceedings of the 2nd International Particle Technology Forum*, San Diego, pp. 607–612.

26. Ramlakhan, M., Wu, C.Y., Watano, S., Dave, R.N., and Pfeffer, R. (2000). Dry particle coating using magnetically assisted impaction coating: modification of surface properties and optimization of system and operating parameters. *Powder Technol.* 112: 137–148.

27. Iwasaki, T., Satoh, M., and Ito, T. (2000). Determination of optimum operating conditions based on energy requirements for particle coating in a dry process. *Powder Technol.* 123: 105–113.

28. Shin-Etsu. (1997). Dry coating: Improved formulation for pellet coating. Technical information, revised edition of No. A-3.

29. Park, J.H. and Chinnan, M.S. (1995). Gas and water barrier of edible films for protein and cellulosic materials. *J. Food Eng.* 25: 497–507.

30. Schubert, H., Herrmann, W., and Rumf, H. (1974). Deformation behaviour of agglomerates under tensile stress. *Powder Technol.* 11: 121–131.

31. Schubert, H. (1974). Tensile strength of agglomerates. *Powder Technol.* 11: 107–119.

8

Fluid-Bed Coating

E. Teunou and D. Poncelet

Département de Génie des Procédés Alimentaires
École Nationale d'Ingénieurs des Techniques des Industries Agricoles et Alimentaires
Nantes, France

CONTENTS

I. SUMMARY

The chapter presents varied descriptions of the fluid-bed process for coating or air suspension coating, with an emphasis on the Wurster system (the most efficient batch fluid-bed apparatus), including a discussion on the application and the efficiency of batch fluid-bed systems and their improvement. An analysis of the performance of the batch fluid bed will show that the continuous fluid bed is the economically suitable solution for coating food powders and the schematic of the ideal continuous fluid bed for food powder coating is given as an indication. Phenomena involved in the process of coating fluidized solid particles are studied and some useful techniques to characterize and to evaluate the quality and the efficiency of the process are given and analyzed. Finally, a short classification of the most used coating materials in food and pharmaceutical industries is presented to give a quick view of what can be done with the technique.

II. INTRODUCTION

Fluid-bed coating is one of the various processes that can be employed for encapsulation and coating of food ingredients or additives such as extrusion, solvent extraction, coacervation, cocrystallization, spray drying, mixing and adhesion in rotating drums, etc. [1]. Its specificity is that it allows to really coat dry solid particles (powders), that is, the engulfing of particles into a coating material. This type of coating process leads to capsules called reservoir systems where the particles are surrounded by a layer (Figure 1[a]) or multiple layers (Figure 1[b]) of coating materials.

The ensuing paragraphs are designed to provide some points on the state of this technology, from the description of the process itself to the requirements regarding the powder to be coated, the coating materials, and the characteristics of the resulting products. A review of the various fluid-bed systems will be done here, showing the specificity of each system and the improvement of the coating efficiency. This last point will underline the necessity of the use of the continuous fluid-bed process in food powder coating.

III. GENERAL KNOWLEDGE ON POWDER COATING AND FLUID BED

Every year, tons of food powders are required with some specific properties that the natural product does not offer. The encapsulation of these products provides an alternative to fulfill this request. It is a process where thin films or polymers (coat or shell) are applied to small, solid particles, droplets of liquid, or gases for a variety of aesthetic and protective purposes. Indeed, encapsulation of food powders can separate the reactive components within a mixture, mask undesirable flavors, protect unstable ingredients from degradation factors, such as heat, moisture, air, and light. It can provide controlled or delayed release and reduce hydroscopicity. It also helps in changing the physical characteristics of the original material, for example, flowability and compression improvement, dust reduction, and density modification [2]. In the food industry, enzymes, vegetal proteins, yeast, bacteria, and aroma are encapsulated in maltodextrine or Arabic gum matrix, film coating of extruded products by lipids, resins, polysaccharides, and proteins.

Note that, the emergence of various encapsulation processes is owing to the fact that no encapsulation or coating process developed to date is able to produce the full range of capsules required by different industries. The specificity of the fluid bed is that it leads to real capsules (Figure 1[a] and [b]) compared to spray drying, which leads to a matrix with the core material randomly dispersed in a polymer (Figure 1[c]).

Fluid-bed technology was quite developed during the 1950s, and was applied for various purposes in chemical industries. Its application to particle and powder coating

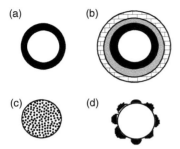

Figure 1 Different types of capsules: (a) reservoir system, (b) reservoir system with multiple layers, (c) matrix system, and (d) imperfect capsule.

Table 1 Comparison of Two Types of Fluid Beds — Batch and Continuous

	Batch (Wurster)	Continuous (horizontal)
Volume (l)	120	120
Flow rate (kg/h)	50	100
Price of the basic equipment in 1999 (€)	1,100,000	610,000
Cost of the coating operation (€/kg)	2.1	0.6
Product quality	Excellent	Passable
	True reservoir capsules (Figure 1[a] and 1[b]) Uniform batch	*Presence of capsules with incomplete layer (Figure 1[d]) heterogeneous product*

Notes: Derived from Glatt Pharmatech data. Glatt Pharmatech S.a.r.l., Parc Technologique — rue Louis Neel — 21000 Dijon. With permission.

is relatively recent and was developed to satisfy the growing demand of pharmaceutical, chemical, agrochemical, cosmetic, and food and feed industries. It is still a batch, expensive, and time-consuming process, which is mostly used in pharmaceutical and cosmetic industries that are able to compensate the cost of the process by the high price of their final product. Its application to food powder coating, which is in unfavorable competition with spray drying (a well-established technology), is actually limited to some high value products, because it is well known that one of the imperative goals of the food industry is to offer foodstuff at low prices, for example, despite the high performance of the Wurster system (see later in the chapter), its use in the food industry is problematic because the actual cost of the final coated powder, as it will be seen later (Table 1), is too high. But this tendency is changing as the technology is currently being upgraded to improve its performance, for example, the continuous fluid-bed process, as it can reduce the cost of production and appears to be an attractive alternative for food powder coating.

IV. THEORY OF FLUIDIZATION AND FLUID-BED COATING

A. Fluidization

The principle of fluidization is to maintain particles in suspension in a closed area by blowing air upward through the powder bed resting on a porous gas distributor plate (Figure 2).

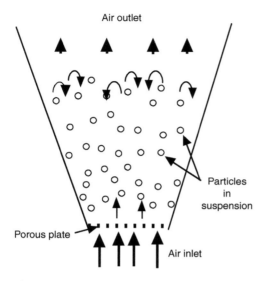

Figure 2   Principle of fluidization.

Many authors [3,4] have described different configurations as a function of air velocity. The state of the fluid bed depends on the air velocity and the powder properties and
can be globally described by two equations:

1. The minimum fluidization velocity U_{mf} (Equations [1] and [2]) is given by

$$U_{mf} = \frac{(\rho_p - \rho_g)^{0.934} g^{0.934} d_p^{1.8}}{111 \mu^{0.87} \rho_g^{0.066}} \quad dp < 100\ \mu m \tag{1}$$

where ρ_g is the gas density,
ρ_p the particle density,
d_p the particle diameter,
d_v the diameter of an equivalent sphere, and μ is the gas viscosity.

$$U_{mf} = \frac{\mu}{\rho_g d_v} \left\{ (1135.7 + 0.0408 Ar)^{1/2} - 33.7 \right\} \quad dp > 100\ \mu m \tag{2}$$

where Ar is the Archimedes number $Ar = \rho_g d_v^3 (\rho_p - \rho_g) g / \mu^2$.
2. The settling or terminal velocity $U_t \cdot U_{mf}$, that leads to a stable fluid bed
   (Equation [3]) is given by

$$U_t = \left[\frac{4 g d_p (\rho_p - \rho_g)}{3 \rho_g C_D} \right]^{0.5} \tag{3}$$

where C_D is the drag coefficient that is a function of the particle Reynolds number,
Re_p. A good correlation between C_D and Re_p for different particle shapes is
given in [5].

In short, fluidization, by its principle, appears to be a segregationist system that must
be well conducted in order to bring a minimum homogeneity. This homogeneity is the

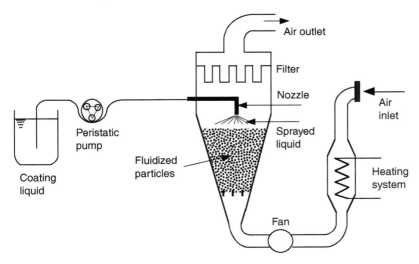

Figure 3   Principle of coating in fluid bed.

real trump that makes this unit operation applicable in various processes such as drying, granulation, agglomeration, pneumatic transport, and of course, coating.

B.   Fluid-Bed Coating

The principle of coating in a batch fluid bed is summarized in Figure 3, which presents a top spray coating system: Particles to be coated (the core) whose temperature and flow rate are variable are introduced into the cell and fluidized by an air current. The coating material is pumped to a nozzle through which the material is sprayed on the core, forming a shell.

During this process, what is expected is a homogeneous layering of the coating material on the particles leading to an onion-like structure (Figure 4[a]). Multilayer coating (Figure 4[b]) is achieved by pumping successively different coating materials. Note that the core can be coated by fine particles with the aid of some binders or plasticizers. A raspberry structure (Figure 4[c]) results from this type of coating, which is in fact a granulation process. Note that the agglomeration process is strictly defined as binding of similar particles to give larger agglomerates with a grape structure (Figure 4[d]). This presents the advantages of reducing the fines or dust and consequently enhances the flowability of the resulting granular material.

There are many phenomena during the coating operation because three phases are present: solid (particles), liquid (liquid coating materials), and gas (fluidizing air). These phenomena are classified chronologically here but most of the time many of them take place simultaneously. They are:

1.   Air suspension of particles in the coating chamber (particles dynamics).
2.   Spraying of coating material as droplets with the objective of increasing the probability of particle-droplets impacts since droplets can easily be dried (heat transfer) before collision with the particle. In this case there is no way for coating to take place.
3.   Spreading of droplets on the particle surface followed by flattening and adhesion of the droplet on the particle (mass transfer). Then, in the best case, by the

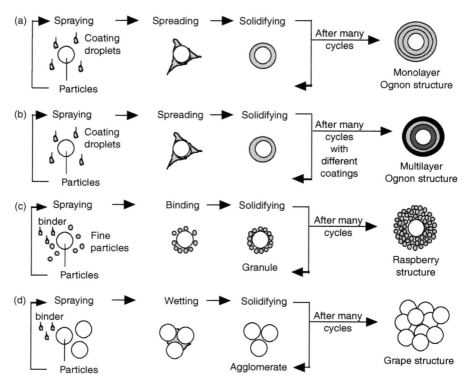

Figure 4 Mechanisms of coating formation and agglomeration in fluid beds. (a) Coating by film layering, (b) coating by film layering multilayer coating, (c) coating by granulation, and (d) as glomeration.

coalescence of droplets on the particle surface before the drying (heat transfer) of the droplets to form a layer.

4. Layering or superposition of different layers of droplets around the particle that results in a homogeneous reservoir system, that is, a real coating. After several cycles of wetting–drying, a continuous film will be formed (Figure 4), with a controlled thickness. It is mainly at this stage that the tendency of agglomeration between two or several particles is high.

Agglomeration is the dreadful phenomenon during the coating process. Indeed, after wetting of the particle surface by the coating material, there is always a competition between continuous layering of the coating material (following the wetting and drying cycle) on the dried particle on the one hand and agglomeration of wetted particles on the other.

The success of the coating operation depends on the spreading of the droplets on the particle surface. This phenomenon is a function of the wettability of particles by droplets and requires a wetting energy [6,7] that depends on the contact angle between the three phases present (solid–liquid–gas) and can be expressed as a wetting coefficient (Wm):

$$Wm = \gamma_{sv} - \gamma_{lv} - \gamma_{sl} \tag{4}$$

where γ_{sv} is the interfacial tension between solid and vapor, γ_{sl} the interfacial tension between solid and liquid, and γ_{lv} is the interfacial tension between liquid and vapor.

A liquid can wet a surface if the wetting modulus is larger than zero ($Wm > 0$).

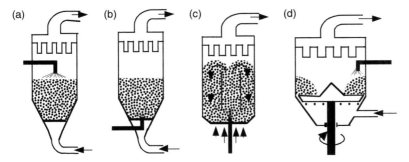

Figure 5 Different types of batch fluid beds: (a) top spray, (b) bottom spray, (c) Wurster, and (d) rotor with side spray.

V. APPLICATIONS

A. Different Types of Fluid-Bed Coaters

1. Top Spray Processing

In the 1950s the coating operation was done in the available top spray granulators (Figure 5[a]), a system where the coating material is sprayed from the top of the product container. Its efficiency in terms of deposited material (spray drying of the coating material) and coating quality (Figure 1[d]) is poor. The resulting capsules present worse controlled release kinetics. So, despite its availability, its big capacity, and the easily accessible sprayer, the top spray is not the one for coating but the system is very useful for agglomeration and granulation.

2. Bottom Spray (Wurster) Processing

This type of processing, where the coating liquid is sprayed from the bottom (Figure 5[b]), was developed by Wurster during the 1960s to significantly increase the collision probability between particles and coating droplets in order to improve the coating material efficiency and bring about a reduction in spray drying (dust reduction). The technique appeared very efficient for larger particles but presented high risk of agglomeration for smaller particles, due to the high concentration of wet particles at the bottom of the container. Wurster decided to put particles in motion using an insert. So, he invented the insert bottom spray coater, also known as Wurster (Figure 5[c]). The circulation of particles increases the drying rate and reduces the potential of agglomeration, leading to a great homogeneity in the coating quality (particles are surrounded by a smooth and continuous coating material). This apparatus is adapted for particles with various sizes and is used for moisture or oxygen barrier, for enteric or aesthetic coating with prolonged controlled release.

3. Rotor or Tangential Spray Processing

The rotor processing system (Figure 5[d]), was developed to produce coated particles with higher spherical shape and density. As is shown in Figure 5(d), the rotor reactor consists of a disk rotating in the fluidizing chamber. The combination of the rotation and the airflow provides the above specific properties. The coating film quality is similar to those obtained with a Wurster reactor. The main limitation of this design is the high agitation in the reactor that limits its application to coating materials that are not too crumbly or friable.

A comparison of different fluid-bed coating systems [8] shows that the type of process will most likely influence performance characteristics. But what must be kept in mind is that each system can be used successfully with an understanding of its advantages and limitations.

B.   Improvement of the Performance of the Fluid-Bed Coating Process

Fluid-bed coating is still an empirical unit operation, which requires, after formulation, some feasibility trials and the theory of its scale-up is yet to be established. Some major investigations have been carried out by various research groups to master this technology. These investigations are favored by the development of technology, which allowed the development of a new generation of sprayers, nozzles, air filters, and air distributor plates, with high performance. The scope of our research includes working on the development of effective criteria of performance to assess the success of any improvement approach. It appears that any coating process can be well-assessed by three efficiency criteria, which are:

- E_c, the material efficiency, generally named coating efficiency [9]
- E_e, the energetic or thermal efficiency [10]
- E_q, the quality efficiency, which definition cannot be generalized since it is related to a required or adequate property.

There are many ways to improve a given fluid-bed coating system but the best ways are based on the process itself (design and modeling followed by automation) and on the coating material.

1.   Optimization by Material Efficiency Improvement (Hot Melt and Dry Coating)

For most of the fluidized bed, the coating solution must have a low viscosity in order to be pumped. For this, the coating material is generally dissolved at a low rate (2 to 10%; this rate can be increased to 30% for dispersion in a latex or pseudolatex system) in the adequate solvent, which is not always suitable. During the coating process, a lot of energy is consumed for the evaporation of the large amount of solvent and the global material efficiency is low. The hot-melt coating can be an alternative to such situations. It can take place in a top, bottom, or rotor reactor, though the top spray process is the process most adopted for hot melt. Its specificity is such that, the coating material, melted by heating and sprayed on particles, is directly solidified by cold air rather than by drying. This confers to hot melt several important production advantages: (1) sprayed liquid = 100% coating agent, (2) short processing time, (3) no drying step required, and (4) no solvent used, that is, low cost, flexible and consistent, the recovery of the active ingredients is obtained by temperature release. Its main limitation is that it is not suitable for coating heat-sensitive (biological product) products where the dry coating system looks promising. Dry coating (which is explained in detail in another chapter) presents the same features as hot-melt coating with the difference that the product is not wetted and the coating is achieved by depositing powder on the core using a reduced volume of plasticizer.

2.   Optimization by Design and Modeling

The fluid-bed process has been permanently upgraded, through the different types as mentioned above, and the Wurster looks like a finalized batch system from the design point of view for coating. The next step is the development of a continuous fluid bed that will effectively impose coating in the food industry. But this requires the optimization of the batch process operation and a few researchers [11–15] have carried out some significant investigations for this purpose by modeling. Prior to modeling, a number of works are in progress to understand the process, characterizing by means of in-line measurements [16], the aerodynamics and hydrodynamics of air and particles in the system. All this leads to some significant models, that is, those that can be considered as closely correlated to experimental data, from the droplet characterization, the particle velocities in different areas of the fluid bed, to the heat and mass transfer. These models are not all unique and some of them may be found in another form, the tendency being to derive simple models that can be used for process control, that is, the ability to automatically launch and conduct the fluid bed just by entering data concerning particles and coating material properties.

a. Coating Droplet Size. The first important characteristic is the droplet size produced from pneumatic nozzles. It may be predicted by the following correlation (Equation [5]), [17]. Note that this equation depends on the type of spray nozzle.

$$d_g = \frac{585.10^3 \cdot \sqrt[3]{\sigma}}{V_{rel} \cdot \sqrt{\rho}} + 597 \left(\frac{\mu}{\sqrt{\sigma\rho}} \right)^{0.45} \left(\frac{1000.Q_{sol}}{Q_a} \right)^{1.5} \tag{5}$$

where σ is the fluid surface tension (N/m), ρ the fluid density (kg/m^3), μ the fluid viscosity (mPa/sec), Q_{sol} the fluid volumetric flow rate (m^3/sec), Q_a the air volumetric flow rate (m^3/sec), and V_{rel} is the relative velocity $\approx$ outlet air velocity (m/sec).

b. Evaporating Time. The evaporating time is given by Equation (6) [18] (neglecting vapor pressure and low Reynolds number).

$$t_{total} = \frac{\lambda \rho_\ell \cdot (d_0^2 - d_c^2)}{8 K_d (T_a - T_s)} + \frac{\lambda d_c^2 \rho_p (X_c - X_f)}{12 K_d \Delta T_{av}} \tag{6}$$

where d_c is the droplet diameter at critical point (m), d_0 the capillary diameter (m), K_d the thermal conductivity, ρ_ℓ the density of liquid (kg/m^3), ρ_p density of particle (kg/m^3), X_c critical moisture content (%), X_f final moisture content of the dried particle (%), T_a air temperature (°C), T_s droplet surface temperature (°C), T_{av} average temperature (°C), and λ is the latent heat of vaporization.

c. Mass Flow Rate. The flow rate (Q_{ms} [kg/sec]) is approximated by the minimum spout flow rate (Equation [7]) [19].

$$Q_{ms} = \rho_p \cdot 5.92 \cdot 10^{-5} \left[\frac{d_p}{\phi D_c} \right]^{0.05} \left[\frac{D_i}{D_c} \right]^{-2.6} \tag{7}$$

where d_p is the particle diameter (m), D_i the equivalent diameter of the bed (m), D_c the equivalent diameter of the air slot (m), and ϕ is the sphericity of the particle.

There are various other parameters, not mentioned here, whose model strongly depends on the type of fluid bed and the design, for example, particle velocities, the residence time, etc. [20].

C. Coated Powder Characterization

Among the three types of efficiency mentioned earlier, quality efficiency is probably the most difficult to define as it is related to required or adequate properties of the coated particles. These characteristics can give some information about the effectiveness of coating and its assessment and about the functional properties of the coated powder. Some useful characteristics of coated particles are described below.

The release behavior is the most important characteristic of the capsules. Capsules, which are protective barriers for various purposes (to oxygen, moisture, etc.),

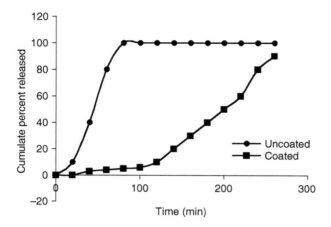

Figure 6   Released curve of a chocolate powder coated with starch.

are designed to release the core slowly (diffusion) or suddenly (shell dissolving
or breakage) by various mechanisms: heating, dissolution, mechanical or chemical rupture, shell modification by pH, etc. There are many ways to measure
the barrier properties of a shell material in accordance with the purpose of this
barrier and the trigger event. Most of the time, the coated particles are placed
in an environment whose properties are chosen to correspond with the application conditions; then the diffused core or the dissolved shell materials in the
medium are quantified as a function of time. Figure 6 presents an example of
a released curve obtained by dissolution of a coated chocolate powder in water
under gentle agitation.

The size of capsules during coating is the most measured characteristic of particles. Its
use in coating assessment requires very sensitive measurement methods as the
final layer of the coating is only about 25 μm. Using a microscope is the best
measurement method for this purpose but it is still a cumbersome technique and
it is here that the light scattering technique of size analysis can help efficiently.

The morphology of the capsule can be a quality criterion of a coating operation
when it is directly measured in parallel with the size. This is the case of the
external capsule structure. The measurement of the internal structure requires
the cutting or breaking of a number of particles to examine the inner surface
of the capsule shell. This is generally done by imbedding the capsules in an
appropriate medium that can solidify easily, then the solidified medium is cut
into slices on which some cut capsules can be seen and analyzed using optical or
scanning electron microscopy. There is always a correlation between the internal
structure and the capsules' properties that are targeted, especially stability and
release behavior.

The strength of the capsule is a mechanical property that is required to predict the
behavior of the coated food powders during their storage or, the process in which
they are involved. This parameter is crucial if the mechanical breakage is the
trigger event for release. There are a number of methods to characterize the
strength of a powder. Most of them are based on axial compression of a single
capsule or the bulk powder [21], leading to the determination of the maximum
strength that can be supported by the capsules without any damage. Others,
called friability or abrasion tests are based on higher agitation (in a mixer)

or percussion for a given time (corresponding to the handling and processing time), followed by an analysis of the agitation effect (size analysis, release test, etc.). They are all easy to apply where coarse particles are concerned.

Sensorial analysis of the capsule gives useful information about different aesthetic and sensorial properties of the coated powder such as the color, the shape, the taste, the flavor, and the roughness. The method consists in presenting different samples of the coated powder to a well-trained human panel (twelve or more persons). After testing the capsule in the desired conditions, the panel will select one or two samples that satisfy the objective. The tendency today is to develop some robots that can replace the human panel and give accurate information about those properties. This has been done with relative success since there are a number of robots that can identify or characterize individual aesthetic or sensorial properties but human intervention is still vital as the selection is based on the combination of various properties.

D. Coating Materials

It has been mentioned earlier that the role of coating is to fulfill a variety of aesthetic and protective purposes. The most commonly used coating materials in food and pharmaceutical industries are generally a reproduction of what is used in spray drying or cooling, for example, hydrocarbons, proteins, and lipids [22]. They must be in liquid form during the process in order to be pumped and for this they are dissolved or dispersed in a solvent. Their use also requires formulation and feasibility studies where the additives to be added and the optimum concentration are determined to keep the apparatus in good working order, for a good layering, and suitable final properties. Organic solvents are used in some specific cases, but water is preferred in the food industry (and more and more in the other industries) for three reasons: water is an edible, available, and low-priced solvent. It is easy to handle compared to other solvents. The new international regulations are becoming more and more severe about solvent release in the atmosphere and their concentration in food, drugs, and cosmetic products. The cheapest coating material is also preferred, as the food industry has become very competitive. The predominantly used materials to render the coating process economically feasible are:

1. For taste masking, carbohydrates (sugars [23], maltodextrins, starch, cellulose derivatives, and gums [24]), proteins (hydrolyzed gelatin), various hot melts (lipids and wax) and polymers (shellac).
2. For enteric coating, various polymers and hot melts [25,26], starch and cellulose derivatives.
3. For controlled release, cellulose derivatives, various hot melts and polymers [27,28].
4. For stability, starch and cellulose derivatives, hot melts, Arabic gum, and shellac [23,29–31].
5. For encapsulation of microorganisms for fermentation purposes, alginates and pectinates [27,28].

Note that there are various additives that are mixed to the initial formulation to homogenize (emulsifiers), and stabilize (thickeners) the mixture or to modify the rheological behavior of the coating solution (plasticizers) [32].

E. Guidelines for a Coating Operation

The role of the process manager is to create in the fluidized chamber propitious conditions for wetting and coating while keeping the whole layer in motion by flow-through gas. He can benefit from following some basic rules for a successful coating operation.

Most of the time, the engineer has to make an optimization in order to enhance the process performance. But in the case of fluid-bed coating, the efficiency of the process strongly depends in addition on the preparation phase, a step-by-step procedure that is the basis of successful coating:

- The first step is the description of the objectives of the coating operation — is it to protect the core, to mask a taste or flavor [29], to change the color or to favor a particular type of exchange, to improve some functional properties or some commercial aspects? etc. [30,33]. This step is vital for choosing the coating material and the type of capsule (matrix or reservoir system).
- The second step is to figure out how the capsules will be used or how the core will be released. This step is very important to define the shape, the particle size, size distribution, and the strength of the capsules [34].
- The formulation step is a critical step where compromises must be made with the composition of the coating solution (coating material, solvent, and additives such as plasticizer, stabilizers, texturizer, emulsifier, binders, flow enhancers, etc.) with regard to steps 1 and 2, the legislation, and the process operation. Preliminary tests of wettability and adhesion are sometime necessary to finalize this stage.

Thus, the relation between the objectives, the nature and the properties of particles, and the coating solution after formulation are determinants of the type of coating apparatus that will be chosen among a large variety of fluid beds. The choice of continuous fluid beds for food powder coating is indispensable.

VI. CONTINUOUS FLUID BED FOR FOOD POWDER COATING

The use of the continuous process in the food industry is widespread for coarse food material coating such as nuts, dried fruits, corn flakes, puffed grain, and so on. Conveyor belt processes and coating–blending drums are used for this purpose. The necessity to coat fine food particles continuously was pointed out in the 1970s with the emergence of the widespread use of spray drying and its derivatives: spray chilling, spray cooling, and prilling. These types of processes produce the matrix system (Figure 1[c]) with poor coating quality where parts of the core material are exposed to air and are not protected. They are limited to small particle size ($<100~\mu$m) [22].

For higher coating quality of fine particles, the continuous fluid-bed process appears to be an attractive alternative for food powder coating with low cost production. Indeed, a comparison of the characteristics of a batch fluid bed (Wurster) and a horizontal continuous one (Table 1) shows that a continuous process is the only alternative to allow application (low price) of fluid-bed coating for food powders. But, as it is mentioned in Table 1, the quality and the homogeneity of the final product is still much below par from the ones obtained with a Wurster system. A number of continuous fluid beds have been developed with more or less success.

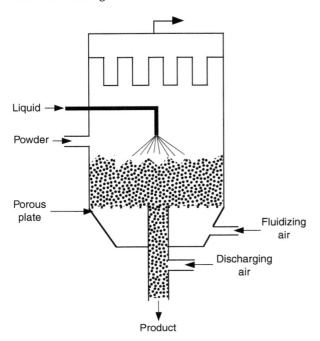

Figure 7 Monocell continuous fluid bed (top spray).

A. Single Bed Continuous Fluid Bed

This type of fluid bed operates along the same lines as the batch fluid-bed granulator, the difference being that particles with a predefined size are automatically discharged via the discharge pipe located in the center of the processing chamber (Figure 7). The other advantages are high output, compact design, predeterminable grain size, dust free and narrow particle size distribution, and low energy consumption. The main limitation of this method is that it is very difficult or impossible to meet contradictory specifications for agglomeration and solid bridges (layering) in a single cell.

B. Horizontal Continuous Fluid Bed

The real horizontal fluid bed (Figure 8) refers to a fluid bed with no obstacles in the system. Powder is admitted in the fluidizing chamber at one end from which it moves slowly during the desired process until it gets out at the other end. These types of fluid beds exist and are actually used in the food industry for various purposes. They are in general vibrating fluid beds with conveying belt and have been designed with a reasonable length since the 1970s. They were especially developed for granulation (instantizing) of powders after spray drying.

The main disadvantage of the real horizontal fluids bed is their length (several meters), which is necessary to allow the required residence time for the process. This disadvantage is worsened when the process involved is coating. A detailed analysis of the advantages and disadvantages of a horizontal fluid bed was done by Teunou and Poncelet [35], including comparison with the Wurster system and characteristics of an ideal fluid-bed coater.

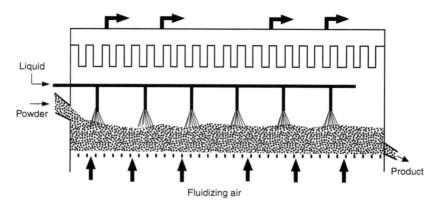

Figure 8   Horizontal continuous fluid bed (top spray).

C.   Multicell Continuous Fluid Beds for Coating

Increasing the residence time in the fluid-bed chamber using fluid-bed properties [36] can reduce the length of the horizontal fluid bed. Rümpler made some progress [37] by developing a multicell fluid-bed coater, which is a horizontal top spray fluid bed divided into four or five compartments connected by regulating flaps. This apparatus leads to imperfect capsules where the coating material does not recover the total surface of core material (Figure 1[d]). However, these capsules are better than those obtained by spray drying and are adequate for physical protection (e.g., taste masking).

Another type of multicell fluid bed (presented by Leuenberger [38]) consists of several minibatches connected to each other with a transport, dosage, and classification system. This apparatus was promoted by Glatt [39], in order to combine advantages of batch and continuous process in the pharmaceutical industry. There is no published information about the quality of coating obtained by this system but it can be noted that this is an expensive installation.

VII.   CONCLUSION

There is real demand for food powder coating to fulfill various purposes and these objectives are actually reached by spray drying or granulation with, in many cases, unsatisfactory results. This may explain why the actual volume of coated food powder is still relatively small compared to the huge potential that exists. Different types of fluid-bed coating have been proved not only to bring an adequate answer to these coating objectives but also to be too expensive techniques that cannot meet the main requirement of foodstuff, that is, low prices. The challenge today is to focus on further equipment evolution (instrumentation and control) that may result in the improvement of the continuous fluid-bed coating and its application to food powder coating.

REFERENCES

1.   Arshady, R. Microcapsulation for food. *J. Microencapsulation* 4: 413–435, 1993.
2.   Dezarn, T.J. Food ingredient encapsulation. Ed. Plach SJ. ACS Symposium Series-Washington, DC pp. 74–86, 1995.

3. Geldart, D., Ed. Single particles, fixed and quiescent beds. In: *Gas Fluidisation Technology*. Chichester: John Wiley & Sons, 1987.

4. Jones, D.M. Air suspension coating for multiparticulates. *Drug Develop. Ind. Pharm.* 20: 3175–3206, 1994.

5. Geankoplis, C.J. *Transport Processes and Unit Operations*. 3rd ed. New Jersey: Prentice Hall Inc., 1993.

6. Briant, J. *Phénomènes d'interface — agent de surface — principe et mode d'action*. Paris: Technip, 1989.

7. Yvon, J., Thomas, F., Villieras, F., and Michot, L.J. Surface-activity of water. In: *Handbook of Powder Technology. Vol. 9: Powder Technology and Pharmaceutical Processes*. Eds. Chulia D., Deleuil M., and Pourcelot Y. Netherlands: Elsevier, 1994.

8. Jones, D.M. Considerations for various fluid bed techniques. Proceedings of Technological training center on current practices in fluid bed drying granulating and particle coating technology. Basel, 2001, Sec. 4.

9. Dewettinck, K. and Huyghebaert, A. Top-spray fluidised bed coating: effect of process variables on coating efficiency. *Food Sci. Technol. lwt* 31: 568–575, 1998.

10. Filkova, I. and Munjundar, A.S. Industrial spray drying systems. *Handbook of Industrial Drying*. Vol I. In: Ed. A.S. Mujumdar. New York: Marcel Dekker, 1995, pp. 263–307.

11. Alden, M., Torkington, P., and Strutt, A.C.R. Control and instrumentation of a fluidised-bed drier using the temperarture-difference technique. I. Development of a working model. *Powder Technol.* 54: 15–25, 1987.

12. Diego, L.F., Gayan, P., and Adanez, J. Modelling of the flow structure in circulating fluidised beds. *Powder Technol.* 85: 19–27, 1995.

13. Dewettinck, K. and Huyghebaert, A. Fluidised bed coating in food technology. *Trends Food Sci. Technol.* 10: 163–168, 1999.

14. Fyhr, C. and Kemp, C. Mathematical modeling of batch and continuous well-mixed fluidized bed dryers. *Chem. Eng. Process.* 38: 11–18, 1999.

15. Kusharski, J. and Kmiec, A. Hydrodynamics: heat and mass transfer during coating of tablets in a spouted bed. *Can. J. Chem. Eng.* 61: 435–439, 1993.

16. Masters, K. *Spray Drying*. London: Leonard Hill Books, 1979, p. 702.

17. Masters, K. Drying of droplets/sprays. In: *Spray Drying Handbook*. New York: John Wiley & Sons, 1988, pp. 298–342.

18. Mörl, L. and Drechsler, J. Using Lasentec FNRM in-process particle characterisation in fluid bed application – Preparing for Process Control. *Lasentec Technical Abstract*, M-2-007 (http://www.lasentec.com/M-2-007_abstract.html).

19. Rocha, S.C.S., Taranto, O.P., and Ayub, G.E. Aerodynamics and heat transfer during coating of tablets in two-dimensional spouted bed. *Can. J. Chem. Eng.* 73: 308–312, 1995.

20. Valenti, C. Etude d'un rocédé d'enrobage en lit fluidisé: le procédé Wurster. Thèse. Nancy: Université Henri Poincaré- I, 1998.

21. Shubert, H. Food particle technology. part II. Properties of particles and particulate food systems. *J. Food Eng.* 6: 1–32, 1987.

22. Jacson, L.S. and Lee, K. Microencapsulation in food industry. *Lebensm.-Wiss. U. technol.* 24: 289–297, 1991.

23. Buckton, G. and Darcy, P. The influence of additives on the recrystallisation of amorphous spray dried lactose. *Int. J. Pharm.* 121: 81–87, 1994.

24. Thevenet, F. Acacia gums stabilizers for flavor encapsulation. *Am. Chem. Soc.* 590: 37–44, 1988.

25. Baldwin, E.A., Nisperos, M.O., Hagenmaier, R.D., and Baker, R.A. Use of lipid in coatings for food product. *Food Technol.* 54(6): 56–64, 1997.

26. Dian, N.L.H.M., Sudin, N., and Yusoff, M.S.A. Characteristics of microencapsulated palm-based oil as affected by type of wall material. *J. Sci. Food Agric.* 70: 422–426, 1996.

27. Gareth Leach, G. Production of a carotenoid-rich product by alginate entrapment and fluid bed drying of dumalielle salina. *J. Sci. Food Agric.* 76: 298–302, 1998.

28. Vilstrup, P. *Microencapsulation of Food Ingredients*. 1st ed. Surrey: Leatherhead, 2001.
29. Brake, N.C. and Fennema, O.R. Edible coatings to inhibit lipid migration in confectionary product. *J. Food Sci.* 58(6): 1422–1425, 1993.
30. Guilbert, S. and Gontard, N. Prolongation of the shelf-life of perishable food product using biodegradable films and coatings. *Lebensm-Wiss. U. Technol.* 29: 10–17, 1994.
31. Keogh, M.K. and O'Kennedy, B.T. Milk fat microencapsulation using whey proteins. *Int. Dairy J.* 9: 657–663, 1999.
32. Rizzotti, R. Les agents de texture épaississants. gélifiants. *Stabilisants.* IAA 563–573, 1994.
33. Janovsky, C. Encapsulated ingredients for baking industry. *Cereal food World* 38: 85–87, 1993.
34. Wan, L.S.C., Heng, P.W.S., and Chia, C.G.H. Spray drying process for microencapsulation and the effect of different coating polymers. *Drug Develop. Ind. Pharm.* 18: 977–1011, 1992.
35. Teunou, E. and Poncelet, D. Batch and continuous fluid bed coating: review and state of the art. *J. Food Eng.* 53: 325–340, 2002.
36. Dumon, R. *Les Applications Industrielles des Lits Fluidisés*. Paris: Masson, 1981.
37. Rumpler, K. Continuous fluid bed lipid coating of particles. Proceedings of the 9th International Glatt Symposium, Weimar, 1999.
38. Leuenberger, H. Pseudo-continuous fluid bed processing: The Glatt multi cell concept. Proceedings of the 10th International Glatt Symposium on Process Technology, Prague, 2000.
39. Glatt International (2002). Spray granulation dryers (AGT system) in the ceramic industry. Technical information (http://www.glatt.de).

Powder Handling and Analysis

9
Particle Size Analysis of Food Powders

Patrick O'Hagan, Kerry Hasapidis, and Amanda Coder
Particle Sizing Systems
New Port Richey, Florida

Heather Helsing and Greg Pokrajac
Particle Sizing Systems
Langhorne, Pennsylvania

CONTENTS

I. SUMMARY

Various food components are increasingly being utilized in powder form, both for processing as well as end use. This allows for reduced shipping costs, increased stability, and ease of use for processing. One of the most important physical parameters of powders with regard to handling them is particle size. Particle size can influence flow out of storage bins, the blending of different components, compaction, and the segregation of a mixture. Detailed information about particle size can determine how to design process or storage equipment, which raw materials to use, how to mix components together, and how long they can sit without separation or caking. Not only is particle size important to powder handling, but it also significantly influences important properties essential to food products in general: taste, smell, texture, and appearance. For example, many food products are processed into emulsions, like spreads and beverages. Particle size can be used to determine the stability of these emulsions. Foams, an important characteristic of beer and coffee drinks, are air particles (bubbles) in a liquid matrix. The particle size of the bubbles will determine how the foam forms and how long it lasts. Particle size can influence the dissolution rates of powdered soups and in turn influence the taste of such reconstituted products. Particle sizing is even used to monitor process end points (grinding, homogenization, or other size reduction processes) and to quality control final products.

So with its major influence on the properties of powders, final food products, and food processing, it is important, then, for food scientists and engineers to have an understanding of how particle sizing is done, what techniques are commonly available, what particle sizers actually measure, and how the results are communicated and interpreted.

II. BASIC PARTICLE SIZING CONCEPTS

A. Spherical Equivalent Diameter

While the idea of particle size seems simple enough, especially for spheres, consider any one of the particles depicted in Figure 1. They could be particles of starch, spice, or other food materials. It should be clear that real particles are not spheres at all but objects with very complex shapes. What are the sizes of these particles? A full description of these particles will entail knowledge of the identity and arrangement of every atom that makes up the particle. Such information cannot be obtained easily and would be cumbersome to use in any case. Add to that the almost infinite variability that will exist over the billions of particles that make up the powder. Thus, the field of particle technology has created many concepts over the years to describe the general shape of an ensemble of particles. Such concepts have been detailed elsewhere [1] but include various shape and size definitions. One of the most important particle technology concepts is that of the spherical equivalent diameter. A spherical equivalent diameter is used to describe the complex shape of a particle with a single number. This number is the diameter of a sphere (diameter is all that is needed to completely describe a sphere) that has the same physical property that the actual particle does. By physical property, it is meant a property that is related in some degree to particle size. For example, consider a real particle of food material that is water insoluble and has a specific gravity greater than one (it is denser than water). If it is dropped into a container of water, it will sink to the bottom or more appropriately, it will sediment. The speed at which this particle sediments or sinks is related to its size and shape. We know for certain that there will be a sphere of similar composition, which will sediment at the same rate. So rather than try to describe the complex shape of the particle with many parameters, we can describe it by saying that it sediments like a sphere of diameter, d_s. Thus, we can say that the spherical equivalent diameter of the particle is d_s. This concept can be broadened to many physical properties as can be seen in Figure 2.

Each physical property of the particle can be ascribed a spherical equivalent diameter. And as this figure indicates, these diameters will not be the same. They are differentiated with names that describe the physical property they relate to. Thus, the particle can be said to have a hydrodynamic diameter of 0.5 μm (it diffuses like a spherical particle of that size)

Figure 1 Depictions of "real" particles.

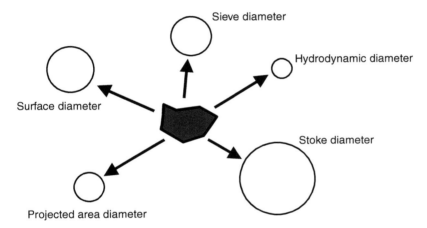

Figure 2 Different spherical equivalent diameters for the same particle.

or a Stokes diameter of 0.7 μm (it sediments like a spherical particle of that size), and so on. Taking the concept of spherical diameter further, if we measure the physical property distribution of an ensemble of particles, for example, the range of sedimentation speeds, then we can use this information to determine the size of the particles that are present. In other words, we can measure the particle size distribution. So, we have a particle size analyzer even though we are not measuring particle size directly. The concept of spherical equivalent diameter thus allowed the development of a range of particle size analyzers that measure the physical properties of a group of particles and relate that to particle size. Some of these techniques will be discussed in detail in a later section.

B. Particle Size Distributions

Particulate systems of interest are made up of billions of particles. This massive amount of particle size information is most often portrayed as a particle size distribution (PSD) with a y-axis in terms of weighted frequency and the x-axis in terms of spherical equivalent diameter. To further simplify this information, PSDs are often reduced to various statistical parameters: means, variances, and ranges. PSDs are commonly displayed in two forms: differential and cumulative. The cumulative distribution is the mathematical integral of the differential distribution. Figure 3 displays both the differential and cumulative forms of the same particle size information.

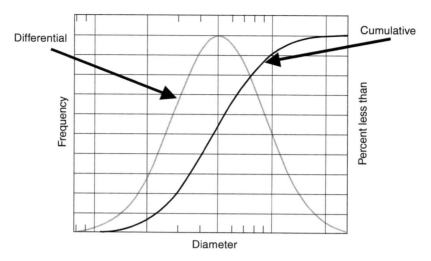

Figure 3 Differential and cumulative particle size distribution curves.

Table 1 Cumulative Particle Size Results from a Homogenization

Pass	Particle Concentration > 0.5 μm (p/ml)	Volume fraction %
First	2.38×10^{10}	1.30
Second	3.52×10^{9}	0.22
Third	1.63×10^{9}	0.19
Fourth	1.48×10^{9}	0.58

Various particle-sizing techniques produce raw data in one form or the other. So a measurement that produces a differential PSD must be transformed mathematically to a cumulative distribution if that is what is desired. Particle-sizing information from sieves is differential in raw form but is most often used in cumulative form, which is derived from the raw differential data. It is usually easier to compare PSD information of different samples when the PSD has been reduced to one or two statistical parameters. For example, the particle concentration greater than a certain size can be used to monitor a size-reduction process like homogenization. This often provides clearer information than directly comparing the entire PSD. Consider the data displayed in Table 1.

From this table, we can clearly see that the concentration of particles greater than 0.5 μm decreases with subsequent passes through a homogenizer. Thus, one cumulative parameter is used to provide important information about the homogenization efficiency. A significant reduction in the particle concentration occurs in the first three passes. The fourth pass does not produce a significant reduction and might be considered as adding unnecessary cost to the production of this emulsion.

For differential distributions, it is common to use means, modes, and medians as statistical parameters while for cumulative distributions, D numbers, like D90 (90% less than) or D50 (50% less than) are used (see Figure 4). Another important aspect of particle size distributions is *weightings*. With regard to particle size distributions, a weighting is the

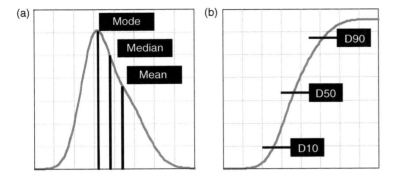

Figure 4   Common statistical parameters used for (a) differential and (b) cumulative distributions.

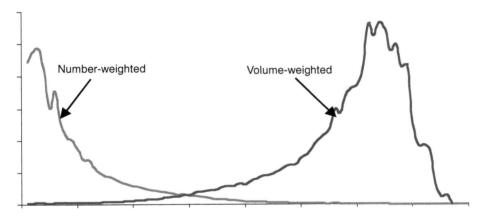

Figure 5   Number-weighted distribution versus volume-weighted distribution.

power to which the diameter term is raised. To illustrate this, consider a PSD expressed in discrete form as follows:

$$f(d) = \Sigma n_i d_i^a \tag{1}$$

where a is the weighting, n_i is the number of particles of size d_i and $d_1, d_2, \ldots, d_i$ represent all the particle sizes present. When diameter is raised to the zero power ($a = 0$), the distribution is said to be number-weighted since the y-axis depends only on the number particles in each class. If diameter is raised to the first power ($a = 1$), the distribution is length-weighted. If the diameter is raised to the third power ($a = 3$), the distribution is volume-weighted. In Figure 5, the same particle size data is expressed as both a number-weighted and a volume-weighted differential distribution. Clearly, the volume-weighted PSD emphasizes the large particles while the number-weighted PSD emphasizes the fines. Depending on the particulars of the application, one or the other weighting might be more appropriate. It should be clear that the statistical parameters used to describe these two distributions would be different. The mean diameter of the number-weighted PSD in Figure 5 is 2.2 μm while the mean diameter of the volume-weighted PSD is 26 μm. Both are equivalent ways of describing the same particle size data. So when evaluating particle size information, for example, and a supplier of a raw material specifies a product by its mean particle size, it is important to consider the weighting of the distribution from which the mean was derived. Furthermore, just as various particle-sizing technologies can produce raw data in differential or cumulative form, so also can this raw data be in different weightings. So image analysis produces raw data that is number-weighted while laser diffraction produces raw data that is volume-weighted.

III. PARTICLE SIZING METHODS

In this section, an overview of the most common particle-sizing technologies will be given. There are literally dozens of technologies that are used to measure particle size and PSDs [2]. Table 2 contains a brief list of particle-sizing techniques in use. Most do not measure particle size directly but measure a physical or optical property related to particle size. As stated in the previous section, each technology produces raw data that is weighted in a certain way. Some measure only particle size parameters whereas most provide information about the entire distribution. These distributions might be cumulative or differential. Some techniques only measure powders (dry dispersal), some only dispersed particles, and others can measure both. It is important to appreciate that there are so many different techniques in use for measuring particle size because there are so many applications. Some techniques provide certain types of information (e.g., particle counts or molecular weights) while others handle certain materials best. So the suitability of a certain technology must be evaluated with regard to the material to be tested as well as the information required. In light of this, the following discussion will emphasize capabilities such as resolution, sensitivity, and dynamic range of each technology that can impact the usefulness of the obtained information. The type of materials that can be measured and whether the technique can be used in the lab or online will also be discussed.

A.   Light Scattering Methods

Light scattering methods are those that derive particle size information from the way that light interacts with a dispersion of particles. There are two main techniques: static light scattering (also known as laser diffraction) and dynamic light scattering. They are also known as ensemble methods because the interaction of light is derived from millions of particles present simultaneously in the volume illuminated by the light source. So the signal that reaches the detector is a supposition of scattered light rays from many particles. Both are in common use and have large dynamic ranges. As a matter of fact, outside of sieves and image analysis, light scattering instruments are the most often found in laboratories. This is because they can test a wide range of materials and are easy to use. They provide accurate distribution information but not with high resolution or high sensitivity due to the ensemble nature of the signal. And the distributions are not quantitative, in the sense that concentration information is not derived from them.

Laser diffraction analyzers consist of a hemispherical array of photo detectors arranged around a light interaction cell. Light, most commonly from a laser, is passed through a dispersion of particles in the interaction cell. Light is scattered in all directions by the particles. The detector array measures the intensity of light at all angles relative to the incident beam. The resulting data, a scattering pattern, is analyzed by the application of Fraunhofer and Mei scattering theories. This is seen in Figure 6.

Laser diffraction can be used to perform both wet and dry analyses and because of the simple optical arrangement, it can be used in the laboratory as well as for online monitoring. It is easy to operate within the sample preparation limits of the materials to be tested and accurate PSDs can be obtained in less than 1 min. The dynamic range is large if both Mei and Fraunhofer theories are used, usually from 0.1 μm to several thousand microns; however,

Table 2   Partial List of Common Particle Sizing methods

Technique	Physical property	Attributes	Size range μm
Single particle optical sensing	Light obscuration	Wet or dry Differential Lab or online High resolution, counts	0.5–2500
Electrozone	Volume exclusion	Wet only Differential Lab only High resolution, counts	0.5–1200
Laser diffraction	Light scattering	Wet or dry Differential Lab or online Wide dynamic range	0.1–2000
Dynamic light scattering	Brownian motion	Wet only Differential Lab or online Nanoparticle sizing	0.001–5
Optical microscopy	Optical contrast	Wet or dry Differential Lab or online Particle shape	1–200
Sieves	Particle size	Wet or dry Differential Lab only Mass-weighted PSDs; large sample size	20–5000
Acoustic spectroscopy	Acoustics	Wet only Differential Lab or online High concentrations	0.01–10
Field flow fractionation	Brownian motion/size exclusion	Wet only Differential Lab only Direct particle size	0.01–50
Centifugal sedimentation	Sedimentation	Wet only Differential or cumulative Lab only High resolution	0.01–30
Gravity sedimentation	Sedimentation	Wet only Differential or cumulative Lab only Large sample sizes	0.5–100
Capillary hydrodynamic fractionation	Size exclusion	Wet only Differential Lab only High resolution	0.01–1
Time of flight	Time of flight	Dry only Differential Lab only Sizes fine powders	0.2–200

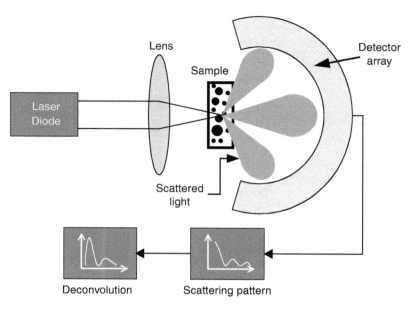

Figure 6   Schematic of a laser diffraction particle size analyzer.

the ability to measure the large particles requires a more spread out optical arrangement and such an instrument will have a larger footprint. In general, laser diffraction instruments are based on first principles and thus do not need to be calibrated. Once the experimental parameters, like the wavelength of the incident light and the position of the detectors, are set, the size is derived directly from the theory.

While laser diffraction instruments can provide complete distribution information, there are some significant limitations. The resolution, the degree to which differences between features in a distribution can be detected, is low. This can be seen in the distributions in Figure 7.

A wet dispersion of a ground pharmaceutical powder was tested on a laser diffraction instrument (circles) and a method known for high-resolution (single particle optical sensing, SPOS) (squares) [3]. Clearly, the width of the distribution measured by laser diffraction was wider and is evidence that its resolution is much lower than that from the SPOS instrument. That this difference can be attributed to the resolution capabilities of the two instruments is supported by the fact that the single particle counter agreed with sieve measurements made on the dry powder with regard to the absence of particles greater than 40 μm. The low resolution of laser diffraction is a consequence of the ensemble nature of the technology. Laser diffraction instruments also have low sensitivity. Sensitivity is the smallest detectable concentration of material. This capability is important since there are many applications, for example, the monitoring of filtering or grinding processes that require the ability to detect the absence of particles down to a low concentration (parts per million or billion). Figure 8 contains a perturbation analysis [4] of data [5] obtained from spiking an emulsion with small amounts of large particles to simulate the growth of oversized particles as an emulsion destabilizes.

As the graph indicates, the detection limit for oversized particles greater than 5 μm in an emulsion for laser diffraction is anywhere from a 0.7 to 7% volume fraction while for the SPOS technique it is 0.005%, more than 100 times better than the best laser diffraction

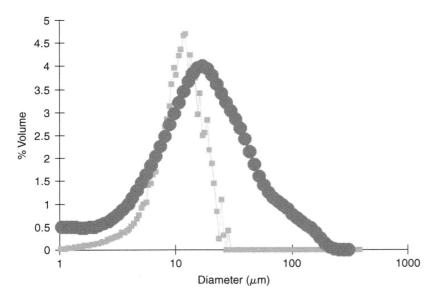

Figure 7   Comparison of particle size data of pharmaceutical powder obtained from SPOS (squares) and laser diffraction (circles) techniques.

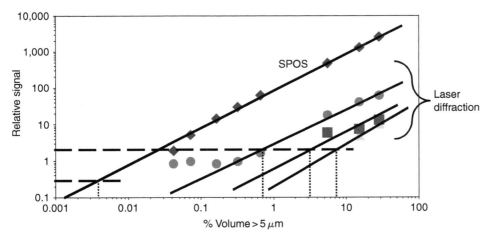

Figure 8   Perturbation analysis of emulsion particle size data comparing the sensitivity of SPOS and laser diffraction.

unit. This clearly indicates that laser diffraction is not suitable when the application requires the detection of small concentrations of particles.

The other main type of light scattering method is dynamic light scattering (DLS), also known as photon correlation spectroscopy (PCS). The experimental setup is shown in Figure 9. Like laser diffraction, DLS particle sizers scatter light (also from a laser) off a dispersion of particles. Instead of an array of detectors, a DLS analyzer captures light from one angle only. At any one angle, the scattered light fluctuates with time around an average value. The frequency of these fluctuations is related to the diffusional properties of the suspended particles. The diffusion of small particles suspended in a liquid is also

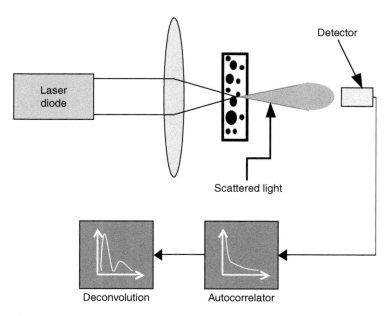

Figure 9   Schematic of dynamic light scattering particle size Analyser.

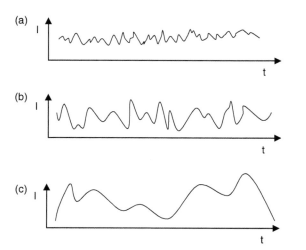

Figure 10   Intensity versus time profiles for (a) 1–100 nm particles (b) 100–500 nm particles (c) 500–1000 nm particles.

called Brownian motion and is caused by molecules of the suspended medium colliding with the particles. Brownian motion is random and is related to particle size. Small particles diffuse faster than larger ones. The relationship is described by the Stokes–Einstein equation seen below:

$$D = kT/6\pi\eta R \tag{2}$$

where D is the diffusional coefficient, R is the particle radius, k is the Boltzmann's constant, η is the viscosity of the diluent, and T is the temperature in kelvins. In Figure 10, a series of

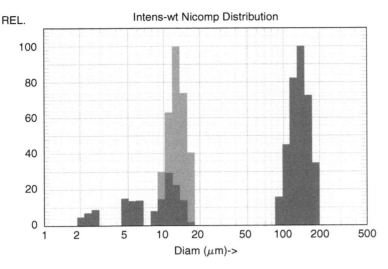

Figure 11   Dynamic light scattering results from two different proteins.

light intensity time profiles for different PSDs are shown. The profiles with high frequency fluctuations represent suspended particles with a small average particle size while the profiles with low frequency fluctuations represent particles with a higher average particle size. There are two main approaches for obtaining the diffusion information from the intensity time profiles measured by the detector. The most common way is to use an autocorrelator to obtain time constants. It is also possible to use Fourier transforms to obtain the frequency components but this involves the mixing of two light beams (the scattered light with a reference beam), which is difficult and makes the measurement vibration sensitive. The monotonically delaying function (referred to as a correlation function) obtained from the autocorrelator is mathematically deconvoluted to produce a PSD.

The main advantage that DLS has is its ability to measure the PSD of particles that are well below 100 nm in size, about the lower range limit of laser diffraction. With sufficient scattering intensity, particles as small as 1 nm can be sized. Proteins, micelles, and other macromolecules are in this range. An example of a measured size distribution of small protein particles is shown in Figure 11. The size of the native protein particles is about 4 nm. Other advantages of DLS are that it is based on first principles and is easy to use. Unlike laser diffraction, DLS does not require prior knowledge of the optical properties of the scattering particles to obtain an accurate PSD. The technique can be applied to online monitoring as well. Unfortunately, because DLS is based on the measurement of Brownian motion, it can only be used for particles suspended in liquids. This also restricts the technique to application on primarily submicron distributions as above-micron particles tend to sediment. Like laser diffraction, DLS is not a high resolution or high sensitivity technique. So both light scattering techniques are able to measure a general simplified representation of the real PSD, which can provide accurate mean diameters and a measure of polydispersity.

B.   Fractionation Methods

Fractionation techniques of particle size use some physical method to divide a dispersion of particles into "fractions" based on size. A detector is used to determine the relative amount of material in each of these fractions. These techniques produce, in general, PSDs that are of higher resolution than the light scattering methods. Together with this, there is a fractionation technique that covers every size range from the deep submicron to the hundreds of microns in size, but each specific technique itself has a limited dynamic range.

The most common fractionation techniques are based on sedimentation. Sedimentation is the process by which suspended particles respond to the force of gravity by separating from the suspending media. The velocity by which they do this is in part influenced by particle size. In general, the larger sized particles separate faster than the smaller ones. This difference in rates means that the particles not only separate from the suspending media but also from each other. Thus, by the time they approach the detection zone, the original mixture of particle sizes is separated into layers or fractions of differently sized particles. Thus, the first particles to enter the detection zone are the large particles followed by the smaller ones. Once the sedimentation velocity of the particles is measured, the particle size of each fraction is determined by the application of Stokes' law:

$$V = \Delta\rho D^2 g / 18\eta \tag{3}$$

where V is the sedimentation velocity, $\Delta\rho$ is the difference in density between the particle and diluent, D is the Stokes' diameter, g is the acceleration due to gravity, and η the liquid viscosity of the diluent. A gravity sedimentation instrument usually consists of a long tube or column containing a suitable liquid.

There is a light source (optical or x-ray) and detector situated some distance down from the top of the column of liquid. The light source illuminates a band in the column. The sample is introduced into the top. The sufficiently dense particles begin to sediment down the column. The longer the distance traveled, the better the separation into fractions.

Each fraction passes through the illuminated band of light. The detector measures the obscuration of the light as each fraction passes through. This can be seen in Figure 12. The main advantages of sedimentation techniques based on gravity is that they are of higher resolution than light scattering methods. Like light scattering methods, they are also based on first principles. Gravity sedimentation is easy to use and relatively large amounts of material can be tested. The particles must be, however, of known density and must be suspended in a liquid. The density must be uniform or else errors will occur and sufficiently large or else the measurement time will be too long. Optical corrections might have to be

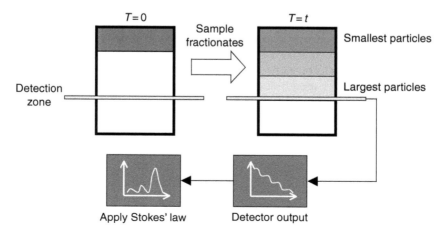

Figure 12   Schematic of gravity sedimentation particle size analyzer.

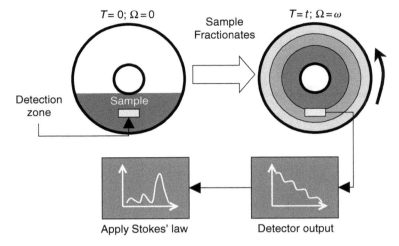

Figure 13 Schematic of centrifugal sedimentation particle size analyzer.

made for some materials. This technique has never been used for online monitoring. As it stands, gravity sedimentation can only be applied to particles larger than 1 μm.

However, a variation of this technique is available for obtaining the PSD of much smaller particles. The main problem posed by smaller particles is the long sedimentation times. This can be overcome by the application of a centrifugal field. Such a field will simulate higher gravitational forces, speeding up very slowly moving particles. Such devices consist of a cylindrical container of liquid, which can be spun at high rates. As in gravity sedimentation, there is a light source and detector situated some distance away from the axis of rotation. Sample is introduced through an opening near the axis. As the container with the sample is spun, particles will move away from the axis toward the outer circumference. As they do this, they will pass through the detection zone. This can be seen in Figure 13. Disk centrifugation, as the technique is known, can produce very high-resolution submicron PSDs down to 10 nm. It is based on first principles like the gravity method (Stokes' equation) and requires knowledge of the particle density, which must be sufficiently larger than that of the diluent and uniform. It does not do well with broad distributions as at the rotational speeds necessary to get small particles to sediment, the larger particles will fall out too quickly to be properly separated into fractions.

Another fractionation method, which can produce high-resolution submicron distributions, is capillary hydrodynamic fractionation or CHDF. CHDF makes use of the parabolic flow properties of a liquid forced through a capillary tube.

This is illustrated in Figure 14. This instrument consists of a long capillary tube. An injection valve at the beginning of the tube introduces sample and a detector at the end will measure the relative amounts of each size fraction that exits. The fractionation occurs due to the parabolic flow properties of the liquid as it flows through the capillary. Particles near the edge of the capillary tube travel at speeds slower than those in the middle of the tube where the liquid velocity is the fastest. Large particles are not able to approach the sides of the capillary due to their size and thus have the fastest velocity. This means that they are the first to exit the capillary tube. The elution time must be calibrated using particles of known size. As stated earlier, this technique can provide high-resolution PSDs in the submicron regime but cannot size particles much larger than a few microns as they tend to clog the capillary tube. The PSD is directly obtained meaning that no mathematical algorithms or

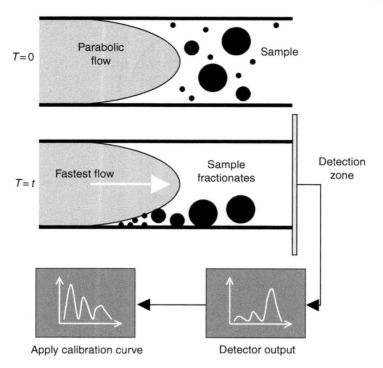

Figure 14 Schematic of capillary hydrodynamic fractionation particle size analyzer.

other conversions are required — the PSD comes directly from the detector response. The sample requirements are limited since the eluant (the mobile phase) is a special mixture (in order to obtain a useful parabolic flow profile) and the particles must be compatible with it. This technique has never been used online. Like the sedimentation techniques discussed earlier, the measurement time is much longer than the light scattering methods.

Field-Flow Fractionation (FFF) is another technique that makes use of the parabolic flow properties of a liquid flowing through a tube. This technique has many variants by which different "fields," applied perpendicular to flow, are used to fractionate the particles. These fields will primarily fractionate based on size but also on other properties [6], such as mass, surface potential, or chemical affinity. This gives the technique a great deal of flexibility in terms of the type of materials tested, the size range, the measurement time, and the resolution achieved [7]. FFF is arranged much like CHDF except that instead of a capillary tube, the flow path consists of a short rectangular channel with high aspect ratio [7]. The exact construction of this channel will depend on what type of force field is applied perpendicular to flow. Usually, the sample is introduced into the liquid at one end of the rectangular channel. The sample is carried into the channel and then the flow is stopped. The perpendicular force is applied, fractionating the sample by particle size or another property, and then the flow is restarted. As with CHDF, the particle fraction in the middle of the channel is forced out first because the liquid velocity is fastest in the middle. Depending on the specific force, either the smallest or largest particles can exit first; it all depends on how the field fractionates the particles.

This is illustrated in Figure 15. Consider an FFF device utilizing a centrifugal cross-force (known as SFFF — Sedimentation Field-Flow Fractionation). The centrifugal force will fractionate the particles in such a way that the large particles are closest to the channel

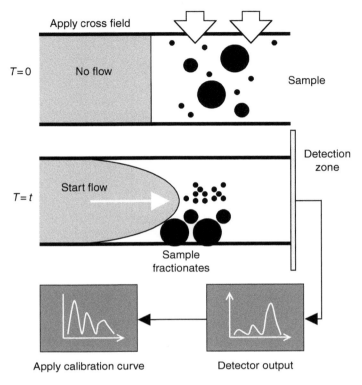

Figure 15 Schematic of field flow fractionation particle size analyzer.

wall (they sediment the fastest) and the smaller particles (which sediment slower) are in the middle of the channel. In this arrangement, the smaller particles exit first, followed by the large ones. When using another liquid flow such as the cross-force (SFFF), all the particles are forced to the channel wall. The particles will then diffuse away. The particles with the largest diffusion coefficient (generally, the smallest) will be farther from the wall and closer to the part of the channel with the fastest flow. These particles will elute first. FFF's flexibility means that it can size particles over many different size ranges. SFFF can size particles in the 10 nm to 1 μm range [8], but knowledge of the sample density is required. FFF can size particles 10 nm to 50 μm. Since the elution time is related to the diffusion coefficient only, the PSD is directly obtained [9] from the application of the Stokes–Einstein equation (eq. 1). The same is true for SFFF. Like CHDF, FFF techniques generally have longer run times and are for wet analysis only. They have never been applied to online monitoring.

C. Single Particle Methods

Single particle sizing techniques involve the sizing and counting of particles, one at a time. These techniques provide excellent resolution and sensitivity in the above-micron range. Because of these capabilities, they provide information and details of distributions that light scattering methods cannot. They can achieve this, in general, because they do not rely on fitting algorithms or ill-defined mathematical transformations. This additional detailed information can be useful in the determination of grinding, homogenization, or other size-reduction process endpoints [10]. These techniques are not limited only to testing materials

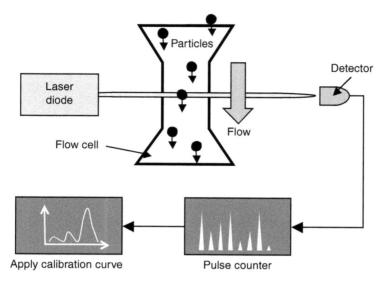

Figure 16   Schematic of single particle optical sensing particle size analyzer.

made up of larger particles. They can provide useful information about primarily submicron distributions as well [11]. For such materials, it is useful to be able to quantify the amount of large or oversized particles that are several standard deviations from the mean diameter but that can still affect the performance [12]. These particles are often present in concentrations that cannot be seen by light scattering techniques [13].

Single particle optical sizing (SPOS) is an optically based single particle method. It involves passing particles through a small optical flow cell. A laser light source illuminates a thin section of the flow path in such a way that particles must pass through this detection zone on their way through the flow cell. This is shown in Figure 16. As is indicated in this figure, detectors are arranged to measure both the transmitted light and the scattered light. The scattered light detector is used to increase the sensitivity for particles <2 μm. As a particle passes through the view volume, the detector situated directly across from the light source will register a decrease in the intensity of the light passing through the flow cell while the scattering detector will register an increase. The response of both these detectors is related to the size of the particle. Since the particles flow through the view volume only once, the response measured by the detectors is a pulse. The height of the pulse is related to the size of the particle. This is illustrated in Figure 17. A multichannel analyzer (MCA) is used for the pulse height analysis. The resulting pulse height distribution, obtained from passing several hundred thousand particles through the flow cell, is converted to a PSD using a calibration curve.

SPOS can count and size particles in the range of 0.5 to several thousand microns although the individual sensors have limited dynamic ranges. It can be used to test dry powders or particles dispersed in liquids. It not only measures particle size but it can also provide particle concentration information. SPOS has been used for online monitoring of particles dispersed in liquids [14]. As stated earlier, SPOS has excellent resolution and sensitivity compared to other techniques designed to size larger particles [3,4]. As a matter of fact, the sensitivity of SPOS is such that it can be used to determine the level of particle contamination in clean liquids [15]. The biggest difficulty of implementing SPOS is the dilution required of many samples in order to pass the particles through the sensor one at

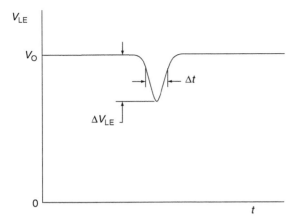

Figure 17 SPOS signal.

a time. Another issue is sampling. SPOS measurements can size several hundred thousand particles per minute but this represents a small amount of material. While not a problem with samples where the goal is to quantify the amount of oversized particles in primarily submicron distributions, users must take care to properly sample the material to be tested in order to insure representative results when using the SPOS method. SPOS sensors must be calibrated.

Another well-used single particle technique is the electrozone method also known as the Coulter method. This involves flowing particles dispersed in a liquid through an aperture. An electric field is applied across the aperture. This generates an electrical current that passes between the electrodes and through the aperture. By flowing liquid through the aperture, the particles are carried through. So long as the particles are composed of a nonconducting material, every particle that passes through the aperture will cause the current between the electrodes to drop since the particle blocks the electric field lines between the electrodes in an amount related to the volume of the particle. Since the particles are flowing, they have a short residence time in the aperture and thus the drop in current is momentary. So every particle that travels through the aperture produces a pulse, the height of which is related to the excluded volume and thus the size of the particle. This is shown in Figure 18.

The most common arrangement is to keep the current constant. This means that when a particle passes through the aperture, the voltage required to maintain the current at a constant level increases. This voltage jump is related to the size of the particle passing through the aperture. So like SPOS, the raw data from the electrozone method is a voltage pulse–height distribution, which is converted to a PSD via a calibration curve. And just as it does for SPOS, pulse counting makes the electozone method one that has high resolution and sensitivity; it can size and count particles in the range of 0.5 to about 1500 μm. Multiple apertures, each of which has a limited dynamic range, cover this range. This small dynamic range means that clogging of the aperture is a common problem. So samples with broad distributions will have long measurement times as various apertures will have to be switched in and out. The electrozone method is limited to particles dispersed in liquids only, because the medium must be conducting. Furthermore, from a practical point of view, this limits this method to particles that can be dispersed in water since it is difficult to make organic solvents conduct (though it can be done). As with SPOS, only small amounts of sample are tested and this technique has never been used online.

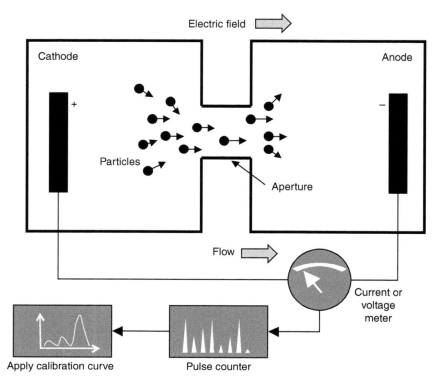

Figure 18 Schematic of electrozone particle size analyzer.

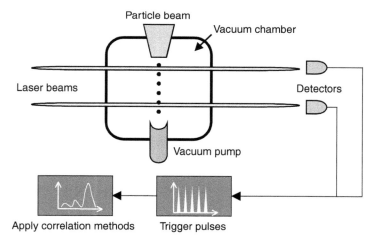

Figure 19 Schematic of time-of-flight particle size analyzer.

There is one more optical single particle method worth mentioning, since in its commercial form, it makes measurements on dry powders only. Time-of-flight instruments create a beam of particles. This is shown schematically in Figure 19. This beam is created by introducing the particles into a vacuum chamber. The velocity of each particle is related to its particle size. The velocity of the particles is measured by using two laser beams, separated by a known distance, that intersect the path of the moving particle. When the particle passes through the first laser beam, a clock is started. Passing through the second laser beam stops the clock. The time measured by the clock is the time it took that particle to travel the distance between the beams. From this information, the particle velocity is determined. The smaller particles have larger velocities.

In theory, this method can be used as a single particle counter like SPOS and electrozone, but in practice it is not. In order to increase the amount of material sampled, autocorrelation techniques must be used to determine which start pulses and stop pulses go together. While this only affects the resolution slightly, it affects the sensitivity significantly. So time-of-flight instruments do not have the same sensitivity as SPOS or electrozone and cannot provide accurate count or particle concentration information.

D.   Sieves

Sieving is probably the oldest particle-sizing technique still in use. A sieve consists of a screen made from metal wire with well-defined spacing between the wires. Only particles with a dimension less than the openings can pass through. All the larger materials remain in the sieve. A sieve analysis is usually done with several screens having different sized spacings. The screen with the largest spacing is on top. After the smallest screen, a pan is placed to catch the material that is so small that it passes through all the above screens. Each sieve, the pan, and the sample to be tested is weighed on a balance. The sample is introduced into the top sieve. Mechanical energy is applied to the stack of sieves. This energy keeps the large particles from clogging the spacings and preventing the small particles from passing through. The sieves and pan are reweighed and the mass fraction of the sample in each sieve is determined. Sieves still have a unique place in powder analysis, compared with the other methods described above, sieves can test much larger amounts of sample. This means the statistical accuracy of sieve data is much better. No other technique can approach the amount of material that sieves can analyze. Laser scattering methods use micrograms while sieves use tens of grams or more. Sieves are easy to use and do not require complicated sample preparation schemes. Sieving can be done wet or dry and can size particles in the range of 20 to 10,000 μm. Sieves are much less expensive than the more modern techniques described earlier. Sieves directly provide mass-weighted PSDs, something that none of the above techniques can do. Unfortunately, the use of sieves is very labor intensive. Because of this, the accuracy of the results can be very dependent on the skills of the operator. While the dynamic range is large, submicron particles cannot be sized and end up in the pan. The availability of screen sizes is limited; and sieving is a low-resolution technique.

IV.   WHY MEASURE PARTICLE SIZE?

As stated earlier, particle size can come into play in many aspects of food powders. It is important in processing, handling, shelf life, and finally eating. Depending on the specific food item, particle size can figure into the taste, color, texture, and smell of final food product. These are the characteristics of foods that the customer cares about the most and will determine whether a product becomes economically successful or not. The manufacturing of food products involves the transport of the various ingredients, the mixing/processing of these ingredients, and the transport and packaging of the final product. Particle size can play a role in all these steps as well. It is with this in mind that specific examples of the analysis of various samples are presented.

A. Process Monitoring

Powders are often classified using sieves/screens or hydrocyclones. This classification is done for the purpose of achieving a certain particle size range. The particle size might be important for flowability or to achieve optimum blend or make sure there is no significant segregation after the powder is blended with another. Figure 20 shows some volume-weighted PSDs for various size fractions of a vitamin A powder used as a nutritional supplement. It is clear from the data that screens do not provide an absolute size cutoff. The 60 mesh fraction, for example, should contain particles between 60 and 45 mesh or 250 and 355 μm. Instead, the actual PSD has a range of 100 to 650 μm and a volume-weighted mean of 384 μm. The presence of particles larger and smaller than expected could have a significant effect on properties like flowability or whether this powder will segregate or cake while stored in a silo or train car.

This example illustrates the importance of using particle size analysis to quality control incoming or classified ingredients to ensure they are within specified limits. It also illustrates an important issue with using screens for particle size analysis. Sieves do wear out. In other words, the spacings may change or become nonuniform across the entire sieve. This may have contributed to larger than expected particles to be in the 60 mesh sieve. Sieves must be qualified with material of known size (like glass beads) on a routine basis and replaced if not within specifications. Moreover, standard operating procedures must be adhered to in order to achieve a repeatable separation. Variations in the applied vibrational energy might have contributed to the fact that smaller than expected particles remained in the screen.

Granulation is another processing technique used in the food industry. Granulation is a process by which fine particles are aggregated into larger particles. This is done for a variety of reasons http://www.fitzmill.com/food/dry_granulation/dry_granulation.html including the removal of fines, prevention of segregation by making the PSD more uniform, and to improve flowability. It is important to monitor the progress of the granulation to determine when the end point is reached. Figure 21(a) contains volume-weighted cumulative curves taken with SPOS. Each curve represents a different period of time during the dry granulation of the powder. The left-most curve is the earliest. As the process continues, the curves move to the right. The D50 starts at 150 μm and grows to 450 μm at the end of the process.

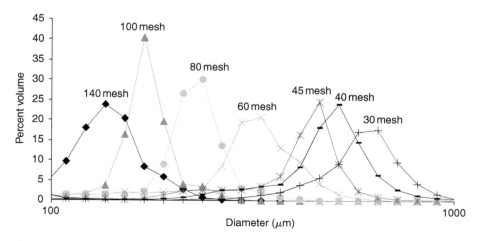

Figure 20   Particle size distributions of screen classified vitamin A powder.

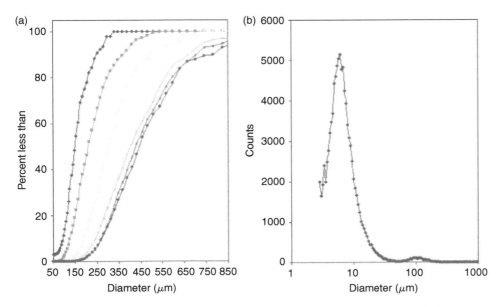

Figure 21 (a) SPOS volume-weighted cumulative curves used to monitor a granulation process; (b) SPOS number-weighted differential PSD of powder during granulation process.

Several points have to be made regarding the use of particle size analysis from this example. First, the PSDs are shown in cumulative form. Converting the curves from differential to cumulative makes it easier to view the changes that are occurring and to monitor the growth of the median particle size (D50). Second, the number-weighted raw data of the SPOS technique was converted to volume-weight. Figure 21(b) contains the number-weighted differential distribution from one of the analyses. Clearly, the distribution is bimodal; the sample consists of a relatively small number of granulated particles in a sea of fine particles. On a number-weighted basis, the fines dominate the distribution. Because of this, the number-weighted D50 does not change much at all. Since the goal is to monitor the growth of the granules, using the volume-weighted data is the preferred approach. On the other hand, since single particle techniques are better suited than light scattering methods in quantifying these fines, they have the advantage of providing better information on potential dust or flowability problems owing to the presence of these smaller particles.

While it is sometimes required to make fine particles larger, it is probably more common to make large particles smaller. One of the methods used for this is homogenization. Homogenization is a process by which particles dispersed in a liquid are forced under high pressure through a small orifice. Through a combination of shear forces and impact forces, large particles and aggregates are reduced to smaller particles. Homogenization is used to make stable oil-in-water emulsions or disperse solid particles for beverages. Consider Figure 22, which contains volume-weighted DLS data obtained from the same oil and water mixture homogenized at different pressures. In the case of this specific formulation, the result of increasing the pressure from 2,500 to 25,000 psi is to steadily decrease the mean diameter and the width of the distribution. The emulsion made at the highest pressure will probably be the most stable since it has the lowest mean diameter and the narrowest distribution.

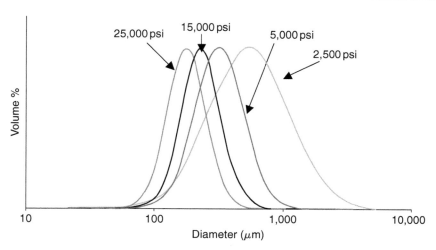

Figure 22   Volume-weighted DLS data for emulsions homogenized at different pressures.

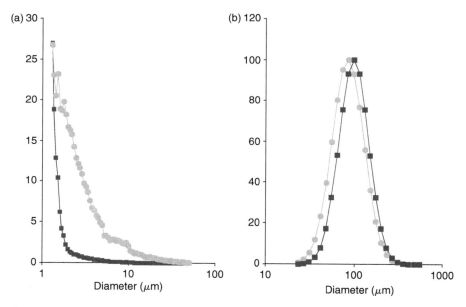

Figure 23   Comparison of (a) SPOS and (b) light scattering methods for oversized particle determination.

As it turns out, one cannot always infer stability from light scattering particle size data [16]. This is evident from the results displayed in Figure 23. The two samples were solid suspensions made by homogenization. This suspension is used by the semiconductor industry to polish silicon wafers. Large particles, caused by the loss of stability or contamination, can scratch wafers with the effect of decreasing chip yield. And it does not take a lot of particles to bring about a significant scratching problem. Of these two samples, one was observed by microscopic methods to scratch wafers and the other did not. To the light scattering method, both samples appear quite similar, with the so-called "bad" sample having a slightly lower mean diameter. When tested with a method that has the sensitivity to detect

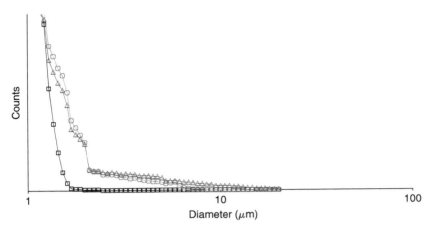

Counts

Diameter (μm)

Figure 24 SPOS particle size distribution data for beverage emulsion after first (triangles), second (circles), and sixth (squares) passes through homogenizer.

small amounts of large or oversized particles, like SPOS, the true difference is apparent. What makes one sample perform within specification and one operate out of specification is an extremely small amount of large particles, most likely aggregates, that are present in concentrations below the detection limit of light scattering methods. The presence of such large particles in oil-in-water emulsions has been shown to be a predictor of emulsion stability [17].

Figure 24 contains SPOS data of a beverage emulsion after various passes through a homogenizer. The SPOS technique sized particles greater than 1 μm. Particles greater than 1 μm in size were 2 to 3 standard deviations away from the mean of the main peak, which was at about 0.35 μm [3]. So the SPOS technique was looking at a very small part of the overall distribution. But it is in this range that colloidal stability is either achieved or not [16,17]. It took six passes through the homogenizer to reduce all the particles to below 2 μm. It is worth noting that light scattering data indicated that from passes 1 through 6, the mean diameter did not change significantly [3]. The manufacturing of emulsions is an instance where utilizing multiple particle size measurements is the best approach. First, it is important to monitor the homogenization process such that the specified mean diameter is achieved. Wide dynamic range particle sizers of the light scattering type are appropriate tools for this. After the correct mean diameter is achieved, a single particle method like SPOS can be used to make sure the number of passes were sufficient to remove oversized particles that act as sites of nucleation for further agglomeration, in other words, to stabilize the emulsion. As the middle section of this chapter indicated, there is no technology available that can measure the mean diameter of an emulsion and at the same time provide quantitative information on the presence of small numbers of large droplets. Two separate measuring tools are required to do this.

Grinding or milling is another method used to reduce the size of particles. Figure 25 contains volume-weighted cumulative SPOS results from a black pigment powder [11] at various times as it was being milled. During the course of hundreds of minutes, the particle size was reduced as reflected in the fact that the D50 went from 1.5 to 0.65 μm. This particular set of data is unique because the overall change in particle size before and after milling is about 1 μm. In order to see differences, the particle size analyzer used will have to have exceptional resolution, unlike the granulation example mentioned above where the shift in particle size was hundreds of microns that could be monitored by light scattering

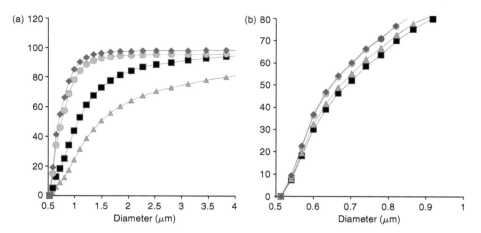

Figure 25 (a) SPOS cumulative volume-weighted curves from a milling process at start (triangles), 15 min (squares), 336 min (circles), and 420 min and after (diamonds); (b) SPOS cumulative volume-weighted curves from a milling process at 420 min (squares), 468 min (triangles), 564 min (circles), 618 min (diamonds).

very easily. As shown by this example, SPOS can differentiate a change of as little as 0.2 μm. Only a single particle method could be used to monitor this specific process.

B. Powder Handling

One of the most important characteristics of powders with regard to handling them is flowability. Flowability impacts mass flow and thus controls the choice of storage equipment like hoppers and silos. It will determine how much energy it takes to convey powdered ingredients from one stage of processing to another, how well they will blend with other powders, and what happens to them when they are stored (segregation or caking). Flowability can be measured and this should be done when developing and designing a process that uses powders. But after the process is online, the powdered ingredients must retain the flow properties that work optimally with all the conveying, blending, and storage equipment that make up the process. This is particularly important with the use of food powders, which are often derived from natural sources, and will vary in terms of properties over time and source. This requires regular testing of incoming ingredients.

Generally, the flowability of powders depends on the cohesive forces between the particles making up the powder [18]. These forces are influenced by many environmental conditions like storage temperature and humidity [19]. The inherent properties of powders also come into play like particle size and surface chemistry. The environmental conditions can be controlled but the inherent properties might vary by source or time of year. This suggests that an important way of determining whether incoming powdered ingredients will present flow problems or not is to make sure that they meet certain particle size specifications. In general, the finer the particle size, the stronger the cohesive forces between particles, and the more resistant to flow will be the powder [19]. Thus, powders consisting of smaller particle sizes tend to flow less easily, are more likely to cake, and altogether be difficult to handle.

Figure 26 contains the flow functions for two powders derived from egg yolks. These powders are used in food products like mayonnaise, dressings, and sauces as well as in baking. The flow functions are derived from flowability measurements performed with

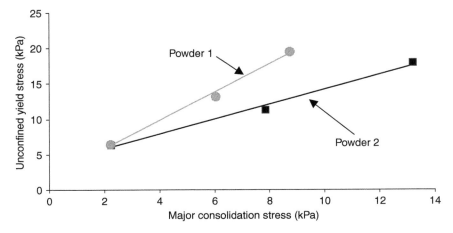

Figure 26 Flow functions of two egg yolk powders.

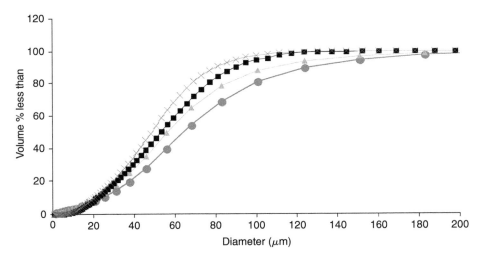

Figure 27 Volume-weighted cumulative curves of egg yolk powders measured by SPOS and laser diffraction, Powder 1: LD (triangles), SPOS (circles); Powder 2: LD (squares), SPOS (X's).

different consolidation stresses [19]. The inverse of the slope of the best-fit straight line is called the flowability or the flow index. The higher the slope, the lower the flowability. The measured flow indices of these powders are both quite low and suggest that they are cohesive and will not flow well, but clearly, powder 2 has a higher flowability than powder 1. Figure 27 contains the cumulative PSDs as measured by SPOS and laser diffraction for these powders. The particle sizing results correlate with the flowability data. Clearly, the coarser powder flows better, as expected. It is worth mentioning that while the overall mean diameter, in this particular case, predicts flowability and can be used as a quality control specification for powdered ingredients, it can be useful and important to be able to measure the fines content of powders. These fines can affect flowability as mentioned earlier. The differential PSDs measured by SPOS for the egg yolk powders are shown in Figure 28. It is clear that there is a substantial number of fines below 10 μm for both powders and that the amount of fines are different for each.

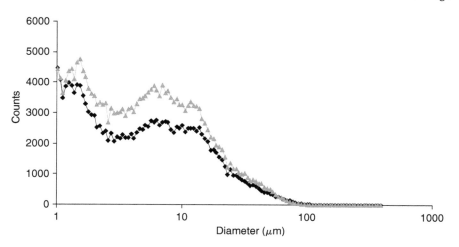

Figure 28   Number-weighted PSDs obtained from SPOS; Powder 1 (squares); Powder 2 (triangles).

Whether these differences play a role in the flowability requires additional measurements, but not all handling issues are related to mass flow. The hazards of airborne particles are well established [20]. Currently, Occupational Safety and Health Administration (OSHA) sets acceptable levels of airborne particles less than 10 μm. Thus powders, which have substantial populations of these particles, require special handling (masks, ventilated work areas). Therefore, it is important to measure the presence of these particles. Consider the PSDs in Figure 29. As can be seen, the number-weighted measurement is better suited to quantify fines than a volume-weighted measurement. It follows that techniques like the single particle methods, which provide number-weighted raw data, are better suited for this task. A volume-weighted method that does detect the fines because of a volume contribution below their detection limit will not see them even if the data is converted to number-weight. Hence, it is important to consider the attributes of various particle-sizing techniques in relation to the aspect of the PSD most important to the application of interest.

C.   Product Quality

There are many qualities of food products, like taste, color, texture, and smell that determine its success in the marketplace. These qualities must be controlled and tested in order to ensure a uniform product over time and place. Particle size does play an important role in many of these qualities. An example would be coffee. Coffee is processed from beans to powdered form by grinding. All types of coffee are graded and priced according to particle size (usually by sieve analysis). The reason for this is that the particle size of the ground beans determines the degree to which certain chemicals, particularly acids, are extracted. The acids are known to give coffee its characteristic bitter flavor. As a matter of fact, over 30 different types of acids have been identified in brewed coffee (www.coffee.org). Most, but not all, of these acids were observed to increase in concentration for a given brew time as the grind became finer. This can be explained by the fact that the surface area-to-mass ratio increases with decreasing particle size. Increased surface area-to-mass means more contact with the extracting medium, and thus more extracted chemicals.

Figure 30 contains the volume-weighted PSDs of three different types of coffee. As can be seen, all the PSDs consist of a single peak with a tail of fines. Clearly, the mean diameter of each peak is different. Coffee 1 has a mean of 270 μm and is the finest. Based

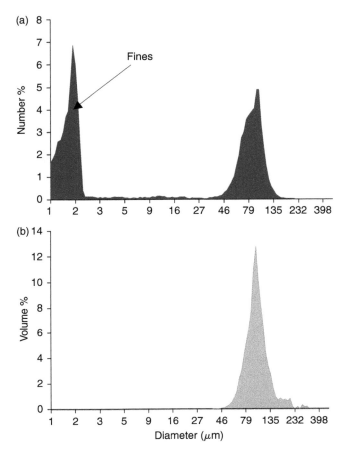

Figure 29 (a) Number-weighted and (b) volume-weighted PSDs.

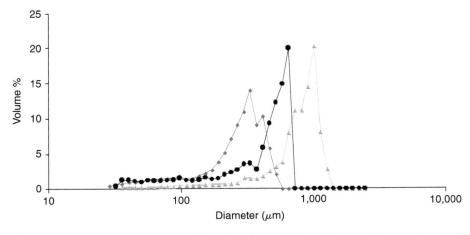

Figure 30 Volume-weighted PSDs from the dry analysis of three coffee samples: Coffee 1 (diamonds), Coffee 2 (circles), Coffee 3 (triangles).

on this particle size information, it should produce the bitterest coffee drink. As a matter of fact, coffee 1 is for making espresso, one of the strongest tasting coffees. The other two coffee samples were made for drip brewing, each representing a different grade. The finer one, coffee 2, is a lower grade and is sold for less money. It is often used in automatic coffee machines, which require a grind that will brew faster. It is appropriate to classify coffee particle sizes in terms of mass-weighted information since the total amount of flavor chemicals is related to the mass of coffee particles. The smaller the size, the more flavor chemicals that can be extracted. While the use of sieves for sizing coffee dates back to a time when this was the only method available, it is still the most useful since it provides a mass-weighted PSD directly and can sample large amounts of material. Laser diffraction is sometimes used but provides volume-weighted distributions. An assumption (often wrong for coffees) of uniform density must be made. Even so, as Figure 29 shows, particle size analyzers other than sieves can still provide useful information and do so much faster than a sieve measurement. Often, a light scattering particle sizer is used to monitor the particle size during the grind and sieves are used for the final grading.

Odor or smell is another food property influenced by particle size. Odors occur from the release of gaseous chemicals. The concentration of gas phase chemicals is related to vapor pressure, which is, in turn, related to intermolecular interactions. These interactions tend to decrease with particle size. Thus, the vapor pressure above finer particles tends to be higher making them appear to smell more strongly. This can be understood with the aid of Figure 31. The left drawing might represent the upper few layers of molecules in a relatively large particle. The radius of curvature is so large that for sufficiently short distances the surface is almost flat. For a surface molecule to escape into the gas phase, it most overcome the attractive forces between it and all the molecules directly adjacent to it. The total intermolecular force on a particular molecule will be the sum of the forces between this molecule and all the molecules surrounding it. As the radius of curvature decreases, the number of molecules adjacent to the surface molecules decreases. In this case, the attractive forces are lower and it is easier for a molecule to escape into the gas phase and contribute to the sensation of smell. So it follows that molecules on the surface of a smaller particle (with a smaller radius of curvature) will have less attractive forces keeping it in the particle.

Figure 32 contains the number-weighted PSDs of two different grades of ground mustard. These products are used to impart mustard flavor and as emulsifiers in mayonnaises and salad dressings. Although used primarily for taste, mustard has a particularly strong distinctive odor. In the case of the above products, it was observed that the finer powder had the stronger mustard odor. This was due in part to the reduction in particle size but the fact that fine particles are more likely to become airborne and enter the nasal passages might also have played a role. Again this is an example of the utility of number-weighted methods to quantify fines.

Milk, when separated and dried, can be an important source of many nutritional components. There are many powdered dietary supplements on the market that consist of

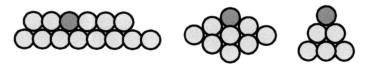

Figure 31   Illustration of decrease in intermolecular forces as radius of curvature and thus particle size decrease from left to right.

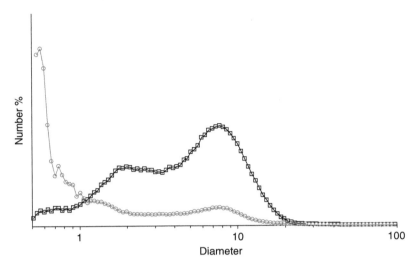

Figure 32 Number-weighted PSDs of fine and coarse ground mustard seeds obtained by SPOS: fine (circles); coarse (squares).

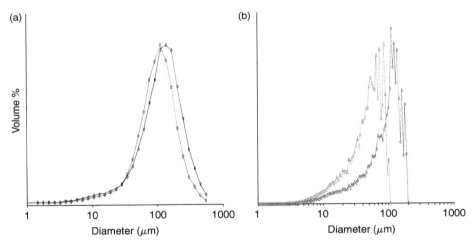

Figure 33 Volume-weighted PSDs for milk protein powders: (a) from laser diffraction; generic (diamonds), name brand (circles) (b) from SPOS; generic (diamonds), name brand (triangles).

almost pure protein derived from milk. These are used to increase protein intake without the fat that comes from eating meat as a source of protein. These supplements are in powdered form that is usually dispersed in another liquid like water and consumed. Even though the main purpose is as a nutritional supplement, such powders, to be successful in the market, should result in a beverage that is appealing to the customer. One of the problems with beverages derived from reconstituting or dispersing powders is a "gritty" or "chalky" texture imparted by large particles that did not break up or dissolve. The human tongue can feel particles as small as 20 μm. Thus, to produce a beverage from a powder that will be perceived as a true liquid, particles large enough to be felt by the tongue must not remain. Even small numbers of these particles can cause the consumer to reject the product. Figure 33 contains the volume-weighted PSDs of two protein nutritional supplements, one generic and the other a name-brand product. The ability to detect small amounts of large particles is the main goal of the analysis. The light scattering measurements are quite similar for both samples with just a slight shift to larger mean diameter for the generic product. The SPOS results indicate a larger shift, which primarily comes about owing to the sensitivity of the measurement to small populations of oversized particles. The number-weighted SPOS results were almost identical. So the difference in the two products is not in the overall distribution, which is just about the same, but in the amount of large particles. The larger concentration of aggregates or oversized particles might impart a texture, which will be rejected by the consumer. The manufacturer of the name-brand product either grinded the powder more efficiently or classified it in some way.

ACKNOWLEDGMENT

The authors would like to thank Micro Measurement Laboratories for making some of the SPOS and LD measurements used in this work. Mike Hawes of Belovo, Inc provided samples of the various egg powders that his company makes, and Greg Martiska of Sci-Tec, Inc., made the flowability measurements.

REFERENCES

1. Kaye, Brian. *Characterization of Powders and Aerosols*, Weinheim, Germany, Wiley-VGH (1999), Chap. 1.
2. Allen, Terence. *Particle Size Measurement*, Vol. 1, 5th ed., Chapman & Hall, London (1997).
3. O'Hagan, P., Hasapidis, K., and Pokrajac, G. *J. Powder/Bulk Solids Technol.* June: 31–36 (1998).
4. Litchy, M., Nicholes, K., and Grant, D. Comparison of Instruments Used for Measuring Concentrations of Large Particles (≥ 1 μm) in CMP Slurry, in *Proceedings of the 5th International Conference on CMP for ULSI Multilevel Interconnection (CMP-MIC)*, (Tampa, FL, 2000) pp. 570–573.
5. Driscoll, D., Etzler, F., Barber, T. et al. Physicochemical assessments of parenteral lipid emulsions: light obscuration versus lases diffraction, *Int. J. Pharm.* 219: 21–37 (2001).
6. Seon, S.J. and Schimpf, M. Thermal Field-Flow Fractation of Colloidal Particles, in *Particle Size Distribution III*. Provder, T., Ed. ACS Symposium Series 693, American Chemical Society, Washington DC (1998), Chap. 13, pp. 182–195.
7. Giddings, J., Caldwell, K., and Jones, H. Measuring Particle Size Distribution of Simple and Complex Colloids Using Sedimentation Field Flow Fractionation, in *Particle Size Distribution: Assessment and Characterization*. Provder, T., Ed. ACS Symposium Series 332, American Chemical Society, Washington DC (1987), Chap. 15, pp. 215–230.
8. Allen, Terence. *Particle Size Measurement,* Vol. 1, 5th ed. Chapman & Hall, London (1997), p. 212.
9. Giddings, J. Field-Flow Fractionation: Analysis of Macromolecular, Colloidal, and Particulate Materials, *Science* 260: 1456–1465 (1993).
10. Turbitt, C., O'Hagan, P., Hasapidis, K., Nicoli, D., and Pokrajac, G. Are we finished yet? Effective end point determination using SPOS, *Powder Bulk Eng.* 15: 27–38 (2001).
11. O'Hagan, P., Hasapidis, K., Nicoli, D., and Pokrajac, G. SPOS, A Unique Tool for Sizing Aggregates, in *Fine Powder Processing International Conference Proceedings*, State College, PA (1999) pp. 78–86.
12. Nicholes, K., Litchy, M., Hood, E., Easter, W., Bhethanabotla, V., Cheema, L., and Grant, D. Analysis of Wafer Defects caused by Large Particles in CMP Slurry Using Light Scattering

and SEM Measurement Techniques, in *2003 Proceedings of the 8th International Conference on CMP for ULSI Multilevel Interconnection (CMP-MIC)*, Marina Del Ray, FL (2003).

13. Nicoli, D., O'Hagan, P., Pokrajac, G., and Hasapidis, K. High-Sensitivity particle size analysis of colloidal, *Am. Lab.* 32: 18–22 (2000).

14. Nicoli, D. et al. Ultra-Sensitive Online Monitoring of CMP Slurries by Single-Particle Optical Sensing (SPOS), in *Proceedings for the Japan Conference* (November 1999), Sec. I, pp. 1–9.

15. Particulate Matter in Injections, in *The United States Pharmacopeia*, United States Pharmacopeial Convention, Rockville, MD (2000), Vol. XXIV pp. 1971–1977.

16. Turbitt, C., O'Hagan, P., Hasapidis, K., and Nicoli, D. A Multi-Tool Approach to Colloid Stability: SPOS and Separation Analysis, in *2001 Fine Powder Processing International Conference Proceedings*, State College, PA (2001) pp. 123–132.

17. Driscoll, D., et al. *Am. J. Health-Syst. Pharm.* 52: 623–634 (1995).

18. Schulze, D. *Powder Bulk Eng. Int.* 5: 32–39 (2002)

19. Marinelli, J. Factors that Impact a Bulk Solids Flowability, on http://www.solidshandlingtech.com/articlefactors.htm (2001).

20. Gregg, W. *Proceedings of Powder and Bulk Solids Conference* (1998) pp. 201–206.

10

Food Powder Flowability

John J. Fitzpatrick
Department of Process Engineering
University College
Cork, Ireland

CONTENTS

I. INTRODUCTION

A. Importance of Powder Flowability

Many industrial operations involve powders including food powders. These operations can be categorized as powder production, storage, transport, and processing, and the common

Table 1 Common Unit Operations Involving Powders

Category	Unit operations
Powder production	Crystallization/precipitation/particle separation, drying, comminution
Storage	Bins, silos, hoppers, bunkers
Transport	Feeders: screw, belt, rotary valve
	Conveyors: screw, belt, bucket elevator, pneumatic conveying
Processing	Mixing, fluidization, agglomeration/granulation, forming, coating, packaging

unit operations are presented in Table 1. In a general sense, powder flowability is all about the movement of powders. The movement of powders can be broken down into two classes: packed or fluidized. Packed movement is where the particles are not suspended in a fluid and are in close contact with each other whereby there is significant friction and cohesion forces resisting flow. Fluidized or suspended movement is where the particles are fully suspended or supported by a fluid, for example, in fluid beds and dilute phase pneumatic transport. Considering the operations presented in Table 1, it can be seen that powder flowability is important in many of these operations. For example, in drying, the manufactured powder must flow out of the dryer; likewise particulate material must flow in and out of comminution equipment. Flow must also occur out of silos, hoppers, in screw feeders, and pneumatic conveyors. Most of the processing operations in Table 1 require the flow of powders including mixing and packaging. Overall, the flow behavior of powders is very significant throughout operations involving powders.

B. Flowability Problems in Powder Storage and Handling

The focus of this chapter is on packed movement of powders, for example, out of hoppers and silos and in screw feeders. A hopper is the converging section at the bottom of a silo, and a feeder is usually placed at the bottom of the hopper to control the flow rate of the discharging material. One of the major industrial powder flow problems is in obtaining reliable and consistent flow out of hoppers and feeders without excessive spillage and dust generation [1]. These problems are usually associated with the flow pattern inside the silo. Typical flow patterns are illustrated in Figure 1. The worst-case scenario is no flow. This can occur when the powder forms a cohesive arch across the opening, which has sufficient strength within the arch to be self-supporting. A device, such as a sledgehammer, is required to try and initiate flow, and evidence of this can be too often seen by the hammer marks left on silos that is commonly referred to as "bin rash." Mass flow is the ideal flow pattern where all the powder is in motion and moving downward toward the opening. Funnel flow is where the powder starts moving out through a central "funnel" that forms within the material, and then the walls of powder at the top collapse and move through the funnel. This process continues until the silo empties or flow stops. Most flow problems are caused by a funnel-flow pattern and can be cured by altering the pattern to mass flow [2].

In some funnel-flow silos, the powder outside the funnel may be able to support itself and form a stable rathole, as illustrated in Figure 1(d). This is another no flow scenario and a flow promotion device is required to force the collapse of the rathole and initiate flow. Other problems with funnel flow include the sudden collapse of a chunk of powder into the funnel placing extra mechanical load on the hopper and feeder, which may cause failure. Collapse of fine powders into the funnel may result in mixing with air causing fluidization of the powder. This will flood the feeder resulting in fluidized powder spilling out of the feeder and generating dust. This is often referred to as flooding. Another problem is erratic

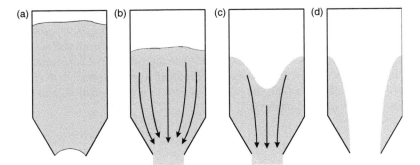

Figure 1 Flow/nonflow patterns in hoppers and silos: (a) bridging or arching; (b) mass flow; (c) funnel flow; (d) ratholing.

flow due to intermittent collapse of arches and ratholes. Powder segregation may also be a problem. During silo filling with a powder mix of different size particles, the bigger particles tend to roll toward the silo wall while the finer particles tend to remain in the center. During funnel-flow discharge, the center will discharge first; thus, the finer particles discharge first followed by the larger particles resulting in demixing or segregation of the mix. In general, most flow problems in hoppers are caused by lack of flow at the hopper wall creating funnel flow, while most problems with feeders occur because they obstruct flow over part of the hopper outlet and thus prevent flow at the hopper walls [3].

C. Powder Flowability and Food Powders

So far, this chapter has dealt with powder flowability in general without any reference to food powder flowability. This is because food powders are powders first. They are discrete particles like any other powder. They have particle properties, such as size distribution and densities, just like other powders. Even the mechanisms for surface interactions resulting in cohesion between particles are mostly the same type of mechanisms that occur with other powders. Thus, powder flow property measurement techniques, design procedures, problems, and problem solving procedures for the flowability of powders in general can be equally applied to the flowability of food powders. What is essentially different about food powders is their composition, which typically consists of proteins/peptides, sugars/carbohydrates, fats/lipids, minerals, and other biological components. These components and how they are influenced by storage conditions and water can have a major influence on the flowability of the powder, as discussed later in the chapter. This has an important impact on food powder flowability in light of the fact that many food powders are used directly at home or in food outlets. Food powders can be considered as dynamic systems whose physical and chemical properties may change over time [4]; hence, the flowability of the powder may alter from the time it leaves the manufacturing plant to time it is consumed.

II. METHODS OF ASSESSING POWDER FLOWABILITY

A. Importance of Consolidation

There are many different flow properties and methods for measuring these properties. Flow property measurements for packed powders will be different from fluidized powders, as the former have some degree of consolidation while the latter have none. This section will focus only on packed powders. Consolidation refers to the pressure or stress acting on the powder causing it to compact. It is important because it influences the strength developed within the powder. Typically, if a powder is consolidated more, then a greater force is required to make the powder fail and flow, that is, the powder has developed a greater strength to resist flow. Thus, this strength developed within the powder depends on consolidation, and this should be represented in any test and analysis.

B. Empirical Methods

There are many empirical tests used to assess flowability, including measurement of angles of repose [5], Hausner ratio, Carr indices [6], flow through a defined opening, and compression testing [4]. These tests are not useful in the design of hoppers and silos for reliable flow. However, they may be useful in quality control, where a change in a measured value may be indicative of change in the flow behavior of a given material.

C. Shear Cell Techniques and Application to Hopper Design

In the 1950s and 1960s, Andrew Jenike pioneered the use of shear cells for determining flow properties of powders that could be directly applied in the design of hoppers and silos for mass flow. He developed a mathematical analysis using two-dimensional stress analysis and soil mechanics principles to evaluate the minimum hopper angle from the horizontal required for mass flow and the minimum hopper opening size required to prevent cohesive arching [7]. The analysis required measurement of the following three flow properties: flow function, effective angle of internal friction, and angle of wall friction. The flow function is a measure of how the unconfined yield strength developed within the powder varies with maximum consolidation stress, as illustrated in Figure 2. Jenike used the flow index (inverse slope of the flow function) to classify powder flowability, and this was extended by Tomas and Schubert [8] and is presented in Table 2. The effective angle of internal friction is an index of the friction that develops between powder particles as they move over each other under pressure. The wall friction developed between the powder and the hopper wall will be a major influence in determining whether or not the powder will slide along the wall [9]. The angle of wall friction is an index of the adhesion developed between the powder and

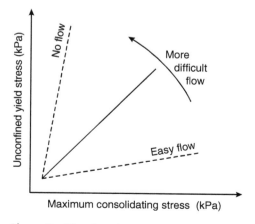

Figure 2   Flow functions.

Table 2 Classification of
Powder Flowability Based on
Flow Index (ff_c)

$ff_c < 1$	hardened
$1 < ff_c < 2$	very cohesive
$2 < ff_c < 4$	cohesive
$4 < ff_c < 10$	easy flow
$10 < ff_c$	free flowing

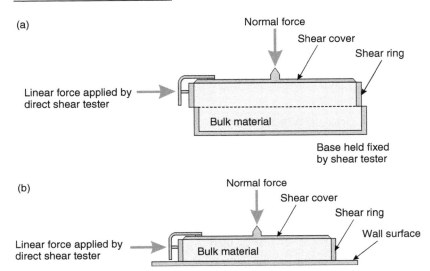

Figure 3 Jenike shear tester modes of operation for measuring: (a) flow function and effective angle of internal friction; (b) wall yield locus.

the hopper wall, under pressure. This angle, along with the hopper angle, will be the major influence in determining whether the flow pattern in a hopper or silo is mass or funnel flow.

Jenike measured the first two properties in his Jenike shear cell, illustrated in Figure 3(a). The Jenike shear cell has a cylindrical base, upon which is placed a ring, that can slide horizontally over the base. The ring and base are filled with powder. A lid is placed on top of the powder and normal loads are then applied to the powder through the lid. The testing consists of measuring a number of yield loci, usually four. A yield locus is obtained from measuring the shear force applied to the ring to cause powder failure under different normal loads for a given consolidating load. The flowfunction and the effective angle of internal friction can then be obtained from the yield loci data. The wall friction characteristics of a powder can be measured using a Jenike shear cell whereby the cylindrical base is replaced by a flat plate of the hopper wall material, as illustrated in Figure 3(b). The test consists of measuring the horizontal stress required to make the powder fail as a function of normal pressure. Angle of wall friction (ϕ_W) is defined as the angle formed with the horizontal by a line drawn from the origin to a point on the curve, as illustrated in Figure 4. This angle may vary with normal pressure.

Obtaining useful reproducible data has always been a difficult task [10], because it is not just a powder that is being tested, but it is also a powder/air mixture, and the method of packing, consolidating, and shearing will greatly affect the measurements. As a result, the European Federation of Chemical Engineers (the Working Party on the Mechanics of

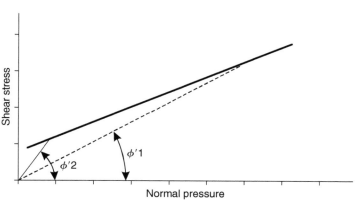

Figure 4   Wall yield locus and angle of wall friction.

Particulate Solids) produced a detailed report on the Standard Shear Testing Technique (SSTT), and this is the recommended guide to anyone conducting Jenike shear tests [11]. Recently, an American standard (ASTM standard D-6128: Jenike's shear tester) has also been produced and is based on the SSTT. There is a fine powder called limestone CRM-116 that is available, which is used as a standard powder to check one's ability in performing shear tests correctly [12,13]. It can be obtained from the Community Bureau of Reference of the European Union (BCR). It has certified yield loci and flow function, which were obtained from Jenike's shear cell tester.

In addition to the Jenike shear cell testers, there are annular or ring shear cell testers that have a rotational as opposed to a translational displacement [14]. They have a number of advantages over the Jenike cell, including unlimited shear strain and ease of operation. Schulze developed a new ring shear tester that gives good agreement with the certified results of the BCR limestone CRM-116 obtained using Jenike's shear cell tester [15–17]. Overall, both the Jenike shear tester and the Schulze ring shear tester can be used in measuring the values of flow properties that can be applied in the design of hoppers using Jenike's design method.

III.   FACTORS AFFECTING THE FLOWABILITY OF FOOD POWDERS

A.   Particle Size and Shape

Particle size has a major influence on powder flowability. A powder may be considered having a particle size less than 200 μm, and as the size decreases below this, the flowability gets worse. One may not notice a major change in flowability as size is reduced from say 80 to 60 μm, however, there should be a noticeable change in flowability if the powder size is reduced by an order of magnitude, for example, from 100 to 10 μm. The rationale behind this reduction in flowability at smaller particle sizes is due to the increased surface area per unit mass of powder. There is more surface area or surface contacts available for cohesive forces, in particular, and frictional forces to resist flow. Particle size usually refers to the mean particle size; however, another important aspect is the particle size distribution, which is usually presented as a mass or volume fraction size distribution. If there is a significant amount of fines in the distribution, then the fines may have a dominant influence on flowability, and powder flowability may be worse than that expected from the mean particle size. Intuitively, one would expect particle shape to affect flowability, as shape will influence the surface contacts between the particles; however, there is not much reported work on the influence of shape on powder flowability. One recent paper describes the results of work that investigated how powder particle shape affected minimum hopper angle and outlet size required for mass flow [18].

B. Surface Interactions

Surface interactions resisting powder flow can be categorized as internal friction and cohesion. Internal friction is simply the frictional resistance of one particle moving over another under normal pressure. When no normal pressure exists, there may still be forces of attraction between the particles that can resist flow, and this is called cohesion. For dry powders, these cohesion forces can be classified as van der Waals, electrostatic, and magnetic forces [19]. van der Waals forces of attraction are usually the most important. They occur on a molecular level due to the polarization of atoms and molecules resulting in attraction between atoms on the surfaces of each of the particles. The smaller the particle size, the greater the contact surface area per unit mass of powder, and thus the greater the combined van der Waals attractions between the particles. Another important aspect of the van der Waals attraction is that it rapidly decreases with increased distance between the particles; thus, as particles are compacted closer together, the van der Waals attraction will be much greater and may result in a compact, such as a tablet.

   Another important contribution to cohesion is the presence of liquid, usually water or oils. Liquid on the surface of the particles will increase the contact area between particles. Above a certain liquid content, liquid bridges will form between the particles at the contact points and this will produce capillary forces as a result of surface tension. The capillary force will bind the particles together in the same way as two glass plates are bound together if there is a layer of liquid between them. The van der Waals interaction between particles will be greatly reduced by the presence of the liquid; however, the capillary forces are usually a lot stronger than the van der Waals forces.

C. Moisture Content and Caking

Food powders, like many other powders, are susceptible to absorbing moisture from an ambient atmosphere if the powder particles are in contact with the ambient atmosphere. The equilibrium relative humidity at 20°C is low for most food powders, usually below 25%. As most ambient relative humidities are higher than this, there is a driving force for moisture sorption. However, the diffusivity of water vapor into a bed of powder is very small; therefore, the powder may only absorb moisture at an exposed surface, and it may take a very long period of time for moisture to diffuse into the bulk of the powder. On the other hand, if a powder is disturbed and there is intimate contact between the powder particles and air, a powder may absorb a significant amount of moisture within a short period. Powder moisture content usually has a significant impact on powder flowability, with increasing moisture content leading to reduced flowability due to the increase in liquid bridges and capillary forces acting between the powder particles [20]. In addition, this may also lead to severe flowability problems due to powder caking.

   Caking occurs when the powder particles stick together to form a hard crust on an exposed powder bed surface, or lumps in the powder bulk, or in the worst-case scenario, solidification of a bed of powder. The causes of caking can be broadly classified as being due to (1) very strong cohesion and (2) formation of solid bridges between powder particles. Very strong cohesion is produced by the cohesive forces described above. For example, if a powder is strongly compacted, the van der Waals forces will interact more strongly and this may produce a cake. Another example is a very hygroscopic fine tea powder, which readily picks up moisture resulting in conversion from a powder form into a solid plug [21]. Heating of food powders to temperatures greater than their sticky point temperature can greatly increase the cohesion between particles as the particles become more plastic.

Caking of food powders due to the formation of solid bridges may be caused by a number of mechanisms. Melting and solidification of fats may also cause solid bridges whereby increase in temperature causes solid fats to melt producing a liquid that can redistribute itself between the particles. If the temperature is subsequently reduced, the liquid fat will solidify and form solid bridges between the particles. Absorption of moisture by soluble powders may cause surface components to solubilize and form a solution between the particles. A change in air conditions that causes drying of these particles will result in water being removed and solid bridges remaining between the particles, forming a hard cake. Another mechanism of interest to many food powders is the formation of solid crystal bridges between particles. Many food powders containing sugars in their amorphous state may crystallize producing solid crystal bridges between the particles. Measurement of glass transition properties using differential scanning calorimetry is a method for detecting the existence of amorphous sugars. Crystallization will only commence if the powder temperature is greater than its glass transition temperature (T_g), whereby the molecules have sufficient mobility to initiate crystallization [22]. T_g is usually well above the storage temperature for most dry powders. However, sugars in their amorphous state are very hygroscopic and will readily absorb moisture from ambient air, and this increase in moisture will cause a significant reduction in T_g. This will initiate crystallization if T_g is reduced below the powder temperature. The rate of crystallization is controlled by the difference in temperature between the powder and its glass transition temperature, which is dependent on moisture content and relative humidity [23,24]. At higher moisture contents, the amorphous sugars will readily dissolve and convert into a thermodynamically more stable form, which makes them less hygroscopic, resulting in the liberation of water to air and reduction in moisture content. This produces supersaturation resulting in crystallization and solid crystal bridge formation between the particles, which can cause severe caking. Finally, many sugars, such as lactose, exist in numerous polymorphic and isomeric forms, which may have a significant influence on powder flow and caking characteristics [25].

D. Storage Conditions

Storage conditions include storage temperature, exposure to relative humidity of air, storage time, and consolidation. In general, varying the storage temperature from above freezing to 30 or 40°C does not usually have a major impact on powder flowability [26], provided no melting of components occurs or no component exceeds its glass transition or sticky point temperatures, as discussed above. For powders containing solid fats, an increase in temperature may cause melting of fats, which may produce viscous liquid bridges leading to increased cohesion. If the powder experiences a temperature below freezing, some of the water may freeze forming ice bridges between the powder particles, resulting in caking. As mentioned in the previous section, the relative humidity of ambient air is usually a lot higher than the equilibrium relative humidity of most food powders; thus, the powder can readily absorb moisture provided it is in intimate contact with air during handling, and this can lead to increased cohesion and even caking.

Time consolidation is another storage condition whereby a powder is left under its own weight for a given storage time. This can often lead to increased strength within the powder and increased adhesion between the powder and the hopper wall, which in turn may lead to arching in a hopper or conversion of a mass-flow hopper to a funnel-flow hopper or caking. Time-consolidation effects are usually caused by (1) increased bulk density owing to the powder consolidating over time, and- (2) physical and chemical changes occurring during the storage time [21]. Powder consolidation over time may cause an increase in bulk density, which leads to the powder particles being pushed closer together. This results in increased van der Waals interaction leading to greater cohesion. Physical and chemical changes often require time, such as the migration of liquids or the crystallization of sugars, which lead to increased cohesion or even caking over time.

A very important aspect of food powder storage conditions is the storage conditions that the powder encounters after it has left the manufacturing plant up until the time it is finally used by the consumer, which might be at home or in a food outlet. A powder may be exposed to varying temperatures and relative humidities and this may have a significant influence on the powder flowability observed by the user, which is a critical quality parameter. Oftentimes, control of storage conditions is out of the control of the manufacturer and thus flow additives/anticaking agents are sometimes used to help prevent potential problems.

IV. OVERCOMING FLOWABILITY PROBLEMS

A. Proper Hopper Design

Mass flow is the preferred flow pattern for consistent reliable flow, as there are, potentially, many problems associated with funnel flow as described above. The main disadvantage of mass flow is that steeper hopper angles are required that require taller silos to be constructed, which is more costly. Funnel-flow design may be acceptable for cases where the material is coarse (>500 μm), free-flowing, nondegrading, and where segregation is not a problem. The starting point for mass-flow hopper design is measurement of the flow properties using a Jenike or ring shear tester, and a compression test to measure bulk density as a function of normal pressure. This information can then be applied in Jenike's analysis to calculate the minumum hopper angle to the horizontal required for mass flow and the minimum hopper opening size for conical- and wedge-shaped hoppers. A number of laboratories worldwide are capable of conducting these tests. An important aspect of the testing is that the powder sample and the testing conditions are representative of what is happening in the plant. For example, particle size and moisture content of the powder sample must be representative along with the storage temperature and time.

In the case of an existing hopper that is giving flow problems, companies can resort to retrofitting as opposed to buying new equipment [2]. The first part of retrofitting is identifying the problem, for example, arching, problems caused by funnel flow, and limited discharge rate. If the problem is arching, then a larger hopper outlet is required. If the problem is because of funnel flow (ratholing, flooding, erratic flow, or segregation), then there are a number of strategies that can be considered to produce mass flow. Modifications to an existing hopper may be as simple as installing a low friction liner, for example, TIVAR 88 (an ultrahigh molecular weight polyethylene). Traditionally, engineers chose 2B stainless steel as the hopper wall material for new construction. A new alternative is to line the hopper with TIVAR 88, which has been shown to have lower wall friction and superior flow performance than 2B stainless steel in many applications. Wall friction tests must be carried out to evaluate the wall friction characteristics with the liner and verify reduction in wall friction.

If the hopper is a conical hopper, then it could be replaced with a wedge-type hopper, as this requires 10 to 12° less of a hopper angle to produce mass flow [27]. A cylindrical

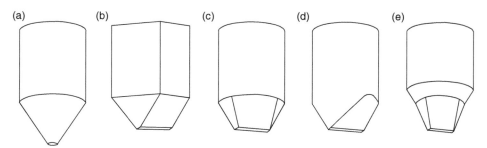

Figure 5   Hopper shapes: (a) conical; (b) wedge; (c) transition; (d) chisel; (e) expanded

silo can be connected to a wedge-type hopper using transition- or chisel-shaped hoppers, as illustrated in Figure 5. Another consideration for hoppers experiencing funnel-flow difficulties is to install an expanded flow hopper, whereby a portion of the hopper is cut away and a properly designed mass-flow hopper is inserted instead, as illustrated in Figure 5(e). The silo will still flow mainly in funnel flow except in the mass-flow hopper where the material will discharge in mass flow. The flow-will be reliable and consistent, as long as the size of the top of the mass flow hopper is large enough to prevent a stable rathole from forming in the funnel-flow region (i.e., it is greater than the critical rathole diameter). Expanded flow design is not cost-effective for silo diameters less than 8 m. In all the above retrofits, powder property measurements will be required for the design calculations.

B.   Proper Feeder Design

A feeder is a device at the bottom of the hopper that controls the powder discharge rate leaving the silo, for example, a screw or belt feeder. Proper feeder design is as important as proper hopper/silo design in achieving reliable consistent flow. An incorrectly specified feeder can turn a mass-flow hopper into a funnel-flow hopper with all the problems associated with funnel flow, thus the feeder must be specified in unison with the hopper [28]. For example, if a screw feeder has a constant pitch and channel depth, then powder is withdrawn preferentially from the back of the screw, as illustrated in Figure 6(a). The constant pitch flights do not allow any increase in capacity in the direction of feed. Therefore, the last flight fills with material and there is no more capacity to take material over the entire outlet length. Any modifications made to the hopper to ensure mass flow are now rendered useless. The improperly designed screw feeder creates a preferential flow channel that enforces a funnel-flow pattern with its resulting problems of ratholing, erratic flow, flooding, segregation, etc. The key to proper feeder design is to ensure that the feeder increases in capacity in the direction of discharge as, for example, with a tapered screw feeder, illustrated in Figure 6(b). Similarly, for belt feeders, the feeder must increase in capacity in the direction of discharge to ensure fully active mass flow across the entire cross section of the hopper.

C.   Use of Mechanical Aids

Many hoppers/silos in the industry operate in funnel flow, and this may be acceptable as long as arching and ratholing do not occur. If problems occur, then the proper route to obtain consistent reliable flow is by redesigning or retrofitting a hopper/feeder system to produce mass flow. Sometimes this may be considered too expensive or available headspace may prove limiting. In these situations, it may be necessary to consider alternatives like

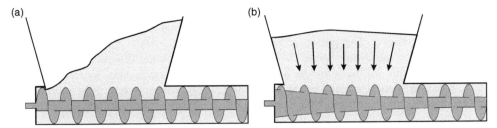

Figure 6 Screw feeder designs: (a) improper screw design causing preferential flow channel; (b) tapered shaft screw feeder.

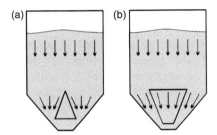

Figure 7 Hopper inserts: (a) inverted cone; (b) hopper in hopper.

mechanical aids. These aids can be categorized as vibrating discharge aids (such as internal and external vibrators), mechanical extractors/dischargers (such as sweep augers), aeration devices (such as air cannons), and hopper inserts. More detailed information about these aids is given in [29].

Care must be taken when using mechanical flow aids, because, if they are used incorrectly or for the wrong powder, they may have no effect at all or may even make the problem worse. Considerable literature exists describing various types of dischargers, mainly written by vendors; however, there is little useful guidance provided to select and implement a reliable discharger and how this is related to powder properties [30]. Vibrators and air cannons are usually most successful when required to encourage a bulk solid to commence flow after a period of storage at rest. Care must be taken in the positioning of these devices so that they are most effective in destabilizing arches or ratholes. However, if the powder is very cohesive, having a steep flowfunction, then increasing pressure on it may only serve to increase its strength and make the problem worse. Mechanical extractors/dischargers, such as sweep augers, should only be used for the worst of the worst or extremely cohesive materials. Generally, they should only be considered where the flowability of the bulk solid is so poor as to make other approaches impossible or uneconomical [31]. Hopper inserts can be used to modify or alter flow in a silo. One of the first inserts was an inverted cone suspended in the hopper cone above the outlet, as illustrated in Figure 7(a). A more recent design is the hopper in hopper design, as illustrated in Figure 7(b). Inserts are often useful in improving bin discharge. However, they have limited use with cohesive materials. Once again, care must be taken with the design and implementation of inserts, because if they are used incorrectly, they may make the problem worse.

Air permeation devices can be applied to overcome limiting discharge rates of fine powders ($<100\ \mu$m) from hoppers [2]. As powder moves from the top of the silo toward the outlet, the pressure exerted on it first increases, thus increasing its bulk density and squeezing out air. However, the pressure then decreases as it nears the outlet of the hopper, resulting in a reduction of bulk density and creation of a partial vacuum between the powder particles in the hopper. This creates an airflow inward from the outlet counter to solids flow. As the discharge rate is increased, the airflow rate in the counter direction also increases and eventually this will limit the discharge rate. Using an air permeation system to supply air to break the partial vacuum can increase the limiting discharge rate. It is important to realize that the air requirement for these systems is very low and injecting excessive air could fluidize the powder leading to flooding.

D. Flow Additives/Anticaking Agents

Flow additives are inert chemicals with small particle size (around 1 to 4 μm) that are mixed into the bulk powder [32]. They act by placing themselves between powder particles and disrupting cohesive interaction between particles, such as van der Waals interaction and liquid bridge formation. In the food industry, it is common to add flow additives to render the product more free-flowing. Applications include salt, spices, seasonings, powdered drinks, gravy mixes, cheese powders, and food flavors. A limited number of flow additives is approved for use in foods, including silicon dioxide and calcium silicate. Food powders that are directly handled by the consumer need to flow freely and must not agglomerate. Left to themselves, many food powders have a tendency to agglomerate and cake over time, which is very unsatisfactory for the consumer; thus, there is a need for flow additives to prevent this from happening. In particular, most food powders have a tendency (some more than others) to pick up water from the ambient atmosphere over time, which will reduce flowability and may eventually result in caking. Flow additives help in preventing this from occurring; however, as the powder picks up more and more water, this may eventually overwhelm the flow additive.

E. Control of Particle Size, Moisture, and Temperature

There may be opportunities to improve powder flowability by controlling powder particle properties and processing conditions. For example, increasing powder particle size and reducing moisture content tends to reduce cohesion and make the powder more free flowing. For example, sugar is dried to a low moisture content to prevent it from caking. Agglomeration/granulation can be applied to increase particle size and thereby reduce the powder's cohesive strength. Controlling the temperature may influence powder flowability, as described above. It may be necessary to control process temperature and relative humidity when processing very hygroscopic powders. For example, it is necessary to control the relative humidity and temperature of air during the mixing of a fine hygroscopic ceybrite tea powder, because the powder readily adsorbs water from ambient air, making it very cohesive and difficult to mix.

REFERENCES

1. Knowlton, T.M., Carson, J.W., Klinzing, G.E., and Yang, W.C. The importance of storage, transfer and collection. *Chem. Eng. Prog.* April, 44–54, 1994.
2. Purutyan, H., Pittenger, B.H., and Carson, J.W. Solve solids handling problems by retrofitting. *Chem. Eng. Prog.* April, 27–39, 1998.
3. Johanson, J.R. Troubleshooting bins, hoppers and feeders. *Chem. Eng. Prog.* April, 24–36, 2002.

4. Peleg, M. Flowability of food powders and methods for its evaluation — a review. *J. Food Process Eng.*, 1, 303–328, 1978.

5. Teunou, E., Vasseur J., and Krawczyk, M. Measurement and interpretation of bulk solids angle of repose for industrial process design. *Powder Handling Process.* 7(3), 219–227, 1995.

6. Carr, R.L. Evaluating flow properties of solids. *Chem. Eng.* 72, 163–168, 1965.

7. Jenike, A.W. Storage and flow of solids. Bulletin 123, Engineering Experiment Station, University of Utah, 1964.

8. Tomas, J. and Schubert, H. Particle characterisation. Partec 79, Nurnberg, Germany, 1979, pp. 301–319.

9. Prescott, J.K., Ploof, D.A., and Carson, J.W. Developing a better understanding of wall friction. *Powder Handling Process.* 11(1), 25–36, 1999.

10. De Silva, S.R. Characterisation of particulate materials — a challenge for the bulk solids fraternity. *Powder Handling Process.* 12(4), 355–362, 2000.

11. IchemE. *Standard Shear Testing Technique for particulate solids using the Jenike shear cell.* IChemE, UK, 1989.

12. Akers, R.J. The certification of a limestone powder for Jenike shear testing — CRM 116. Loughborough University of Technology, UK, BCR/163/90 (Community Bureau of Reference, 1990)

13. Community Bureau of Reference (BCR) of the European Commission, Directorate for Science, Research and Development: Reference Materials, 1994.

14. Schwedes, J. Testers for measuring flow properties of particulate solids. *Powder Handling Process.* 12(4), 337–354, 2000.

15. Schulze, D. Measurement of flow properties of particulate solids in food and pharmaceutical technology using a new automated ring shear tester. PARTEC 98, First European Symposium on Process Technology in Pharmaceutical and Nutritional Sciences, Nürnberg, 1998, pp. 276–285.

16. Schulze, D. Flowability and time consolidation measurements using a ring shear tester. *Powder Handling Process.* 8(3), 221–226, 1996.

17. Schulze, D., Heinrici, H., and Zetzener, H. The ring shear tester as a valuable tool for silo design and powder characterization. *Powder Handling Process.* 13(1), 19–24, 2001.

18. Bumiller, M., Carson, J., and Prescott, J. Particle shape analysis of powders and the effect of shape on powder flow. Fourth World Congress on Particle Technology, Sydney, Australia, 2002.

19. Shamlou, P.A. Handling of Bulk Solids — Theory and Practice. Butterworths, London, UK, 1988.

20. Scoville, E. and Peleg, M. Evaluation of the effect of liquid bridges on the bulk properties of model powders. *J. Food Sci.* 46, 174–177, 1981.

21. Teunou, E. and Fitzpatrick, J.J. Effect of relative humidity and temperature on food powder flowability. *J. Food Eng.* 42(2), 109–116, 1999.

22. Roos, Y.H. *Phase Transitions in Foods.* Academic Press, San Diego, CA, 1995.

23. Jouppila, K. and Roos, Y.H. Glass transitions and crystallization in milk powders. *J. Dairy Sci.*, 77, 2907, 1994.

24. Jouppila, K., Kansikas, J. and Roos, Y.H. Glass transition, water plasticization and lactose crystallization in skim milk powders. *J. Dairy Sci.* 80, 3152–3160, 1997.

25. Listiohadi, Y.D., Hourigan, J.A., Steele, R.J., Johnson, R.L., and Sleigh, R.W. Lactose — polymorphism, caking behaviour and use in pharmaceuticals. Fourth World Congress on Particle Technology, Sydney, Australia, 2002.

26. Teunou, E. and Fitzpatrick, J.J. Effect of storage time and consolidation on food powder flowability. *J. Food Eng.* 43(2), 97–101, 2000.

27. Marinelli, J. and Carson, J.W. Solve solids flow problems in bins, hoppers, and feeders. *Chem. Eng. Prog.* May, 22–28, 1992.

28. Marinelli, J. Choosing a feeder that works in unison with your bin. *Powder Bulk Eng.* December, 43–54, 1996.

29. Institution of Mechanical Engineers (IMechE, UK). *Hopper and Silo Discharge — Successful Solutions*. Published by Professional Engineering Publishing Ltd for IMechE, London, UK, 1999.

30. Arnold, P.C. Some observations on the relevance of flow properties in the selection of bin dischargers. *Powder Handling Process.* 12(4), 371–374, 2000.

31. Bradley, M. Strategy for selecting the solution. In *Hopper and Silo Discharge — Successful Solutions*. Published by Professional Engineering Publishing Ltd for IMechE, London, UK, 1999, pp. 95–110.

32. Peleg, M. Physical characteristics of food powders. In *Physical Properties of Foods*. Peleg, M. and Bagley, E.B. (Eds.) API Publishing Co., Westport, CT, USA, 1983, pp. 293–324.

11

Phase Transitions During Food Powder Production and Powder Stability

Bhesh R. Bhandari
University of Queensland
Australia

Richard W. Hartel
University of Wisconsin
Madison, Wisconsin

CONTENTS

I.   INTRODUCTION

Dry solid food materials are produced in various forms and dimensions, ranging from pieces to very fine micron size powders. The forms of dehydrated product are determined by the original state prior to processing, pretreatments, and water removal methods. Two methods of water removal are usually employed to produce the dry products, separation by crystallization/sublimation (freeze drying) and water removal by vaporization (various drying techniques). Depending on the physical states of the original (raw) material, the drying processes can be classified as solid or liquid drying. Although both types of raw products may have a similar amount of initial moisture (more than 75%), solid products (i.e., fruits and vegetables) have defined shape and size and do not flow, contrary to liquids (e.g., fruit puree). However, solids with high moisture content are characterized by viscoelastic behavior and may be prone to disintegration, even with a small shear force.

In general terms, drying is a process during which there is a transition of viscoelastic or liquid material to solid form. The dried material may be either a crystalline solid or an amorphous solid. Formation of crystalline powders requires sufficient time to allow the molecules to form into a crystal lattice. Under rapid drying conditions or in foods where no single component crystallizes, the molecules generally solidify in an amorphous or random form. In some cases, this amorphous solid may be in the glassy state, where the equivalent viscosity is greater than 10^{12} Pa sec. The temperature at which this liquid to glassy solid (or solid–liquid) transition takes place is called the glass transition temperature (T_g). This transition takes place over a range of temperatures, but normally a midpoint temperature is taken as glass transition temperature. To achieve maximum shelf stability of the product, drying should continue until the product is converted to a glassy state, since the molecular mobility and reactivity of the molecules are minimized or even halted in this state. In some cases, a mixture of crystalline and amorphous solids may be formed within a powdered food.

Many dried products are found in powder form. Both granular products with a dimension in the order of millimeters and fine powder products with average size less than 100 μm are treated as powders. These powders are generally free-flowing, can be used in very small portions (in milligrams), and are easy to transport in pipes pneumatically. In many processing situations, the powder forms are essential, such as in mixing and dissolution. The two main methods of conversion of liquid to powder form are drying and crystallization. Other processes such as grinding and milling also contribute to powder production, but these are purely size reduction processes. The particle size, distribution, and shape, and the density of the powders are highly variable and depend on both the characteristics of the raw materials and processing conditions during drying. These parameters contribute to the functional properties of powders, including flowability, bulk density, ease of handling, dust forming, mixing/segregation, compressibility, and surface activity (Peleg, 1983). Powders have a large surface area per unit volume and may be hygroscopic (e.g., high degree of moisture absorption). The stability of a powder, in terms of physical and chemical properties, is usually impaired by increased moisture sorption.

II. FOOD POWDERS AND THEIR MICROSTRUCTURE

Food powders may be amorphous, crystalline, or mixed (semicrystalline). Some examples of food powders in these state are listed in Table 1. Powders in crystalline state possess defined molecular alignment in the long-range order (Figure 1). The molecules in crystalline form are tightly packed; therefore, only radical or functional molecular groups on the external surface of the crystals can interact with external materials, such as water (absorption). Thermodynamically, the crystalline form is in the lowest energy level or stable equilibrium state (Hartel, 2001). Molecules in the amorphous state are in anarchy, tangled, more open, and porous; therefore, an individual molecule possesses more sites for external interactions, and, as a result, an amorphous structure can absorb water easily. The microstructure of an amorphous solid may consist of short-range order and regions of high and low densities (Figure 1) and have higher entropy than the corresponding crystals. The short-term order in amorphous glasses signifies the order of molecules over a small range in spatial dimension (Hartel, 2001). Amorphous glasses can undergo crystallization or structural relaxation to achieve an equilibrium condition (Yu, 2001). Mixed powders (coexistence of amorphous and crystalline structure) have both amorphous and crystalline regions. This can occur during processing when molecular motion ceases due to insufficient solvent or reduced temperature. For example, lactose crystals are embedded in an amorphous lactose glass when precrystallized whey is dried. Grinding of castor sugar to produce icing sugar destroys the aligned molecules on the surface and creates amorphous regions. The presence of amorphous regions in the starch granule is another example of a mixed structure. A microcrystalline structure can also be achieved by micronization of powder particles. This may not even be detected by x-ray diffraction.

Table 1   Physical States of Various Food Powders

Forms	Examples of powders
Amorphous	Milk, some whey powders, instant coffee and tea, spices, cheese, protein, coffee whitener, cocoa, spice mixes (gravy, soup, etc.)
Crystalline	Refined sugar, organic acids, polyols, salts
Mixed	Some whey powders, starch powders, ground icing sugar

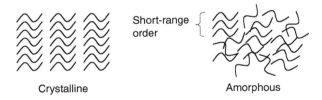

Figure 1   Schematic representation of crystalline and amorphous molecular structure.

A. Importance of Physical State

The physical state of the powder influences the functionality and handling properties. The physical state also determines the particle shape to some extent, particularly in the case of crystalline powders. Flow characteristics are very important properties of powder for transport, handling, mixing, and instantization. Amorphous powders are more hygroscopic than crystalline powders. Therefore, they tend to be sticky when they absorb moisture, which has an adverse effect on flowability of powder. Amorphous powders also tend to be more porous and bulky than crystalline powders. Hardness, cohesiveness, and compressibility of amorphous powders are different from those of crystalline powders. The particle density and hardness of crystalline powders are high and the powder may be less compressible under stress. Amorphous powders deform and compact more, and tend to cake under stress. For example, a bag of powder at the bottom of a bin or stack in the bulk storage room is more prone to caking.

Rehydration properties such as solubility and rate of dissolution can be influenced by the form of the powder. Crystalline powders typically dissolve slowly because dissolution occurs only at the outer surface exposed to the solvent. The particles are impermeable to the solvent because the molecules are bonded tightly and the density is high; therefore, the dissolution proceeds from outside to inside. In contrast, amorphous powders can dissolve rapidly because the particles are porous and hygroscopic. However, some vitrified powders (least porosity and structure fully relaxed) can be as tough to dissolve as crystals. Amorphous powders (such as gum and milk powders) tend to lump during dissolution due to particle swelling. The particle size is also responsible for this lumping and rate of dissolution. The amorphous state also positively influences the melting behavior of the mixed form powder, as in starch granules, where the absorption of moisture by amorphous starch enhances the melting of the crystalline fraction. The heat of dissolution of the amorphous phase differs from the heat of dissolution of the crystalline phase. The dissolution of the amorphous phase essentially involves solid–liquid transition and heat of mixing. Dissolution of crystalline material involves heat of dissociation of packed molecules into solvent and heat of mixing. Therefore, the heat of solution of the crystalline material is higher than for the amorphous phase dissolution.

Encapsulation is a process whereby the active and noncompatible ingredients (such as flavor, vitamins, enzymes, and fat) are entrapped within a matrix of a carrier material. Amorphous powders are often used for encapsulation. The amorphous form is essential for the entrapment of the active or noncompatible component, an example being fat in whole milk powder, which is entrapped within the amorphous matrix of lactose and protein. The crystalline form, however, excludes any other component. Since crystallization is a process in which molecules of the same species are tightly bound, they do not accommodate foreign molecules. During phase change of powder from amorphous form to crystallized form, such as during storage, the encapsulated material is ejected and the quality is degraded. Another form of encapsulation involves the agglomeration of tiny crystals that entrap other noncompatible molecules within the interparticulate space. This method is employed during the encapsulation of active components by cocrystallization of sugars.

The stability of the crystal form is superior to that of the amorphous form. The amorphous structure, being in a thermodynamically nonequilibrium state, always tends to relax or crystallize if given the opportunity. The absorption of moisture accelerates this process in food powders. This normally results in caking, loss of flavor, and degradation of other noncrystalline components present in the system. The crystalline form is stable at any temperature below the melting point, whereas the long-term stability of the amorphous form is attained well below T_g. The melting point temperature of food crystals is at least two times higher than the amorphous form transition temperature. The glass transition temperature property is strongly influenced by the presence of water in the amorphous powder (every 1% of moisture present can depress the T_g by 10°C in amorphous powder).

III. FORMATION OF FOOD POWDERS

The structure of food powder formation depends on the type of materials, other components present, and the process conditions. Some materials, such as those having large molecular size, only form amorphous structures. Examples are hydrocolloids, starches, and proteins. Other materials, such as salts, form only crystalline powder because they cannot withstand high supersaturation and have low energy of crystallization. Low molecular weight carbohydrates, such as lactose, glucose, sucrose, and organic acids and polyols, can either be in crystalline or amorphous states. Their structure depends on the processing conditions and other components or impurities present in the system. Sugar-based powders present an interesting example of moisture management to promote formation of desired forms.

Materials are also classified into good glass formers and poor glass formers based on glass forming ability (Yu, 2001). Some materials need to undergo significant conformational change for crystallization. For some materials, the conformational equilibrium has also to be maintained, affecting the crystal forming ability. Poor glass formers can still exist in amorphous form due to adverse processing conditions or impurities.

The form of the powder, whether crystalline, amorphous, or mixed, can be detected in the x-ray diffraction pattern, melting endotherm in differential scanning calorimetry (DSC), and water sorption trend analysis. In a system with a small amorphous fraction, water sorption analysis can be the most sensitive technique (Hancock and Zografi, 1997). This is due to a marked difference of sorption behavior in crystalline and amorphous structures (Figure 2).

A. Crystalline Food Powders

Crystallization is a phenomenon of separation of solute from solvent or other components. Supersaturation in a solvent or liquefaction to provide molecular mobility is the primary

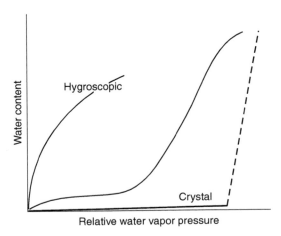

Figure 2   Moisture sorption property of crystalline and amorphous food powders.

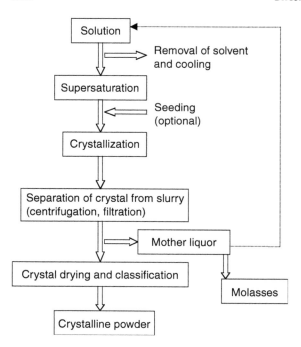

Figure 3 Schematic representation of a typical crystallization.

requirement for crystallization to take place. A crystalline powder form can be obtained by crystallization from solution or the melt by holding at a temperature above T_g. Examples of common powders found in crystalline states are salts, sugars, and organic acids. Crystalline powders are nonhygroscopic, stable, and easy flowing. A typical processing method for producing crystalline sugars from a solution is described in Figure 3. In this process, crystallization involves concentration of the solute above supersaturation by removal of solvent through evaporation and by cooling. Tiny crystal particles can also be added to the supersaturated solution for controlled crystallization (seeding). In the case of sucrose manufacture, supersaturation during crystallization is maintained by evaporation of water under vacuum, whereas in lactose crystallization, supersaturation is maintained by evaporation and controlled cooling during crystallization. Maintenance of the conformation equilibrium is also important in some sugars. Lactose solution exists in conformation of two (α and β) forms at a proportion of around 1:1.5 at 50°C. As the α-lactose crystallizes in the monohydrate form, the β-lactose changes into the α-lactose (a reaction known as mutarotation) to maintain the equilibrium in the solution.

The crystals obtained from crystallization from solution (Figure 3) are dried to remove residual surface moisture. The surface may contain a small amount of amorphous mother liquor with some impurities, which behaves as an amorphous solid after drying. The presence of this amorphous layer is undesirable because of its high hygroscopicity and the fact that it may cause clumping and caking of the crystalline powder during storage. Impurities in the solution can influence crystal growth and may cause some disorientation of the crystal shape. Sometimes milling is needed to meet the particle size specification. Milling causes amorphization (or disordering) of the crystalline structure particularly on the surface of the micronized crystals. Therefore, milled sugar exists in mixed form (crystal + amorphous) and is prone to caking. Anticaking agents, such as cornstarch, are sometimes added to icing sugars to avoid caking.

Figure 4   Maltitol powder (a) crystallized from solution (b) crystallized from extrusion melt.

Some special crystalline powders are also manufactured from melts. Examples are cocrystallized sucrose powder encapsulated with flavors, fruit solids, peanut butter, or honey (Chen, 1994; Bhandari et al., 1998). Spontaneous crystallization of sucrose at high temperature and low moisture under continuous agitation results in rapid crystallization and agglomeration of tiny crystals. The amorphous liquid acts as a liquid bridge between the crystal particles. The encapsulating materials are entrapped with the agglomerates. The product is in granulated form and may need further removal of moisture.

Maltitol powders may be manufactured by crystallization from solution or by extruding the melt (Figure 4). The purity of the product depends on the method of crystallization; in this case 99.7% purity is achieved by crystallizing from solution and 92 to 94% from extrusion. The method of crystallization and residual moisture content can influence the melting property of the crystals. For example, maltitol has a DSC melting peak at 152.1°C (moisture 0.08%), whereas it is at 144.7°C (moisture 0.39%) for crystals prepared by the extrusion method.

The crystalline form may also be prone to agglomeration and caking. Temperature and humidity variation in the environment can result in absorption of moisture, and dissolution of the crystal surface forming a thin amorphous layer. This layer eventually cements adjacent particles due to its sticky property. Recrystallization may also occur in this layer. Consequently, a free-flowing crystalline powder will develop caking when held at humid conditions.

B.   Amorphous Food Powders

Amorphous structure exists in many important food powders such as high or low molecular weight carbohydrates and proteins. The amorphous state is a nonequilibrium state, and its formation also originates from a nonequilibrium condition. Amorphous powders are obtained from two material systems: solution and dry crystalline solids. Process conditions directly influence the formation of amorphous powders. A condition of rapid cooling and rapid removal of solvent does not allow enough time for the molecules to align themselves in a low energy crystalline state. Therefore, rapid solvent removal and cooling produces amorphous powder. Drying is one of the most common methods used to produce amorphous powders (Figure 5).

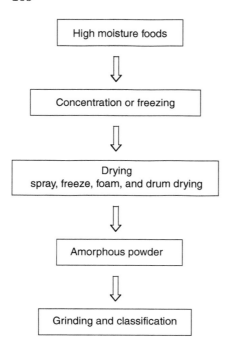

Figure 5 Amorphous powder formation during drying.

A process of melting the crystalline solid, rapid cooling, and grinding can also give an amorphous powder. Partial or complete amorphous structures can also be formed from crystalline solids by milling or grinding them. A micronization process can virtually convert nearly all the crystalline arrangement to a nonequilibrium amorphous structure, though some micocrystallinity may always be present.

A material with slow crystallizing tendency promotes amorphous structure formation; for example, fructose does not crystallize easily from a supersaturated solution due to its high solubility. The presence of impurities or large molecules can virtually prevent crystallization and promote glass formation. Addition of impurities is also used to obtain amorphous structure of poor glass formers, such as the addition of sorbitol to mannitol (Yu et al., 1998). Different components inhibit crystallization to different extents (Hartel, 2001). Inhibition of crystallization may arise from mass transfer influences, where the diffusivity of solute molecules is decreased in the presence of other components, or from specific adsorption effects, where the impurity molecule interacts with an existing crystal surface and impedes further molecular deposition (Hartel, 1987; Hartel and Shastry, 1991). This may create a situation of completely amorphous or mixed structure.

C. Mixed Powders

As described earlier any processing situation that slows or inhibits crystallization promotes formation of mixed structure (crystalline + amorphous) powders. Any solid products made in the presence of impurities, by medium or fast cooling or solvent removal, or a low level of supersaturation results in mixed structures. Some natural materials, such as starch granules or powders, have some degree of mixed structure, though the proportion of amorphous fraction can normally be much lower than the crystalline one. In starch granules, the straight-chain sections of amylopectin form crystalline regions, whereas the branched-chain sections of the amylopectin, along with amylose, make up the amorphous regions (Oates, 1997). Crystallization of powders in prior or postprocess conditions also gives a mixed powder, for instance, partial crystallization of lactose in skim milk or whey before or after spray drying. This is employed to obtain a powder with low hygroscopicity.

IV. PHASE TRANSITIONS IN THE PRODUCTION OF FOOD POWDERS

A. Glass Transition

A phase transition is involved in the manufacturing of food powders. In this context, phase transition refers to the transition from liquid to solid or vice versa (Figure 6). The liquid can be a solution or a melt (no solvent). The solid in this case can be either amorphous or crystalline. All the phases are reversible and the state change mechanisms are time and temperature dependent. An amorphous liquid (or solid) state passes through a highly viscous rubbery state before transition to solid glassy (or liquid state). The temperature at which the transition to the rubbery state occurs is defined as the glass transition temperature (T_g). The T_g and melting point temperature (T_m) of some food materials are shown in Table 2.

Amorphous glasses are sometimes referred to as "the fourth state of matter" because they have the structure of a liquid but the properties of a solid. They may also be called a "solid solution," a "glass," or "vitrified solid." They behave as an extremely viscous liquid having a typical viscosity above 10^{12} Pa sec (Downton et al., 1982). In this state, molecular movement is highly limited, which is necessary for an orderly alignment of molecules to crystallize. The viscosity of the rubbery state is less than 10^{12} Pa sec, although few measurements have attempted to quantify conditions in this region. Downton et al. (1982) measured the sticky point of sugar-based materials at approximately 10^8 Pa sec. Undoubtedly, the nature of the rubbery transition into the sticky zone is time dependent.

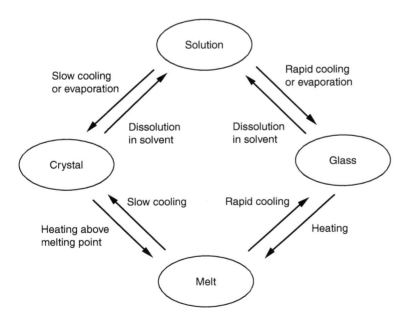

Figure 6   Schematic diagram indicating phase transition of a material.

Table 2 The Phase Change Properties of Some Individual Anhydrous Components (Unless Specified Otherwise) and Foods

Food material	$T_g(°C)$	$T_m(°C)$[a]	Ref.
Fructose	16	108	Roos (1993); Truong et al. (2002); Raemy and Schweizer (1983)
Glucose	36	143 (85)	
Galactose	38	165	
Sucrose	67	192	
Maltose	92	165 (102)	
Lactose	101	223 (201)	
Citric acid	12	153 (135)	Maltini et al. (1997); Truong (2003)
Malic acid	−16	130	
Tartaric acid	18	206 (206)	
Lactic acid	−60	17	
Maltodextrins			Roos and Karel (1991a)
DE 36 (MW 500)	100		
DE 25 (MW 720)	121		
DE 20 (MW 900)	141		
DE 10 (MW 1800)	160		
DE 5 (MW 3600)	188		
Starch	243		
Ice-cream	−34.3		[e]Donhowe et al. (1991)
Honey (moisture 15–18% w/w)	−38 to −46		Sopade et al. (2002)
Dried strawberry ($a_w = 0$, anhydrous)	25		Khalloufi et al. (2000)
Partially dried strawberry ($a_w = 0.75$)	−65		Roos (1987)
Dried apple ($a_w = 0$, anhydrous)	4.5		Welti-Chanes et al. (1999)
Partially dried apple ($a_w = 0.756$)	−79		
Apple juice (freeze dried, $a_w = 0.231$)	−26		

[a] Values in bracket are for the hydrated form of crystals.
[e] T_g of maximally concentrated solution.

The glass transition temperature, T_g, is a measured transition point of a glass from its solid to liquid state, or vice versa. Unlike the first-order transition of crystalline solid to liquid by melting, the glass transition is a second-order phase change and is characterized by abrupt changes in viscosity, specific heat, mechanical relaxation, and free molecular volume (Roos, 1995b). In simple terms, above T_g, matter is in a liquid state, whereas below T_g it is in a solid-like state. Texture and physicochemical stability of a material are related to T_g. Below T_g, the physical status of the material remains virtually unchanged; therefore, the shelf life is at a maximum.

A phase/state diagram is normally used to highlight the solid–liquid behavior of a material in relation to temperature and concentrations of solute in water. Each food material has its own phase/state diagram. A change in the concentration of solute is achieved either by removal of water (drying) or by freezing (water crystallization). The physical state of sucrose as a function of solvent concentration and temperature is shown in Figure 7. The required conditions for glassy states, such as temperature and concentration in water and presence of crystal solid below the solubility curve, are also depicted in this figure.

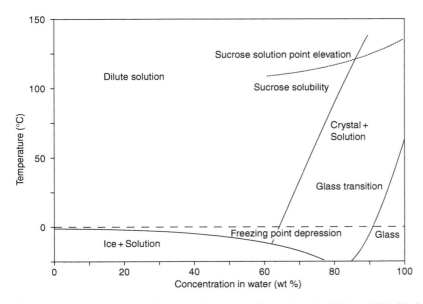

Figure 7 Simplified state diagram of sucrose. (From Hartel, 2001 — With kind permission of Springer Science and Business Media).

B.   Phase Transitions during Drying of Foods for Powder Production

The most common way of obtaining food powders is by drying liquid food materials. Three techniques, namely freeze, drum, and spray drying, are typically used. The principles of drying for these three techniques differ and they also produce powders with different shapes, dimensions, and functional properties. The important factors influencing the physical state properties of the material are formulation and processing conditions. The formulation factors include the properties of the primary components, such as poor or good glass formation and glass transition curve, and the effect of the presence of other components on the T_g curve and glass forming properties. The processing factors include the temperature in the dryer, relative humidity of the air, and the drying technology employed.

The material undergoing drying can be a mixture of low or high molecular weight solids and water. The T_g of starches, proteins, and gums (high molecular weight materials) is higher than that of the sugars, organic acids, and polyols (low molecular weight materials). The T_g of water is the lowest at around $-135°C$. Therefore, the influence of water in depressing the T_g of the mixture is significant especially in the early stages of drying. The T_g of materials increases as the drying progresses due to the removal of water. The increased T_g values of some food materials as a function of moisture content in their low moisture range are shown in Figure 8. Materials with high glass transition temperature are easy to dry, whereas those with low T_g may be impossible to dry since they remain in the fluid state under normal drying conditions.

1.   Freeze Drying

Freeze drying involves the removal of frozen water by sublimation. The method necessitates high vacuum and the drying rate is slow and therefore, expensive. Its application is limited to high value products, such as meat, coffee, fish, and some seafoods, and fruits. The freeze drying process consists of three stages: freezing, primary drying (mainly frozen water and some sorbed water), and secondary drying (sorbed water). The product must be frozen prior

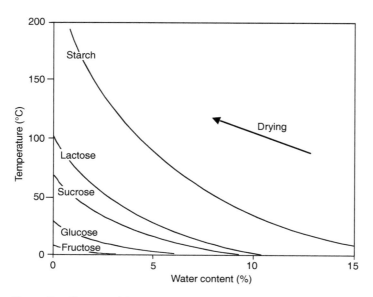

Figure 8   Glass transition temperature of some food materials as a function of moisture content.

to drying. During freezing, depending on the composition of the food material, 65–90% of the initial water is in the frozen state and the rest, 10–35%, of water is in the unfrozen (sorbed) state. The rate and amount of ice formation during freezing is dependent on the composition, viscosity of unfrozen fraction, and isothermal annealing between its glass transition temperature and the melting point of ice (Roos, 1995a). If the water concentration in the product is low, all the water remains in the unfrozen state. Water remained unfrozen in partially dried strawberry when the water activity of the materials was below 0.75 (Roos, 1987) and water in honey (moisture ≈ 16% w/w, a_w ≈ 0.6) was unfreezable (Sopade et al., 2002). This suggests that all the water present is strongly influenced by the sugars and organic acid constituents in these food systems preventing ice crystallization.

   During freeze drying, the heat of sublimation is supplied by conduction and radiation. The drying front recedes from the surface toward the inside. The sublimation of pure ice is possible only below the triple point at which all three states of matter exist together. The triple point of water is at around 610.8 Pa absolute pressure and 0.0098°C. But food contains solid matter and therefore, requires drying at lower temperature. Generally, freeze drying is done at −10°C or lower and at absolute pressures of about 270 Pa or less (Liapis and Bruttini, 1995). After the frozen fraction of the water is sublimed, the temperature of the chamber is normally increased and evaporation of sorbed water takes place purely under vacuum.

   Due to the presence of solute and the freezing point depression by the solutes present in the food systems, freeze drying should be undertaken below the eutectic point to avoid melting and eventual collapse of the structure. If there is no real eutectic point, the drying should take place below the glass transition temperature of the amorphous phase. This refers to the T_g of the concentrated unfrozen phase. For carbohydrate-based matrices, it has been found that the maximally freeze-concentrated state contains around 80% solids (Roos and Karel, 1991a; Roos, 1995a). Products with moisture content as low as 10% can collapse during freeze drying if the chamber temperature exceeds the T_g. Collapse is a dynamic process and therefore, a time-dependent property. The rate of collapse depends on $T - T_g$ (To and Flink, 1978; Levi and Karel, 1995). Collapse occurs due to viscous flow and therefore, relaxation time of such flow can be predicted using the WLF equation (Levi and Karel, 1995).

$$\log_{10} \frac{\tau}{\tau_g} = \frac{-C_1(T - T_g)}{C_2 + (T - T_g)} \tag{1}$$

Here, τ is the relaxation time, μ_g is the relaxation time at glass transition temperature T_g, T is the temperature, and C_1 and C_2 are "universal" constants (17.44 and 51.6 K respectively) that represent a best fit for a wide range of materials (which can be adjusted for some systems). For various food liquids during freeze drying, the collapse temperature can vary between -5 and $-60°C$ (Bellows and King, 1973) depending upon their composition. Foods that are high in sugars, such as fruit juices, have lower collapse temperatures. If the collapse temperature is lower than the practical range of operation of the drier, it is impossible to freeze dry the material without deformation. The collapse temperatures can be raised by the addition of high molecular weight materials such as maltodextrin (Tsourouflis et al., 1976).

Freeze drying involves removal of ice without an intermediate phase change; thus, the space occupied by the ice remains empty. Ideally, since the product is in a frozen solid state, it retains its shape and size. The empty space left in the product makes it very porous. The final product obtained is normally the same size as the initial frozen product. To convert this freeze-dried matrix to a powdery form, it needs to be ground to a given particle size range. Due to the porous nature, freeze-dried products have very low density and therefore, take up a lot of space in packaging. The flakes are brittle and irregular in shape. Being highly porous, they also can be very hygroscopic. Freeze-dried powders typically dissolve rapidly and retain almost all of the nutritional property, flavor, and aroma. The structural stability of the material during freeze drying is very important in retaining the porous structure. In any adverse event (e.g., water sorption), the product will collapse leading to a dense structure, quality degradation, and difficulty in dissolving the product.

Freeze drying is the only method of drying where the liquid–solid phase transition of the material to be dried actually takes place during freezing and where drying continues (under sub-T_g conditions) until the T_g of the product reaches well above room temperature.

2. Roller Drying

Roller drying is a process where the majority of the water evaporation takes place by boiling, normally at atmospheric pressure. The product is smeared as a thin film on to a steam-heated revolving drum. The temperature of the product rises rapidly above 100°C. In the first few seconds, the water evaporates by boiling and during the rest of the contact time, the sorbed water evaporates due to the difference in vapor pressure between the interface of the product film and the boundary layer. Roller driers can also operate under vacuum, but these are expensive to operate (Moore, 1995). The dried product, in the form of flakes, is scraped off the drum by a knife after one revolution. The residence time of the product on the drum depends on the revolution speed of the drum, and is usually more than 5 sec. This is, in fact, the oldest liquid drying process commercially applied in the food industry. Because of quality degradation due to high temperature and long residence time, roller drying has limited application and has been gradually replaced by the spray drying process over past decades. Current applications of roller drying are for drying animal feeds, some vegetable purees, apple-sauce, precooked breakfast cereals, dry soup mixtures, and special milk products such as confectionery and baby foods. Roller dried powders are in quickly rehydratable flake form. Roller dried milk powder is preferred in chocolate manufacture due to its high free fat content contributing to low viscosity of the mix (Twomey and Keogh, 1998). In contrast, roller dried caseinate produces a high viscosity in solution, which is a desirable property when preparing meat products, such as fermented and cooked sausages. This property of the caseinate is due to some degree of denaturation of protein. Starch gelatinization (also a phase transition phenomenon) also occurs during roller drying because of the high temperature of the product during drying.

Prior to roller drying, the product is usually concentrated to reduce the drying time, increase the throughput and allow better adhesion of the product on to the drum surface. The liquid–solid transition takes place by the end of a drum rotation or after scraping and cooling the product. The product during scraping can be in a rubbery state due to the high temperature of the drum. The real transition from rubbery to solid state may take place during cooling after scraping away from the drum surface. The transition converts the product from a flexible rubbery state to a brittle glassy state. Further drying may be carried out, such as by using microwaves, to remove the moisture from unevenly dried particles on some spots of the drum, mainly caused by bubbling and uneven thickness of the film. The product is recovered in the form of a thin sheet, but is structurally very weak. It is relatively porous, the porosity being created by vapor bubbles produced during boiling. The thin sheets or flakes are broken into a coarse powder in a conveying auger. Final grinding to a required size is done in a hammer or crushing mill.

Drying of sugar-rich products such as fruit juices and whey are difficult during roller drying due to the low glass transition temperatures of the solids present in these foods. Drying of cheese whey can be done, but the product comes as a thermoplastic sheet and lactose may crystallize in monohydrate form if enough water is present and the temperature is above the glass transition temperature. Crystallization of lactose renders the product less hygroscopic and more stable during storage. Lactic acid present in the whey depresses the T_g because the T_g of lactic acid is $-60°C$ (Table 2). This worsens the thermoplasticity and sticky nature of the product. Drum-dried acid whey needs immediate cooling and further drying in a fluidized bed. Sensitive water-insoluble ingredients, such as fat, are not effectively encapsulated in the amorphous matrix of carbohydrates or protein during drum drying. This leads to a high free fat content and susceptibility to oxidation. The stability of the powder will be more influenced by this, rather than by any phase change (crystallization) taking place during storage. Powders obtained by roller drying are easy to reconstitute due to the irregular, relatively porous, and larger particle size.

3. Spray Drying

Spray drying is the most commonly applied method of liquid to solid conversion. The drying time is very short, in the order of a few seconds, and the temperature of the product rarely reaches the temperature of the outgoing air. Besides the superior quality of the product, the possibility of a wide range of drying capacity has made it popular in the food, pharmaceutical, and chemical industries. The capacity of the spray drying unit can vary from a few grams of powder to several tons per hour. The mean particle size obtained from the spray drying process can range from 30 to 200 μm. There are normally single- and two-stage drying systems. Most of the large industrial spray dryers are integrated with a fluidized bed dryer (Figure 9). The two-stage drying system reduces production cost, produces agglomerated powder with improved reconstitution properties, minimizes loss of fines and, to some extent, the stickiness problem is better handled by introducing cool low humidity fresh air.

Wall deposition is a regular occurrence in the spray dryer. This may be caused by (1) coarse droplets contacting the wall before sufficient drying has occurred, (2) thermoplasticity of the product (low T_g), (3) surface dusting on the wall due to surface geometry, roughness, and electrostatic forces, and (4) localized buildup as a result of poor insulation

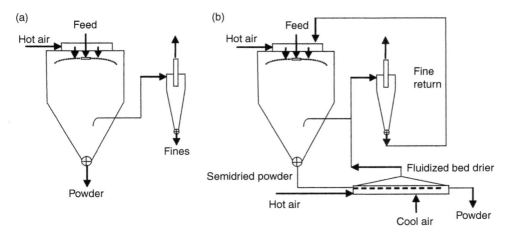

Figure 9 Schematic diagram of (a) single- and (b) two-stage spray drying plants (agglomeration and drying takes place in the second stage).

(heat loss) and condensation of humid air (Bhandari and Howes, 1999; Masters, 2002). Above all, stickiness due to thermoplasticity of the material is the major problem.

Rapid drying of droplets in the spray drying process causes a correspondingly rapid increase in the viscosity of the material, which prolongs the time required for the realignment of the molecules. Thus, the dried droplet is in an amorphous form. Figure 10 depicts the major changes taking place during the spray drying of a liquid product. The physical state changes as it passes through the dryer from solution to syrup, and finally to solid form. If the viscosity of the dried product is below a critical level of about 10^6 to 10^8 Pa sec, it may stay as a syrup or in a rubbery state even at low moisture levels (Downton et al., 1982). Depending upon the product characteristics and composition, and drying conditions, the surface of the drying droplets may remain plastic, resulting either in them sticking on the dryer wall or remaining as a sticky powder rather than turning into a relatively free-flowing powder. The problems of stickiness during spray drying of sugar-rich and acid-rich foods such as fruit juices, honey, some starch derivatives (glucose syrups/maltodextrin at higher dextrose equivalent values), whey, sugars, and hydrolyzates of proteins and sugars have been well recognized. During drying of these products, they may either remain as syrup or stick on the dryer chamber wall (Bhandari et al., 1997a). There is also the problem of unwanted agglomeration in the dryer chamber, fluidized bed, and conveying system. This leads to lower product yields and operating problems.

It has also been reported that the critical viscosity of stickiness is reached at temperatures 10 to 20°C above the measured T_g value of the material (Roos and Karel, 1991a; Hennigs et al., 2001). The stickiness curve at 10 to 20°C above the glass transition curve is depicted schematically in Figure 11. Flow dynamics within the dryer can also influence the stickiness property of the drop. Particles that are tangentially directed toward the dryer wall can impact and stick easily onto the wall even at their glass transition temperature. On the other hand, the inertia of mobile particles and the kinetic energy of the moving air can loosen the contact point of particles from the wall, which eventually results in an increase in the sticky temperature. Overall, the dominating effects of the film-forming properties of the material along with the dynamic conditions in the dryer can locally elevate the stickiness temperature above the glass transition temperature.

To avoid stickiness, process conditions must be such that the product (or the surface of the droplets) is at least below the sticky zone at the time it contacts the dryer wall.

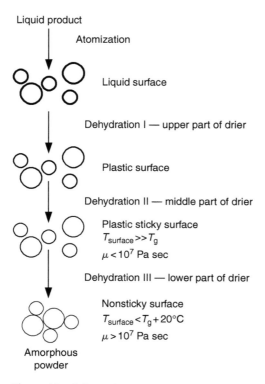

Figure 10 Schematic representation of physical changes in droplets during the spray drying process. (Dehydration I, II, III represent only the arbitrary stages of dehydration, μ = viscosity, T_g = glass transition temperature, $T_{surface}$ = surface temperature of drying particle.) (From Bhandari, B.R., Datta, N., and Howes, T. (1997b) *Drying Technol*. 15: 671–684. With permission.)

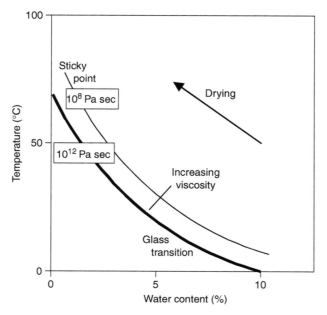

Figure 11 The glass transition temperature and stickiness behavior of the product during spray drying.

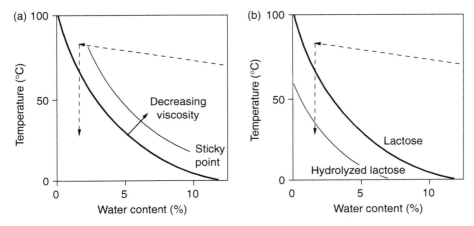

Figure 12 (a) Indicative sticky and glass transition curves for lactose and (b) glass transition curve for lactose and partially hydrolyzed lactose. (The dotted line is the droplet surface temperature.)

So it is important that the process conditions match the product being dried. The inlet and outlet conditions should be balanced to obtain appropriate humidity conditions at the exit of air. Use of high temperature at the exit reduces the humidity but this may increase the particle temperature well above the sticky zone. Too low a drying air temperature will increase the moisture in the product and thus, result in stickiness behavior. A trade-off in drying conditions is therefore necessary to minimize the stickiness behavior. Normally, a relatively low inlet air temperature is used in drying sticky products with a low T_g. Some products with very low T_g cannot be dried by spray drying.

For example, spray drying of whey is presented schematically in Figure 12. In whey, lactose is the principle component that dominates T_g (Joupilla and Roos, 1994) because salts and proteins have little effect. Anhydrous lactose has a T_g of 101°C, as shown in Table 2. Because of its sufficiently high T_g, spray drying of whey is not difficult in reasonably low drying temperature conditions (Figure 12[a]). However, the drying conditions should still be well controlled as the powder is very hygroscopic. Partial or complete hydrolyzation will convert the lactose into monosaccharide units (glucose and galactose), which have substantially lower T_g than lactose (Table 2). This means that the sticky curve for this product is much lower (Figure 12[b]) and it will not be possible to spray dry the completely hydrolyzed lactose because the average T_g of the glucose and galactose units will be lower than 35°C.

Honey is a mixture of sugars, 90% of which are glucose and fructose. The glass transition temperature of honey is around −45°C (≈16% moisture w/w). Considering the T_g of the principal components of honey (glucose and fructose) the glass transition temperature of anhydrous honey may lie between the T_g of fructose (16°C) and glucose (32°C). This means that spray drying of honey is not possible. Addition of carrier materials such as maltodextrin at levels up to 50% is necessary to dry this honey. The indicative sticky curve with or without maltodextrin is shown in Figure 13. Addition of maltodextrin substantially raises the T_g of the mixture. Sometimes a drying index is used to calculate how much maltodextrin is needed to permit spray drying (Bhandari et al., 1997b). The addition of a carrier means a dilution of the honey by 50%, but in terms of honey flavor it may not matter much if a strong honey is selected for spray drying. Spray dried honey with maltodextrin is commercially available and is used mainly in premixes and for flavoring purposes.

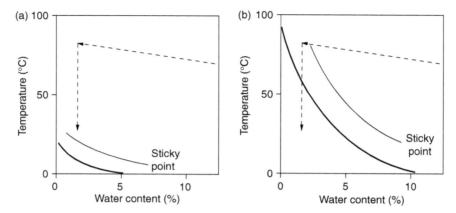

Figure 13   Sticky point curve of (a) honey and (b) honey and maltodextrin mixture. (The dotted line is the droplet surface temperature.)

V.   STICKINESS EVALUATION TECHNIQUES

Stickiness is attributed to the adhesion of food particles to the contact surface. The contact surface may be similar (cohesion) or dissimilar (adhesion). For example, adhesion takes place between drying droplets and the dryer wall, whereas cohesion is responsible for caking of the powder during processing or storage. Sensory stickiness correlates well with viscosity and surface tension of a material (Kilcast and Roberts, 1998). The factors rendering the particles adhesive or cohesive may be similar, but higher cohesion is not necessarily associated with higher adhesion or vice versa (Papadakis and Bahu, 1992; Adhikari et al., 2003).

The mechanisms of interaction of two surfaces can be divided into four major groups: intermolecular and electrostatic forces, liquid bridges, solid bridges, and mechanical interlocking. The type of stickiness mechanism related to drying or caking is the liquid bridge. These bridges can be pendular, funicular, or capillary. Further, the adhesion of semisolid foods to dissimilar surfaces is not possible unless they are characterized by good wettability or spreadability to the adherend. Good wettability means the food and the adherend have a strong mutual affinity and are likely to adhere well (Michalski et al., 1997).

The nature of the surface is the most important aspect to be considered in stickiness evaluation. Nowakowki and Hartel (2002) analyzed the stickiness development of amorphous candy due to moisture absorption and found that the bulk T_g does not relate to the surface stickiness. A peak in stickiness was observed at about 10 to 20°C below the bulk T_g. This was attributed to moisture at the surface being higher than in the interior and to the diffusion of moisture toward the interior being slow. On the other hand, during moisture desorption (drying), Adhikari et al. (2003) demonstrated that film formation in the droplets influences the stickiness property of the material because a surface crust solidifies earlier than the internal part of the droplet. Film formation is a material specific property.

Various methods have been used to characterize powder particles in relation to their stickiness behavior. Several researchers have attempted to devise instrumentation based on the properties of food materials, such as resistance to shear motion, glass transition temperature, viscosity, and optical properties. These techniques can be classified as stickiness evaluation techniques for drying droplet and powder stickiness.

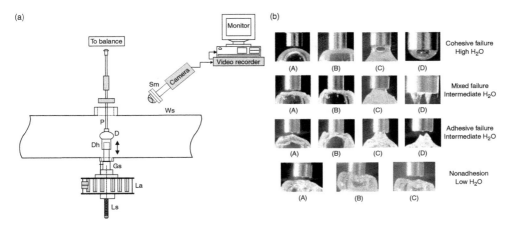

Figure 14 (a) Schematic diagram of stickiness testing device. P = stainless steel probe, D = sample drop, Dh = sample holder, Gs = guide shaft with peripherals for linear motion, La = linear actuator, Ls = lead screw, Ws = glass working section, Sm = stereomicroscope system with video camera. (b) Sequence of events during adhesive failure — (A) = approach, (B) = contact (C) = separation, and (D) = complete separation, in case of maltodextrin (DE 6) at $95 \pm 2°C$, 2.0% RH, and 30 mm/min probe withdrawal speed. (From Adhikari, B., Howes, T., Bhandari, B., and Truong, V. (2003). *J. Food Eng.* 58(1): 11–22. With permission.)

A. Stickiness Testing of Drying Droplet

Surface properties of a droplet, such as film formation and dried wall material properties, can greatly influence stickiness. The film formation property once again is a function of drying conditions. Adhikari et al. (2003) developed a rig for testing the *in situ* stickiness behavior of a drying droplet (Figure 14[a]). There are two setups in this rig — one for measuring stickiness and the other for studying the drying kinetics. Both the setups are housed inside a glass chamber supplied with hot air with controlled flow, temperature, and humidity. The linear actuator with an appropriate step size is used to achieve the forward and backward movements of the probe. The tensile and compression force during stickiness testing and air temperature is continuously logged. The temperature history of the sample is recorded using microthermocouples. The image of the bonding, debonding, and failure modes during the testing is recorded using a color camera and recorded on a personal computer through a frame grabbing card. The same image capturing system is also used to monitor the drying droplet.

During drying, the failure can be adhesive (between probe and drop) or cohesive (drop divides). At the initial stage when the drop is liquid, cohesive failure occurs (Figure 14[b]). Tensile force (stickiness) or cohesiveness increases to the maximum as drying progresses due to an increase in viscosity and interparticulate force. After a certain cohesive force is reached, the adhesive failure starts prevailing. An illustration of the sequence of the events during the semiadhesive failure, which is more common during the later stage of drying, is presented in Figure 14(b). Results show that at the initial stage of the semiadhesive failure, the cohesive mode is dominant but very rapidly the adhesive failure dominates the separation process.

The quantitative results on the evolution of stickiness during the dehydration of droplets of fructose and fructose–maltodextrin mixture are shown in Figure 15. In the case of fructose, it can be seen in this figure that the maximum tensile pressure required to separate the probe does not change significantly. This means that the fructose remains

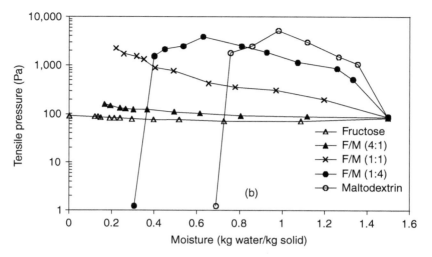

Figure 15 Evolution of stickiness (tensile pressure) of droplets of fructose (F), maltodextrin (M), and their mixtures during drying (drying at 63°C, 2.5% RH, 0.75 m/sec air velocity, and 30 mm/min probe withdrawal speed). (From Adhikari, B., Howes, T., Bhandari, B., and Truong, V. (2003). *J. Food Eng.* 58(1): 11–22. With permission.)

liquid at this temperature even at complete dryness. This can be related to the low glass transition temperature of fructose as presented in Table 2. The drop of pure maltodextrin solution, in contrast, does not remain spatially homogeneous during drying and the surface undergoes remarkable changes.

The rapid rise in cohesive strength (Figure 15) of a maltodextrin drop associated with the decrease in average moisture content is due to the fact that the outer surface of the drop of maltodextrin forms a skin soon after the onset of drying. As drying progressed, the skin became thicker. Subsequently, the outermost layers were nearly at a glassy state while the majority of the maltodextrin drop within remained as a viscous solution. It appeared that the success of maltodextrin as a drying agent was associated not only with its high T_g, but also with the formation of skin and its growth in thickness. This suggests that glass transition temperature of the average mass of droplets alone does not necessarily explain the stickiness property of film-forming materials. In other words, stickiness is a surface property of the material. The formation of a surface film during spray drying is also affected by the air/surface temperature difference, drying rate, initial moisture content, and spray drying characteristics such as droplet size and drop velocity.

B. Stickiness Testing Methods for Powder

1. Shear Cell Method

The Jenike shear cell is a common method used for quantification of cohesion and adhesion of granular materials including food powders. In this method, the powder is consolidated with the maximum applicable consolidation load. A plot of shear stress versus consolidation normal stress provides a yield locus curve. Although this method provides useful data on the cohesive force between particles, it is not usually used to characterize stickiness of powders. This method is commonly used for studying the flow behavior of powders through chutes and hoppers.

2. Propeller Driven Method

Originally developed by Lazar et al. (1956), this method has been used by several researchers with or without modifications to evaluate the effects of temperature on powder stickiness (Brennan et al., 1971; Downton et al., 1982; Wallack and King, 1988; Pasley and Haloulos, 1994). The tester basically comprises a test tube containing powder with known moisture content. The test tube is immersed in a water bath. A machine-driven impeller stirs the powder. When the temperature of the powder is slowly raised by increasing the water bath temperature, at the sticky point a maximum force of stirring is recorded. The sticky point temperature of powders can be determined at various moisture contents (Hennigs et al., 2001). Ozkan et al. (2002) developed a viscometry technique based on the measurement of the torque required to turn a propeller inserted into powders. The stickiness property determined by these analytical techniques is more related to cohesion between particulates than adhesion between particles and other dissimilar surfaces (like the dryer chamber wall). This method can only evaluate already dried powder. Moreover, a temperature lag during heating in a large sample will influence the stickiness measurement.

3. Optical Probe Method

A method based on the changes in optical properties of a free-flowing powder was recently reported by Lockemann (1999). The motion of the powder in a constantly rotating tube is observed with a fiber-optic sensor. The tube and sensor are all immersed in an oil bath to maintain the temperature. A sharp rise in reflectance of a freely flowing powder is observed at its sticky point. This method has been tested for colored food, such as carotene, but has not been applied to particulate materials that tend to be transparent on softening or melting. However, this seems to be a promising method for evaluation of already dried powders.

4. Blow Test

Paterson et al. (2001) attempted to develop a blow test for measuring the stickiness of powders. This method measures the velocity of air needed to blow a channel into a packed bed of powder and the stickiness of powder is classified based on the air velocity range. The apparatus consists of a multisegmented circular distributor (sample holder) where the preconditioned sample is packed in the distributor. Air at a given temperature and humidity is blown from a 45° angle on to each segment with increasing flow rate (maximum 22 l/min) until a channel is formed. This method is claimed to measure the time-dependent bond formation between the particles held at various temperatures above glass transition temperature of the powder. A significantly faster particle bonding was observed at around 10°C above T_g. The results from this method would represent further the caking behavior of the powder. This method is physically demanding and several steps are required before the measurement.

5. Fluidization Method

Bloore (2000) described a small fluidized bed setup to study the stickiness property of the powder at different humidity and temperature conditions. The positive point of this method as compared to other tests is that the particulates are in a dynamic condition, which is closer to the spray drying and fluidized bed drying situations. The stickiness observed by this method is dominated by the cohesive property of the particles. Compared to other methods, the data obtained from this method can represent conditions in spray and fluidized bed dryers.

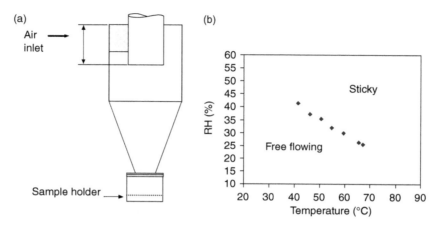

Figure 16 (a) Design of the cyclone chamber and sample holder. (b) Stickiness curve of skim milk powder as a function of temperature and relative humidity measured using cyclone setup. (From Boonyai, P., Bhandari, B., and Howes, T. (2002). Proceedings of the International Conference on Innovations in Food Processing Technology and Engineering. December 11–13. Jindal, V.K., Noomhorm, A., Rakshit, S., and Khan, I., (Eds.). Asian Institute of Technology, Bangkok, Thailand, 809–816. With permission.)

6.   Cyclone Method

A cyclone technique was described by Boonyai et al. (2002) to investigate stickiness behavior of food powders as a function of temperature and moisture conditions, simulating the dynamic condition in a spray drying system. The powder particles are individually in contact with a preconditioned air stream and hence, a rapid simultaneous heat and moisture transfer occurs at the surface. The cyclone consists of a detachable sample holder at the bottom (Figure 16[a]). A few grams of the sample is put in the sample holder for the test. Stickiness is observed within 1 to 2 min when particles become cohesive and stick to each other and some adhere to the chamber wall due to adhesive force. If a longer time span is allowed, all particles become completely immobilized. The testing time may also depend on hygroscopicity of the material and particle size. Small particles of very hygroscopic powder may become sticky immediately at the sticky condition. Theoretically, stickiness should be observed instantaneously as soon as the thin surface layer of the particles attains equilibrium with the air condition (relative humidity [RH] and temperature). The stickiness is observed visually; thus, there might be personal variation on the judgment of the exact sticky point. The stickiness property of a skim milk powder measured by this technique at various RHs is presented in Figure 16(b).

VI.   PHASE TRANSITIONS DURING INSTANTIZATION OF FOOD POWDERS

The rate at which a powder attains its original state in water is influenced by its reconstitution properties. The reconstitution properties are important functional physical properties of food powders. These properties include wettability, sinkability, dispersibility etc. of the powder while trying to dissolve in water. Powders with particle size less than 100 μm are difficult to wet because of their small interparticulate space. If the particles are fine and compact, the bulk surface gets hydrated and swells, preventing further water penetration so that a lump develops with a dry surface inside covered by a hydrated surface outside.

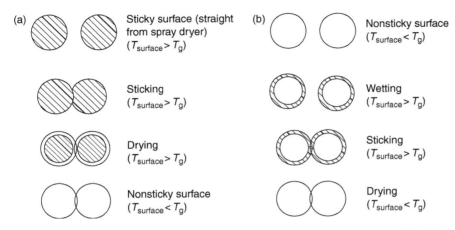

Figure 17 (a) Instantization by spray drying to a sticky state followed by fluidized bed drying. (b) Instantization by rewetting and fluidized bed drying.

This is most common in powders made with high molecular weight carbohydrates, gums, and proteins. To improve the wettability of a powder, the particle size is increased by agglomerating a number of particles to form a granular size between 500 and 3000 μm. This agglomeration process is termed instantization. Instant powder gets wet quickly and disperses in water within 10 sec or less without any lumping.

There are various methods of instantizing powders. The first method is removal of the powder from the spray dryer at higher moisture content (in thermoplastic state) allowing the particles to stick in a fluidized bed at elevated temperature (Figure 17[a]). The second method of instantization involves rewetting the surface of individual particles, allowing them to come in contact and stick together, and then drying to remove water and cause individual particles to stick together as agglomerates (Figure 17[b]).

Instantization of powder involves liquid–glass (Figure 17[a]) or glass–liquid–glass (Figure 17[b]) transition. The sticky property of the surface of the particle depends on the temperature; thus, it is necessary to appropriately select a temperature and humidity condition of the air for optimum instantization. Longer residence time in the instantizing equipment can result in fusion and caking. After optimum particle agglomeration, the air temperature should be reduced below the T_g of the particle to halt further agglomeration and caking. Dehumidified air should be used to compensate for the effect of reduced temperature on the drying rate. Therefore, careful control of the operating parameters is essential to obtain a specific particle size of the agglomerates.

VII. STABILITY OF POWDERS

Stability of powder means maintaining the physical, chemical, and microbial characteristics unchanged during handling and storage. The major factor that influences stability is moisture content. Water being a strong plasticizer, a small amount of it will depress the glass transition temperature significantly to influence the molecular mobility of the matrix in the condition of storage. At a microscopic level, food materials can exist as a phase separated system, resulting in variation in water content, reactant concentration, and transition temperatures within the microstructure. This is often an important factor influencing the chemical stability of an amorphous food system (Lievonen et al., 1998).

A. Crystallization and Caking

All amorphous products are metastable and therefore, can potentially crystallize (for crystallizing species) over time during storage. The rate of crystallization is a function of the extent the product is held above $T_g (T - T_g)$, with increasing crystallization rates for higher temperatures (Senoussi et al., 1995; Jouppila et al., 1997).

Crystallization is also encouraged by water absorption. The extent of water absorption by powder depends on its sorption property. If a local portion of the product in a package picks up moisture, the glass transition temperature is locally depressed for that particular portion and crystallization rate there is accelerated. Formation of the lattice during crystallization generally excludes water molecules (except for hydrates) and the excess moisture is lost to the environment. Absorption of this ejected moisture at the surface of neighboring particles creates interparticulate liquid bridges resulting in "caking" (Peleg and Hollenbach, 1984). Surrounding particles that absorb moisture are also crystallized and crystallization proceeds as a chain phenomenon. Caking might be accompanied by discoloration, loss of nutrients (such as lysine), and lipid oxidation. Water migration and accumulation in a spot can also serve as a potential site for mold growth. Melting of free fat in a powder (such as in whole milk powder) can also contribute to caking, but the strength of the interparticulate bond is weaker (Ozkan et al., 2002).

Crystallization of sugars is delayed by other ingredients present in the concentrated powder system by the same mechanism as in the supersaturated solution (Roos and Karel, 1991b; Hartel, 2001). The crystallization process rejects impurities including volatiles. Senoussi et al. (1995) found a loss of diacetyl as a function of rate of crystallization of lactose during storage. They found that when the lactose was stored at 20°C above T_g, the amorphous product went through immediate crystallization and practically all diacetyl was lost after six days. Levi and Karel (1995) also found an increased rate of loss of a volatile, 1-N-propanol, in an amorphous sucrose system as a result of crystallization. The crystallization of amorphous sugars (such as lactose) is a diffusion-controlled process and the temperature dependence of the crystallization time can be predicted by the WLF relationship (Equation [2]) (Roos and Karel, 1991b). In other research (Joupilla et al., 1997), the Avrami equation (Equation [3]) was used to predict the crystallization time of lactose in a skim milk powder system during storage at a given temperature condition.

WLF equation:

$$\log_{10} \frac{\theta_{cr}}{\theta_g} = \frac{-C_1 (T - T_g)}{C_2 + (T - T_g)} \tag{2}$$

where θ_{cr} = crystallization time at temperature T and θ_g = crystallization time at T_g.

Avrami equation:

$$\theta = 1 - e^{-kt^n} \tag{3}$$

or

$$\frac{I_f - I_t}{I_f - I_0} = e^{-kt^n} \tag{4}$$

where θ is the crystallinity, t time, k rate constant, n the Avrami exponent, I_f a leveling off value of peak intensity or peak area in x-ray diffraction patterns at maximum amount of crystallization, I_t peak intensity at time t, and I_0 is the peak intensity at 0 time.

The caking of an amorphous powder is a time-dependent phenomenon. The rate of caking due to the viscous flow property of the material is a function of $T - T_g$, and the kinetics of caking can again be determined by using a WLF-type relationship (Aguilera et al., 1995). The coalescence of the particles at early stages due to viscous flow driven by surface energy can be approximately predicted by Frenkel's mechanistic model (Wallack and King, 1988; Downton et al., 1982).

$$\left(\frac{x}{a}\right)^2 = \left(\frac{3}{2}\right)\left(\frac{\sigma t}{a\mu}\right) \tag{5}$$

where x is the interparticulate bridge radius, a the initial particle radius, t the time dimensionless proportionality constant of order unity, t is the contact time (sec), σ the surface tension, and μ is the viscosity. By approximating the values of $x/a = 0.1, a = 30$–$40\ \mu$m and $t = 1$–10 sec for coffee extract with $\sigma = 70$ mN/m, a particle viscosity range of 10^5 to 10^7 was found from Equation (5) (Wallack and King, 1988). This equation demonstrates that raising the viscosity, increasing the powder particle size, and decreasing the powder surface energy can reduce stickiness.

This model relationship is particularly important in relation to the stability of the dried product during the drying process and storage, as sticking and structural collapse phenomena have been found to take place at a critical viscosity range of 10^6 to 10^8 Pa sec (Downton et al., 1982; Wallack and King, 1988). It is important to note that the caking event is time dependent, whereas sticking is a relatively instantaneous phenomenon (say $t = 1$ sec). Longer contact times increase the tendency toward sticking and caking, all other things being equal. Thus, a dried product with a relatively free-flowing property immediately after drying could also cake in a collection or packaging container over a period of time if the surface viscosity is still relatively low due to higher temperature or moisture levels. For this reason, the dried product needs to be cooled immediately to an appropriate temperature before packaging. Temperature changes and moisture migration in the bags during travel through different climate zones and consolidation pressure can cause undesirable caking in the powders (Ozkan et al., 2002).

There are various methods employed to characterize the degree of caking by analyzing the change in the basic properties of the powders, such as flowability, angle of repose, inter-particulate cohesion, size distribution, and particle morphology (Table 3).

Table 3 Methods Employed to Measure the Degree of Caking

Method	Principle
Flowability	Discharge mass flow rate from a bin or funnel
Angle of repose	Heap angle along the horizontal plane ($<40°$ for free-flowing powder)
Cohesion	Negligible shear stress for a free-flowing powder (Jenike shear cell method)
Caking index	Weight fraction retained by a mesh with an opening size of maximum particle size of the powder (Equation [6] in text)
Microscopical attributes	Ratio of interparticulate bridge diameter to particle diameter (microscopic techniques)

Source: From Aguilera, J.M., de Valle, J.M., and Karel, M. (1995). *Trends Food Sci. Technol.* 6: 149–155. With Permission.

Aguilera et al. (1995) described an empirical technique to characterize the caking kinetics:

$$1 - \frac{\phi(t)}{100} = \exp\left(-\frac{t - t_d}{\tau}\right) \tag{6}$$

where $\phi(t)$ = caking index (fraction of particles larger than the given screen size), t = time, t_d = delay time factor for the onset of caking phenomenon (i.e., heating time, time required for the moisture to diffuse into the outer surface, etc.), τ = relaxation time for caking.

Equation (6) was validated with caking of a protein hydrolyzate powder (74 to 125 μm size). Caking was induced by exposing the fish hydrolyzate in petrie dishes to a given humidity and temperature. The samples were sifted mildly through a sieve larger than the particle size (125 μm). The retained sample was recorded as the caking index.

B. Diffusion of Volatiles and Permeability to Oxygen

Glass transition temperature is directly related to the molecular mobility of not only the matrix, but also to water and other components entrapped in the matrix. Volatile components in a food system have very limited mobility in a glassy matrix. Diffusion through the matrix is primarily through the pores of the matrix. Diffusivity of volatiles when the temperature exceeds T_g is greatly increased, and continues to increase with an increase in temperature (Roos and Karel, 1991a). The rubbery matrix above the glass transition temperature accelerates diffusion due to the increased free volume and change in transport and solution behavior of volatiles (Whorton, 1995). To and Flink (1978) found that the loss of volatiles was very high at collapse temperatures during freeze drying or during storage of the dried product. Levi and Karel (1995) found an increase in the rate of release of a volatile, 1-N-propanol, in an amorphous sucrose : raffinose system stored at temperatures above T_g.

Permeability of glass matrices to oxygen is also limited, which protects sensitive flavors against oxidation. Gejl-Hansen and Flink (1977) reported that the oxygen absorption into a freeze-dried, lipid–carbohydrate system was increased significantly at the collapse temperature, which is related to T_g.

C. Chemical Changes

Chemical changes in dry products occur very slowly. But, because of the increase in internal mobility of reactants and diffusivity of oxygen, various chemical reactions are accelerated in dried products if stored above T_g. Karmas et al. (1992) and Buera and Karel (1993) found an increased rate of nonenzymic reactions in dried vegetables stored at above T_g. The rate of oxidation of sensitive compounds in a dry matrix is also enhanced above T_g due to crystallization, which forces the encapsulated materials from the system to the surface. An increase in diffusivity of oxygen through the matrix above T_g also accelerates deteriorative reactions, but the consequence of crystallization on oxidation is significantly higher because of the release of encapsulated sensitive compounds. Shimada et al. (1991) reported that unencapsulated methyl linoleate underwent rapid oxidation, whereas encapsulated oil remained unoxidized, in a lactose-based food model. Moreau and Rosenberg (1996) also found rapid oxidation of milk fat that had migrated to the surface as a result of lactose crystallization from a whey protein–milk fat matrix. In another study, Saltmarch et al. (1981) found the highest rate of Maillard reaction in a whey powder at a critical water activity of 0.44. At this condition, the amorphous lactose began to crystallize, releasing water, and consequently mobilizing reactants for the browning reaction.

D. Structure Relaxation and Stability of Amorphous Products

When a liquid is supercooled (below its melting point) without crystallization, there is a transition of liquid to solid state with a discontinuity in the enthalpy line as depicted in Figure 17. The point of discontinuity represents the glass transition temperature (T_g). It has been widely claimed previously that the reactivity of a material is halted below the glass transition temperature, indicating the material's greatest stability. However, it has recently been shown that although molecular mobility in amorphous material is greatly reduced below the T_g, it is not completely ceased. Therefore, the stability of materials is not perfect even below the T_g. This means that mobility in a glass is dependent on more than just $T - T_g$ (Champion et al., 2000).

The amorphous glassy state is thermodynamically a nonequilibrium state. When a supercooled liquid is cooled, the viscosity increases and relaxation of molecules within the liquid become slower. If the relaxation is slower than the timescale of the cooling rate, the system departs from the equilibrium state and consequently, it converts to a solid at higher temperature than the expected temperature if the relaxation is at least equal or faster than the cooling rate. When this nonequilibrium glass is held at temperatures along the transition region or below, but in close vicinity to the T_g, an excess thermodynamic quantity relaxes toward a thermodynamic equilibrium state (Figure 18). This occurs as a result of internal rotation of the main molecular chain to reduce configuration energy (Yoshida, 1995). This phenomenon is expressed in various terms: enthalpy relaxation, molecular relaxation, annealing, or physical aging. The rate of relaxation accelerates with increasing temperature; therefore, it is highest in the temperature region close to the glass transition temperature. Smaller molecules have higher relaxation rates (Kim et al., 2001).

The knowledge of weak or strong dependence of relaxation time on temperature below T_g is very important in interpreting the stability of the material. The relaxation rate can be a few hours to as long as several years. Relaxation rate depends on the thermal history (state of nonequilibrium) during manufacturing and the storage temperature (provided there is no change in moisture). A material with a weak relaxation time will change its property during prolonged storage; therefore a strong dependence on temperature is desirable (Hancock and Shamblin, 2001). However, such material will be sensitive to a small fluctuation in temperature. The temperature dependence of relaxation time is normally determined by measuring the temperature dependence of viscosity or structural relaxation time at or above

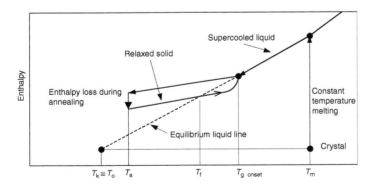

Figure 18 Schematic diagram representing the enthalpy relaxation of supercooled liquid during annealing at a temperature T_a (T_m = melting temperature, $T_{g\ onset}$ = onset glass transition temperature during cooling, T_f = fictive temperature, T_a = annealing temperature, T_k = Kauzmann temperature at which the configurational mobility becomes zero — configurational energy of amorphous solid intersects with that of crystals $\cong T_o$).

T_g and by applying the Vagel–Tamman–Fulcher (VTF) equation:

$$\eta = \eta_0 \exp\left(\frac{DT_0}{T - T_0}\right) \tag{7}$$

where D is the strength parameter, indicating the extent of deviation from the Arrhenius law (typical values for D are 3–7 for very fragile liquids and 30–∞ for strong liquids), η_0 is taken for normal liquids to be 10^{-5} Pa sec, and T_0 is the ideal glass transition temperature, which usually lies about 20 to 50°C below the experimentally measured T_g (Lu and Zografi, 1997). At a given temperature, enthalpy relaxation values $\Delta H(t_a, T_a)$ are a function of aging time and the maximum extent of enthalpy relaxation $\Delta H(t_\infty, T_a)$. The decay of relaxation has been found to follow the Kohlrausch–Williams–Watts (KWW) equation in pharmaceutical systems (Yu, 2001).

$$\begin{aligned}\Phi(t_a) &= 1 - \frac{\Delta H(t_a, T_a)}{\Delta H(t_\infty, T_a)} \\ &= \exp^{-(t_a/\tau_{\text{eff}})^\beta}\end{aligned} \tag{8}$$

Here, $\Phi(t_a)$ is a decay function (proportion of glass that has not relaxed) and t_{eff} is the effective relaxation time (average) of the system, which depends on the aging temperature T_a and also on the departure of the system from the final equilibrium state. β (stretched exponential power) describes the extent to which the relaxation process is nonexponential (width of the relaxation time distribution).

Relaxation processes in amorphous systems are nonexponential, in which case β takes on values less than one. The $\Delta H(t_a, T_a)$ can be obtained from DSC experiments by integrating the enthalpy overshoot area or by integrating the difference in specific heat (ΔC_p) between aged and nonaged samples. $\Delta H(t_\infty, T_a)$ is commonly obtained by the extrapolation of the C_p line of the liquid and C_p curve of the unaged samples within the integer limit of T_a (annealing temperature)

$$\Delta H(t_\infty, T_a) = \Delta C_p(T_g - T_a) \tag{9}$$

Enthalpy relaxation has been reported for various inorganic, organic, and polymeric materials. Following enthalpy relaxation, mechanical properties, such as stress relaxation, creep relaxation behaviors, mechanical stress, dynamic viscoelastic properties, dielectric constant, and dielectric loss of materials are influenced (Schmidt and Lammert, 1996). Due to the densification of the material during relaxation (Lourdin et al., 2002), molecular porosity, and consequently the hygroscopicity and water sorption properties of the material can also be altered. Physical aging implies a progressive change (time dependent) not only in the thermodynamic properties such as enthalpy, specific volume, and refractive index but also in the macrolevel physicomechanical and texture properties (such as hardness, brittleness). Therefore, these changes are of relevant practical importance as they concern the stability and shelf life of dried food products. The chemical reactivity of the susceptible component is obviously facilitated by the physical movement of the internal structure. It has been reported that long-term physical and chemical stability can be achieved only if an amorphous product is stored below T_0 (where there is zero molecular mobility), whose value has been approximated at 50°C below its calorimetric T_g. T_0 has been found to be similar to the Kauzmann temperature (T_k), which is the temperature where the entropy of a supercooled liquid equals the entropy of the crystalline form. T_k is the lower limit of T_g at maximum enthalpy relaxation (Figure 18). The presence of additives can delay, accelerate, or erase the structural changes of the materials below T_g. Water acts as a deplasticizer below glass transition temperature, which means that the enthalpy relaxation is accelerated by water below the T_g of the material (Seow et al., 1999; Kiekens et al., 2000).

The concepts of relaxation time and molecular mobility outlined above have been successfully applied for predicting the stability of pharmaceutical formulations (Lechuga-Ballesteros et al., 2002). Thermal analysis of some pure food compounds, such as sucrose, fructose, glucose, and maltose, has indicated the occurrence of enthalpy relaxation (Wungtanagorn and Schmidt, 2001; Truong et al., 2002). This concept has not been widely researched in food materials, though the importance of this concept has been highlighted in some cereal products. In a recent study, Kim et al. (2001) described the enthalpy relaxation of amorphous potato starch powder produced by ball milling of native starch. Chung et al. (2002) studied the glass transition and enthalpy relaxation of native and gelatinized rice starches using the DSC technique. Lourdin et al. (2002) demonstrated the structural relaxation of sorbitol-plasticized amorphous starch during storage. By measuring the mechanical properties of the material over various timescales, they found that the starchy material became stiffer and more brittle after annealing the samples at sub $- T_g$ temperature to allow relaxation. More research still needs to be done to relate the glass transition concept to the absolute stability of dried food materials.

VIII.   CONCLUSIONS

In terms of molecular structure, there are three forms of powders: crystalline, amorphous, and the mixture of the two. Rapid removal of water, melting, and rapid cooling result in transition from liquid to amorphous solid forms. Crystallization is a time-dependent phenomenon; therefore, the transition from liquid to solid occurs if there is sufficient time for realignment of the molecules. The materials can also be classified as poor or good glass formers based on their ability to retain their amorphous structure. The powders produced in many drying processes are in amorphous form, though it is possible to produce crystalline or semicrystalline powder by altering the process parameters or introducing another processing step prior to or after drying. The crystalline powders are thermodynamically least hygroscopic, and thus stable. Many sugar- and acid-rich products have solid components with very low glass transition temperatures. Due to this, the transition from liquid to solid structures may not take place in conventional drying conditions. When there is no transformation from liquid to solid state even after removing moisture, the product exhibits stickiness during drying. Stickiness is a very problematic issue in powder production or during further processing and packaging of the powder. A similar phenomenon also results in caking during storage. Amorphous structure can absorb water, which depresses the glass transition temperature below the storage temperature. The regaining of molecular mobility can provoke crystallization, loss of flavor, volatiles, and chemical degradation. Glass transition temperature was conventionally thought to be the threshold for stability of the dried product; however it has been found that structural relaxation takes place even below the glass transition temperature.

The phase/state transition property of the material is an important issue pertaining to processing and storage of solid state (dehydrated or frozen) products. Knowledge of the phase/state transitions is required to optimize the shelf life and processibility of the product. Traditionally, moisture content or water activity has been considered as the principle factors influencing the physical and chemical stability of food powders. Studies on the glass transition property of materials have been providing more and more evidence that the stability of dehydrated systems can also be predicted by relating to the glass transition temperature of the system. Due to the complex nature of food systems and possible interactions between the components, the prediction of the glass transition and related phenomena are also complex. Another area that needs more attention is the structure relaxation of the dehydrated amorphous food systems. Many unanswered stability issues may be revealed by systematically investigating this event.

REFERENCES

Adhikari, B., Howes, T., Bhandari, B., and Truong, V. (2003). *In situ* characterization of surface stickiness of sugar-rich foods using a linear actuator driven stickiness testing device. *J. Food Eng.* 58: 11–22.

Aguilera, J.M., de Valle, J.M., and Karel, M. (1995). Caking phenomenon in amorphous food powders. *Trends Food Sci. Technol.* 6: 149–155.

Bellows, R.J. and King, C.J. (1973). Product collapse during freeze-drying of liquid foods. *AICHE Symp. Series* 69: 33–41.

Bhandari, B.R. and Howes, T. (1999). Implication of glass transition for the drying and stability of dried foods. *J. Food Eng.* 40: 71–79.

Bhandari, B.R., Datta, N., and Howes, T. (1997a). Problems associated with spray drying of sugar-rich foods. *Drying Technol.* 15: 671–684.

Bhandari, B.R., Datta, N., and Howes, T. (1997b). A semi-empirical approach to optimize the quantity required to spray dry sugar-rich foods. *Drying Technol.* 15: 2509–2525.

Bhandari, B.R., Datta, N., Rintoul, G.B., and D'Arcy, B.R. (1998). Co-crystallization of honey with sucrose. *Lebensmittel-Wiss. u-Technol.* 31: 138–142.

Bloore, C. (2000). Development in food drying technology — overview. International Food Dehydration Conference- 2000 and Beyond, Food Science Australia, Melbourne, pp. 1.1–1.5.

Boonyai, P., Bhandari, B., and Howes, T. (2002). Development of a novel testing device to characterize the sticky behavior of food powders — a preliminary study. Proceedings of the International Conference on Innovations in Food Processing Technology and Engineering. December 11–13. Jindal, V.K., Noomhorm, A., Rakshit, S., and Khan, I., (Eds.). Asian Institute of Technology, Bangkok, Thailand, pp. 809–816.

Brennan, J.G., Herrera, J., and Jowitt, R. (1971). A study of some of the factors affecting the spray drying of concentrated orange juice, on a laboratory scale. *J. Food Technol.* 6: 295–307.

Buera, M.P. and Karel, M. (1993). Application of the WLF equation to describe the combined effects of moisture and temperature on nonenzymatic browning rates in food systems. *J. Food Process Pres.* 17: 31–45.

Champion, D., Le Meste, M., and Simatos, D. (2000). Towards an improved understanding of glass transition and relaxations in foods — molecular mobility in the glass transition range. *Trends Food Sci Technol.* 11: 41–55.

Chen, A.C. (1994). Ingredient technology by the sugar cocrystallization process. *Int. Sugar J.* 96: 493–495.

Chung, H.J., Lee, E.J., and Lim, S.T. (2002). Comparison in glass transition and enthalpy relaxation between native and gelatinized rice starches. *Carb. Polym.* 48: 287–298.

Donhowe, D.P., Hartel, R.W., and Bradley, R.L. Jr. (1991). Ice crystallization processes during manufacture and storage of ice cream. *J. Dairy Sci.* 74: 3334–3344.

Downton, G.E., Flores-Luna, J.L., and King, C.J. (1982). Mechanism of stickiness in hygroscopic, amorphous powders. *Indust. Eng. Chem. Fundam.* 21: 447–451.

Gejl-Hansen, F. and Flink, J.M. (1977). Freeze-dried carbohydrate containing oil-in-water emulsions — microstructure and fat distribution. *J. Food Sci.* 42: 1049–1055.

Hancock, B.C. and Shamblin, S.L. (2001). Molecular mobility of amorphous pharmaceutical determined using differential scanning calorimetry. *Thermochim. Acta* 380: 95–107.

Hancock, B.C. and Zografi, G. (1997). Characteristics and significance of the amorphous state in pharmaceutical systems. *J Pharm. Sci.* 86: 1–12.

Hartel, R.W. (1987). Sugar crystallization in confectionery products. *Manuf. Confec.* October, 67: 59–65.

Hartel, R.W. (2001). *Crystallization in Foods.* Aspen Publishing, Gaithersburg, MD.

Hartel, R.W. and Shastry, A.V. (1991). Sugar crystallization in food products. *Crit. Rev. Food Sci. Nutr.* 1: 49–112.

Hennigs, C., Kochel, T.K., and Langrish, T.A.G. (2001). New measurements of the sticky behaviour of skim milk powder. *Drying Technol.* 19: 471–484.

Jouppila, K. and Roos, Y.H. (1994). Water sorption and time-dependent phenomenon of milk powders. *J. Dairy Sci.* 77: 1799–1807.

Jouppila, K., Kansikas, J., and Roos, Y.H. (1997). Glass transition, water plasticization, and lactose crystallization in skim milk powder. *J. Dairy Sci.* 80: 3152–3160.

Karmas, R., Buera, M.P., and Karel, M. (1992). Effect of glass transition on rates of nonenzymatic browning in food systems. *J. Agric. Food Chem.* 40: 873–879.

Khalloufi, S., El-Maslouhi, Y., and Ratti, C. (2000). Mathematical model for prediction of glass transition temperature of fruit powders. *J. Food Sci.* 65: 842–848.

Kiekens, F., Zelko, R., and Remon, J.P. (2000). Effect of the storage conditions on the tensile strength of tablets in relation to the enthalpy relaxation of the binder. *Pharm. Res.* 17: 490–493.

Kilcast, D. and Roberts, C. (1998). Perception and measurement of stickiness in sugar-rich foods. *J. of Texture Studies* 29: 81–100.

Kim, Y.J., Suzuki, T., Hagiwara, T., Yamaji, I., and Takai, R. (2001). Enthalpy relaxation and glass to rubber transition of amorphous potato starch formed by ball milling. *Carbohydr. Polym.* 46: 1–6.

Lazar, M.E., Brown, A.H., Smith, G.S., Wang, F.F., and Lindquist, F.E. (1956). Experimental production of tomato powder by spray drying. *Food Technol.* 10: 129–134.

Lechuga-Ballesteros, D., Miller, D.P., and Zhang, J. (2002). Residual water in amorphous solids — measurement and effects on stability. In: *Amorphous Food and Pharmaceutical Systems.* Levine, H. (Ed.). The Royal Society of Chemistry, Cambridge, U.K., pp. 275–317.

Levi, G. and Karel, M. (1995). Volumetric shrinkage (collapse) in freeze-dried carbohydrates above their glass transition temperature. *Food Res. Int.* 28: 145–151.

Liapis, A.I. and Bruttini, R. (1995). Freeze drying. In: *Handbook of Industrial Drying*, 2nd ed. Mujumdar, A.S. (Ed.), Marcel Dekker, N.Y., pp. 309–344.

Lievonen, S.M., Laaksonen, T.J., and Roos, Y.H. (1998). Glass transition and reaction rates — nonenzymatic browning in glassy and liquid systems. *J. Agric. Food Chem.* 46: 2778–2784.

Lockemann, C.A. (1999). A new laboratory method to characterize the sticking properties of free-flowing solids. *Chem. Eng. Process.* 38: 301–306.

Lourdin, D., Colonna, P., Brownsey, G., and Ring, S. (2002). Influence of physical ageing on physical properties of starchy materials. In: *Amorphous Food and Pharmaceutical Systems.* Levine, H. (Ed.). The Royal Society of Chemistry, Cambridge, U.K., pp. 88–112.

Lu, Q. and Zografi, G. (1997). Properties of citric acid at the glass transition. *J. Pharm. Sci.* 86: 1374–1378.

Maltini, E., Anese, M., and Shtylla, I. (1997). State diagram of some organic acid-water systems of interested in food. *Cryo-Letters* 18: 263.

Masters, K. (2002). *Spray Drying in Practice.* SprayDryConsult International ApS, Krathusparken, Charlottenlund, Denmark.

Michalski, M.C., Desobry, S., Babak, V. and Hardy, J. (1999). Adhesion of food emulsions to packaging and equipment surfaces. *Colloids and Surfaces A: Physicochemical and Eng. Aspects* 149: 107–121.

Moore, J.G. (1995). Drum dryers. In: *Handbook of Industrial Drying*, 2nd ed. Mujumdar, A.S. (Ed.), Marcel Dekker, N.Y., pp. 249–262.

Moreau, D.L. and Rosenberg, M. (1996). Oxidative stability of anhydrous milkfat microencapsulated in whey proteins. *J. Food Sci.* 61: 39–43.

Nowakowski, C.M. and Hartel, R.W. (2002). Moisture sorption of amorphous sugar products. *J. Food Sci.* 67: 1419–1425.

Oates, C.G. (1997). Towards an understanding of starch granule structure and hydrolysis. *Trends Food Sci. Technol.* 8: 375–382.

Ozkan, N., Walisinghe, N., and Chen, X.D. (2002). Characterization of stickiness and cake formation in whole and skim milk powders. *J. Food Eng.* 55: 293–303.

Papadakis, S.E. and Bahu, R.E. (1992). The sticky issues of drying. *Drying Technol.* 10: 817–837.

Pasley, H. and Haloulos, P. (1994). Stickiness — a comparison of test methods and characterization parameters. In: *Drying'94.* Rudolph, V., Key, R.B., and Mujumdar, A.S. (Eds.). pp. 165–172.

Paterson, A.H.J., Bronlund, J.E., and Brooks, G.F. (2001). The Blow test for measuring the stickiness of powders. Conference of Food Engineering 2001, AICHE Conference. Reno, NV, U.S.A., November 4–9, pp. 408–414.

Peleg, M. (1983). Physical characteristics of food powders. In: *Physical Properties of Foods.* Peleg, M. and Bagley, E.B. (Eds.). AVI Publishing Company, Inc., Westport, CT, U.S.A., pp. 293–323.

Peleg, M. and Hollenbach, A.M. (1984). Flow conditioners and anticaking agents. *Food Technol.* 38: 93–99.

Raemy, A. and Schweizer, T.F. (1983). Thermal behaviour of carbohydrates studied by heat flow calorimetry. *J. Thermal Anal.* 28: 95–108.

Roos, Y. (1987). Effect of moisture on the thermal behavior of strawberries studied using differential scanning calorimetry. *J. Food Sci.* 52: 146–149.

Roos, Y. and Karel, M. (1991a). Water and molecular weight effects on glass transitions in amorphous carbohydrates and carbohydrate solutions. *J. Food Sci.* 56: 1676–1681.

Roos, Y. and Karel, M. (1991b). Plasticizing effect of water on thermal behavior and crystallization of amorphous food models. *J. Food Sci.* 56: 38–43.

Roos, Y.H. (1993). Melting point and glass transition of low molecular weight carbohydrates. *Carbohydr. Res.* 238: 39–48.

Roos, Y.H. (1995a). Glass transition-related physicochemical changes in foods. *Food Technol.* 49: 97–107.

Roos, Y.H. (1995b). *Phase Transitions in Foods.* Academic Press, New York.

Saltmarch, M., Vaginini-Ferrari, M., and Labuza, T.P. (1981). Theoretical basis and application of kinetics to browning in spray-dried whey food systems. *Prog. Food Nutr. Sci.* 5: 331–344.

Schmidt, S.J. and Lammert, M. (1996). Physical aging of maltose glasses. *J. Food Sci.* 61: 870–875.

Senoussi, A., Dumoulin, E.D., and Berk, Z. (1995). Retention of diacetyl in milk during spray-drying and storage. *J. Food Sci.* 60: 894–905.

Seow, C.C., Cheah, P.B., and Chang, Y.P. (1999). Antiplasticization by water in reduced-moisture food systems. *J. Food Sci.* 64: 576–581.

Shimada, Y., Roos, Y., and Karel, M. (1991). Oxidation of methyl linoleate encapsulated in amorphous lactose-based food model. *J. Agric. Food Chem.* 39: 637–641.

Sopade, P.A., Bhandari, B.R., D'Arcy, B.R., Halley, P., and Caffin, N. (2002). A study of vitrification of Australian honeys at different moisture contents. In: *Amorphous Food and Pharmaceutical Systems.* Levine, H. (Ed.). The Royal Society of Chemistry, Cambridge, U.K., pp. 169–186.

To, E.C. and Flink, J.M. (1978). 'Collapse', a structural transition in freeze dried carbohydrates II — effect of solute composition. *J. Food Technol.* 13: 567–581.

Truong, V. (2003). Modelling of the glass transition temperature of sugar-rich foods and its relation to spray drying of such products. Ph.D thesis, University of Queensland, Australia.

Truong, V., Bhandari, B.R., Adhikari, B., and Howes, T. (2002). Physical ageing of amorphous fructose. *J. Food Sci.* 67: 3011–3018.

Tsourouflis, S., Flink, J.M., and Karel, M. (1976). Loss of structure in freeze-dried carbohydrates solutions — effect of temperature, moisture content and composition. *J. Sci. Food Agric.* 27: 509–519.

Twomey, M. and Keogh, K. (1998). Milk powder in chocolate. *Farm Food* 8: 9–11.

Wallack, D.A. and King, C.J. (1988). Sticking and agglomeration of hygroscopic, amorphous carbohydrate and food powders. *Biotechnol. Prog.* 4: 31–35.

Welti-Chanes, J., Guerrero, J.A., Barcenas, M.E., Aguilera, J.M., Vergara, F., and Ovas, G.V. (1999). Glass transition temperature (T_g) and water activity (a_w) of dehydrated apple products. *J. Food Process Eng.* 22: 91–101.

Whorton, C. (1995). Factors influencing volatiles release from encapsulation matrices. In: *Encapsulation and Controlled Release of Food Ingredients.* Risch, S.J. and Reineccius, G.A. (Eds.). ACS Symposium Series 590, American Chemical Society, Washington D.C., pp. 135–141.

Wungtanagorn, R. and Schmidt, S.J. (2001). Phenomenological study of enthalpy relaxation of amorphous glucose, fructose, and their mixture. *Thermochim. Acta* 369: 95–116.

Yoshida, H. (1995). Relationship between enthalpy relaxation and dynamic mechanical relaxation of engineering plastics. *Thermochim. Acta* 266: 119–127.

Yu, L. (2001). Amorphous pharmaceutical solids — preparation, characterization and stabilization. *Adv. Drug Deliv. Rev.* 48: 27–42.

Yu, L., Mishra, D.S., and Rigsbee, D.R. (1998). Determination of the glass properties of D-mannitol using sorbitol as an impurity. *J. Pharm. Sci.* 87: 774–777.

12

Blending, Segregation, and Sampling

Scott A. Clement
Jenike & Johanson, Inc.
San Luis Obispo, California

James K. Prescott
Jenike & Johanson, Inc.
Westford, Massachusetts

Contents

I. INTRODUCTION

Blending is an important aspect of many food production processes. In some cases, blending and packaging make up the entire process, while in others, blending is a minor step in a long and complex process. Some applications call for a simple combining of material streams, while others demand a high degree of homogeneity.

Blending should not be seen as an independent unit process, but rather as an integral part of the larger process. The goals of producing an adequate blend, maintaining that blend through additional handling steps, and verifying that both the blend and the finished product are sufficiently homogeneous are often not as straightforward as initial inspection might lead us to believe.

Mixing and blending are two commonly used terms that are not universally applied in the same way. They are used to describe:

- Combining two or more powdered or granular components.
- Homogenizing the contents of a vessel before discharge.
- Contacting, wetting, or dissolving a dry material with a liquid.
- Combining liquid components.
- Preparing or maintaining a slurry or suspension.
- Kneading a dough or paste.

Distinctions between mixing and blending have sometimes been made, based on the equipment used, the material handled (liquids versus solids), or whether combining streams of different components or simply homogenizing a product. No definition is universally accepted or used. For the purpose of this chapter, we will use the terms mixing and blending interchangeably. Our discussion will be limited to blending of bulk solids, which we define as granular or powdered materials composed of discrete particles. Many of the blenders mentioned are capable of liquid addition, generally by use of spray bars. Any discussion on mixing of liquids alone, or paste, or dough, is beyond the scope of this chapter.

A. Blending Mechanisms

The principal mechanisms through which blending is accomplished are convection, diffusion, and shear.

Convection. Convection is the transfer of a group or body of particles from one location to another. This can occur, for example, as a result of:

- Cascading of material within a tumble blender.
- The action of the blade of a ribbon blender.
- Material movement resulting from gas pressure gradients in a pneumatic blender.

Diffusion. Diffusion is the random redistribution of particles. This is not necessarily in the direction of the principle movement of a body of material, but often at angles to it, and as a result of increased particle mobility. Examples of where diffusion can occur include:

- Fluidization caused by the action of a pneumatic blender.
- Movement of material parallel to the axis of a tumble blender caused by collisions with other particles or the walls or internals of the blender.

Shear/impact. There is some disagreement about how shear in blending is defined. It is sometimes defined as the development of slip planes or shearing strains within a bed of material. This definition would lead to the conclusion that blending within a gravity blender is a result of shear. Likewise, blending that results from cascading of material within a tumble blender would be considered shear.

We consider these to be examples of convection, where the movement of the body of material results from the flow of material, which involves development of one or more slip planes. For our purposes, we define the blending mechanism of shear as high intensity impact or splitting of the bed of material to break up agglomerates, or overcome cohesion. This can be very effective at producing small-scale uniformity, usually on a localized basis. Examples of types of equipment that use shear include:

- Intensifiers (choppers) in a variety of blenders.
- Pin mixers.
- Other high intensity mixers.

II. DEFINING BLEND STRUCTURE

Powders and other bulk solids are made up of discrete particles. The most often stated objective of blending several different powders is "a uniform blend." In order to avoid eventual problems, however, we also must consider two important aspects of uniformity: the scale of uniformity, and the structure of the particles themselves within a blend.

A. Scale of Uniformity

A dry blend of powders consists of individual particles, usually of different particle size or chemical composition. We could state that we have a goal of having every single sample collected from this blend meet some target value, such as percentage of each component. However, since we are working with discrete particles, the size of the defined sample significantly impacts the results of any test used to compare blend uniformity to the target value. Let us assume that we are making pudding mix in a batch blender. We dispense the proper mass of each component and add each to the blender. Before even starting the mixer, we could state that a sample size equal to the entire batch (however impractical) would meet our target value, although clearly the blend within that sample is nonhomogeneous. Unless we are setting a record for the largest bowl of pudding, mixing at a smaller scale is required. Smaller samples taken at different locations will not be the same, thus quantifying the nonuniformity.

At this point it is important to clarify how we are using the term *sample size*. Statisticians use the term *sample* to identify the portion of the whole (or population) that is evaluated in order to represent the properties of the whole. Thus, *sample size* would be the number of samples evaluated, or the sum total of the samples on a mass or volume basis. For our purposes, we use the term *sample size* to describe the physical size of each discrete sample of blended material to be evaluated (usually on a mass basis). Each of these samples represents one data point. For example, we may choose to collect 30 samples with a sample size of 50 g each.

In the above example, let us say that we start the mixer and blend the material. Visually, the pudding blend looks uniform, but is the blend homogeneous? Again, the scale of uniformity must be considered. Just as at the outset, a sample equal to the size of the entire batch meets the same target value. Smaller samples, perhaps equal to the size of a box of pudding mix, may all be at the target value as well. From the consumer's perspective (and hence yours) the blend is adequate. However, we can proceed to even smaller scales, to find that each of the individual components remain as discrete particles, with different chemical compositions; at this scale, the blend is not uniform. Is this of concern? It may be, depending on your application. This effect becomes more pronounced with smaller package (or product) sizes, lower percentage components, and larger particle sizes of these minor components. To address this, it is best to select for analysis sample sizes representative of the size being consumed by the customer. For example, in the case of the pudding mix, the ideal sample size is the weight of the pudding mix in a box, since normally the entire box is made into pudding. Samples smaller than this may have increased variability due to the discrete nature of the particulate system. Samples larger than that consumed would mask variability. For example, in a box of breakfast cereal, a blend of flakes and raisins within a box should contain a certain amount of raisins. The number of raisins from box

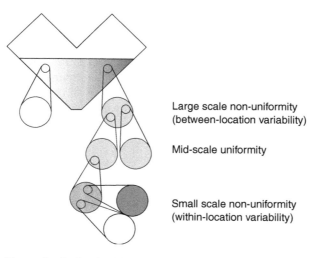

Large scale non-uniformity
(between-location variability)

Mid-scale uniformity

Small scale non-uniformity
(within-location variability)

Figure 1   Scale of uniformity.

to box will be relatively consistent if the process is appropriately designed. However, on a smaller scale, say the size of a spoonful of this blend, some samples may contain no raisins while others may have five raisins, with other samples having any number in between. This level of variability would be unacceptable on the box-to-box scale, although the consumer is used to tolerating spoon-to-spoon variability. While this example is confined to samples being "analyzed" by the consumer, the same behavior will occur when analyzing samples containing large particles or agglomerates. Small samples highlight small-scale variability, while larger samples mask it.

Sample size should represent the quantity of product the customer will use, analyze, or inspect. In this way, the manufacturer is using the same basis the customer uses for quality assessment.

B.   The Structure of the Particles within a Blend

The simplified case of blending two different ingredients, that do not adhere, bind, or otherwise interact with themselves or each other, is often referred to as a *random* blend. In a random blend, all individual particles are free to move relative to each other, so no form of bonding takes place. Dissimilar particles can readily separate from each other and collect in zones of similar particles when forces such as gravity, airflow, or vibration act on the blend. This process is referred to as *segregation*. Random blends are never perfect, and inevitably, even in the best-blended state, there will be variations in concentrations of components, owing simply to chance.

Completely random blends are rarely encountered in most industrial applications. Instead, many particles do have some tendency to interact with one another by way of chemical, molecular, physical, or other means such that these particles end up agglomerating, coating, or bonding to one another. When such interaction occurs, the blend is referred to as an *ordered, structured*, or *interactive* blend. In principle, this behavior, if completed in an ideal manner, could create new particles, each consisting of the individual ingredients. This superparticle is sometimes referred to as a *unit particle*. A perfect unit particle would consist of the desired percentage of each ingredient. This concept of combining ingredients to create new, chemically homogeneous particles is one of the goals of granulation methods (wet or dry), which are not discussed in this chapter. However, even with dry blending, in many applications, unit particles are created during blending. The formulator and engineer must consider this when developing a product and selecting equipment associated with blending and blend handling.

A blend of perfect unit particles of identical size will not segregate after discharge from the blender. However, if the unit particles are not monosized, then segregation by size may occur. Even *if* the blend is chemically homogeneous this segregation may cause problems because it may affect other blend properties such as solubility or bulk density. In reality there is often some variability in chemical composition that is related to particle size, which can result in compositional variations as a result of size-based segregation. Therefore, segregation generally should not be ruled out as a possible concern, even if chemically perfect unit particles are created.

There are several ways to test whether or not unit particles have been created. Microscopy may give a quick visual indication as to whether or not particles have adhered to each other. Microscopic images may be challenging to interpret, however, as it may not be obvious whether particles have actually bonded or are simply adjacent to each other. A second way to evaluate unit particles is to sieve the blend and subsequently assay each sieve cut individually. The ideal case is uniform composition across each sieve size. However, the results of this method will be a function of the screen sizes selected and the work imparted on the screens to effect sieving. The greater the forces and duration of sieving, the more opportunity there is for unit particles to degrade back to their individual components. Creating a curve of sieve-cut assay as a function of work imparted on the sieves may provide insight as to the integrity of the bonds within the unit particle. From a practical standpoint, unit particles that are sensitive to variations in sieving will be sensitive to forces imparted during handling; hence, in the production process, these materials should be handled gently after final blending.

Whereas the ideal unit particle is a perfect composite of individual ingredients, the worst-case scenario is when specific ingredients have a tendency to adhere only to themselves, without adhering to dissimilar ingredients. This often happens with fine materials in the creation of agglomerates. The strength and conversely the dispersive ability of agglomerates can vary widely, from soft agglomerates to rock-hard lumps. The tendency for agglomerates to form is a function of chemical composition, the presence of free volatiles, moisture content (which, in turn, is often a function of the humidity to which the material is exposed), temperature history (including temperature cycling), and consolidating pressure acting on the material. In general, agglomerates present in the ingredients added to the blender, or formed during blending, must be destroyed to provide an adequate blend to create an acceptable product.

III. SEGREGATION

A. Definition

A discussion of blending should include a discussion of segregation since the two act against each other. Segregation can be defined as a separation of a mix or blend of particles into zones of particles of similar size, composition, density, resiliency, capacity for holding a static charge, or any number of other physical or chemical properties of the particles.

Segregation can occur upstream of a blender, during its discharge, or in downstream equipment. Within a blender, segregation can work counter to the mechanisms (diffusion, shear, and convection) that would act to produce a uniform mix. As a blender increases particle mobility, it creates a condition where segregation may also occur. Segregation during a blending operation, particularly of coarse, free-flowing materials, is not uncommon. In this case, a state of equilibrium is reached between mixing and segregation, which limits the achievable blend quality [1]. Note that this equilibrium condition may not always correspond to a condition of best uniformity; that is, the best blend (lowest variability) may not be at equilibrium. Additionally, segregation may further reduce the blend quality during discharge of the blender or during subsequent handling of the blend.

Segregation can cause problems in many food processing and handling operations. In the manufacture of packaged foods, a continuous stream or large batches of product are divided into package sizes that are assumed to be the same in terms of bulk density (or inversely, bulk index), fill weight, and composition. In producing ingredients for sale, maintaining and delivering products that are uniform within a shipment and between shipments is important. Variations, even those that may not impact product performance, are frequently perceived as a product quality problem, resulting in customer complaints. Within a process, particle size or compositional changes of an ingredient over time can cause process upsets or rejected product. When handling ingredients from a supplier that vary over time, a plant may experience changes in flowability of the materials, or difficulty in packaging resulting from changes in bulk density. In a volumetric packaging line, package weight variations may force the processor to increase the fill weight, giving away an excessive amount of product. If filling is gravimetric, low bulk density resulting from air trapped in excessive fines may make it difficult to achieve the required package weight. These are a few of the problems attributable to segregation that have been experienced by food processors.

B. Segregation Mechanisms

Four primary segregation mechanisms are of interest in food processing applications. Other mechanisms exist [2], but are less frequently encountered in this industry. The segregation mechanisms of interest to us are:

1. Sifting.
2. Fluidization.
3. Dusting.
4. Trajectory.

Segregation may occur as a result of one of these mechanisms or a combination of several. Whether segregation occurs, to what degree, and which mechanism or mechanisms are involved depend on a combination of the properties of the blend and the process conditions encountered. Material properties that influence segregation include:

- Particle size and particle size distribution.
- Particle density.
- Particle shape.
- Particle resilience.
- Cohesiveness of the bulk material.
- Ability to develop and hold an electrostatic charge.
- Affinity for other ingredients or processing surfaces.

Of all these, segregation based on particle size is by far the most common [1]. In fact, particle size is the most important factor in all the primary segregation mechanisms.

Process conditions that can influence segregation include:

- Interparticle motion within a bed of particles in contact with one another.
- Free falls and impacts.
- Sliding on a surface.
- Fluidization.
- Induced vibration.
- Air currents.

Particular care must be taken at points of storage and transfer since these present the opportunity for such conditions to occur.

1. Sifting (or Percolation) Segregation

This is the most common form of segregation for many industrial processes. Under appropriate conditions, fine particles tend to sift or percolate through coarse particles. For segregation to occur by this mechanism there must be a range of particle sizes. A minimum difference in mean particle diameter of 1.3:1 is often more than sufficient. In addition, the mean particle size of the mixture must be sufficiently large (typically greater than about 100 μm) [3], the mixture must be relatively free flowing, and there must be relative motion between particles. This last requirement is very important, since without it even highly segregating blends of ingredients that meet the first three conditions will not segregate.

Figure 2   Sifting segregation (Photo courtesy of Jenike & Johanson, Inc.)

Relative motion can be induced in a variety of ways, for example: (1) as a pile is formed when filling a bin, (2) as material moves rhythmically over the idlers of a belt conveyor, or (3) as particles tumble and slide down a chute.

One way that relative motion can contribute to a segregation problem involves the flow pattern in a vessel (bin, silo, or blender) during discharge. Two primary flow patterns are possible when discharging a bulk solid from a vessel: mass flow and funnel flow [4]. In mass flow, the entire contents of the vessel are in motion during discharge, while in funnel flow stagnant regions exist (these flow patterns are discussed in more detail in another chapter). In a funnel-flow vessel, relative motion exists at the interface of the flow channel and the stagnant material that surrounds it. This creates the opportunity for the fine particles to sift out of the flowing material and into void spaces in the stagnant material, potentially resulting in a layer of concentrated fines surrounding the flow channel.

The result of sifting segregation in a bin is usually a side-to-side variation in the particle size distribution. The smaller particles will generally concentrate under the fill point, with the coarse particles concentrating at the outside of the pile. Within the converging flow channel of a funnel-flow bin, larger particles can be made to change direction more easily to flow with the streamlines of the flow channel than can smaller particles. The latter therefore tend to move vertically downward into a slower moving or stagnant region.

2. Fluidization (Air Entrainment in a Solids Bed) Segregation

When handling powders that can be fluidized, variations in particle size or density often result in vertically segregated material. Finer or lighter particles often will be concentrated above larger or denser particles. This can occur during the filling of a bin or other vessel or within a blending vessel once the blending action has ceased.

A fine powder can remain fluidized for an extended period of time after filling or blending. In this fluidized state, larger and/or denser particles tend to settle to the bottom and fine particles may be carried to the surface with escaping air as the bed of material deaerates.

Fluidization often results in separate horizontal layers of fines and coarse material. When a hopper is being filled at a sufficiently high rate, the material will become aerated. The coarse particles move downward through the aerated bed while the fine particles remain fluidized near, or are driven towards, the surface. This can also occur after blending if the material is fluidized during blending. Fluidization is common in materials that contain a significant percentage of particles smaller than 100 μm [5]. Fluidization segregation is likely to occur when fine materials are pneumatically conveyed, when they are filled or discharged at high rates, or if gas counterflow occurs. As with most segregation mechanisms, the more cohesive the material, the less likely it will segregate by this mechanism.

Gas counterflow can occur as a result of insufficient venting during material transfer. As an example, consider a tumble blender discharging material to a drum, with an airtight seal between the two. As the blend transfers from the blender to the drum, air in the drum is displaced and a vacuum is created in the blender. If both are properly vented, air moves out of the drum and into the blender, but if not, the air must move from the drum to the blender through the blender discharge. In doing so, the fines may be stripped off the blend and carried to the surface of the material in the blender.

3. Dusting (Particle Entrainment in Air) Segregation

Like fluidization segregation, dusting is most likely to be a problem when handling fine powders (typically with particles smaller than about 50 μm [5] that are made up of a range of particle sizes). If the powder is allowed to become airborne, for example, during belt transfers or loading a surge bin, air currents will carry off some particles.

Settling velocity is a term used to describe the gas velocity required to keep a particle suspended. The lower the settling velocity of a particle, the longer it is likely to remain suspended. The settling velocity of a particle in air can be calculated as follows, based on Stokes equation for aerodynamic drag on a spherical particle:

$$u_t = (1.37 \times 10^5)\rho_s D^2 \tag{1}$$

where u_t is the settling velocity (ft/sec), ρ_s the particle density (lb/ft^3), and D is the particle diameter (ft).[6]

The above equation suggests that particle diameter is much more significant than particle density in determining settling velocity.

Dusting segregation can occur even in a windless area, such as inside a bin, since a flowing stream of material will induce airflow.

For example, consider a mix of fine and large particles that is allowed to fall into the center of a bin. When the stream hits the pile of material in the bin, the column of air moving with it is deflected and sweeps off the pile toward the perimeter of the bin, where it becomes highly disturbed, but generally moves back up the bin walls in a swirling pattern. At this point, the gas velocity is much lower, allowing many particles to fall out of suspension. Because settling velocity is a strong function of particle diameter, the finest particles (with low settling velocities) will be carried to the perimeter of the bin while the larger particles will concentrate closer to the fill point where the air currents are strong enough to prevent the fine particles from settling.

Dusting segregation can also result in less predictable segregation patterns, depending on how the bin is loaded, dust collection use and location, and other segregation mechanisms that may be at work. For example, consider a pneumatic conveying line delivering a powder into a large bin, perpendicular to the bin wall. We might expect that dusting segregation would cause the finest material to fall farthest from the entrance point. However, the resulting pattern in this case is likely to be complex, and, depending on the range of particle sizes involved, may be the result of trajectory segregation (discussed below) as well as dusting segregation.

4. Trajectory Segregation

Particles leaving a conveyor or chute with a given trajectory may land at varying distances from their starting point, resulting in a side-to-side variation in the pile that forms. This is the result of particles having different size, shape, and density, resulting in varying momentum and aerodynamic drag.

For example, consider a mix of particles sliding down an inclined chute surface. The fine fraction is more frictional on the chute surface than the coarse material. If loading of the chute is light, so that all of the particles are in contact with the chute, the coarse particles will accelerate more than the fines, which are retarded by their higher friction angle. This results in the fines leaving the end of the chute at a lower velocity than the coarse particles. If the stream is allowed to fall some distance and form a pile, the coarse material will concentrate farther away from the chute, and the fine material will concentrate closer to the chute.

Similarly, if the chute is fully loaded, a mix of fine and coarse material segregates by sifting as it slides down the surface of an inclined chute. The fines, which are in contact with the chute, move more slowly down the chute than does the coarse material, which tumbles over the compacted bed of fines. The effect is the same. The coarse material exits the chute with a higher velocity than that of the fines.

Even if the particles start at the same initial velocity, there are opportunities for trajectory segregation to occur.

Williams [1] defines *stopping distance* as "the horizontal distance that [a particle] would travel after infinite time," and gives the equation for this distance as:

$$X \propto (v_0 \rho_s D^2)/\eta \tag{2}$$

where X is the stopping distance, v_0 the initial horizontal velocity, ρ_s the particle density, D the particle diameter, and η is the fluid viscosity.

He goes on to conclude that since the stopping distance is proportional to the square of a particle's diameter, a particle that is twice the diameter will travel four times as far for a given initial velocity. This effect is somewhat diminished by the limited mobility of a particle within a flowing stream of particles.

This relationship is limited to relatively large particles as very small particles are easily entrained within turbulent gas currents induced by the stream of particles. The finest particles will be carried much farther than the largest particles, and the distances they cover become more of a function of air velocities and particle settling velocities. For these particles, dusting segregation becomes the predominant mechanism.

IV. ANALYSIS AND PROBLEM SOLVING

A. Sampling

Sampling is essential in determining the state of the blend in the blender and in downstream equipment. Collected samples are analyzed with respect to the variables of interest to the application at hand, for example, particle size, composition, pH, dissolution rate, color, etc. The overall *average* of the sample results represents the average *composition* of the blend, while *variations* from sample to sample reflect the *homogeneity* of the blend. The variability is often expressed as a standard deviation, coefficient of variation, or relative standard deviation, although there are many other mixing indices that can be used to describe variations among samples [7].

There are two major concerns with collecting and analyzing samples when homogeneity is considered: (1) the first is being able to collect a sample that truly represents the state of the blend from where it was sampled, which is a significant challenge owing to the potential for sampling error; (2) the second is being able to process or analyze the data in a meaningful way.

1. Sampling Error and Bias

Sampling from a stationary bed is often accomplished with a sampling thief. A sampling thief is a probe that can be inserted into a bed of material to collect a sample from below the surface. Many designs exist, but in their basic form, they are shaped like a rod, or a lance, frequently with a pointed tip and with some type of handle to aid in insertion. Once the theif is inserted into the bed of material, a small cavity below the surface can be uncovered, allowing material to flow into a sample cavity (or multiple cavities). The cavity is then covered, capturing the sample, and the thief is removed from the material. This practice is very common in the food industry, particularly for sampling blends still in a blender, or sampling from bulk containers. Unfortunately, this approach has severe limitations. Collectively, these limitations lead to sampling error. Some of the more pronounced sources of error are:

- The insertion of a thief disturbs and smears the bed, resulting in a sample that does not represent the material that was there prior to the thief being inserted [8].
- Results from a thief are highly operator-dependent; changes in the angle of insertion, insertion rate, twisting or rocking during insertion of the thief or collection of the sample all contribute to variability that can yield significantly different findings [9].
- Thief results are a function of its depth of penetration, such that even if the blend were uniform, results would be different from top to bottom of the blend [10].
- The lack of a standard thief design adds further uncertainty; it has been shown that merely changing the design of a thief can change blend results from unacceptable to passing. In a real application, it would be difficult to know which thief was giving the "correct" results.

There are three basic forms of sampling error. One type results in increased variability of samples, yielding data, which indicate that the homogeneity of the blend is worse than it actually is. Because this type of sampling error is so common, poor blend results are often blamed on it, regardless of whether data supports this claim.

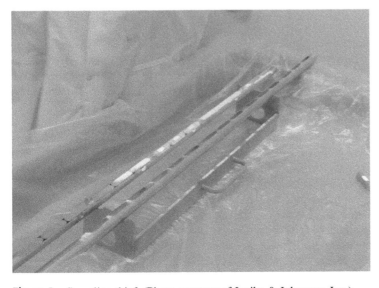

Figure 3   Sampling thief. (Photo courtesy of Jenike & Johanson, Inc.)

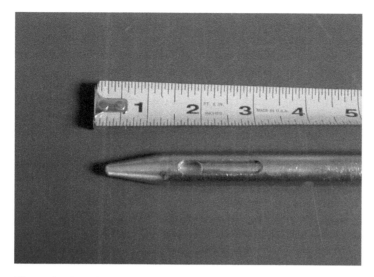

Figure 4   Sampling thief cavity. (Photo courtesy of Jenike & Johanson, Inc.)

A second type of error is an overall shift in the results, whereby the measured average from the samples collected is higher or lower than the anticipated blend composition, possibly without introducing any additional variability in the results. This type of error is often referred to as *bias*. Bias can result if one component adheres to, or is repelled by, the thief, or owing to preferential flow of one component into the thief. Bias can also result if samples are not collected from appropriate locations for example, if samples are collected from a dead zone in the blender that is holding the "missing" material. Therefore, great care must be taken with biased data, and its true cause must be determined.

A third type of error is from a sampling device that yields lower variability than actually present in the blend, perhaps due to smearing of the sample. This type of error, nicknamed "counterfeiting," is hard to detect, in part because statistical tools do not identify it, and in part because the user is generally reluctant to dismiss data that passes acceptance criteria.

It has been said that, as the name implies, a thief is not to be trusted. In recognizing these limitations, the industry has been asking for a "perfect" thief for a long time, so attempts have been made to improve upon the design of the thief to give results that are more accurate. The focus has been on how to collect a sample without disturbing it. Unfortunately, the thief violates the two golden rules of sampling and thus will never be perfect.

Since sampling a stationary powder bed has limitations, a better approach is to follow the two "Golden Rules" of sampling, which are: (1) always collect the sample while it is in motion, and (2) always collect a full-stream sample for a short time period [11]. This eliminates the errors introduced in using a thief. If the sample is collected from the process itself (e.g., upon discharge from the blender while filling a bin), this represents the true state of the material at that point in the process.

If a full-stream sample is collected, the resulting sample is often too large for analysis. In order to reduce it to a smaller size, a subsample must be collected from the larger one. This process, if done incorrectly, can itself induce significant errors. The larger sample collected is immediately prone to segregation as it is placed into a container. Subsampling with a scoop or spoon, or cone-and-quartering techniques, will result in further confusion as to the actual results. Instead, a nonbiased splitting technique is needed. This can be accomplished using

Table 1   Repeatability of Sampling Methods [12]

Method	Std. dev. samples	Est. sample error, %
Cone and quarter	6.81	22.7
Scoop sampling	5.14	17.1
Table sampling	2.09	7.0
Chute riffling	1.01	3.4
Spinning riffling	0.125	0.42
Random variation	0.076	0.25

a sample splitter, which will uniformly divide a sample into two smaller ones, or a spinning riffler, which can separate a sample into many smaller ones (six to sixteen is common). This process can be repeated to get to a sufficiently small sample size. In this manner, each smaller sample represents the larger one. Sample splitting is based on the assumption that the sample to be split is uniform or sufficiently representative of the sampling location of interest; splitting it will mask any variation within that sample.

B. Meaningful Analysis of the Data

Interpretation of the results is equally important. An important consideration is the accuracy of the results, which immediately ties to the often-asked question of how many samples are needed.

1. Number of Samples Required

In measuring homogeneity (or variability) of a blend, the fundamental estimator is some form of standard deviation. To get a good estimate of the standard deviation, a relatively large number of samples are required, whereas comparatively few are needed to get an estimate of the mean value. The probable range of the true mean, based on a limited number of samples, is bracketed by using the student's t-distribution [12]. The probable range of the true standard deviation is bracketed by using the chi-squared distribution. The full treatment on statistics is beyond the scope of this chapter, but many introductory statistics textbooks will cover this in sufficient detail. For example, if 10 samples were collected which yielded a mean of 100 and a standard deviation of 5.0, one could say with 95% confidence that the true mean is between 96.4 and 103.6, a range of ±3.6%. Compare that to the same confidence on the range of the true standard deviation, which would lie between 3.4 and 9.1: a range of −31% to +83%. The actual number of samples required will depend on the actual variability of the blend (less homogeneous requires more samples) as well as the confidence needed in the result, so there is no "right answer": the number of samples required depends on the application. Consider the following application where uniformity of the blend is often used to determine blend time, and where standard deviations from two (or more) datasets are compared.

Thief data from within a blender is useful for determining when a blend is achieved. As blending progresses, the variance of samples collected from within the blender will decrease over time. After a while, this variance will level off, with some "noise" due to sampling error and the statistics of a limited number of samples. When the variance has flattened out, no further blending will be achieved with additional blending time. Note that if blending is too long, the material may segregate within the blender, resulting in overblending and an increase in the variance again. By comparing the results over time, blending time can be optimized.

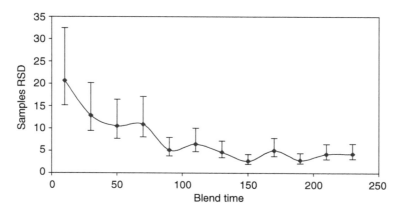

Figure 5 Typical blending trend.

With a limited number of samples, the "noise" from one time point to the next can be quite high. Consider an example with 10 samples collected from two different time points, one sample set from the blender after 10 minutes of blend time, the next set after 14 minutes of blend time. The first set yields a mean of 100 and a standard deviation of 1.9. The next set of data coincidentally yields a mean of 100 but a standard deviation of 3.7. In this example, plant personnel believed that, based on this data, the blend was segregating with additional blend time, and decided it was essential to keep the blend time to 10 minutes. A closer look at the statistics would show that actually these values are not different with statistical significance. In this case, additional samples from each time point, in addition to samples collected from other time points, would help address whether the blend has or has not in fact hit an optimum at 10 minutes.

Although this chapter is not intended to cover aspects of statistics needed to fully analyze these issues, the message to the reader is that statistics must be considered in greater detail than most in the industry are doing now.

2. Stratified Sampling

Stratified sampling of blends and products is being adopted by the pharmaceutical industry [13], and serves as a path forward for many industries to follow. Stratified sampling is the process of selecting multiple units deliberately from various locations within a lot or batch or from various phases or periods of a process [14]. This sampling approach is designed to allow *variance component analysis* [15] to be performed on the data to quantify the variability attributed to the uniformity of the process (across locations) as well as any sampling error or small-scale variability (within each location) that may be present.

An example of stratified sampling is to collect samples from ten locations within a blender, in triplicate. All thirty samples are analyzed. Rather than report the results as a single standard deviation, a variance components analysis is used to separate the variance between locations from the within-location variance. High *within-location* variance, if found, could be attributed to sampling error, improper sample handling or subdivision, or variability of the analytical method used. It could also be attributed to variability of the blend at the small scale as discussed previously, such as owing to agglomerates or other large particles, relative to the sample size. Therefore, high within-location variability of the blend must be investigated relative to the sampling and analytical methods, as well as the formulation itself.

High *between-location* variance is most likely attributable to either poor blending or segregation, although, in principle, sampling bias as a function of sample location can yield between-location variations that do not exist within the blend [16].

It is important to keep in mind that the purpose of sampling is to uncover hidden blend quality issues, not to mask or overlook suspected problems. Samples should be intentionally collected from suspected "hot spots," or regions where the blend may be less uniform, not just from the middle of the blender. Such hot spots may include the surface of the material or regions where the material may have been stagnant during blending.

3. Comparing the Blend with the Product

Powder blend samples are invaluable in determining the state of the blend at various points in the process, and are essential to determine blender and powder handling equipment performance. However, the best way to determine the adequacy of the process as a whole is to sample the product itself once it is packaged or processed to its final form. As with blend samples, a stratified sampling approach should be used. Multiple samples should be collected at the production line at a given instant (for within-location variations), at multiple time points (for between-location variations), from the beginning to the end of a batch, and across a meaningful time period if a process is continuous.

Stratified sampling can give insight into what is occurring with the process (such as segregation), as long as other variables are documented at the same time. For example, if there is a surge bin before the packaging equipment, it is important to note the level in the bin when each sample was collected. For simplicity, it may be best to initially fill the bin from the beginning, then let it discharge completely, while samples are collected at regular intervals. As with collecting samples using a thief, additional samples should be taken from potential "hot spots," such as at the beginning or end of the run. In this example, a plot of particle size versus time would likely reveal any trend of segregation induced by the bin and other upstream equipment.

With an understanding of the flow pattern and segregation mechanisms that may be occurring upstream of the packaging process, one can usually deduce why data variations occur. A full understanding of the physical behavior (segregation mechanisms, flow patterns) of the material is necessary to make meaningful use of sample data.

Stratified sampling of both the blend and the final product provides a powerful tool for diagnosing the root causes of most uniformity problems. With this approach, the data can be interpreted to rank the likelihood that the problems are attributable to blending, sampling, segregation, weight variability, component losses, analytical error, or poor particle size distribution [17].

C. Segregation Testing

In developing a product or designing a process, it is beneficial to know whether the material will be prone to segregation, and, if so, by which mechanism or mechanisms.

When developing a new product, this information can be used to modify the material properties (such as particle size distribution, moisture content, or cohesiveness) to minimize the potential for segregation and to refine ingredient specifications or sources. In developing a process, understanding the potential for segregation can alert the designer to potential risks that may then be avoided. In some cases, significant process steps, such as granulation, may be required to avoid potential segregation problems.

There are two ASTM standards (D6940-03 and D6941-03) on segregation test methods that involve devices developed by Jenike & Johanson, Inc. These testers are designed to isolate specific mechanisms (sifting or fluidization), and test a material's tendency to segregate by that method. A description of these test methods follows.

1. Sifting Segregation Test Method

The sifting segregation test is performed by center-filling a small funnel-flow bin and then discharging it while collecting sequential samples. If sifting segregation occurs during either filling or discharge, the fines content of the discharging material will vary from beginning to end. Samples are collected from the various cups (i.e., the beginning, middle, and end of the discharge) and measured for segregation by particle size analyses, assays, or other variables of interest.

2. Fluidization Segregation Test Method

The fluidization segregation test is run by first fluidizing a column of material by injecting air at its base. After the column is thoroughly fluidized, the air is turned off and the material is allowed to deaerate. The column is then split into three equal sections: top, middle, and bottom, and the resulting samples are measured for segregation.

Segregation tests are useful for identifying which segregation mechanism(s) may be active for a given blend, the general trend that may be observed in the process, and as a comparator between materials. However, the test results have limitations. Most notably, there is no direct way to use the results as a basis for designing a system that will minimize segregation. The results are not scaleable, and are not tied quantifiably to the process. Testing

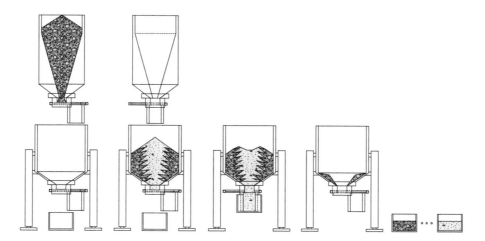

Figure 6   Sifting segregation test method.

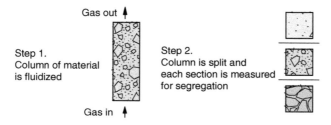

Figure 7   Fluidization segregation test method.

points out, with comparative data, whether a material or blend is prone to segregation by a particular mechanism.

D. Solutions to Segregation Problems

Identifying and correcting a segregation problem is seldom simple; it requires:

- Knowledge of the material's physical and chemical characteristics.
- An understanding of the segregation mechanisms that could be active.
- Knowledge of the process conditions that could serve as a driving force to cause segregation.
- Sufficient sampling to support the hypothesis of segregation (e.g., powder samples and final product samples, samples from the center versus periphery of the bin).
- A full understanding of the fill/discharge sequence, including flow pattern and inventory management, that gives rise to the observed segregation.

Determining which segregation mechanism or mechanisms are at work is an essential starting point. Flow property measurements (wall friction, cohesive strength, compressibility, and permeability) can help to provide understanding of the behavior of the material in storage and the transfer equipment. Testing for segregation potential can provide additional insight about the mechanisms that may be causing segregation.

From the previous discussion about segregation mechanisms, it can be concluded that certain *material properties* as well as *process conditions* must exist for segregation to occur. Elimination of one of these will prevent segregation. It stands to reason then that, if segregation is a problem in our process, we should look for opportunities to either change the material or change the process.

1. Changes to the Material

Often, changing the material is not an option, but the question should always be raised, particularly when developing a new product or process.

Changing the particle size distribution is sometimes the answer. Segregation based on particle size would not be possible if all particles were the same size. In practice, it is seldom practical to achieve this, but reducing the range of particle sizes within the material, or adjusting the mean particle size, may improve the situation significantly. Matching the particle size distributions of the different components is another way to minimize compositional variations due to size-based segregation. This may require purchasing one or more ingredients with a larger size than desired and milling them to achieve the desired size range.

Another possible solution may be to increase the cohesive strength of the material. Particularly in the case of sifting segregation, cohesive materials are much less likely to segregate.

In some cases, granulation can be used to bind different components of a blend together to form "unit particles" that are each composed of all the ingredients of the blend.

2. Changes to the Process and Equipment

Some generalizations can be made when designing equipment to minimize segregation:

- The blending process should generally be located as far downstream in the process as possible. Ideally, blending should occur immediately before the blend is used or packaged.
- Postblend handling of the material should be minimized.

- Surge and storage bins should be designed for mass flow.
- Velocity gradients within bins should be minimized [18].
- Eccentric hoppers should be avoided owing to their inherently large velocity gradients.
- For a given storage capacity, a mass-flow bin with a tall, narrow cylinder will minimize the potential for segregation as compared to that of a short, wide bin. Keeping the material level in the bin high is often beneficial.
- Utilize design symmetry whenever possible. For example, if designing a bin with multiple outlets, all the outlets should be located the same distance from the bin centerline.
- Ensure that any division of the process stream does not result in differences between the streams. Consideration must be given to any potential for segregation upstream of the split.
- Use venting to avoid gas counterflow.

Other specific solutions may be apparent once the segregation mechanism has been identified. For example, if fluidization segregation occurs as a result of filling a bin by pneumatically conveying the material, alternate means to disengage the material from the air stream may be effective in reducing the problem. This may involve a cyclone separator at the bin inlet, or simply a tangential entry into the bin. If material is segregating by sifting when it is loaded into a bin, an inlet distributor may help. An inlet distributor works by breaking up a single incoming stream into multiple streams and distributing them around the bin so that a single central pile does not form.

Mass flow is usually beneficial when handling segregation-prone materials, especially materials that exhibit a side-to-side (or center-to-periphery) segregation pattern, with overall uniformity in the vertical direction. This type of segregation is common when handling coarse, free-flowing materials that contain a range of particle sizes.

It is important to remember that mass flow is not a universal solution; it will not address a top-to-bottom segregation pattern. For example, consider the situation in a tumble blender where fluidization has caused the fine fraction of a blend to be driven to the top surface at the end of a blending cycle. Mass flow would effectively transfer this segregated material to the downstream process, delivering the coarser blend first, followed by the fines.

V. BLENDERS

Blending may be accomplished on a batch or continuous basis.

A. Batch Blending

A batch blending process consists of three sequential steps:

1. Weighing and loading the components.
2. Blending.
3. Discharge.

Product motion within a batch blender is confined only by the vessel, and directional changes are frequent. Unlike a continuous blender, the retention time in a batch blender is rigidly defined and controlled, and is the same for all the particles.

Batch blenders come in many different designs and sizes, and make use of a wide range of blending mechanisms. Blending cycles can take from a few seconds to 30 minutes or more, although properly matching the blending mechanism to the materials should keep the cycle short. Blender discharge may be rapid, or take several hours, particularly if the blender is used as a surge vessel to feed a downstream process. Some batch blenders, such as fluidized zone blenders, are designed for high mixing rates and rapid discharge, making them capable of handling fairly high production rates on a nearly continuous basis. Batch blenders are typically used where:

- Quality control requires strict batch control (weight, composition).
- Ingredient properties change over time and compensation must be on a batch-by-batch basis.
- The blender cannot be dedicated to a specific product line.
- Production quantities are small.
- Many formulations are produced on the same production line.

Advantages of batch blending include:

- Lower installed cost.
- Lower operating cost for small throughputs.
- Lower cleaning costs when product changes are frequent.
- Accurate continuous metering into the blender being unnecessary.
- The option of manual loading.
- Maintaining batch integrity.
- Production flexibility.
- Easy preblending of minor ingredients.
- Positive control of blending time.
- Positive product containment during blending.

B. Continuous Blending

In a continuous blending process, the weighing, loading, blending, and discharge steps occur continuously and simultaneously. Product motion is not completely random, but on average directed from the feed point toward the outlet. Retention time is influenced rather than controlled, with some particles remaining in the blender longer than others, based on factors such as:

- Blender speed.
- Feed rate.
- Blender slope (if any).
- Blender length (in some cases).
- Agitator design.

In order to blend in a single pass, most continuous blenders are more lightly loaded than batch blenders in order to improve the blend quality. Blend quality for a particular blender is a function of retention time.

Continuous blending is typically used when:

- A continuous process is required (often by high production rates).
- Maintaining strict batch integrity is not essential.
- Several process streams are combined.
- The goal is to smooth out or minimize product variations.

A continuous blending system is advantageous because:

- It is easily integrated when the upstream and downstream processes are continuous.
- These is less opportunity for batch-to-batch variation caused by loading errors.
- Automation can sometimes improve quality and reduce operational costs.
- Higher throughputs are often possible.
- No subsequent surge and feed systems are required.
- It usually requires less floor space for a given capacity.

C. Blender Types

1. Stationary Vessel with Internal Agitators

This is a broad classification of blenders; many can be used as either batch or continuous blenders, and can handle most materials. These blenders frequently incorporate spray bars for addition of liquids, usually in relatively small amounts. Rotational speeds can vary significantly, as can the blending mechanisms involved. Generally, convection and shear are the primary blending mechanisms.

These blenders usually do a good job of breaking up agglomerates; some incorporate intensifiers for this purpose. An intensifier is a high-speed agitator (or chopper) that imparts high shear over a small region. It does not perform the bulk of the blending, but is effective in providing some diffusive mixing and in breaking up agglomerates. This can be particularly important when adding liquid to a dry blend where lumps may form. Jacketing for heating or cooling can be added to most designs. Another advantage is that these blenders typically take very little headroom, particularly when compared to tumble blenders of similar capacity.

Energy input is relatively high, so heat buildup and attrition may be a problem with some materials. Clearance between the rotors and vessel walls can result in dead spots that can contain material that is not involved in the mixing process. Some of this material may not discharge easily and may have to be manually removed between batches, depending on the between-batch cleanout that is required.

Segregation can occur during discharge, depending primarily on the material being handled. With some blends, segregation can occur when the blender is brought to a stop.

a. Ribbon, Paddle, and Plough Blenders.   These blenders consist of a cylindrical or U-shaped horizontal vessel with an internal agitator. Several agitator designs are available, making this type of blender suitable for most materials. The blending action can be relatively gentle to aggressive, depending on the agitator design and speed and the use of intensifiers.

Because of their versatility and widespread use, these blenders handle a wide range of products, including baking ingredients and mixes, drink mixes, powdered spices, snack foods, vitamins, coffee, and tea.

The typical agitator designs include:

Ribbon.   One or more helical "ribbon" blades around a central shaft move the product in various patterns to mix by convection. Ribbon blenders are probably the most common agitated blenders used in food processing.

Paddle.   The construction of this type of blender is similar to that of ribbon blenders, but rather than using a helical ribbon, a series of paddles on radial spokes provides the mixing action. The paddles are typically mounted so that they run close to the trough.

Plow.   A plow blender's agitator is similar to that of a paddle blender, but instead of paddles plow-shaped elements are designed to split the material they run through, and often

Figure 8   Ribbon blender. (Photo courtesy of Lowe Industries, Inc.)

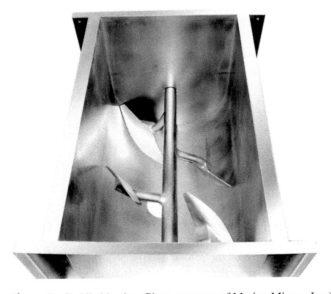

Figure 9   Paddle blender. (Photo courtesy of Marion Mixers, Inc.)

Figure 10   Plow blender. (Photo courtesy of Littleford Day, Inc.)

to throw the material away from the vessel wall toward the axis of the agitator, producing more intensive mixing. Intensifiers are frequently added for breaking up agglomerates, and the vessel shape is usually cylindrical.

A partial list of manufacturers includes:

- Hosokawa Bepex.
- Jaygo Inc.
- Kemutec.
- Littleford-Day.
- Lowe Industries.
- Munson Machinery Company, Inc.
- Patterson Industries.
- Paul O. Abbe.
- Ross Mixers, Inc.
- Scott Equipment Company.
- The Young Industries, Inc.

b. Pin Mixers.   Pin mixers utilize a single rotor turning within a cylindrical shell at high speed to wet a powder, form agglomerates, or break up agglomerates. Pins or small paddles extend radially from a central shaft. Retention time is very short, generally less than a minute.

A partial list of manufacturers includes:

- Hosokawa Bepex.
- Littleford Day.
- Munson Machinery Company, Inc.
- Scott Equipment Company.
- The Young Industries, Inc.

c. Fluidized Zone (Mechanical) Blenders.   These blenders consist of twin counterrotating paddled agitators that mechanically fluidize the ingredients. Rotation is such that the blend is lifted in the center, between the rotors. Mixing is quite intensive, producing intimate blends in a short period of time. Blend cycles are often less than a minute, and bomb-bay doors allow rapid discharge of the blend. These combine to give this blender a high throughput capacity relative to its batch size.

Fluidized zone blenders are used for materials with a wide range of properties, including cohesive materials. Various spray bars and integrated pin mills are available for liquid

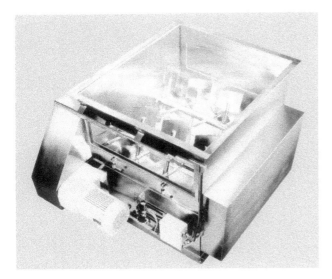

Figure 11 Bella™ twin shaft fluidized zone mixer. (Photo courtesy of Dynamic Air, Inc., St. Paul, MN, USA.)

or fat dispersion and agglomerate breaking. Low rotor speeds can be used to blend friable materials.

A partial list of manufacturers includes:

- Dynamic Air.
- Forberg.

d. Planetary Mixers. This is a varied subcategory, which includes blenders that are used primarily for powders and some that are used primarily for slurries and pastes. One type of planetary mixer commonly used for powders is the planetary screw (sometimes called a conical screw mixer).

The planetary screw is composed of a screw conveyor inside a conical hopper. The screw is located in such a manner that one end is near the apex of the cone and the other end is near the top of the hopper, with the tip of the flights near the surface of the hopper. The screw rotates about its axis while revolving around the axis of the hopper, pulling material up from the bottom. Advantages include the ability to handle a wide range of materials, from free-flowing to cohesive. Potential disadvantages include sifting segregation within the blender (fines at top of screw, coarse material toward the bottom of the pile, resulting in a nonuniform steady state), possible segregation during discharge, and a dead region at the bottom of the cone during blending.

A partial list of manufacturers includes:

- Hosokawa Micron.
- Jaygo Inc.
- Littleford Day.
- Krauss Maffei Process Technology, Inc.
- Ross Mixers, Inc.

e. Vertical Ribbon Blenders. As the name implies, these are similar in appearance to ribbon blenders, except that the axis of rotation is either vertical or steeply inclined. Feed and discharge are through relatively small nozzles on either end, and large doors are often placed on one side to make thorough cleaning possible.

Figure 12   Day-Nauta planetary blender. (Photo courtesy of Littleford Day, Inc.)

Common variations include single-agitator and dual-(parallel) agitator designs, choppers to break up agglomerates, and spray nozzles to add small quantities of liquids. Single agitators provide gentler mixing, while the dual-agitator design provides more intense mixing because of the interaction between the rotors.

The agitators are designed to move the blend upward near the vessel walls, while gravity causes the material to cascade downward along the blender's axis. This action often results in a reduced blending time relative to a traditional ribbon blender. AZO Inc. manufactures vertical ribbon blenders.

f. Vertical Helical Augers.   This design places a tapered ribbon flight inside a conical hopper, with its axes collinear. It has many of the advantages and disadvantages of the planetary screw. The principal difference is that the axis of the flight rotation is vertical and collinear with the axis of the conical vessel. Some designs incorporate a flight design that minimizes stagnant material at the bottom of the hopper during blending.

A partial list of manufacturers includes:

- National Bulk Equipment, Inc.
- Ross Mixers, Inc.

g. Cut Flight Augers.   Cut flight augers act primarily as a conveyor, but provide some mixing as well. They are normally used as a continuous mixer where intensive mixing is not needed, and are available from most screw conveyor manufacturers.

2.   Tumble Blenders

A variety of tumble—blender designs are available for food processing applications; some of the more common types are described below. Tumble blenders as a group can be described as vessels that are rotated about an axis other than their geometric axis. They are typically gentle on the ingredients, although attrition can occur in any blender. They are used for materials ranging from fine powders to coarse materials such as diced vegetables. Because the basic design incorporates no internal moving parts, they are suitable for use where sanitation is important. For the same reason, they handle abrasive materials well. The lack of rotational seals, agitators, etc., also makes tumble blenders easy to clean, depending on the access ports provided.

The primary blending mechanism is convection, but diffusion plays an important role if lateral (along the axis of rotation) mixing is required. This can be significant if the ingredients are loaded into the blender side-by-side rather than one on top of the other. Side-by-side loading is most common when components are loaded into a V-blender from opposite legs of the V. Static internal baffles, asymmetric designs (e.g., differing leg lengths on a V), and off-axis rotation are used in some designs to improve lateral mixing.

Intensifiers are sometimes used to add shear, and spray bars for liquid addition can be integrated, but the basic design with no internal parts is by far the most common design in food processing applications. Tumble blenders are also used for drying, heating/cooling, or chemical treatments in some processes.

Virtually all tumble blenders are batch blenders. One example of a continuous tumble blender with which the authors are familiar is the Zig-Zag® blender by Patterson Kelly. Tumbling here is accomplished by a roughly horizontal zigzag-shaped tube mounted in trunnions. The material is fed into one end of the tube and blended as it progresses through the tube as though through several rotations of a twin-shell blender.

Tumble blenders are best suited for relatively free-flowing materials. The ability to produce an intimate blend of cohesive materials is limited and discharge through a relatively small outlet can create problems when handling cohesive blends.

a. Twin-Shell Blenders (V-Blenders). Twin-shell blenders, or V-blenders, are probably the most common tumble blenders in food processing applications. These blenders usually discharge in a funnel-flow pattern because of the relatively shallow wall angles near

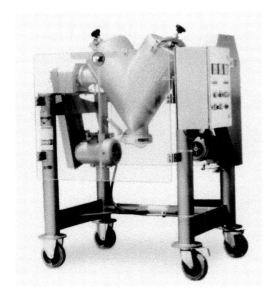

Figure 13 Twin shell blender. (Photo courtesy of Patterson-Kelley, Co.)

the outlet. Some manufacturers modify the geometry of the blender near the outlet, adding a steep-walled cone to prevent dead regions during discharge. Loading may be through one or both legs of the V, or through the outlet (at the tip of the V) with the blender inverted. Twin-shell blenders are quite versatile and are used to blend spices, drink mixes, and soup mixes.

A partial list of manufacturers includes:

- Gemco.
- Jaygo Inc.
- Patterson Industries.
- Patterson-Kelley.
- Paul O. Abbe.

b. Twin-Cone Blenders.   Twin-cone blenders consist of two opposing cones with their large ends mated to opposite ends of a cylinder. In some designs, the axis of the two cones is offset to improve mixing parallel to the axis of rotation, which is perpendicular to the geometric axis of each cone. This improves lateral mixing, which would otherwise be a relatively weak result of diffusive (random) mixing.

Many variations of a twin-cone blenders are available, with features such as intensifiers and spray bars. Gemco offers a model with one removable cone that can be used to transport the blend as a portable container. Spare Porta-Hoppers™ keep the blender in production while full hoppers are transported and discharged. This concept is beneficial in reducing the potential for segregation since each transfer step provides an opportunity for segregation to occur.

A partial list of manufacturers includes:

- Gemco.
- Jaygo Inc.
- Kemutec.
- Patterson Industries.
- Patterson-Kelley.
- Paul O. Abbe.

Figure 14   Porta-Hopper double cone blender. (Photo courtesy of Gemco, Inc.)

Figure 15   Mass-flow BINSERT® tumble blender. (Photo courtesy of Jenike & Johanson, Inc.)

c. Mass-Flow BINSERT® Tumblers.   Materials that blend easily also often segregate easily [2]. Discharging a highly segregating blend from a blender to a downstream process or a surge bin can result in an unacceptable blend, even if the blender is producing a high quality blend, owing to segregation during discharge, as previously discussed. One solution is to provide mass flow and control the velocity gradients within the flowing material.

A BINSERT® tumbler makes use of a hopper-within-a-hopper design on one end of the blending vessel to improve segregation control over that of a typical blending vessel, even one designed to provide mass flow. Internal baffles are often used to improve mixing parallel to the axis of rotation. The BINSERT® tumbler is manufactured by Jenike & Johanson, Inc.

d. In-Container Tumble Blending.   In-container tumble blending is used when transfers after blending must be minimized in order to preserve blend quality and minimize segregation. Blended material may be transported in the container directly to the end use point, for instance to a feed point above a packaging line.

Containers are secured into a stationary blending unit that tumbles them, often at an oblique angle. Multiple portable containers can be used to store preblend and postblend materials so that the blender can be fully utilized. Since cleaning procedures do not involve the blender itself, downtime is not required for routine cleaning procedures. The portable containers are sometimes used to store the product between blending and use, particularly if strict quality control requires release of a blend batch before packaging or use.

To achieve maximum segregation control, the containers should be designed to provide mass flow during discharge. These systems are widely used in the pharmaceutical industry as well as for such products as soup mixes.

Figure 16 In-container tumble blender. (Photo courtesy of Jenike & Johanson, Inc.)

A partial list of manufacturers includes:

- Jenike & Johanson, Inc.
- Matcon.
- Patterson-Kelley.
- Tote Systems.

3. Gravity Blenders

Gravity blenders, also called static in-bin blenders, are specially designed bins or silos that double as blenders by using the flow of the material through the vessel to blend their contents.

Gravity blenders can be designed for either single pass operation or recirculation. Single pass blenders are generally limited to smoothing out process variations or variations between batches, while systems designed for recirculation can produce an intimate blend of the entire contents of the bin. Single pass systems face the inherent problem that the first material loaded into the blender is not involved in the blending activity. For this reason, even single pass blenders are sometimes designed to recirculate a small amount of material prior to discharge.

While gravity blending has a somewhat limited application, it can offer some significant benefits. Gravity blenders often can be integrated into a process that already requires material storage and, since they have no moving parts, little or no additional operation or maintenance costs are incurred.

Successful performance of any gravity blender is dependent upon involving the entire bin contents in the blending process; for this to occur, the blending bin must be designed for mass flow [19].

Two concepts are commonly employed for gravity blending:

1. Material is withdrawn from various parts of a bin and recombined at the discharge point (tube blender).
2. Velocity gradients are used to influence retention time in the bin so that some particles are retained longer than others (BINSERT® blender).

Carson and Royal [20] summarize the requirements of a good gravity blender. A good gravity blender should:

- Be designed for mass flow (no dead regions).
- Allow for a large differential between the time it takes a particle in the fastest flowing region to exit the blender compared to the time of a particle in the slowest flowing region. In addition, particles in the fastest flowing region should start to discharge as soon as possible after they enter the blender.
- Require a minimal amount of recirculation of its contents.
- Provide blending uniformity that is independent of the blender's fill or discharge rate or material level.
- Provide the ability to blend a wide variety of materials (e.g., fine and coarse particles, free-flowing and cohesive materials).
- Cause segregating materials to remain blended as they discharge from the blender.
- Be cost-effective to install and operate.

a. Tube Blenders. Numerous designs have been developed to make use of the concept of withdrawing material from different parts of a bin, through use of either a central tube with multiple inlets, or multiple tubes that terminate at different elevations.

Phillips blenders are one of the more well-known tube blender designs. The Phillips design consists of multiple tubes that extend vertically up through the cylinder of the blending silo. The lower ends of the tubes terminate in the hopper of the blending vessel, just below an insert. Each tube is divided longitudinally into three parts. Multiple openings are placed in each tube segment at different levels so that material flows only into the uppermost hole below the surface of the material.

Another well-known design is the Multi-Port® gravity blender made by Young Industries. The Multi-Port® blender utilizes multiple outlet nozzles that protrude through the wall of the bin at different locations in the cylinder and hopper. These nozzles are connected, through a blending tube, to a blending chamber below the main vessel. The blending chamber is a small hopper where the different streams are collected.

Tube blenders are best applied to materials that are free flowing, since cohesive materials have the potential to arch over the nozzles of the blending tubes. Materials with a high angle of internal friction may form steep flow channels outside the blending tubes, which reduces blending efficiency. Tube blenders also do not handle materials with a wide range of particle sizes or the addition of minor ingredients well [20].

b. Static BINSERT® Blenders. BINSERT® blenders consist of a hopper within a hopper (normally conical) below a vertical cylindrical bin. These blenders are specifically designed for each application based on the properties of the materials to be blended.

Figure 17   Static BINSERT® blender. (Photo courtesy of Jenike & Johanson, Inc.)

Velocity gradients are used to produce the blending action. By varying the geometry of the inner and outer hoppers as well as the configuration of the geometry below, a significant velocity differential can be achieved between the material flowing through the inner hopper and that flowing through the annular space around it. The result is a large difference in retention time between the fastest and slowest particles in the blending bin.

BINSERT® blenders can be used to handle a wide range of materials, including cohesive and highly segregating blends. Since these blenders are designed for mass flow, all the material in the bin is involved in the blending activity, with no stagnant regions. Another benefit is that existing bins often can be retrofitted for use as a BINSERT® blender in a cost-effective manner.

A partial list of gravity blender manufacturers includes:

- Columbian Tectank
- Fuller Solids Handling
- Jenike & Johanson, Inc.
- The Young Industries

4. Pneumatic (Fluidization) Blenders

Pneumatic fluidization can be an effective method of blending fine powders.

A pneumatic blender consists of a static shell with either a fluidizing membrane or an array of nozzles at the bottom. These are used to inject gas (usually air) into the material to produce pressure gradients that cause convective blending. The resulting fluidized material allows diffusive blending to occur as well. In the designs using nozzles, the nozzles are often pulsed sequentially in a circular pattern.

Mixing in these compact blenders can be quite rapid, and they can be combined with a storage vessel. Blending typically can be optimized by adjusting open and closed dwell times, cycle time, and gas pressure. The limitations are that not all materials can be effectively blended, and some materials may segregate as a result of the gas flow through the bed of particles.

Figure 18 Blendcon® air blender. (Photo courtesy of Dynamic Air, Inc., St. Paul, MN, USA.)

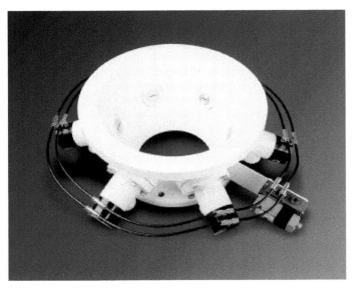

Figure 19 Pneumatic blender. (Photo courtesy of NOL-TEC Systems®, Inc.)

A partial list of manufacturers includes:

- Dynamic Air
- Glatt Air
- Nol-Tec

5. Rotating Vessels

Rotating vessels are well suited to either batch or continuous operations, and include vessels of varying shapes, but generally involve cylinders and cones. Rotation is along the geometric axis of the vessel, and internal flights, lifters, or dividers are usually attached to the vessel's interior surface.

While diffusion is at work in these blenders, the primary blending mechanism is typically convection.

In batch blenders, materials may be loaded and unloaded from one end. In a continuous blender, the ingredients are fed in one end, progress through the vessel, and the blend is discharged from the other end.

Small amounts of liquid additives can be added, but these are low intensity, low shear blenders, so if agglomerates form, they may not be broken up effectively by the blending action. This makes rotating vessels useful in applications where gentle handling of a product is necessary. A particular form of this type of blender is used for coating breakfast cereals and snack chips, which are quite fragile. Other materials handled include grains, starch, coffee, tea, dried vegetables, and spices.

A partial list of manufacturers includes:

- Continental Products Corporation
- Munson Machinery Company, Inc.
- Scott Equipment Company

D. Blender Selection

Blender selection is not a straightforward process and a cookbook approach is not feasible. The reasons for this include the number of variables involved and the fact that there is a lack of understanding of the fundamentals of blending bulk solids. This is not an indictment of those specifying or designing blenders, but rather an observation about the body of knowledge available on the subject [21–23]. Blending is a complex process frequently involving multiple blending mechanisms as well as segregation mechanisms. Analytical methods are generally highly simplified, and usually applicable only to idealized cases.

As a result, systematizing blender selection is not a simple process. In practice, a blender is often selected because of experience with a similar application rather than the rational evaluation of a variety of potential solutions. Not uncommonly, one plant will use a particular type of blender for a given product, while another plant will use another type of blender for the same product.

The best outcome can be reached when you know as much about your products and ingredients as possible early in the selection process.

As a starting point, a quantitative definition of the required blend quality must be developed. Frequently this is stated as XX% material A and YY% material B. As discussed earlier, this is an inadequate definition of blend quality. We need to define:

- Sample size that will be used for evaluation (which should usually relate to package size)
- Target values for composition

- Allowable variation from the target
- Methods to be used to evaluate blend quality

In addition to a definition of the blend quality, the properties of the ingredients and the blend should be accounted for in blender selection or design. For example, determining the arching dimensions for the blend may steer you away from a tumble blender with a small outlet. Knowing that a particular segregation mechanism is likely may steer you toward a blender that will provide mass flow during discharge. The following properties are likely to be helpful in selecting the optimum blender for a given application:

- Compressibility (bulk density as a function of consolidating pressure)
- Wall friction angles
- Cohesive strength
- Permeability
- Potential for segregation
- Abrasiveness
- State of incoming materials (agglomerates, caking, cohesive strength, etc.)
- Variability of incoming materials over time or between batches

In some cases, blender choices may be narrowed by specific process or space requirements. For example, when a blended batch must be evaluated by a time-consuming quality assurance process prior to discharging it, in-container tumble blending may be an attractive approach. Sometimes additional processes, for example, drying, heating, or cooling, can be combined with the blending operation. This may require a stationary vessel so that the blender can be jacketed (although some tumble blenders are used in these applications as well). Large-scale storage of material can often be combined with a gravity blender, reducing both capital and operating costs.

Sometimes ingredient and blend properties will determine the blending mechanism that must be used or how that mechanism will be imparted. For example, if we were interested in blending fragile cereal flakes, we would avoid high intensity shear and look for a blender that will use convection and diffusion to blend: perhaps a rotating vessel. If agglomeration during blending is likely, or if liquid addition is necessary, high intensity shear may be necessary to ensure that lumps are broken and the blend quality is good on a small scale.

Once the choices have been narrowed to a few blender types, scaled blending tests should be performed. In order to evaluate the results of these tests and compare the performance of the blenders, a stratified sampling and analysis plan should be prepared and carried out. The criteria for comparison should be clearly spelled out based on the important characteristics of the product. Analysis may be on a particle size basis, chemical analysis, or some other measurable criterion.

Other considerations in blender selection include material flow, containment, cleaning, and minimizing dead regions. These are discussed in more detail below.

1. Material Flow

Flow of ingredients into the blender and discharge of the blend are important aspects of achieving and maintaining an adequate blend. Reliable flow is the most obvious aspect, as unreliable flow may result in nonuniform product, an interrupted cycle, blender downtime, or plugged equipment that must be manually unloaded.

Segregation is another material flow related issue that has to be considered. Segregation of the ingredients prior to loading the blender may result in bulk density and particle size distribution swings, if not compositional changes. If these properties are not consistent between batches, the blend will not be consistent.

2.   Containment

The need for material containment during blending may not be immediately evident. Most commercially supplied equipment comes with covers and doors that limit access to the blender internals for safety; these same covers are easy to seal and often provide sufficient containment while blending. Containing materials while a blender is filled and discharged is not as easily accomplished. Containment design should address processes immediately upstream and downstream in addition to the blender. Two containment methods frequently employed are closed and open containment systems. Closed containment systems use mechanical barriers limiting particulate escape. An example of an open system is discharging a ribbon blender into a portable container, allowing product to free-fall in open air between the two.

In closed systems, openings are covered and their sizes minimized. Dust collection for a closed system utilizes a low airflow rate and slightly negative internal pressure to minimize dust escape. Less commonly, a closed system is designed for a slightly positive pressure when contamination by room air is not permissible. Because airflow rates are low in a closed system, duct velocities may be low, which can result in settling, buildup, and plugging if the dust collection lines are not properly sized. For this reason, systems that return the material directly to an enclosure, such as a blender-mounted reverse pulse bag unit, are recommended for their simplicity, reliability, low maintenance, and conservation of product.

Open systems make use of air velocity at openings and transfer points to minimize the amount of dust escaping into the production environment. The required airflow rate is related to the number and size of the openings and product particle velocity at the opening. "Capture velocity," or the velocity required to prevent dust escape, must be maintained at each opening. This type of system is typified by vacuum systems with high airflow rates, frequently integrated into an area dust collection network with captured dust from various sources returned to a central bag house collector.

Open systems are typically not as effective in dust control as closed systems, and product collected through an open system is unlikely to be acceptable for use in a final product since contamination of the recovered dust is likely. As a result, the recovered material becomes waste. However, open systems have the advantage of using a central fan or vacuum device to service a large area, limiting initial equipment costs as well as ongoing maintenance requirements.

3.   Cleaning

Requirements for inspection and cleaning of food processing equipment vary greatly, depending on the products and materials handled. A discussion of these requirements is beyond the scope of this chapter; the reader should refer to the appropriate regulatory agencies (USDA/FDA) for assistance in setting up a sanitation program.

The two important goals in cleaning a blender are to maintain sanitary conditions (minimize microbial growth and foreign material) and to prevent batch-to-batch cross-contamination.

Maintaining sanitary conditions is normally the most rigorous standard, and is typically scheduled based on microbial growth rates or Current Good Manufacturing Practices (CGMP) [24]. If cleanout is often, after every batch or several times a day, the cost of cleaning can be significant owing to both labor and lost production. Heating may be necessary to meet sanitation requirements or to minimize drying time. Clean-in-place (CIP) designs can help minimize labor cost and lost production time but can add significantly to capital cost.

Prevention of cross-batch contamination is typically done on an as-needed basis determined by production needs and quality standards. Because blenders are expensive and often have large throughput capacity, a single blender may be used for several different products in one production shift. Use of products that may cause serious allergic reactions or other consumer health concerns requires cleaning of equipment according to extreme cleanliness standards between batches. Products labeled nut-free, for example, must be strictly isolated from those containing nuts. Kosher and organic products also have specific cleaning and inspection requirements.

Many blenders for food handling applications are fabricated from stainless steel, but carbon steel is also common for some products, particularly those that are dehydrated or those that contain oils. The potential for contamination from rust is usually the criterion for material choice. If frequent wet washing of the blender is required, stainless steel should be used.

Nonwelded surface finishes and weld finishes should be defined to meet specific cleaning needs. Gaskets, covers, clamps, shaft seals, shafting, impellers, valves, dust collection components, and all other product contact materials should be reviewed to ensure compatibility with the products handled and compliance with cleaning requirements. The degree of disassembly required to meet sanitation requirements must also be considered in order to minimize downtime and cleaning costs.

4. Dead Regions

Dead spots can occur during blending or during discharge; both cases present potential problems. Dead spots during blending may result from clearances between an agitator and the wall, or as a result of material buildup on an agitator or vessel feature such as trunnion reinforcement. Minimizing dead regions during blending is important to produce a quality blend, since stagnant material will not be involved in the blending activity.

Dead regions during blender discharge result from a funnel-flow discharge pattern, which can lead to segregation problems, as previously discussed.

VI. AN EXAMPLE: PACKAGED FOOD PRODUCT

This example illustrates sampling, analyzing relevant variables, and flow pattern effecting sifting segregation. Manufacturers who sell bulk material in packages must meet minimum weight requirements. Since some variability in package weight is inevitable, packages must be overfilled in order to ensure the minimum is achieved. This excess product is essentially given away, resulting in lost profit to the manufacturer. The cost can run to hundreds of thousands of dollars each year for a single production line.

A manufacturer of a precooked food product realized that its weight variations were high, and that this was related to the product flow through upstream equipment. In this process, there are several similar production lines. In each line, material is accumulated in a surge bin prior to packaging. Although a specific weight is required in each box, filling is on a volumetric basis. Therefore, a consistent bulk density is critical to achieving consistent package weight.

It was observed that when the material level in the surge bin was high, boxes were heavy. When the level was lower, boxes were light. Initially, plant personnel were suspicious that the higher level compacted the material, resulting in a higher density product, and,

Table 2   Measurement of Particle Size for Each Box of a Given Weight

Cumulative size (mesh)	Box weight (g)					
	823	809	804	800	785	769
+6	2.8	2.5	2.8	3.2	3.5	7.1
+8	20.2	24.1	39.2	42.5	34.4	56.4
+10	74.6	72.8	83.2	82.5	85.7	89.2
+12	95.0	95.5	95.9	94.0	97.8	97.9

as the level was lowered, less compaction resulted in a less dense product. However, this hypothesis had two flaws. First, the surge bin was located about 10 feet. above the packaging equipment to allow room for other plant equipment. A long, narrow sloping chute brought product from the bin to the packaging equipment. Unlike a liquid head, any changes in head pressure within the bin would not be transmitted to the packaging equipment, because wall friction within this chute would quickly take up any additional load. The second flaw was that the material was nearly incompressible. Although the bin was only eight feet tall, it would have taken pressure differences in excess of a 100 foot depth to produce sufficient variations in density to give the weight variations that were observed.

Upon realizing that head pressure could not be to blame, segregation was suspected. Although the material was closely sized, attrition in the upstream processes resulted in broken particles. Since the material was relatively coarse (a mean particle size between 8 and 10 mesh) and free-flowing, sifting segregation was suspected. To confirm this, boxes were collected throughout the run and then weighed and screened. The results confirmed that segregation was occurring. Table 2 shows the weight of some boxes along with the particle size determined by sieving. The acceptable weight was 794 ± 20 g.

Flowability tests of the material showed that the surge hopper was discharging in funnel flow. Furthermore, the bulk density of the +8 mesh material was significantly lower than that of the −10 mesh material, easily accounting for the observed variations.

With all this information, the problem was easily identified. The inlet and outlet of the funnel-flow surge bin were on the centerline of the bin. Upon filling the surge bin, material segregated by sifting. This resulted in a concentration of fines in the center of the bin, and a concentration of coarse material at the wall. When the material level was high, the fines were discharged through the flow channel that formed above the outlet; as the level dropped, the predominantly coarse product from the wall discharged. Without a full understanding of what was really happening, the symptoms could easily have been misinterpreted as being caused by the pressure in the bin, rather than by segregation.

To correct the problem, one bin was selected from the most poorly performing production line and retrofitted with an insert to convert it to mass flow. This significantly reduced the weight variations and this line, which had been giving the greatest variations, became the best performing line. Subsequently, other bins on the other lines were also retrofitted with such an insert.

REFERENCES

1. Williams, J.C. The segregation of particulate materials: A review. *Powder Technology* 15, 245–251, 1976.
2. Prescott, J.K. and Carson, J.W. Analyzing and Overcoming Industrial Blending and Segregation Problems. International Union of Theoretical and Applied Mechanics Symposium, Segregation

in Granular Flows, Kluwer Academic Publishers, Rosato, A.D., and Blackmore, D.L. Eds., Dorecht, Netherlands, 2000, p. 89–102.

3. Williams, J.C. and Khan, M.I. The Mixing and Segregation of Particulate Solids of Different Particle Size. The Chemical Engineer. January 1973, 19–25.

4. Jenike, A.W. Storage and Flow of Solids. Rev. 1980. University of Utah, Salt Lake City, UT, 16th Printing, July 1994.

5. Pittenger, B.H., Purutyan, H., and Barnum, R.A. Reducing/eliminating segregation problems in powdered metal processing. Part I: segregation mechanisms. *P/M Sci. Technol. Briefs* 2, 5–9, March 2000.

6. Burton, D.J. Ed. *Hemeon's Plant and Process Ventilation*. 3rd ed. Lewis Publishers, Boca Raton, FL, 1999.

7. Fan, L.T., Chen, S.J., and Watson, C.A. Solids mixing. *Industrial Eng. Chem.* 62, 53–69, July 1970.

8. Harwood, C.F. and Ripley, T. Errors associated with the thief probe for bulk powder sampling. *J. Powder Bulk Solids Technol.* 1, 20–29, Fall 1977.

9. Berman, J. and Planchard, J.A. Blend uniformity and unit dose sampling. *Drug Dev. Ind. Pharm.* 2, 1257–1283, 1995.

10. Berman, J., Schoeneman, A., and Shelton, J.T. Unit dose sampling: a tale of two thieves. *Drug Dev. Ind. Pharm.* 22, 1121–1132, 1996.

11. Allan, T. *Particle Size Measurement*. 2nd ed. London, Chapman & Hall Ltd., 1975.

12. Kepmthorne, O. and Folks, J.L. *Probability, Statistics and Data Analysis*. Iowa State University Press, Ames, Iowa 1971.

13. Boehm, G., Clark, J., Dietrick, J., Foust, L., Garcia, T., Gavini, M., Gelber, L., Geoffroy, J., Jimenez, P., Mergen, G., Muzzio, F., Planchard, J., Prescott, J., Timmermans, J., and Takiar, N. Final blend uniformity working group recommendation. *PDA J. Pharm. Sci. Technol.* 57, 64–74, March–April 2003.

14. *Glossary and Tables for Statistical Quality Control*. ASQC Quality Press, Milwaukee, WI 1983.

15. Davies, O.L. Ed. *Design and Analysis of Industrial Experiments*. Hafner Macmillan, New York, 1960.

16. Prescott, J.K., Ramsey, P., Gladysz, K., and Nowaczyk, F., Bench-Scale Segregation Tests as a Predictor of Blend Sampling Error. Presented at AAPS Annual Meeting, Indianapolis, IN, 2000.

17. Prescott, J.K. and Garcia, T.P. A solid dosage and blend content uniformity troubleshooting diagram. *Pharm. Technol. N. Am.* 25, 68–88, March 2001.

18. Carson, J.W., Royal, T.A., and Goodwill, D.J. Understanding and eliminating particle segregation problems. *Bulk Solids Handling* 6, 139–144, February 1986.

19. Wilms, H. Inserts in Silos for Blending. In: Brown, C.J. and Nielsen, J. Eds. *Silos: Fundamentals of Theory, Behaviour and Design*. E & FN Spon, London, pp. 131–141, 1998.

20. Carson, J.W. and Royal, T.A. Techniques of In-Bin Blending. Presented at Bulk 2000: Bulk Material Handling toward the Year 2000, London, October 29–31, 1991.

21. Adams, J.F.E. and Baker, A.G. An assessment of dry blending equipment. *Trans. I. Chem. E.* 34, 91–107, 1956.

22. Cooke, M.H., Stephens, D.J., and Bridgwater, J. Powder mixing: a literature survey. *Powder Technol.* 15, 1–20, 1976.

23. Bridgwater, J. Fundamental powder mixing mechanisms. *Powder Technol.* 15, 215–236, 1976.

24. Current Good Manufacturing Practice in Manufacturing, Processing, Packaging or Holding Human Food. Code of Federal Regulations, Title 21 — Food and Drugs, Part 110.

13
Powder Flow Properties

Kerry Johanson
University of Florida
Gainesville, Florida

CONTENTS

I. INTRODUCTION

When considering the relevant material properties for powder and granular food products, one should determine the type of equipment needed to process or handle food products and the operating conditions at plant facilities. With this in mind, a brief review of typical unit operations in food industries is in order.

Many food products begin as grain in some form or another. These grains are harvested, placed in silos, processed to remove unwanted parts, and then milled, classified, and finally bagged. Typically the finished consumer product is bags of flour or meal ranging between 5 and 50 lb capacities. During the course of this processing, the basic feedstock may range from coarse free-flowing granules to fine cohesive powders. Often the natural oils in feedstocks cause significant cohesive flow problems or adhesion problems in process equipment. Storage temperature and humidity also cause significant changes in the product and can lead to hang-ups in process equipment. The grains may flow at high rates during gravity discharge and require feeders to control the product rate. However, fine milled powders may flow at either significantly reduced rates or very high rates, depending on the quantity of entrained air present in the material. Typical problems for these types of processes are hang-ups and flow rate limitations or control. Dust control and dust explosion mitigation are also important issues. Segregation is not generally a problem with finer flour materials, but may cause problems with meals and coarser grinds. To quantify these potential problems, a unique set of material properties must be measured.

Other food products start as liquid or slurry feeds that are spray dried to create a powder product. Different particle characteristics are possible depending on the drying rate and liquid throughput. These fine powders or granules are then classified and conveyed to process and handling facilities. They may be mixed with other ingredients and then packaged or they may simply be packaged. Many of the flow problems in these processes arise because of the product moisture content and temperature of the material during processing. Cohesion of these types of material often increases during storage, resulting in very strong agglomerates or cakes. Degradation during conveying and handling is also a significant problem. Segregation of final product mixtures causes many consumer complaints.

Some food products start as a liquid feed and then are crystallized. The solid grains are separated from the liquor through filtration or centrifuges. The product is then dried, classified, and may be "conditioned" in special bins. These bins may cause additional drying or allow sufficient residence time to stabilize the crystallization process. These products can be mixed with other ingredients and finally packaged for consumers. Some of these granular materials are then milled to create fine powder products.

Still other food products are harvested as leaves or sections of a plant. These types of materials make up many of the spices used to entice our palates. These products are dried and then conveyed or manually transported to small feed bins and mixing stations where they are combined to form spice blends, dried soup mixes, flavor enhancers, or dressing mixes. The most important unit operation in this process is the mixing stage. Typically these mixtures contain a combination of coarse particles, fine powders, flakes, and granules. Keeping these products intimately mixed is the primary challenge. In addition, these products often feed small packages requiring small size openings in process equipment. This results in process hang-ups owing to the cohesive or flaky nature of these materials.

It is the goal of this chapter to present to the reader the primary flow property measurements required to predict acceptable process behavior. In order to do this, the effect of several important operational and process parameters must be established.

II. EFFECT OF TEMPERATURE ON FLOW PROPERTIES

Many food products behave poorly at elevated temperatures. There are several reasons for this. Some foods undergo a glass transition where they change from an amorphous solid to a crystalline material. The temperature at which this happens is a function of the material moisture content and the bulk temperature. At elevated temperatures, amorphous particles become soft and solid contact pressures cause particles to bind together. When temperature decreases, the material switches to the crystalline phase, cementing the particles together and forming strong particle bonds. To complicate this issue, ambient moisture in the air can change the glass transition temperature through moisture adsorption into the particles. Understanding these effects allows us to develop test programs that mimic the worst-case process conditions that may be expected at the plant. For example, a material sensitive to glass transition effects will gain significant strength when briefly exposed to humid air and then subjected to temperature cycles during storage. This suggests that testing programs for this type of material should include measuring bulk strength after briefly subjecting the material to humid conditions, inducing moisture pickup, and then exposing it to cyclic temperature swings.

Some food products have significant oil content that becomes mobile when heated to only moderate temperatures. These oils come to the surface of warm food products causing capillary attractive forces between solid particles. This increases the cohesive nature of the product and can cause sufficient strength increase to result in hang-ups. In this case, it may only require one temperature excursion past a critical value to change the material and develop cohesive flow problems.

Temperature effects cause significant increase in bulk strength, which in turn cause stable rathole and arching problems in process equipment. For example, consider the increase in infant formula critical rathole dimension in a 3 m diameter bin as a function of storage temperature (Figure 1). The data presented in Figure 1 is for 7 h storage times and represents the effect that bulk strength has on critical hopper dimensions after nearly one shift storage time. The larger these rathole dimensions the more likely the material will hang-up in process equipment. This figure shows a three-fold increase in potential flow problems with a moderate increase in storage temperature suggesting that storage temperature should be controlled. Infant formula contains significant amounts of soy and warm soy meal creates substantial flow problems.

Storage temperature can cause change in the cohesion of the material, generating hang-up conditions. However, increasing cohesive properties also affects other properties such as density and permeability. Cohesion within a bulk material allows the formation of loose-packed beds. Figure 2 shows the typical void structure for a free-flowing material and a more cohesive material. The cohesive material contains larger void structures.

Attractive forces between particles induce the formation of large void structures in cohesive materials, resulting in bulk materials that are more compressible than less cohesive materials. The formation of these expanded voids also allows more airflow through the bulk material, increasing material permeability. However, applying a consolidation pressure to the cohesive material collapses the loose-packed void structure and significantly increases the bulk density while decreasing the permeability. For example, consider the change in bulk density and permeability material properties when small quantities of oil are mixed with

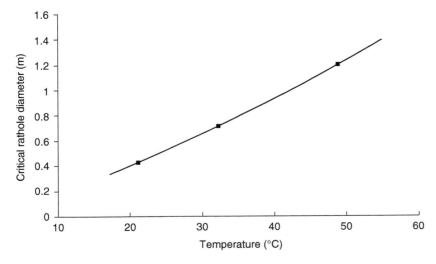

Figure 1 Critical rathole dimension of infant formula containing soy at various temperatures for a 3 m diameter bin.

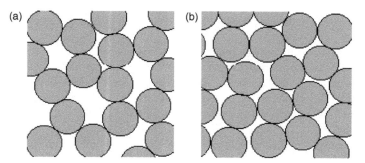

Figure 2 Comparison of (a) loose-packed pore structure caused by cohesion in a granular system, and (b) a more compact void structure existing with free-flowing materials.

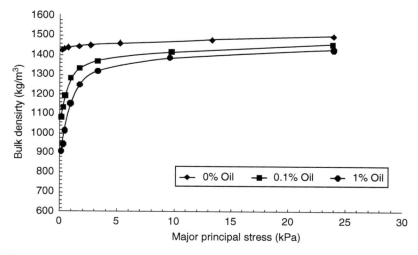

Figure 3 Bulk density of glass beads with oil.

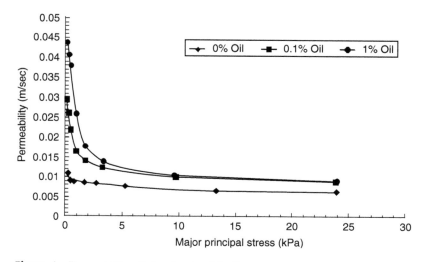

Figure 4 Permeability of glass beads with oil.

closely sized glass beads (Figure 3 and Figure 4). The materials used to generate Figure 3 and Figure 4 have exactly the same particle size distribution. Cohesion was increased by mixing oil with the bulk material creating capillary bonds between individual particles. More cohesive materials have lighter bulk densities at lower consolidation pressures, but will approach the densities of the more free flowing, closely packed material at higher consolidation pressures. The basic mechanism acting to produce this effect is the ability for cohesive bulk material to form stable, loose-packed structures owing to adhesive capillary or van der Waals forces between particles. This same effect is also present in food products.

III. HUMIDITY OR MOISTURE EFFECTS

In bulk materials, moisture absorption generally causes an increase in cohesion. However, sometimes the exact cause of cohesion increase is an indirect effect. For example, consider glass transition of particulate materials. Bulk moisture contents tend to lower glass transition temperatures, allowing moderate temperature fluctuations to force crystalline changes in particles. Changes in temperature cause a softening of the material, followed by creep between individual particles, and eventual crystallization of the surface structure as the temperature decreases. The net effect is to cement particle together causing significant increases in cohesion.

In other cases, the local moisture content is the direct cause of cohesion. For example, crystalline products can cake significantly when exposed to humid conditions owing to the formation and subsequent drying of liquid bridges between particles. The mechanism at work here is dissolution of soluble material and the transport of water vapor away from solid surfaces resulting in the formation of new crystal structures between particles.

The degree of moisture in a bulk solid often determines the flow behavior. Thus, knowing the amount and placement of the water in the powder will help explain flow behavior. A bulk solid system is always three-phase — containing gas, solid, and liquid phases. There are four sources of moisture in a bulk material. First, moisture can be transferred to the solid through the gas phase as water vapor absorbs on the surface of the particles. This presupposes that there is an affinity for the material in question to pick up moisture. Second, the liquid could be retained in the molecular structure of the bulk solid as water of hydration. Third, the liquid phase can also be present in the particle structure and may come to the particle surface through diffusion or as a result of particle breakage. Finally, the moisture may be present in a single component of a mixture and transferred to adjacent particles through cross-species diffusion. This cross-species diffusion is sometimes used to advantage in food systems. Homemakers and restaurant operators know that the addition of rice to sugar containers helps prevent caking by providing selective absorption of moisture during storage.

The rest of this chapter deals with very specific flow properties that determine the behavior of bulk solids material in typical process plant applications. Knowledge of these key properties will aid the practicing engineer in solving flow problems experienced with typical food products.

IV. UNCONFINED YIELD STRENGTH

Process failures often occur when handling bulk powders. Most of these failures are due to hang-ups and stagnant flow conditions that result from cohesive flow properties. Industries handling powders and bulk materials routinely experience unscheduled downtimes.

In fact, it is estimated that these flow stoppages cause 30 to 40% loss in production from typical plants handling powder materials. In addition, plants designed to handle powder materials require three to four times longer start-up times than plants designed to handle only liquid streams.

Many of these flow problems can be avoided with proper system design. However, this requires the measurement and use of primary material properties causing these flow obstructions. Unconfined yield strength is one of the primary properties of interest in predicting and thereby avoiding these flow problems. Yet, it is the least measured and used material property. Often the cohesive nature of material is correlated with material angle of repose, time to discharge through a small funnel, fractal nature of avalanches during pile formation, particle size, and material compressibility.

None of these measurements are primary measurements for cohesive flow obstructions. However, decisions to select proper process equipment are often naively made by measuring these secondary variables and inferring cohesive flow characteristics. This is the wrong approach. Whenever possible, engineers should measure the properties that most closely apply to the given situation. An engineer would not measure fluid thermal conductivity if calculation of time to drain a fluid from a constant temperature tank was the task at hand. The same is true of bulk powder properties. When describing the potential for hang-ups in process equipment, unconfined yield strength is the primary measurement variable. Unconfined yield strength (fc) is defined as the major principal compressive stress exerted on a bulk solid to cause material to fail or yield in shear when first subjected to a known consolidation stress and then exposed to an unconfined state of stress. It is used to determine the arching and rathole tendencies of a bulk material in process equipment.

An examination of typical flow obstructions will help explain why unconfined yield strength is the primary variable that should be used to describe cohesive flow problems. First, these flow obstructions always have a free surface. Hence, the material in the neighborhood of the obstruction is in an unconfined stress state. Second, the powder at this free surface will yield plastically when the stresses within the powder reach some critical value. That critical stress level that produces yield is the unconfined yield strength. Once the initial failure takes place, the stress state acting on the material within the equipment will change to be compatible with a condition of continual deformation without volume change (sometimes called the critical state). If the geometry of the equipment will not allow this unique stress state to form in the entire mass, then the material will yield in such a way as to produce this critical stress state in only a portion of the equipment, resulting in stagnant region formation. Flowing material will exit the process equipment leaving a large stagnant mass. The boundary between this stagnant region and the flow channel is a function of the material strength.

Consider the case of an arch across the outlet of a piece of process equipment. The stress state along the surface of the arch at the point of incipient failure is shown in Figure 5. Note that there are no shear stresses along the arch surface, so this surface is a principal stress surface. The minor principal stress acting on this surface is zero, and the major principal stress acting 90° to this surface equals the critical yield stress for the material (fc). Failure of the arch occurs when the stresses acting within the material exceed the critical yield stress value. Therefore, if unconfined yield stress can be measured, it could be used to predict arches across process equipment outlets. Unfortunately, the matter is complicated by the fact that the strength of a powder sample is a function of the consolidation stresses experienced by the powder — and stress levels in process equipment vary spatially. These stress levels also depend on specific geometry boundary conditions, compaction history of the material within the equipment, and operation parameters. Therefore, prediction of cohesive hang-ups requires knowledge of the stress state in process equipment.

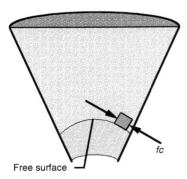

Free surface

Figure 5 Typical arch in process equipment.

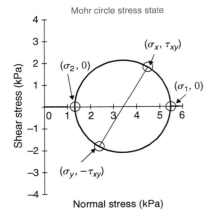

Figure 6 Mohr circle stress state showing two equivalent stresses.

There are four main concepts that the reader should grasp concerning bulk solids stress states to understand the behavior of solids flow in process equipment. First, the magnitude of stress in process equipment is not a linear function of the depth of material in the equipment, but rather a function of the span of the active flow channel. Second, the properties causing hang-ups in process equipment are a function of this consolidation stress and generally increase with stress level. Third, there exists a set of stresses that will cause incipient failure of bulk materials and these stresses are used to predict collapse of hang-ups. Fourth, there exists a set of stresses that are compatible with continual deformation without volume change and these are used to describe flow behavior of bulk materials.

Describing the stress state in process equipment is simplified by the use of two concepts. The first concept is a graphical description of stress states called a Mohr circle. Mohr stress circles can be used to describe a general two-dimensional stress state when those stress states are plotted in normal stress versus shear stress plots (see Figure 6). The circle in Figure 6 represents a single stress state.

Any group of stresses located on opposing sides of the circle will cause the same behavior of the bulk. For example, shearing a bulk material will result in the stress state given by (σ_x, τ_{xy}) and $(\sigma_y, -\tau_{xy})$ acting on two orthogonal planes. However, applying normal pressures defined by $(\sigma_1, 0)$ and $(\sigma_3, 0)$ stress states on orthogonal planes will cause the same shear behavior.

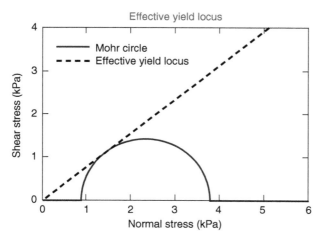

Figure 7 Effective yield locus.

The second concept used to describe stress states in equipment is the stress locus. This concept states that for a given behavior (i.e., flow or yield) there exists a unique relationship between the normal stresses and shear stresses. Consider, for example, the collection of all stress states that cause continual deformation without volume change (flow). Experiments suggest that these stress states can be described by a linear relationship between shear stress and normal stress as shown in Figure 7.

A Mohr circle represents a state of stress and any Mohr circle that is tangent to this line will define a stress state that causes continual deformation without volume change. This line then provides us with a constitutive relationship that describes stress states during slow flow of bulk materials. The slope angle of this line is the effective angle of internal friction. The limiting line itself is the effective yield locus and is used to describe stresses within flowing bulk materials. In general, this locus may have a small intercept on the shear-stress axis. However, in many applications this small shear intercept is ignored when estimating the stresses during flow.

Sometimes food engineers do not want to describe steady flow behavior, but want to determine if a given bulk material will initially flow from a container or process vessel. The desire in this case is to avoid process hang-up conditions. The concept of stress loci can also be used in this situation. For example, hang-ups are characterized by the collection of stress states that will cause yield or failure of the bulk material. The magnitude of this failure depends on the stress history of the sample in the process equipment. Greater compaction stresses result in greater material strength.

Manufacturers of nutraceuticals know that tablet integrity depends on the degree of compaction in tablet presses. Likewise, manufacturers of pet treat briquettes know that the roll gap must be set close enough to create sufficient compaction pressure to hold powder product together. Smaller compaction pressures may not be sufficient to form stable shapes that can be handled. In both these examples, the strength must exceed some critical value for the product to be acceptable. The situation is reversed in gravity feed equipment. Strength causes formation of free surfaces across equipment boundaries when material is placed in the gravity feed process equipment resulting in hang-ups. Avoidance of these hang-ups requires that strength levels be below some critical value. In all these cases, the characteristic stress describing the process behavior is related to a critical yield stress of the bulk material.

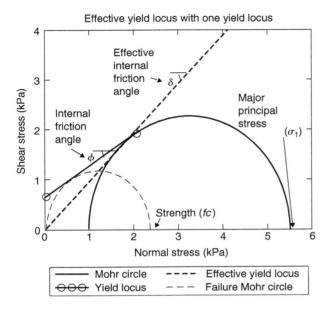

Figure 8   Yield locus concept.

These yield stresses are measured by first preconditioning the material to a pre-scribed state of compaction stress, reducing the stress and then, while at the reduced stress state, determining the failure stresses required to yield the material. This procedure can be repeated for several different failure load conditions to generate a locus of stress states that cause failure at different confining pressures. It is important to note that yield stress states generated in this way depend on the initial stress state imposed on the bulk sample during the preconditioning phase. If these yield stress measurements were all done at the same preconsolidation stress, they form a locus of yield points that describe failure of a bulk sample after first being subjected to a particular compaction state. This compaction stress state is represented by the termination Mohr circle shown in Figure 8. The major principle stress (σ_1) associated with this termination Mohr circle is defined as the compaction stress level.

The yield locus is often described as a line with a slope and an intercept on the shear-stress axis. The slope angle of this line is defined as the internal friction angle (ϕ). The shear intercept is defined as the cohesion value. Figure 8 also contains the effective yield locus (for use with flow conditions) with a slope angle equal to the effective internal friction angle (δ) as described earlier. This is included only for comparison purposes and should not be confused with a yield locus condition. This figure shows a particular yield stress state described by a Mohr stress circle that passes through the point (0,0) and is tangent to the yield locus line. This unique stress state is defined as the unconfined strength stress state. The major principal stress associated with this Mohr circle is the unconfined yield strength (fc). There is one unconfined strength value for each compaction stress state. This implies that there exists a one-to-one relationship between compaction pressure and the unconfined yield stress for any bulk material.

This unconfined yield strength (fc) can be plotted as a function of major principal stress (σ_1) as shown in Figure 9. In general, unconfined yield strength increases with an increase in compaction pressure. The strength plot may or may not go through the origin.

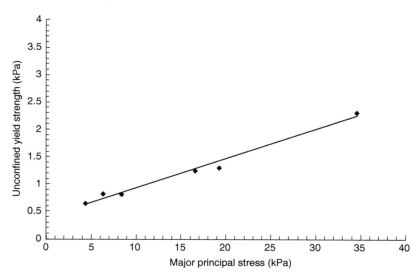

Figure 9   Yield strength versus major principal stress for flour.

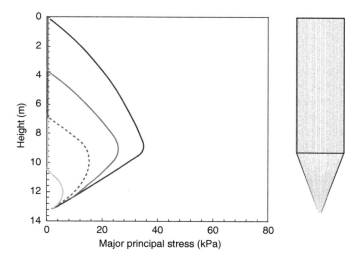

Figure 10   Major principal stress during initial filling process.

Generating these yield loci from shear tests provides information that can be used to determine cohesive hang-up behavior provided we can determine or estimate the stress levels in process equipment. Often new process engineers faced with the task of designing vessels for bulk solid materials will apply fluid flow principles and discover that bulk materials do not behave the same as fluids. Everyone knows that the stress level in a static fluid is a linearly increasing function of depth of the fluid. Applying this assumption to containers full of bulk materials would lead to the erroneous conclusion that the solids stress level is largest at the bottom of the storage vessel. The actual stress profile of bulk materials is quite different [1–3]. Figure 10 and Figure 11 show typical wall stress levels computed from a simple slice model approach for a conical-shaped bin for both filling and unloading conditions.

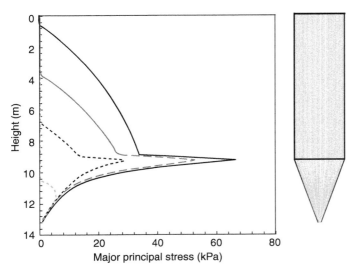

Figure 11 Major principal stress during emptying process. These pressures often persist after flow has stopped.

These two figures show that the stress levels are low at the bottom of the storage bin. The largest pressure occurs at the transition between the cone and the cylinder portions of the storage vessel. In fact, this maximum pressure depends on the mode of operation with the highest pressure occurring during steady flow. Knowing the stress state in a process vessel is critical to understanding the relationship between unconfined yield strength and the hang-ups that may occur in process vessels. Strength is a function of the major principal stress acting on the bulk material, so the strength naturally varies with the position in process equipment. In fact, the strength will increase with increasing stress levels in the bin. The strongest material will be near the transition between the vertical section and the converging hopper. Conversely, strength will be low near the outlet. However, quite often, bulk materials possess some nonzero strength even at zero consolidation stress levels (see Figure 9). This implies that material near the outlet may possess some positive yield strength even if the compaction level is very low. It is these yield strengths near the outlet that cause material to arch across the hopper outlet. Arching will occur whenever the material strength in the neighborhood of the outlet exceeds the stresses causing the arch to fail. An arch gains support from the abutments of the arch which are, in this case, coincident with the hopper walls. Stress levels in converging flow channels and process vessels approach a condition, near the outlet, where the stress magnitude varies linearly with the span of the flow channel or process vessel.

The production of a flow obstruction in the process equipment is governed by a simple principle. A flow obstruction will occur whenever the bulk strength exceeds the stress acting to break the obstruction. Consider the example shown in Figure 12. The stress shows the typical flow loads associated with bulk material in converging bin geometries. The strength of the bulk material is plotted and varies in accordance with the stress level in the bin. Also plotted in this figure is a line indicating the stress level required to overcome arching across the hopper span for normal gravity flow conditions. This required stress level varies linearly with hopper span. Figure 12 shows that bulk material strength is greater than the stress required to break arches when the diameter is less than 1 m. Therefore, for this material, in this configuration, the critical arching diameter is 1 m and conical-hopper

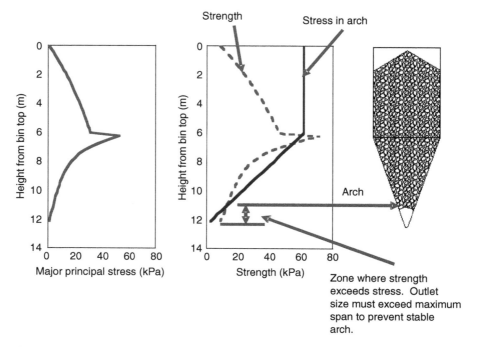

Figure 12 Expected strength values in typical bins leading to arching after imposing an initial flow, then storing the material for a period of time.

outlets should be designed larger than this to prevent arching. It is obvious that the arching criterion depends very heavily on the stress level in the bin or process equipment. Stress levels, and consequently strength values, will be different after flow and just after initial filling. In fact, an examination of Figure 10 and Figure 11 indicate that stress levels near the outlet just after filling are considerably larger than stress levels present during flow. This implies that arching dimension calculations should be different for these two conditions.

We will now consider only the case of arching after initial flow conditions have been established. Arching that occurs after initial filling will be considered later. Numerically, the flow condition necessary to prevent arch formation would imply that major principal stresses acting at the outlet would be greater than the unconfined yield stress of the material (Equation [1]).

$$\sigma_1 \geq fc_{\mathrm{crit}} \tag{1}$$

Jenike [4,5] and others were able to compute the major principal stress acting on the bulk material in conical- and wedge-shaped hoppers that undergo flow. Jenike further postulated that there exists a critical ratio of major principal stress in the hopper and unconfined yield strength that will cause arch formation. This gave birth to the concept of a flow factor (ff) (Equation [2]).

$$ff = \frac{\sigma_1}{fc_{\mathrm{crit}}} \tag{2}$$

The flow factor is a relationship between the major principal stress and the unconfined yield strength used to describe stress states during hang-ups. It is a function of both hopper geometry and material properties. Typically flow factors range between 1.1 and 1.3 for most

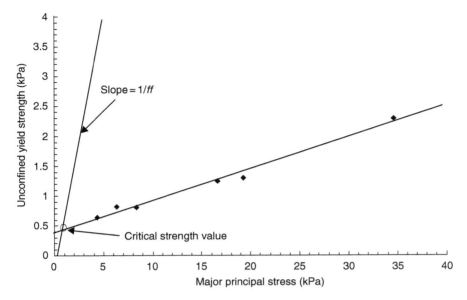

Figure 13 Plotting the flow factor on the flow function plot.

materials flowing in mass-flow bins. Values of 1.7 are used to describe funnel-flow situations. In general, an intercept is generated when the flow factor relationship is plotted with the experimentally measured strength versus major principal stress curves. Since the flow factor represents the critical relationship between the stress and strength for arching, the intersection with the measured property curve represents the expected stress and strength value, at the arch surface, for the particular material when placed in specific hopper geometries. Inherent in the calculation is the fact that these flow factors are based on stresses resulting from material flowing through a hopper. This flow factor arching criteria approach will then apply only to the case where material is placed in a bin, flow is initiated, and then the material is stored for a period of time. During this storage time the material may gain strength.

An example of such calculations may help the reader internalize this concept. Consider the strength of the flour material given in Figure 9. Assume that this material is placed in a conical hopper with a hopper slope angle of 20° measured from the vertical, which is 3 m in diameter at its widest spot with a 0.25 m diameter outlet. Assume the flow factor (ff) for this bin geometry is 1.2. This flow factor is plotted on the experimentally measured strength versus major principal stress plot, yielding a critical value of stress. This stress is used to compute the arching potential for flour in conical bins.

The critical strength value found from the intersection of the flow factor line with the measured strength data yields a value of $fc_{crit} = 0.42$ kPa for this material. Incidentally, this critical strength level corresponds to a critical stress level σ_{crit} of about 0.5 kPa. The critical arching dimension is found by using this critical strength with Equation (3) to compute the critical arching dimension (AI)

$$AI = \frac{H(\theta) \cdot fc_{crit}}{\gamma(\sigma_{crit}) \cdot g} \tag{3}$$

The bulk density (γ) term is evaluated at the consolidation pressure associated with the intersection of the flow factor and the measured strength data. This corresponds to the

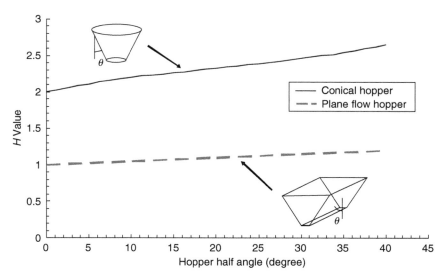

Figure 14 $H(\theta)$ function for various hopper geometries.

major principal stress level near the arch of approximately 0.5 kPa for this example. For the time being, let us assume that material bulk density evaluated at this consolidation pressure is 510 kg/m³. The term $H(\theta)$ is a geometry function that takes into account the fact that arches that form in conical bins are more stable than arches that form across plane flow channels. This is due to the difference in stress states when material converges in only two dimensions versus the increase in arch support when material converges in three directions as in conical-hopper geometries. Figure 14 shows the $H(\theta)$ values as a function of hopper geometry ($H(\theta) = 2.3$ for this example).

The critical arching dimension is the maximum span of the hopper outlet that will form a stable arch. Hopper outlets must be designed larger than this span to prevent hang-ups. The computed conical arching dimension for this example case is about 0.19 m. The conical-hopper outlet for this example is 0.25 m, so this material will not form stable arches across this hopper outlet.

The reader should be reminded that critical arching dimension is computed assuming the material within the hopper was placed in the bin, flow was initiated, and then material was stored for a period of time. Alternate arching dimensions can be developed for the case of initially filling the bin. This alternate method is presented below. The basic flow-no-flow concept is still used. In other words, there exists some critical value of strength that will cause flow hang-ups. This critical strength depends on the stress level. However, pressures near the hopper outlet are computed assuming the bin is initially filled. These initial filling loads are difficult to compute accurately. In addition, they will include the effect of impact pressure caused by the free fall of material into an empty bin. However, an empirical equation for the approximate calculation of initial stress levels in bins and hoppers is presented below (Equation [4]). It is designed to provide an estimate of the initial compaction pressures near hopper outlets in typical bins. If the reader has a more accurate estimate of these pressures near hopper outlets they could be used instead.

$$\sigma_{\text{crit}} = 1.5 \cdot \sqrt{\frac{D_{\text{bin}}}{3.05 \text{ m}}} \cdot \gamma(\sigma_{\text{crit}}) \cdot g \cdot D_{\text{out}} \qquad (4)$$

Figure 15   Typical rathole in funnel-flow bin.

This critical pressure depends on the diameter of the bin (D_{bin}), the hopper outlet diameter (D_{out}), and the bulk density (γ). It should be noted that, in general, bulk density is a function of the stress level applied to the material. Thus, the calculation of the critical stress level may require some iteration to determine the value near the hopper outlet. Once this critical stress level is found, it is used with the flow function plot to compute the critical strength value near the hopper outlet. Assume for the time being that the bulk density of flour at this critical stress condition is about 570 kg/m^3. The calculation of arching dimension proceeds as before using Equation (3) except that the critical value of strength is replaced with the value computed at a stress level given by equation 4. In this example, the critical stress was computed as 2.1 kPa, which yields a critical strength of 0.54 kPa. The critical arching dimension computed in this manner is 0.22 m, which is slightly larger than the arching condition computed previously and close to the outlet size. This suggests that placing flour in this example bin design results in conditions close to arching.

Ratholes are another form of hang-up common to process equipment. These hang-ups are actually a circumferential arch, which produces the formation of a stable pipe down through the material (see Figure 15). Typical rathole formation generates a condition where the material empties from the center of the bin leaving a larger mass of material clinging to the side of the bin. It should be noted that ratholes can occasionally form along one side of the bin and not in a central flow channel. These off-center ratholes are often less strong than those formed in the center of the bin. In either case, this flow obstruction is responsible for significantly reducing the active storage in bins and hoppers. It is, perhaps, the reason for most cohesive flow problems in process equipment. It should be pointed out that ratholes cannot form if the bin or process equipment produces flow along hopper walls. Ratholes gain their strength from the hoop (azimuthal) stress that forms within the stagnant material around the open channel. Failure of this "circumferential arch" or rathole will occur if this stress at the surface of the rathole exceeds the strength of the material or if the material continually deforms across the hopper cross-section. For any given bin geometry, which does not allow flow along hopper walls, there exists a critical strength value that will produce stable ratholes.

Thus, there is a critical rathole dimension for a specific bin configuration. This critical rathole dimension is defined as the maximum size of a stable rathole that could form in a funnel-flow bin. The active flow channel must be larger than this value to prevent stable rathole formation. It is important to stress that stable ratholes are only possible in bin configurations that do not cause flow along hopper walls. One method of rathole elimination is to design hoppers steep enough to produce active flow channels that exceed the critical rathole dimension. Other methods of rathole remediation include the use of flow-aid devices such as air blasters, vibrators, and mechanical agitators. These flow-aid device methods have limited success. If they are to work, they should be placed close to regions where stable ratholes would form in bins and hoppers. These flow-aid devices have a limited range of effectiveness. The failure of these devices to prevent rathole formation is often because of either incorrect placement and lack of enough devices to accomplish the task or to the fact that the bulk material, when subjected to excessive consolidation pressures induced by flow-aid devices, will increase strength more than the induced stress can overcome.

The placement of these devices or the size of the active flow channel is determined by the critical rathole dimension. The procedure for computing this critical dimension is outlined below. The flow-no-flow concept also applies to this condition. Ratholes will be stable if the strength of the material surrounding the rathole is larger than the stress causing failure of the material in this circumferential arch. As was discussed earlier, the pressures in the bin that is initially filled are given by those in Figure 10. This figure indicates that the largest stress level occurs at the transition of the cylinder section and the converging hopper section and it is this stress level that determines the magnitude of the strength used in critical rathole calculations. An estimate of these stress levels can be found from Equation (5).

$$\sigma_{bin} = \frac{\gamma(\sigma_{bin}) \cdot g \cdot D_{bin}}{4 \cdot k_v \cdot \tan(\phi_{wall})} \cdot \left(1 - \exp\left(\frac{-4 \cdot k_v \cdot \tan(\phi_{wall})}{D_{bin}} \cdot H_{bin}\right)\right) \tag{5}$$

This critical stress level approximation depends on the bulk density (γ), maximum bin diameter (D_{bin}), the height of the vertical section above the converging hopper (H_{bin}), the wall friction angle (ϕ_{wall}), and the Jansson k-ratio ($k_v = 0.4$ to 0.65). Once the critical rathole stress is computed, the calculation proceeds in a manner similar to the arching criteria. The strength at the critical stress condition is computed and then used to compute the critical rathole dimension from Equation (6).

$$RI = \frac{G(\phi) \cdot fc(\sigma_{bin})}{\gamma(\sigma_{bin}) \cdot g} \tag{6}$$

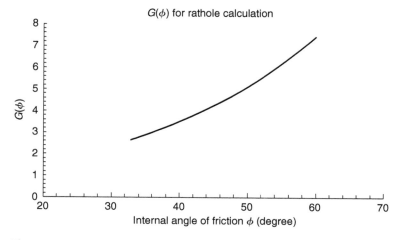

Figure 16 $G(\phi)$ values as a function of internal friction angle (ϕ).

The $G(\phi)$ term relates the internal friction angle (ϕ) to the major principal stress close to the rathole surface and is used to determine the effect that the internal friction angle (ϕ) has on the stability of ratholes. The $G(\phi)$ term is between 3.0 and 4.0 for typical materials. Exact $G(\phi)$ values can be found from Figure 16. If the internal friction angle is not specifically known, then the value of 3.5 is typically assumed for the $G(\phi)$ term. This corresponds to a value of 40° for the internal friction angle.

For example, consider the flour example outlined previously. The strength values are found in Figure 9. Assume that this material is placed in a conical hopper with a hopper slope angle of 20° measured from the vertical which is 3 m in diameter at its widest spot with a 0.25 m diameter outlet and a 6 m tall vertical section. Assume that the internal friction angle is about 40° giving a $G(\phi)$ term of about 3.5. Assume also that the wall friction angle for this bin is 18° measured from the vertical, the bulk density at the critical bin stress level is 715 kg/m³, and the Jansson k-ratio is 0.4. Equation (5) can be used to compute the critical stress for rathole calculations yielding a critical stress of 26.1 kPa. Figure 9 can be used to determine the critical strength yielding a value of 1.8 kPa. Finally, Equation 6 can be used to compute the critical rathole dimension for material in this example bin. The critical rathole computed from this analysis equals 0.9 m. This indicates that the example hopper must be steep enough to cause flow along hopper walls up to 0.9 m in diameter to prevent stable rathole formation.

In summary, cohesive flow problems in process equipment are caused by a material property called unconfined yield strength. This strength is a function of consolidation pressure and will vary spatially in a nonlinear way. The basic approach in overcoming these cohesive flow obstructions is to compute both the unconfined yield strength in the process equipment near the flow obstruction and the stress required to break the flow obstruction as a function of process geometry. Situations where the strength exceeds the stress to break obstructions lead to process hang-ups and should be avoided. This may mean designing larger outlets to overcome arching or designing containers such that flow occurs in large enough diameters to destabilize ratholes. Ratholes are unstable if material flows along bin walls. The ability of a given material to flow along hopper walls depends on another material property called wall friction angle.

V. WALL FRICTION ANGLE

As discussed earlier, some stress conditions in bulk materials lead to failure of cohesive structures. Other stress states cause continual deformation without volume change. This second stress state is called flow. However, sometimes the boundary conditions in a given piece of process equipment are not capable of inducing stress states compatible with flow in the entire equipment or bin. When this occurs, stagnant regions form and a smaller active flow channel develops in a portion of the equipment. This flow channel exists only in regions where stress states are compatible with this unique flow stress state. Bulk solids have two distinct flow patterns in process equipment [6–8]. One pattern is characterized by material flow at the wall during discharge. This flow pattern is called mass flow. The second flow pattern is characterized by the formation of stagnant regions in the process equipment during discharge and is called funnel flow.

The flow properties of wall friction angle and the effective internal friction angle govern the type of flow in process equipment. However, wall friction angle has the strongest effect on flow pattern in gravity feed process equipment. If the process equipment is conical, then there exists a limiting relationship between the conical-hopper half angle and the wall friction angle for mass-flow conditions. If hoppers are sufficiently steep and smooth, the

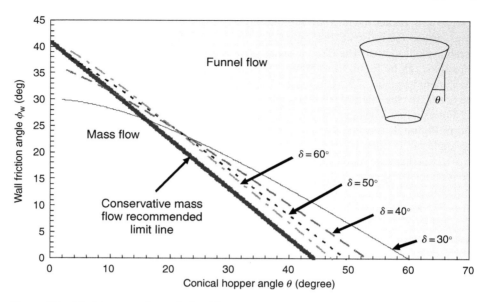

Figure 17   Limiting mass-flow relationships.

entire mass within the bin flows when any material discharges from the bin. This flow profile is called mass flow and simply means flow along the walls. If sloping hopper walls are sufficiently frictional, then the material flowing in a bin will develop an active central flow channel or funnel-flow behavior. In funnel flow, the material around the perimeter of the bin remains stagnant during discharge. Ratholes cannot form in bins where flow along the walls exists. One remedy to rathole formation is to modify the bin to achieve mass flow. The theory [9] behind the mass-flow limit line will not be presented here. However, the mass-flow limitation resulting from the application of this theory will be presented to provide the reader with the means of predicting the shape of conical bins that will produce flow along hopper walls during operation. It should be noted that continual deformation at constant stress is characterized by Mohr circles tangent to the effective yield locus. These stress states are a function of the effective angle of internal friction. Therefore, it should come as no surprise that mass-flow profiles are also a function of the effective internal friction angle. Figure 17 shows the limiting lines between mass-flow and funnel-flow operation as a function of wall friction angle (ϕ_{wall}) and effective internal friction angle (δ).

If the internal friction angle is not known, then the conservative mass-flow angle shown in Figure 17 could be used to provide an estimate of mass for typical materials. It is also recommended that hopper angle for reliable mass flow should be at least 4° inside the limiting lines shown for each internal friction angle (δ). It is important to note that the recommended line shown above is an unconservative estimate for conditions where the internal friction angle is small and the hopper is very steep. However, in this case these types of materials are not typically cohesive and stable ratholes are not generally a problem. In addition, steep conical hoppers are not very practical designs and are seldom used. In general, mass flow in plane-flow (wedge-shaped) hoppers can occur if the hopper angle is about 10° flatter than typical conical hoppers. This information will provide the reader with some basic design constraints for gravity flow in common process equipment shapes.

VI. BULK DENSITY

At first glance bulk density appears to be a straightforward material property. However, there are more methods to measure bulk density and more empirical fits than any other bulk property. Bulk density is defined as the mass of the solid particles and included moisture, divided by the total volume occupied by the particles, surface moisture, and surrounding voids. In its most complex form, bulk density is a function of both the stress and the strain imposed on the material. Overconsolidated material dilates as it shears, increasing the local solid porosity. However, underconsolidated material will compact during shear. Applying compaction stress to a bulk powder material will result in a rearrangement of local particles to achieve a denser packed condition, but vibration will most likely be required to achieve optimum packing conditions. This is owing to the fact that friction exists between particles and vibration reduces the frictional component acting on the assembly of particles. Vibration may also provide some limited local separation of particles to free up locked or jammed particles, allowing them to move and rearrange the discrete particle matrix. Many density measurement procedures differ by the amount and type of vibration applied to the bulk material. For example, loose-packed density is the density that occurs when a container of a given size is filled with a bulk powder material and scraped off level. Tapped density is the density that results when a specific container filled with material is subjected to a controlled number of vibrations at a given amplitude. This is usually characterized by dropping a weight a given distance, a prescribed number of times. Bulk density is also measured by placing material in a container and applying a consolidation load to the material within the container, then measuring the axial deformation at each uniaxial load and computing the density. Finally, density can be measured in a biaxial or triaxial stress–strain apparatus where the density behavior depends on the history of the material within the cell and the amount of shear imposed on the material. Loose-packed density, tapped density, and uniaxial density measurements are the most frequently used parameters to characterize bulk powders.

   Loose-packed density is used as an estimate of a density required for flow rate calculations. Packed density is typically used when computing bin volumes and other process equipment volumes. The uniaxial method of obtaining bulk density results in bulk density as a function of a compaction pressure applied and provides enough details to describe flow problems with typical bulk materials in process equipment. The normal pressure applied

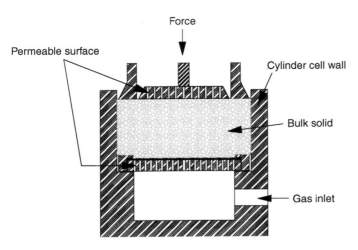

Figure 18   Uniaxial bulk density and permeability measurement device.

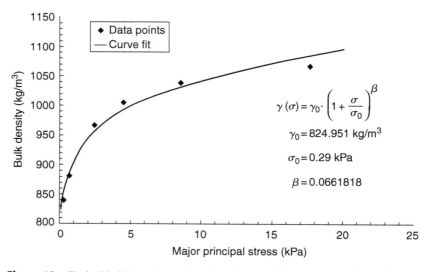

Figure 19   Typical bulk density as a function of consolidation pressure for baking soda.

is assumed to be equal to the major principal stress and uniaxial bulk density test cells are
shallow to reduce the friction effects of the wall (see Figure 18). Material is placed in the
uniaxial cell in a loose-packed condition and scraped off level with the cell. A light lid
is placed on top of the material in the cell and the initial height of material in the cell is
recorded. A weight hanger is then placed on the sample and weights of increasing magnitude
are added to the sample. The deformation as a function of weight applied is monitored and
recorded. The resulting deformation data are used to compute bulk density as a function of
normal stress (principal stress) applied to the material. Figure 19 shows a typical plot of
bulk density as a function of consolidation pressure. The form of the empirical fit shown in
Equation (7) gives a reasonable approximation to typical food product densities in gravity
flow process equipment.

$$\gamma = \gamma_0 \cdot \left(1 + \frac{\sigma}{\sigma_0}\right)^{\beta} \tag{7}$$

where γ_0 is the bulk density at zero consolidation pressure, σ_0 is the stress curve fit parameter,
and β is the compressibility exponent.

Bulk density measurements generated in this manner can be used in a variety of
process equipment where fluidization does not occur. The stress levels in process equipment
can be computed and the bulk density estimated from the data measured with the above
procedure. This suggests that density may be a function of position in process equipment.
In fact, densities can change as material flows through process equipment. Densities exiting
a gravity feed device will be low and are used to determine mass-flow rate from process
equipment. The calculation of active volume in process equipment requires an estimate of
the complete stress distribution in equipment.

This changing density can give rise to flooding and flushing problems from process
equipment if the residence time does not allow for the dissipation of excess gas generated
from these density changes. The dissipation rate is a function of another related property
called permeability. Generally, the finer the material, the more compressible the material
is and the more impermeable the material. Likewise, in general, the more polydispersed a
material is, the more compressible and impermeable the material is. Fine compressible

materials are sensitive to uncontrolled flow rate problems in process equipment. The exact cause of these behaviors is discussed below, but depends on the compressibility and permeability of a bulk material.

VII. PERMEABILITY

The ability of a given bulk material to allow a gas to pass through its pores depends on the permeability of the bulk material. The resistance to gas flow is dependant on the size of the local pores. Therefore, permeability is a strong function of material particle size. Polydispersed systems often fill available voids with particles of smaller diameter creating a structure with much smaller pores than would be found in monodispersed systems of any comparable particle size. Consequently, it is not uncommon for polydispersed systems to have lower permeabilities than monodispersed systems of comparable particle size. Cohesion tends to result in packed states with large open void spaces. Therefore, cohesive materials tend to exhibit high permeabilities at low consolidation pressures and low permeabilities at high consolidation pressures where stresses are large enough to collapse the local cohesive void structures.

   In process equipment, permeability governs how quickly entrained gas within solids voids can dissipate. These gas pressures within bulk solids arise from sudden compaction of material creating a compression of gas within the bulk solid pores. The permeability of the bulk material provides a resistance to gas flow and the gas pressure gradient provides the driving force for gas dissipation. Thus, if these compressive events occur faster than the material's permeability will allow dissipation, the excess gas pressure remains in the bulk solid system and can cause flow rate control problems from volumetric flow devices. These flow rate problems will be discussed below.

   Measurement of permeability is done in the cell described in Figure 18 above. Gas is passed through a bulk material that is in a compacted state in the test cell. The velocity of gas passing through a bulk material is related to the solids permeability through Darcy's law, which is expressed in Equation (8).

$$\vec{U} = \frac{\vec{\vec{K}} \circ \nabla P}{\gamma \cdot g} \cdot \frac{\mu_0}{\mu} \tag{8}$$

   In general, materials can have varying permeabilities in different flow directions. Consequently, in its most general form, permeability is tensorial in nature. However, for many applications the scalar version of Darcy's law is applicable as shown in Equation (9).

$$U = \frac{K \cdot (\partial P / \partial z)}{\gamma \cdot g} \cdot \frac{\mu_0}{\mu} \tag{9}$$

Measuring the pressure drop across a bed of bulk material ($\partial P / \partial z$), the superficial gas velocity relative to the solids (U), and the solid bulk density (γ) will allow calculation of the permeation velocity through the bulk material. It is important to note that the permeability defined in this way has units of velocity (Figure 20). This is not the traditional definition of permeability that would include the viscosity of the gas. However, this definition has some benefits when computing solids flow rate due to gas permeation limitations. This will be addressed below.

$$K = K_1 \cdot \left(\frac{\gamma}{\gamma_a}\right)^\alpha \tag{10}$$

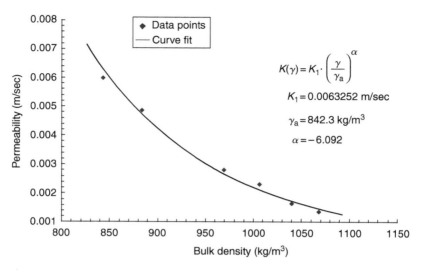

Figure 20   Permeability of baking soda as a function of density.

where K is the material permeability, K_1 the curve fit permeability parameter, γ the material bulk density, and γ_a is the reference density parameter.

VIII.   FLOW RATE

Flow rates of fine powdered foods through process equipment are often difficult to predict. The reason for this is that bulk solids are a three-phase system. They contain solid particles that behave as a columbic solid. Moisture can be carried on the particle surface, creating additional cohesive forces sticking particles together. Finally, air pressure gradients within the solids' pores can influence the behavior by modifying the body forces acting on the collection of particles thereby providing a temporary storage space for compressible gas. It is the storage and movement of gas within bulk solid pores that causes most of the flow rate problems with fine food powders. These fine materials exit process equipment at either very fast rates or extremely slow rates. Often operation oscillates between these two rates in an uncontrolled manner, resulting in very erratic flow rates. The reasons for these flow rate problems are discussed below.

A.   Acceleration Limited Flow

Consider the solids stress levels in typical bins and hoppers. Figure 21 shows the spatial variation in solids stress state. The highest pressures occur near the transition between the hopper section and the cylinder section. This causes a local material consolidation in the bin as material flows through it, resulting in an increase in bulk density (see Figure 22).

  An increase in bulk density causes a resulting decrease in the bulk solid pore volume. Gas is compressible and will increase pressure when subjected to decreasing volume conditions. Conversely, gas pressures will decrease if the volume occupied by the gas increases. Most powder particles are not significantly compressible at the solids contact stresses in most handling systems and storage containers. However, the collection of solid particles within a powder can rearrange to form a denser packing, resulting in a reduction of porosity,

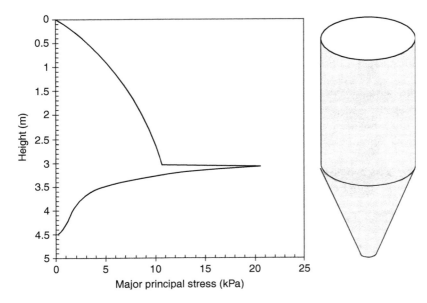

Figure 21 Solids stress level in bin during flow. The highest pressure is at the transition between the cylinder and cone.

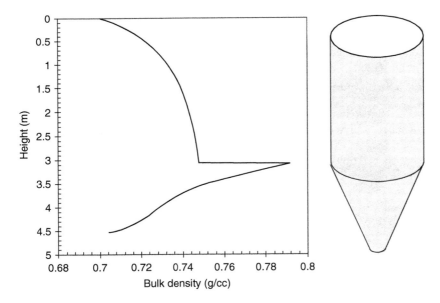

Figure 22 Bulk density in bin.

when subjected to consolidation stresses. This reduction in porosity can increase the gas pressures acting within the interstitial pores. If gas trapped within the pores can escape quickly, then little or no pressure buildup occurs. However, if there is a larger resistance to gas flow through bulk material, then significant gas pressure can build up within the voids, causing temporary insipient fluidization thereby creating aerated conditions. Discharging aerated powder material from process equipment results in very fast flow rates. Fluidic material exits the bin or process equipment as fast as gravitational acceleration allows.

It is important to note that these large flow rates do not require bubbling fluid-bed behavior. Slightly aerated material will flow at acceleration-limited flow rates. Adding additional gas will create pressure-induced flow and result in flow rates faster than those possible by gravitational conditions. In either case, convergence of the flow channel determines the maximum velocity and solids flow rate Qs_{max} obtainable through a given outlet of diameter D_{out} and cross-sectional area A_{out} (see Equation [11]). Using this equation to estimate the acceleration-limit flow rate requires computation of the stress state at the hopper outlet (σ_{out}).

This can be done by noting that the stress near the hopper outlet depends linearly on the hopper span as given in Equation (12). However, this stress depends on the bulk density requiring an iterative solution of Equation (12) and using the empirical density fit given above where γ_0, σ_0, and β are bulk density curve fit parameters.

$$Qs_{max} = \gamma(\sigma_{out}) \cdot A_{out} \cdot \sqrt{\frac{D_{out} \cdot g}{2 \cdot a \cdot \tan(\theta)}} \tag{11}$$

$a = 1$ for wedge-shaped flow channels

$a = 2$ for conical-shaped flow channels

$$\sigma_{out} = \frac{1}{a} \cdot \gamma(\sigma_{out}) \cdot D_{out} \cdot g = \frac{1}{a} \cdot \gamma_0 \cdot \left(1 + \frac{(\sigma_{out})}{\sigma_0}\right)^{\beta} \cdot D_{out} \cdot g \tag{12}$$

Material exits the hopper at a bulk density $\gamma(\sigma_{out})$ compatible with the stress condition near the discharge outlet. Since the flow rate depends on the convergence of the flow channel, anything that changes this convergence will affect the flow rate. Increasing the flow channel slope will increase the flow rate of a material that is close to this maximum acceleration-limited flow rate. The type of hopper geometry also affects acceleration-limited flow. Conical channels will result in lower velocities through the outlet than wedge-shaped flow channels for the same outlet area. The reason for this is that conical channels converge faster than wedge-shaped flow channels, resulting in greater material accelerations through the hopper or flow channel. This acceleration limit flow rate also applies to coarse granular materials.

B.   Permeability Limited Flow

Another possible flow rate limitation exists in process equipment. This limitation results in very slow flow rates and requires that flow through the container be slow enough to allow excess gas pressure to dissipate. Under these conditions, small mass-flow rates can occur when increasing stress levels squeez gas from the material during consolidation. This excess air leaves the system through the top material surface. However, as the compacted material flows through the hopper section, the stress levels decrease and the bulk solid expands — creating a deficit of gas in the bulk solid pore structure. The lack of gas in the bulk solid's pores results in negative gas pressure near the outlet, inducing gas counterflow at the equipment outlet. This creates an upward acting gas pressure gradient that retards the flow of powder through the system.

Bulk solid gas permeability is the limiting factor here. It is not uncommon for these permeability-induced limiting flow rates to be several orders of magnitude lower than the

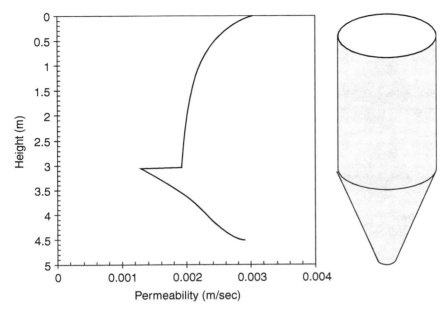

Figure 23 Permeability in bin.

fully aerated acceleration-limited flow rates. Bulk density and gas permeability are key factors in describing this flow rate behavior of fine powder materials. Figure 23 shows the expected permeability values in a typical, conical surge bin. This permeability corresponds to the solids contact pressures shown in Figure 21. These permeabilities vary by a factor of three, depending on the solid stress in the bin. Lowest permeability occurs at the transition between the cylinder and the hopper. This provides a pinch point for gas permeation through the system. Any gas traveling with the material that passes this pinch point will find it easier to permeate downward through the bulk than up through this impermeable zone. Likewise, gas squeezed from the material during flow will find it difficult to pass downward through the impermeable layer of material. Thus, any air required by negative gas pressures within material in the hopper section must enter the bulk mainly through the outlet.

High or low gas pressure within the bulk must be dissipated by one of two mechanisms: either a local rearrangement of particles, requiring solids flow to follow gas pressure gradients, or by local gas permeation. The fact that the bulk solids are constrained by wall-friction conditions suggests that the arrangement of local particles may not be an active mechanism for relieving flow-induced gas pressures. In fact, local particle arrangement in typical process equipment can only occur in the direction of gravity. Upflow of bulk solids never occurs in normal gravity flow situations and always requires an external body force. Thus, for the case of local negative gas pressures near the outlet, the only way to dissipate gas pressures is to rely on the permeability of the bulk material. If the material is impermeable, this dissipation process will be slow. Therefore, the pressure equalization process will be slow and the flow rate from the outlet will be significantly reduced.

Flow-induced gas pressures can result in significant negative pressure in the area above the outlet. The following example (Figure 24), shows the expected gas pressure profile in a bin that is 1 m in diameter with a 15 cm diameter outlet. This pressure profile was generated

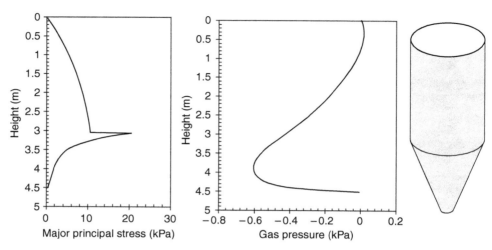

Figure 24 Solids and gas pressures in a bin at a flow rate of 54 kg/min through a 15 cm outlet and a 1.2 m diameter bin.

for a condition where the solids stress level at the bottom of the bin was nearly zero and the flow rate was 54 kg/min — the maximum flow rate possible with this fine material in this bin configuration. It should be pointed out that the corresponding acceleration-limited flow rate is 780 kg/min, which is 14.4 times faster.

Equation (13) can be used to estimate the permeability limiting flow rate from conical- and wedge-shaped flow channels. This requires knowing the measured permeability $K(\sigma_{out})$, and the bulk density $\gamma(\sigma_{out})$, at solids contact stresses close to the outlet. In addition, the outlet area A_{out} and the bulk density $\gamma(\sigma_{bin})$ at the maximum solids contact stresses in the process equipment must be known.

$$Qs_{min} = \frac{\gamma(\sigma_{out}) \cdot A_{out} \cdot K(\sigma_{out})}{1 - (\gamma(\sigma_{out})/\gamma(\sigma_{bin}))} \tag{13}$$

$$K(\sigma_{out}) = K_1 \cdot \left(\frac{\gamma(\sigma_{out})}{\gamma_1} \right)^{\alpha} \tag{14}$$

The bulk density $\gamma(\sigma_{bin})$ at the hopper transition can be estimated by solving Equation (15) for σ_{bin} with bulk density curve fit parameters γ_0, σ_0, and β. It also uses bin geometry parameters such as the diameter of the bin, D_{bin}, and the height of the cylinder section, H_{bin}. An iterative procedure is needed to solve this nonlinear equation. Equation (15) also assumes cylindrical bin geometry.

$$\sigma_{bin} = 2 \cdot \left(\frac{\gamma_0 \cdot (1 + \sigma_{bin}/\sigma_0)^{\beta} \cdot g \cdot D_{bin}}{4 \cdot k_v \cdot \tan(\phi_{wall})} \right) \cdot \left(1 - \exp\left(\frac{-4 \cdot k_v \cdot \tan(\phi_{wall})}{D_{bin}} \cdot H_{bin} \right) \right) \tag{15}$$

The term in the denominator of Equation (13) represents the percent change in material porosity as bulk material flows through the process equipment. This suggests that the limiting flow rate is inversely proportional to the change in the volume of gas squeezed out of the material during flow. This is a function of the bulk material compressibility as indicated by the density ratio of the process equipment $(\gamma(\sigma_{out})/\gamma(\sigma_{bin}))$. This flow-rate limitation is also directly proportional to the material permeability at the hopper outlet $K(\sigma_{out})$. Limiting flow rate is directly dependent on the size of gas volume that must be pressure-equilibrated and the gas flow rate through the outlet required to accomplish this task. Small limiting rates arise from low permeability values and highly compressible materials.

Most bulk materials have limiting flow-rate conditions between the two extremes given in Equations (11) and (13). These two equations can be used to compute expected flow rate of a general material from a hopper of a given outlet diameter and capacity. Equation (16) shows the way to combine these two flow-rate limit equations to give a general limiting flow-rate equation that will work for all materials.

$$Qs_{limit} = Qs_{max} \cdot \left[-\frac{1}{2} \cdot \frac{Qs_{max}}{Qs_{min}} + \frac{1}{2} \cdot \sqrt{\left(\frac{Qs_{max}}{Qs_{min}}\right)^2 + 4} \right] \tag{16}$$

Numerically the computed value of Qs_{min} could be larger than Qs_{max} provided the material permeability is large enough. This equation handles that situation. If Qs_{min} is larger than Qs_{max} (very permeable materials), then Qs_{limit} equals Qs_{max}. If Qs_{min} is small, then Qs_{limit} reduces to Qs_{min}. This equation can be derived from the equation of motion describing flow of bulk materials in a converging geometry, but that derivation will not be presented here. Qs_{limit} can be used as an index to quantify the flow-rate behavior of bulk materials. It combines measured properties in the correct manner to describe flow-rate limits in process equipment. It can also be used to estimate the time required for powdered materials to lose entrained air.

Equation (17) shows a settlement index (SI) that is proportional to the deaeration time for a given material and could be used to rank the fluidizing or flooding behavior of a bulk powder material placed in a container with a total volume, Vol_{tot} and a free surface, $A_{surface}$ available for gas dissipation.

$$SI = \frac{\gamma(\sigma_{out}) \cdot A_{out} \cdot Vol_{tot}}{Qs_{limit} \cdot A_{surface}} \tag{17}$$

Materials with large SI values will lose retained air slowly and may remain partially fluidized during handling. This can be a problem in situations where the process equipment is designed to work in a volumetric feed mode. Often these volumetric feed devices are not equipped to handle fluid-like materials. If the SI exceeds the residence time, there is a possibility of poor flow control through volumetric feed equipment. Thus the SI can provide a means of predicting flow-rate problems.

IX. SUMMARY OF IMPORTANT FLOW PROPERTIES

We have discussed some of the key properties required to predict flow behavior in powder food processes. These basic properties can also be used to describe common product behavior. The following tables summarize the material properties and their uses. Table 1 gives the general uses of these properties. Table 2 shows the calculation summary of important indices used to characterize any bulk material. It is important to point out that, even though these indices are derived with hopper flow in mind, they can be applied to other processing situations. They constitute a set of numbers that describes the basic behavior of a bulk material from a flow point of view. It is naive to think that one number could effectively correlate with all flow problems in processes that handle bulk powder materials. It is recommended that all the numbers be computed for a detailed characterization of any bulk solid.

Table 1   Review of Material Properties

Basic material property	Process variables that effect it	Use for process equipment predictions	Use for product characterization
Bulk density (usually measured as a function of the major principle stress applied to the material)	Temperature, moisture content, particle size, particle shape, size distribution	Process volumes, mass-flow rates, arching dimensions, rathole dimensions, compressibility, solid vessel loads, floodability, flow-rate consistency	Porosity, particle surface area, compatibility, fluidization, tablet weight consistency
Unconfined yield strength (usually measured as a function of the major principle stress applied to the material)	Temperature, moisture content, storage time, chemical content, particle size, particle shape, size distribution	Arching dimensions, rathole dimensions, fluidization ability, blender operation, segregation tendency	Green tablet strength, tablet capping, caking, package filling ability
Permeability (usually measured as a function of the bulk density of the material)	Temperature, moisture content, particle size, particle shape, size distribution gas viscosity	Flow-rate limitations, settlement times, floodability, drying times, flow-rate consistency	Caking, tablet capping, dustiness, fluidization
Wall-friction angle (usually measured as a function of the normal wall stress applied to the material)	Temperature, moisture content, particle size, particle shape	Flow rates, flow profiles, solids vessel loads, rathole dimensions, wear control, chute velocities, mass-flow angles, segregation tendency	Abrasion, coating ability

Table 2 Flow Indices Describing Process Behavior

Index	Definition	Calculation
Arching index (AI)	The minimum size of hopper outlet that will prevent arch formation	$\sigma_{\text{crit}} = 1.5 \cdot \sqrt{\dfrac{D_{\text{bin}}}{3.05 \text{ m}}} \cdot \gamma(\sigma_{\text{crit}}) \cdot g \cdot D_{\text{out}}$ $\text{AI} = \dfrac{H(\theta) \cdot fc(\sigma_{\text{crit}})}{\gamma(\sigma_{\text{crit}}) \cdot g}$
Rathole Index (RI)	Minimum size of active flow channel that will prevent stable rathole formation in hopper	$\text{RI} = \dfrac{G(\phi) \cdot fc(\sigma_{\text{bin}})}{\gamma(\sigma_{\text{bin}}) \cdot g}$
Recommended mass-flow hopper index ($\text{HI} = \theta_c$)	Maximum conical-hopper slope angle measured from the vertical that will cause flow along hopper walls	$\text{HI} \approx \dfrac{41° - \phi_{\text{wall}}}{0.93}$
Feed density index ($\text{FDI} = \gamma_{\text{out}}$)	Density of the material leaving the hopper	$\text{FDI} = \gamma(\sigma_{\text{out}})$
Bin density index ($\text{BDI} = \gamma_{\text{bin}}$)	Density of material in the bin	$\text{BDI} = \gamma(\sigma_{\text{bin}})$
Flow rate Index ($\text{FRI} = Qs_{\text{limit}}$)	Flow rate of bulk material leaving a conical hopper	$Qs_{\text{max}} = \gamma(\sigma_{\text{out}}) \cdot A_{\text{out}} \cdot \sqrt{\dfrac{D_{\text{out}} \cdot g}{2 \cdot a \cdot \tan(\theta)}}$ $Qs_{\text{min}} = \dfrac{\gamma(\sigma_{\text{out}}) \cdot A_{\text{out}} \cdot K(\gamma(\sigma_{\text{out}}))}{1 - (\gamma(\sigma_{\text{out}})/\gamma(2 \cdot \sigma_{\text{bin}}))}$ $\text{FRI} = Qs_{\text{max}} \cdot \left[-\dfrac{1}{2} \cdot \dfrac{Qs_{\text{max}}}{Qs_{\text{min}}} + \dfrac{1}{2} \cdot \sqrt{\left(\dfrac{Qs_{\text{max}}}{Qs_{\text{min}}}\right)^2 + 4} \right]$
Settlement index (SI)	Relative time required for deaeration of material in conical bin	$\text{SI} \approx \dfrac{\gamma(\sigma_{\text{out}}) \cdot A_{\text{out}} \cdot \text{Vol}_{\text{tot}}}{Qs_{\text{limit}} \cdot A_{\text{surface}}}$

Notes: These indices are specifically developed for use in conical bins and depend on the size of the bin. They are referenced to a maximum bin diameter D_{bin} and a bin outlet diameter D_{out}.

$$\sigma_{\text{bin}} = \frac{\gamma(\sigma_{\text{bin}}) \cdot g \cdot D_{\text{bin}}}{4 \cdot k_v \cdot \tan(\phi_{\text{wall}})} \cdot \left(1 - \exp\left(\frac{-4 \cdot k_v \cdot \tan(\phi_{\text{wall}})}{D_{\text{bin}}} \cdot H_{\text{bin}}\right)\right)$$

$$\sigma_{\text{out}} = \frac{\gamma(\sigma_{\text{out}}) \cdot g \cdot D_{\text{out}}}{2} \quad \text{(for cone)}$$

NOMENCLATURE

a	geometry index for cone- or wedge-shaped geometries
AI	arching index
A_{out}	cross-sectional area on the equipment outlet
$A_{surface}$	area of the surface available for gas dissipation
BDI	average density in the bin
D_{bin}	largest diameter of the bin
D_{out}	diameter of the hopper outlet
fc	unconfined yield strength of the bulk material
fc_{crit}	critical unconfined yield strength for arching
FDI	density at the hopper outlet
ff	flow factor for the Jenike arching analysis
FRI	flow-rate index for a bulk material in a bin of diameter D_{bin} and outlet of D_{out}
g	gravitational acceleration
$G(\phi)$	friction angle factor for the Jenike rathole analysis
$H(\theta)$	geometry arching factor
H_{bin}	height of the vertical section in the bin
HI	recommended conical hopper angle for mass flow
K	material permeability
K_1	material permeability curve fit parameter
k	Jansson k-value
k_v	Jansson k-value for a cylinder
Qs_{limit}	combined flow rate limit
Qs_{max}	acceleration limited flow rate
Qs_{min}	permeability limited flow rate
RI	rathole index
SI	settlement index
U	superficial gas velocity
Vol_{tot}	total volume of material in the bin

Greek Symbols

γ	material bulk density
∇P	gas pressure gradient
$\partial P/\partial z$	gas pressure gradient in the z-direction
α	permeability curve fit exponent
β	bulk density curve fit exponent
δ	effective friction angle
ϕ	internal friction angle
ϕ_{wall}	wall-friction angle
γ_a	reference bulk density for the permeability curve fit
γ_0	minimum bulk density parameter for the bulk density curve fit
μ	gas viscosity
μ_0	reference gas viscosity
θ	conical hopper half angle
σ_1	major principal stress
σ_{bin}	maximum stress level in the bin
σ_{crit}	stress level at the arch

σ_{out} stress level at the outlet
σ_x minor principal stress
σ_x normal stress in the x-direction
σ_y normal stress in the y-direction
σ_0 minimum stress parameter for the density curve fit
τ_{xy} shear stress in the y-direction on the x face of bulk solid stress element

REFERENCES

1. Benink, E.J., *Flow and Stress Analysis of Cohesionless Bulk Materials in Silos Related to Codes*, Ph.D. thesis (1989).
2. BMHB, *Silos: Draft Design Code for Silos, Bin, Bunkers and Hoppers* (1987).
3. Cleaver, J.A.S., Ph.D. thesis, University of Cambridge (1991).
4. Jenike, A.W., Gravity Flow of Bulk Solids, Bulletin 108, Utah Engineering Experiment Station (1961).
5. Jenike, A.W., Flow and Storage of Bulk Solids, Bulletin 123, Utah Engineering Experiment Station (1967).
6. Ebert, F., Dau, G., and Durr, V., *Blending Performance of Cone-in-Cone Blenders: Experimental Results and Theoretical Predictions*, Proceedings of The Fourth International Conference for Conveying and Handling of Particulate Solids (2003).
7. Johanson, J.R., Controlling flow patterns in bins by use of an insert, *Bulk Solids Handling*, 2, 495–498 (1982).
8. Nedderman, R.M., *Statics and Kinematics of Granular Materials*, Cambridge University Press (1992).
9. Jenike, A.W. and Johanson, J.R., Stress and Velocity Fields in Gravity Flow of Bulk Solids, Bulletin 116, Utah Engineering Experiment Station (1967).

Food Powder Ingredients

14

Functional Properties of Milk Powders

David Oldfield
Institute of Food, Nutrition and Human Health
Massey University
Palmerston North, New Zealand

Harjinder Singh
Riddet Centre
Massey University
Palmerston North, New Zealand

CONTENTS

I. INTRODUCTION

Milk is a secretion of the mammary gland of female mammals and is often the sole source of food for the very young mammal. Milk has high nutritional value, which also makes it a good medium for microbial growth and other spoilage organisms. In order to preserve the nutritional quality of milk and enhance its utilization and safety, a number of processes have been developed over the years. Liquid milk can be heated, for example, pasteurization and ultrahigh temperature (UHT) treatment; or milk can be concentrated, for example, evaporated milk and sweetened condensed milk; or milk can be made into other dairy products, for example, cheese and yoghurt; and milk can also be dried to make powders.

Milk powders are defined by the Codex Standard 207-1999 [1] as, "milk products which can be obtained by the partial removal of water from milk." Milk powders are typically dried in a spray dryer (refer to Chapter 15) to a maximum moisture content of 5%. Furthermore, fat and protein content can be standardized according to the Codex. Thus milk powders are differentiated on the basis of their fat content into skim milk powder (SMP, also known as nonfat dry milk [NDM]), partially skimmed milk powder, and whole milk powder (WMP).

The functional properties of milk powders depend on:

- Raw milk composition and standardization.
- Processing conditions.
- Storage conditions.
- How the powder is used in a particular food system.

This chapter describes the various applications of milk powders and how the functional properties of milk powders are affected by various composition and processing conditions.

II. MILK PRODUCTION

Milk is produced by mammals and contains a wealth of nutrients for the growing young. Cows, buffalos, goats, and sheep are milked; however, cow's milk is predominantly used for manufacturing milk powders.

Milk is collected during the lactation cycle of a cow, which typically lasts ten months, from the time the calf is born (calving) to when milking is stopped (drying off) to allow the cow to recover before the next lactation cycle. This is followed by a two-month dry period, when no milking occurs, after which a new calf is born and the lactation cycle and milking begins again [2].

During the lactation cycle, the milk composition and the volume of milk collected from the cow change. This results in issues for the manufacturer processing milk at certain times of the lactation cycle. For example, late lactation milk (characterized by high protein and low lactose levels) is considered unsuitable for manufacture into powders used in recombined evaporated milk and in applications where high-heat treatment is used.

In addition to lactation, other factors can add to variability in milk composition and the amount of milk collected: age of the animal, breed, condition of the cow, and changes in feed. Milk production and composition can be affected by seasonality, where the changing seasons determine the type of feed (grass, hay, silage, etc.) and condition of the cow.

III. MILK COMPOSITION

Milk is a complex mixture of fat, protein, lactose, milk salts, and minor components (nonprotein nitrogen, vitamins, etc). The functional properties of milk powders are dependent both on the milk composition and on how the different components are modified by manufacturing conditions.

As milk composition varies throughout a milking season (Table 1) it is routinely analyzed to determine payouts to farmers, plant economics, and processing conditions required to produce a consistent product. Although variation in composition can be quite large from country to country, variations within a country are less. For example, Jelen [3] reports that the average protein content of milk (3.3–3.4%) shows wide variations internationally (2.8–4.2%); however, within countries the variation is ±0.5% or less.

A. Milk Fat and Fat Globule Membrane

Milk fat is mainly (98%) in the form of triglycerides, which are formed when three fatty acids bond (esterification) to a glycerol molecule. The main fatty acids are: myristic (14:0), palmitic (16:0), stearic (18:0), and oleic (18:1) (approximately 72% of the total fatty acid content by weight). Milk fat also consists of small amounts of (about 2% of total fat) diglycerides, monoglycerides, cholesterol, phospholipids, free fatty acids, cerebrosides, and gangliosides.

The composition of milk fatty acids varies with season/lactation and is influenced by the cow's diet. Changing the cow's feed to certain vegetable oils can be used to manipulate the nutritional characteristics of milk fat (more unsaturated fatty acids), while

Table 1 Typical Raw Milk Composition

Component	Level in milk %(w/w)	
	Typical value	Range[a]
Water	86.5	
Fat	4	2.98–6.23
Protein	3.28	2.75–4.16
Casein protein	2.6	2.3–3.0
Whey protein	0.75	0.53–0.82
Lactose	4.6	4.5–5.0
Milk salts (ash)	0.85	0.7–0.9

[a] Range due to country, breed, and lactation.

Source: From [25, 66–70] With permission.

maintaining the quality (may be prone to oxidation) and functional properties (lower melting point — softer milk fat) [4]. There is also interest in the nutritional benefits of certain milk fat components, such as conjugated linoleic acid [5] and phospholipids [6].

In milk, nearly all the milk fat is present as fat globules ranging in size from 1 to 9 μm in diameter with an average diameter of 2 μm. Each globule is stabilized by a milk fat globule membrane (MFGM), which is composed of phospholipids and glycoproteins. The MFGM makes up only 1% of the total weight of the fat globule and is less than 0.01 μm thick.

The MFGM prevents oxidation and enzymic lipolysis of the fat that causes rancid off flavors in the powder. On the one hand, damage to the MFGM can occur during processing, by pump cavitation, high shear, and foaming. On the other, homogenization is used to break the milk fat globules into smaller fat globules, which prevents cream separation in milk. Because the original MFGM is unable to cover the additional surface area created by homogenization, milk proteins are adsorbed onto the newly created fat surface. Heating milk above $\sim$70°C causes denaturation of the membrane protein and interaction with serum proteins [7].

B. Milk Proteins

Milk proteins are important nutritionally (essential amino acids) and provide a wide range of functional properties to milk powders. There are two groups of milk proteins. On acidification of milk to pH 4.6 at 20°C, about 80% of the total milk protein precipitates out of the solution; these proteins are called caseins. The proteins that remain soluble under these conditions are referred to as whey proteins. Both the casein and whey protein groups are heterogeneous. Several reviews and monographs on the structures and properties of milk proteins have been published [8–12].

In milk, the caseins associate together to form spherical colloidal structures called casein micelles with diameters ranging from 50 to 300 nm with an average diameter of 150 nm [13]. The micelle is composed of casein (92%) and inorganic salts (8%), which are largely in the form of colloidal calcium phosphate (CCP). A number of models of the casein micelle structure have been proposed over the years; these include coat-core models [14], submicelles models [15,16], and porous network structures [17]. A common factor in all the models is that most of the κ-casein appears to be present on the surface of the casein micelles. The hydrophilic, C-terminal part of κ-casein is assumed to protrude 5 to 10 nm from the micelle surface into the surrounding solvent, giving it a "hairy" appearance. The highly charged flexible "hairs" physically prevent the approach of other micelles thus providing the micelles with their stability.

Casein micelle structure is relatively heat stable. Milk can withstand heating for 14 to 19 min at 140°C before coagulating. This property is useful for applications where solubility and heat stability are required, for example, recombined evaporated milk, hot beverages, and soups. On the other hand, casein micelle stability is easily destroyed by the addition of acid or by enzymatic hydrolysis (rennet). Acid dissolves the CCP and causes dissociation of individual casein proteins from the casein micelle; enough acid causes casein to precipitate out at pH 4.6, which is the basis of yoghurt and fresh cheese manufacture. Addition of rennet destabilizes the micelle by removing the charged portion of κ-casein responsible for micelle stability. This subsequently allows the micelles to aggregate and a curd is formed (cheese production).

Whey proteins do not form micelles; instead they exist in solution as globular proteins. The main whey proteins are: β-lactoglobulin (50% by weight), α-lactalbumin (20%), bovine serum albumin (5%), and immunoglobulins (10%). β-lactoglobulin is the major whey protein and it tends to govern the general behavior of the whey proteins as a whole. In milk, whey proteins are stable under acid conditions; however, they are sensitive to heat denaturation above ~65°C. Heating above 70°C, causes β-lactoglobulin to aggregate with itself and/or with other whey proteins. Furthermore, β-lactoglobulin can also interact with κ-casein at the micelle surface. These reactions modify properties of the milk system, such as heat stability and acid gelation [18,19].

Nutritionally milk proteins, which are highly digestible, provide a rich source of essential amino acids. Bioactive peptides derived from partially digested milk proteins have a wide range of health benefits [20,21]. Peptides are produced either by digestion in the gut, by microbial enzymes in traditional foods such as cheese and yoghurt, or in the manufacture of ingredient powders such as milk protein hydrolyzate powders.

C. Lactose

The main carbohydrate found in milk is lactose, a disaccharide containing β-D-galactose and α-D-glucose. In comparison with other sugars, lactose has a low solubility and a low level of sweetness. Lactose in solution exists as two isomers; α- and β-lactose. During milk powder production, supersaturation of lactose occurs, however, instead of crystallizing, the rapid drying process causes lactose to form an amorphous state, which is highly hygroscopic. This causes powder stickiness and caking when a combination of moisture absorption and high storage temperature occur during powder storage [22].

Lactose is a reducing sugar, and can react with a free amino group of milk proteins under certain conditions. This reaction is called the Maillard reaction or nonenzymatic browning and it increases with the severity of the heat treatment, and an increase in pH. In milk powders, the Maillard reaction is also dependent on water activity of the powder. The detrimental effects of this reaction are a loss of essential amino acids, especially highly reactive lysine, and discoloration of the powders and milk through the formation of brown pigments. However, in some applications heat-induced lactose reactions are desirable, for example, in powders used for confectioneries, where flavor development and color are desired. In this case, high-heat treatments are used during manufacture to obtain the desired properties.

Lactose is a source of energy; however, some people lose the ability to digest lactose (lactose intolerance), which results in digestive problems. Special powders are available where the lactose has been enzymatically hydrolyzed in the milk prior to powder manufacture.

D. Milk Salts

Milk salts are a relatively minor group in terms of concentration, however, they have a large bearing on the functional properties of the milk proteins. Interactions between milk salts and the milk proteins are important for powder functional properties such as heat stability, solubility, and gelation [23].

Milk contains the cations: calcium, sodium, potassium, and magnesium; and the anions: chloride, phosphate, and citrate. All these ions exist in solution, but calcium and phosphate are supersaturated. The excess calcium and phosphate form insoluble complexes (CCP) in the casein micelle, which helps to stabilize it. About half of the phosphate, two-thirds of the calcium, one-third of the magnesium, and a small proportion of the citrate are in the colloidal phase. The ash content approximates to the content of milk salts, however, organic salts (citrates) are lost during the process [24].

IV. OVERVIEW OF THE PHYSICOCHEMICAL CHANGES DURING MILK POWDER MANUFACTURE

The steps involved in the manufacture of various dairy powders are outlined in Chapter 15. The following section will focus on the effects of different processing steps on the functional properties during the manufacture of WMPs and SMPs.

A. Milk Standardization

After collection from the farms, raw milk is stored in silos and then standardized prior to evaporation and drying. Standardization is carried out to produce powders that have a consistent composition. Both fat and protein components of milk can be standardized, thus minimizing variations in composition caused by different breeds of cow, types of feed, and as a result, of lactation.

Standardization involves the mixing of approved dairy streams in order to achieve the required composition in the powder, and is governed by the Codex for milk powders and cream powders [1]. The following dairy streams are permitted: milk, cream, milk retentate from ultrafiltration, milk permeate from ultrafiltration, and lactose. Furthermore, approved additives (certain anticaking agents, emulsifiers, antioxidants, stabilizers, etc.) can be used to enhance powder functional properties.

The fat content of milk can be standardized using a combination of milk, cream, and skim milk, depending on the type of powder required (Table 2).

For protein standardization of milk powders, the only requirement is that the minimum protein content is 34% on a milk solids–nonfat (SNF) basis (w/w). Furthermore, the whey protein to casein ratio cannot be altered and must remain the same as that of the raw milk. Comparing the average protein contents of milk powders between countries, the range is 35.5 to 39.6 (SNF basis), and the variation within a country is typically ±6% [25]. However, extreme variations can occur with values as low as 33.5% and as high as 41.0% on an SNF basis.

If the natural protein content of milk can produce a powder with a protein content above 34% then down-standardization of milk (addition of lactose or permeate from milk ultra-filtration) at the start of powder manufacture is economically beneficial. Addition of lactose and milk permeate will increase the amount of powder produced, and thereby increase profitability as lactose or permeate are not as expensive as protein. For milk powder with a protein content below 34% up-standardization (addition of retentate from milk ultrafiltration) of the milk prior to drying will allow powders to conform to Codex standards and therefore be sold as milk powders.

Down-standardization can be useful if the protein concentration is naturally high. High protein concentration can cause problems in drying and powder functional properties

Table 2 Different Fat Specifications for Milk Powders

Powder	Fat content
Cream powder	minimum 42% (w/w basis)
Whole milk powder	minimum 26%, <42%
Partly skimmed milk powder	>1.5%, <26%
Skimmed milk powder	1.5% or less

Source: [1]

such as increased viscosity of recombined evaporated milk and reduced solubility of some types of milk powders [26].

Although the effect of protein standardization on the functional properties of milk powders has not been reported in any depth, a number of studies have been carried out on milk. Adding skim milk permeate slightly increases the heat stability of skim milk or evaporated milk [27,28], and does not greatly affect the sensory quality of standardized milk [29].

B. Preheating

Standardized milk is heated at a set time/temperature combination prior to evaporation (known as preheating). The main purpose of preheating is to change the functional properties of the powder. Preheating is also used to raise the milk temperature high enough to ensure that milk boils in the first effect of the evaporator.

A number of different preheating configurations can be used. A typical example is where milk first passes through spiral tube heat exchangers located in the vapor/liquid separator and caldaria sections of the evaporator. This heats the milk up to around 60 to 70°C. These heat exchangers use heat from the evaporators for greater thermal efficiency. Then, if a high temperature preheat treatment is required (above 70°C) a direct heating system (direct steam injection or steam infusion) is used to rapidly raise the milk temperature. Milk then passes through a holding tube and then through one or more flash vessels, which are under vacuum. Some of the water in the milk flash vaporizes and this rapidly cools the milk to just above the first effect temperature.

The main chemical changes that occur in milk during preheating are:

- Denaturation and aggregation of whey proteins.
- Interaction of denatured whey proteins with casein micelles.
- Precipitation of soluble minerals onto the casein micelles.
- Decrease in pH.

Preheating also affects powder solubility properties and shelf life. Generally, the higher the preheat treatment the poorer the powder solubility and the better the oxidative stability of WMPs [30].

C. Evaporation

The purpose of evaporation is to remove as much water as possible without detrimentally affecting product quality. Evaporators are far more efficient at removing water than spray dryers. However, as the milk becomes more concentrated it becomes increasingly more viscous, making it difficult to remove water. Thus evaporators are limited to concentrating milk up to ~50% total solids (TS); higher solid concentrates produce powders with poor functional properties.

Falling film evaporators remove ~80% of the water from milk, concentrating the milk from 9–14% TS up to 44–50% TS. The falling film evaporator is designed to minimize heat damage, by having a low residence time in the evaporator tubes (3–6 min) and operating under vacuum so that boiling temperatures range from 40–70°C.

Whey protein denaturation during evaporation is minimal as the temperatures remain below 70°C. In addition, as the milk solids become more concentrated the heat stability of the major whey protein (β-lactoglobulin) increases [31], probably the result of increasing lactose concentration stabilizing the proteins. Other changes that occur are the precipitation of calcium and phosphate, which are already saturated in milk, onto the casein micelle as the milk becomes more concentrated. This contributes to an overall decrease in the milk pH from 6.7 in raw milk to around 6.2 for 48% TS concentrate.

D. Concentrate Heating

Before spray drying, milk concentrate is heated at a temperature in the range of 65 to 80°C to reduce concentrate viscosity prior to spray drying. This is to optimize atomization of the concentrate in the spray dryer, thereby improving spray drying efficiency and powder properties, for example, the solubility index and coffee sediment [32,33]. An additional benefit is that heat sensitive microorganisms are destroyed by the concentrate heating step.

In addition to concentrate heating, whole milk concentrates are homogenized before spray drying to reduce free fat in the powder. Homogenization can occur before or after concentrate heating although it is not known what effect this may have on subsequent powder properties.

E. Spray Drying

Spray drying removes the remaining water in the concentrate to produce a powder with 2–5% moisture. Milk concentrate is atomized into a fine spray that is dried in a chamber by incoming hot air (150–250°C). Initially, milk droplets dry rapidly at their wet bulb temperature (45–60°C); however, as the droplets progress down the drying chamber they become drier and the particle temperature eventually approaches the air temperature at the dryer outlet (70–90°C).

There are a number of different dryer configurations, with or without internal and external fluidized beds. Two-stage and multistage dryers are more energy efficient and less damaging on milk powder properties than single stage dryers. (For more details on spray dryers see Chapter 15.)

Drying conditions influence powder properties, such as moisture content, solubility, bulk density, and powder flowability [22]. The outlet temperature has a strong effect on powder solubility — increasing the outlet temperature results in poor powder solubility [22]. However, little is known of the chemical changes that occur during spray drying. Under typical drying conditions whey proteins are not denatured to any great extent. However, high outlet temperature conditions (>80°C) cause increased lactosylation of milk proteins (part of the Maillard reaction between proteins and lactose), decreasing their nutritional value [34].

Roller dryers can also be used to produce milk powders, but they are limited to producing WMPs for confectioneries, especially chocolate manufacture [35]. Roller dried WMPs have a high level of free fat and low solubility, which make them ideal for use in chocolate.

F. Powder Storage

After drying, the powder is transported to a packaging machine. Packaging comes in a variety of different forms, from multi-ply paper bags with a polyethylene liner to cans, and bulk bin containers. After filling, the powder may be flushed with an inert gas (carbon dioxide and/or nitrogen) to remove oxygen and increase shelf life; the packaging is then sealed. Powders are stored indoors away from direct sunlight and in cool dry conditions ready for consignment. Milk powder bags (25 kg) are typically stacked on pallets and shrink-wrapped to prevent the bags from shifting and falling off during storage or transportation. Overall,

Table 3 Changes that Occur to Milk Powders During Storage

Detrimental effect	Chemical reaction	Changes to powder properties
Fat oxidation	Unsaturated fatty acids react with oxygen and form hydroperoxides, which breakdown to form rancid flavors and odors	Rancid flavors and odors
Maillard reaction	Aldehyde group of lactose reacts with free amino group of milk proteins	Discoloration (browning) of the powder, loss of protein nutritional value and flavor defects
Powder caking	Amorphous lactose above its glass transition temperature (T_g) flows and forms sticky bridges between powder particles Free fat coating the surface of powder particles can cause caking	Reduction in flowability of the powder, unsightly lumps

packaging is designed to: identify the product; keep the product away from detrimental effects, for example, light, heat, oxygen, moisture, animals, dirt; and facilitate transport, storage, and customer use.

A number of detrimental reactions occur in powder over time, which limits its shelf life (Table 3). Shelf life of milk powders can vary depending on composition, manufacturing conditions, type of packaging, and storage conditions. SMPs have a typical shelf life of around 12 to 18 months from the time of manufacture. In contrast, WMPs have a shelf life of approximately six months.

The fat content of WMP is susceptible to oxidation during storage, which produces rancid off flavors in the powder and greatly limits their shelf life. To minimize oxidation, flushing with nitrogen before sealing the packaging and selective preheat treatment during powder manufacture can be used. Medium to high preheat treatments cause protein denaturation and Maillard reactions in the milk, and these reactions form antioxidants, which slow down fat oxidation. Medium-heat treatments are preferred as high-heat treatments cause other detrimental reactions, for example, loss of solubility. Processing conditions should also ensure that minimal free fat is present in fat containing milk powders. Free fat is more easily oxidized, and can also cause poor flowability in powders. Moderate two-stage homogenization conditions, moderate drying conditions, and gentle handling of powder minimizes free fat [22].

Off flavors can also develop during milk processing and handling. Damage to the MFGM by pump cavitation, high shear, and foaming can cause an increase in enzymatic lipolysis, resulting in rancid flavors. These flavor defects carry through into the powder.

Milk powders are hydroscopic and will adsorb moisture from the air. Moisture absorption speeds up the Maillard reaction and increases powder caking. Increased moisture content of the powder (increased water activity) lowers the glass transition temperature (T_g) of lactose so that powder can cake at lower temperatures. On the other hand, low water content (2.4% compared with 3.0%) in WMPs increases the peroxide value, which is a measure of deteriorative fat reactions leading to rancid flavors [36].

The rate of fat oxidation, Maillard reaction (nonenzymatic browning), and caking are all increased by raising the storage temperature [37]. Milk powders are relatively stable at storage temperatures up to 20°C; however, above 30°C deteriorative reactions proceed rapidly [38].

Powder solubility decreases with storage time and is influenced by factors such as preheat conditions, moisture content of the powder, and storage temperature [30,36,38]. The initial solubility of high-heat powders tends to be poorer when compared with low and medium heat powders; furthermore the solubility of high-heat powders becomes worse more rapidly during storage [38].

V. MILK POWDERS AND THEIR USES

There are two principal commercial milk powders, that is, SMP and WMP that are further classified as either regular (noninstant) or instant. Commercial SMPs are commonly classified as low-, medium-, and high-heat powders on the basis of their whey protein nitrogen index (WPNI). The WPNI classifications are a gross indicator of the SMP's suitability for different applications (Table 4). For example, high-heat SMP is suitable for use in bakery products as high levels of heat treatment denature whey proteins that otherwise would act as loaf depressants, reducing loaf volume. However, the same powders are unsuitable for manufacture into recombined cheese products as heat-induced protein reactions are detrimental to renneting and cheese properties. Hence low-heat powders are used [39].

As WPNI is only a gross indicator with certain limitations [26] more specific end-user tests should be applied to determine the suitability of the powder before use. For typical compositions of milk powders see Chapter 15.

Milk powders are used in the following applications:

Recombining.   Components of milk (e.g., skimmed milk powder and anhydrous milk fat) are mixed together by dissolving them in water and then homogenized. The products that can be made include recombined evaporated milk, recombined milk, recombined sweetened condensed milk, and recombined butter. Recombined products can be eaten or drunk by consumers or used by manufacturers for processing into other food products, for example, recombined cheese and yogurt.

Ingredients.   Manufacturers can add milk powder as a functional ingredient to a wide variety of foods, for example, chocolate, bakery products, beverages, confectionery, and yoghurt.

Table 4   Uses of Different SMPs Based on Their WPNI

WPNI class (mg undenatured whey protein nitrogen/g powder)	Typical preheat treatments	Some applications
Low heat, >6	70–80°C × 15–50 sec	Recombined cheese, recombined fluid milk
Medium heat, 1.5–5.9	90–100°C × 30–50 sec	Recombined sweetened condensed milk, ice-cream, confectionery
High heat, <1.5	115–125°C × 1–4 min	Bakery products, recombined evaporated milk

Consumer powders.   Instead of using recombined milk or fresh milk, consumers can reconstitute milk at home by mixing milk powder and water. Alternatively powder can be used as a whitener by adding it directly to hot coffee or tea. For household use the powders are instantized during spray drying by lecithination and agglomeration, so they dissolve rapidly in water at room temperature.

Selection of a milk powder for a given application is based on a number of factors, which may include: required fat content of the powder, WPNI value, whether the powder is instant or not, and how the powder performs in selected end-user tests.

VI.   FUNCTIONAL PROPERTIES OF DAIRY POWDERS

The functional properties of a milk powder are determined by the physical and chemical properties of its primary constituents, namely proteins, fat, and lactose both individually and in combination. The milk fat component in the powder contributes to a wide range of functional properties depending on the application (Table 5). In addition, milk fat can be delivered using other dairy ingredients such as high fat (cream) powder, butter milk powder

Table 5   Functional Properties of Different Components in Milk Powders

Desired functional property	Application (food system)
Milk fat	
Whitening ability	Coffee whiteners, beverages, soups and sauces, confectionery, and chocolate
Smooth texture/mouthfeel (modify viscosity)	Cheese, soups and sauces, confectionery, chocolate, bakery products, coffee whiteners
Creamy dairy flavor	Dairy beverages, milk chocolate, soups and sauces, ice-cream, and desserts
Whipping and foaming	Ice-cream, desserts, and whipped toppings
Protein	
Water binding	Confectionery, meat products, bakery products
Thickening (modify viscosity)	Confectionery, soups and sauces
Emulsification	Coffee whiteners, meat products, soups and sauces
Heat stability	Soups and sauces, recombined evaporated milk
Foaming and whipping properties	Ice-cream, desserts, and whipped toppings
Solubility	Beverages
Color and flavor development	Confectionery and chocolate
Gelation	Yogurt, cheese
Lactose	
Filler, bulking agent	Free flowing powders
Color and flavor development	Confectionery (caramel) and chocolate
Humectant	Bakery products
Body and mouthfeel (modify viscosity)	Coffee whitener, confectionery

(BMP) and butter powder — and in nonpowdered form: butter, cream, milk fat fractions, and anhydrous milk fat.

Whey and casein proteins are an important component of milk powders. The functional properties that they contribute are shown in Table 5. Functional properties of the proteins are largely determined by the extent of heat treatment applied during powder manufacture.

As milk proteins are highly functional, there are a number of dairy powders with concentrated levels of whey and casein proteins. Correspondingly these powders have reduced levels of fat, and lactose, and milk salts. The following dairy powders contain a mix of whey and casein proteins: milk protein concentrate (MPC), coprecipitates and total milk protein (TMP); solely whey protein: whey protein concentrate (WPC) and whey protein isolate (WPI); or solely casein: caseinates (sodium and calcium) and casein (lactic, acid, rennet).

Lactose is a major component of milk powders comprising up to ~50% of the powder by weight. Functional properties that lactose can provide are shown in Table 5.

In the following section, some important functional properties of milk powders are described.

A.   Heat Stability

The heat stability of milk powder is important in applications such as recombined evaporated milk, soups, and sauces. In fact it is important in any application or process where milk powder is subjected to high temperature conditions; for example, as a part of cooking, sterilization, or when powder is added to hot water or to a hot beverage. The definition of heat stability very much depends upon the application the powder is to be used for. Many studies have examined the heat stability of milk at its original concentration by observing the time taken for the onset of coagulation. In contrast to unconcentrated milk, concentrated milk products, for example, recombined evaporated milk, show very different heat stability characteristics. The heat stability of milk powder is affected by composition of the original milk from which the powder is made, processing conditions, additives, and the composition of the food system the powder is used in. There are a number of published reviews on heat stability [18,40,41].

1.   Heat Stability Tests

For SMPs used in recombined evaporated milk (REM), heat stability is the property of proteins to withstand commercial sterilization without adversely affecting the product (i.e., coagulating, or a large increase in viscosity). For a description of the REM process see Chapter 15. As the heat stability of SMPs can be quite variable, they are routinely tested. The sterilization conditions (typically 120°C for 12 min) and the high concentration of milk solids (26 to 31% TS) are a severe test of the powder's heat stability.

Tests can be carried out under lab-scale conditions or on a small batch of samples made in the plant. A number of samples are usually tested with different levels and types of stabilizing salts added before heat sterilization. This ensures correct selection of stabilizers to maximize heat stability. Both subjective and objective lab-scale test methods have been used to assess heat stability of SMPs for REM.

One subjective method [42,43] involves reconstituting SMP to 20% TS in distilled water. A set volume is added to a small glass tube, allowing some head space in the tube. The tube is sealed with a stopper and placed in an oil bath set to 120°C. The tubes are rocked gently to simulate the passage of cans through a commercial sterilizer. Heat stability is recorded as the time for the sample to show the first visible signs of coagulation; this is called the heat coagulation time (HCT). The test has a number of limitations. The visual "end-point" where coagulation first appears is very subjective. Furthermore, there is no fat inclusion and homogenization, both of which greatly affect heat stability of REM. Inclusion of fat clouds and smears the inside of the glass tube and makes end-point determination extremely difficult. Thus, HCT of SMP correlates poorly with that of commercial samples [44] and the test will only give a general indication of the heat stability of SMP when used in REM.

An alternative is an objective lab-scale method, described by Kieseker and Aitken [45]. The method aims to simulate both the composition of REM and also the processing and sterilization conditions. This test, however, takes longer than the subjective test. SMP is reconstituted in distilled water at about 45°C, and the fat mixed in and the mixture homogenized. The REM is then poured into a stainless steel tube, which is then fitted with a pressure cap. The tube is rocked gently in an oil bath at 120°C for a set period, defined as the sterilization time, typically anywhere from 12 to 25 min. Viscosity of the REM is measured after the tube has been removed from the oil bath and cooled in 25°C water. Suitable viscosity is typically in the 12 to 20 mPa sec range. Visible lumps begin to appear in the REM above 40 mPa sec. The test gives viscosity results that are in satisfactory agreement with viscosity results obtained from commercially produced samples [45]. In addition to assessing heat stability, viscosity is also an important attribute for REM. If the viscosity is too high, REM will not pour smoothly; this may also be due to lumps of precipitated material, which is also a visual defect. Very low viscosity makes REM appear watered down and not give the appearance of full fat evaporated milk.

Testing can also be carried out at the manufacturing site. A small test sample is put through the manufacturing process, so the heat stability of the powder can be observed under actual processing conditions. However, this is time-consuming, expensive, and manufacturing plants more often do not have the facilities for analyzing samples. Usually, any testing in the plant is carried out on the finished product by visual assessment, and the milk powders are tested by the supplier, using lab-scale methods, and certified as "heat stable." Visual assessment looks for signs of fat separation and sediment, as well as excessive thickening (viscosity increase) in the finished product and after accelerated storage tests at elevated temperatures, for example, 37°C.

Heat stability tests are usually only of interest when powders are used for concentrated products (e.g., REM). Powders reconstituted to normal milk solids (9 to 14% w/w) and subjected to UHT-type heat treatments or even sterilization heat treatments (120°C) are unlikely to coagulate [46]. Although the temperature for UHT is 140°C, the short heating times (a few seconds) and the low solids concentration compared to concentrated products means the milk usually has a sufficient HCT to withstand the heat treatment.

2. Heat Stability of Milk and Milk Powder

The HCT of raw milk at 140°C is ~14 to 19 min. Heat stability of raw milk varies during the year, caused mainly by changes in milk composition resulting from lactation and seasonal effects. Heat stability is affected by the natural concentration of urea [47] and soluble calcium concentration of milk [48].

Heat stability is also affected by pH. A plot of HCT versus pH typically shows a maximum at around pH 6.6–6.7 and a minimum in the pH region 6.8–6.9 (Figure 1). The heat stability increases again at higher pH values. The significance of the heat stability curve is that, for optimum milk heat stability the natural pH of milk should be positioned as close as possible to the maximum. However, the pH of the maximum can vary depending on the processing conditions, additives, and milk composition, and is entirely different for concentrated milk products such as REM. The overall heat stability of concentrated milk is much lower than that of unconcentrated milk. The heat stability maximum usually occurs

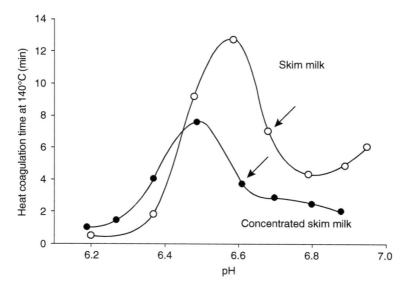

Figure 1 Typical heat stability versus pH curves at 140°C. Skim milk (unconcentrated) (o) and concentrated skim milk (20% TS, high preheat treatment) (●). Arrows show the natural unadjusted pH. In general, the position of the natural milk pH occurs either at the peak of the HCT/pH curves or very close on either side. (Unpublished data, courtesy of D.J. Oldfield and P.M. Kelly. Dairy Products Research Centre, Teagasc, Moorepark, Fermoy, Ireland.)

in the pH range 6.4 to 6.6, and the stability remains very low even at higher pH values (Figure 1).

As long as minimal heat damage occurs during processing (e.g., pasteurization 72°C for 15 sec), the heat stability of reconstituted milk powder should resemble that of the original milk (both at the same solids level). However, high-heat treatments (115–125°C for 1–3 min) cause a decrease in heat stability of milk and milk powders reconstituted to 9–13% TS level [49]. In contrast, high-heat treatment of milk prior to concentration increases the heat stability of the resulting milk concentrates. Therefore, the preheating conditions that are used to enhance the heat stability of SMP in concentrated products detrimentally affect the heat stability of unconcentrated milk, and can adversely affect other functional properties, such as coffee stability [33].

3. Heat Stable Milk Powders

Production of heat stable SMP for use in concentrated milk products (e.g., REM) is primarily controlled by providing a sufficient preheat treatment, before evaporation, to denature the whey proteins. High-heat treatments are typically in the order of 115–125°C for 1–4 min. However, high-heat treatments are not a guarantee of producing heat stable SMP. In countries with seasonal variations in milk composition (e.g., New Zealand, Australia, Ireland) milk from the middle of the season is used, as milk at the beginning and end of the season (late lactation milk) tends to have lower heat stability [48]. It is possible to make heat stable SMP from late lactation milk [50] by carefully screening the powders and controlling the processing conditions [26].

Heat stable powders are used in the manufacture of REM. BMP can be used to replace up to 10% of the SMP in REM. Phospholipids in the butter milk improve heat stability, contribute to the creamy, flavor, and help to emulsify fat [51].

Sodium and potassium salts of orthophosphate and citrate are used to improve the heat stability of REM. Care should be taken as overuse of salts can cause the heat stability to become worse. Thus, heat stability tests are used to select the right combination and level of salts. Food regulations limit the amount of stabilizing salts that can be added — besides the limit of not producing a salty product. For stabilizing salts to work, the overall heat stability maximum needs to be increased and more importantly the pH of the REM should also be positioned on or close to this maximum. For example, di-sodium hydrogen orthophosphate (DSP) and mono-sodium di-hydrogen orthophosphate (MSP) are commonly used to improve heat stability. Addition of either salt shifts the heat stability curve to a more acidic pH and increases the heat stability maximum. The sodium salts also change the pH: DSP increases the natural pH, whereas MSP reduces the natural pH. Thus salt selection depends on whether the pH needs to be increased or decreased to reach the maximum. Often small pH shifts are required to position the pH of REM at the maximum, so combinations of MSP and DSP are used. Addition of MSP or DSP on their own can result in large pH shifts potentially causing the pH of REM to "overshoot" the maximum.

Two-stage homogenization pressures are typically 14–17.5 MPa (2000–2500 psi) first stage and 3.5 MPa (500 psi) second stage [52]. A higher first stage pressure reduces fat globule size and decreases fat separation, but overall heat stability is reduced.

Water quality is important, especially the water hardness (calcium and magnesium ions). High level of calcium and magnesium in water used for reconstitution adversely affects heat stability [52].

B. Coffee Stability of Milk Powders

Whole milk powder has found widespread use as a coffee whitener in countries where fresh milk is in limited supply. Milk powders upon dissolving in coffee solution should remain stable, that is, there should be no observable precipitation. The precipitation is called *feathering* and is composed of milk proteins and fat. Besides coffee stability, milk powders should also have the following attributes:

- Whitening ability.
- Provide body and mouth feel (modify viscosity of the beverage).
- Impart a rich creamy dairy flavor.

1. Coffee Stability Test

Although there is no standard coffee stability test a method developed by the NZDRI has found popular use [53]. The NZDRI coffee stability test measures the amount of precipitated and undissolved powder that is formed when milk powder is stirred into a hot coffee solution (80°C). After standing for 5 min at room temperature the precipitate is sedimented at 164 g in ADMI (American Dry Milk Institute) solubility index tubes. The volume of sediment is read off using graduations on the tube. As a general guideline coffee unstable powders result in sediment volumes ≥ 1.0 ml [53].

The test results are greatly affected by water hardness (calcium and magnesium ions in water) and temperature of the coffee solution into which the powder is added. As the test uses distilled water, results obtained may not accurately represent coffee stability as observed by the consumer. Furthermore different powders, which have the same coffee stability under standard test conditions, can show great variations when tested in coffee made with hard domestic water. The amount of precipitate formed is also extremely sensitive to coffee temperature (above 80°C), and care must be taken to accurately control the coffee temperature prior to milk powder addition.

Precipitated material observed at the surface are called *floaters* and those that sink to the bottom are called *sinkers*. The sinkers sediment easily, and thus can be measured. In contrast the floaters may be difficult to sediment. Furthermore, the test does not record the amount of floaters. Floaters are more of a problem as the consumer sees this type of precipitation first.

Additional test methods involve visual observation and subjective assessment [54,55], and measuring L color value [56].

2. Manufacturing Coffee Stable Milk Powder

During the drying process the powder is agglomerated and lecithinated to improve solubility properties. This will help the powder to dissolve rapidly in coffee and minimize unsightly precipitation (feathering). Agglomeration increases the particle size of the powder, allowing the particles to sink into the coffee, while lecithination coats the particle surface with a wetting agent (lecithin) to reduce surface tension when the powder first contacts the coffee, thus improving wettability.

The preheat treatment that the milk receives has a significant effect on coffee stability. Powders with a medium- or low-heat WPNI are preferred as high-heat treatments tend to produce powders with poor coffee stability [33,57] (Figure 2). Concentrate heating temperature in WMP manufacture is also important as too high or too low a temperature adversely affects coffee stability [33]. Thus, there appears to be an optimum concentrate heating temperature; however, this temperature may vary from plant to plant.

The fat content of the powder has a significant effect on coffee stability. Reduction of the fat content of WMP below the standard 26% (weight of fat/weight of WMP) reduces the amount of feathering [58]. Thus partly skimmed powders (milk fat concentration between 1.5 and 26%) would have better coffee stability and are still likely to provide adequate whitening. Conversely high fat milk powders (>26% fat content) will have poorer coffee

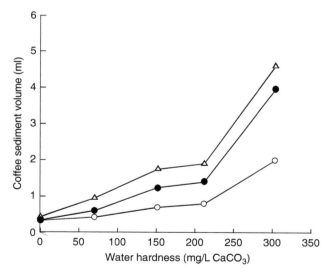

Figure 2   Effect of preheat treatment on the coffee stability of WMP in coffee solutions as a function of water hardness. Preheating conditions for 30 sec at 75°C (o), 94°C (●), and 120°C (△). (From [33]. Reprinted from *Int. Dairy J.*, 10: 659–667, 2000. Copyright 2000, with permission from Elsevier.)

stability. These high fat powders can be reconstituted in warm water before addition to coffee to minimize any feathering problems [56].

3. Formation of Precipitated Material

The combination of hot temperature, low pH, and the high concentration of milk components at the moment of dissolution contribute to the formation of coffee sediment [53,59]. Thus good instant properties (ability to dissolve rapidly in hot coffee solution) and buffering capacity (to raise the pH away from the isoelectric point of the proteins) are critical for WMP. Formation of feathering, if it does occur, happens almost immediately after the powder has been added. The pH of coffee can vary, although it is typically around 4.8 [53]. This is close to the isoelectric point of casein proteins (pH 4.6), which would cause the casein to precipitate. Dissolution of the powder upon mixing allows the buffering action of the powder constituents to raise the pH of the final coffee/milk powder solution to a pH of around 6.2. Interestingly, rapid dissolution (high speed mechanical mixing) greatly reduces the level of coffee sediment of WMP with poor coffee stability [53].

Fat in addition to protein is found in the precipitated material. During homogenization casein micelles are adsorbed onto the fat globule surface and thus stabilize the newly created surface area. However, in hot acidic conditions, casein micelles at the fat globule's surface start to aggregate, combining fat globules into the precipitate. Electron microscopy of coffee sediment shows protein–fat clusters, where the casein micelles have spread and fused with adjacent micelles attached to other fat globules, forming a protein matrix with embedded fat globules [60] (Figure 3). Liquid coffee creamers have a similar mechanism for feathering [55].

4. Effect of Water on Coffee Stability of Milk Powders

The performance of coffee whitener powders will vary greatly depending on the hardness of the water (calcium and magnesium ions) used to make the coffee. Increased water hardness

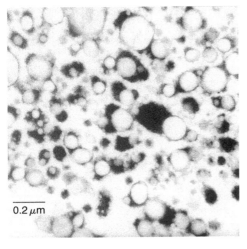

0.2 μm

Figure 3 Transmission electron micrograph of coffee sediment produced from whole milk powders. (From [60]. Reprinted from *Scanning* 21: 305–315, 1999. Copyright 1999. With permission from the Foundation for Advances in Medicine and Science, Inc.)

causes greater amounts of feathering [33,53]. Calcium ions promote protein aggregation reactions through the formation of salt bridges and reduce the electrostatic repulsion between proteins, leading to more protein aggregation.

5. Effect of Compounds in Coffee

Coffee, tea, and cocoa all contain phenolic compounds, which have been shown to improve heat stability (at 140 and 120°C) of milk and concentrated milk [61,62]. The effect of phenolic compounds on whitener powders is likely to be small, as coffee temperatures are lower. However, when milk is used in UHT and retort sterilized coffee beverages the phenolic compounds in coffee will have a stabilizing effect.

Although the mechanism is to be fully explained, O'Connell and Fox [62] have suggested that chelation of the calcium is partly responsible for improving the heat stability of milk. Interactions between the milk proteins and the phenolic compounds are also likely to be important.

The ability of phenolic compounds to stabilize milk is highly dependent on pH [61]. Thus phenolic compounds are only likely to improve heat stability when stabilizers such as potassium and sodium phosphate are added. At pH above 6.8 (pH of imitation whiteners and coffee solution and retort sterilized coffee beverages >7) there is a large improvement in heat stability. In contrast, below pH 6.8 (pH of WMPs and coffee solution <6.4) there is probably little or no improvement.

6. Additional Uses of Milk Powders in Beverages

Milk powders can be used as an ingredient in the manufacture of beverages ranging from coffee to carbonated drinks. Typically skim milk powder (SMP, also known as nonfat dry milk NDM) is used, as it has a low fat content and long storage life for convenience of

Table 6   Milk Powder Used in Different Beverages

Beverages type	Description
Yogurt drinks	A combination of SMP, WPC, culture, lactose, and sweetener can be used
Flavored milk drinks	Popular flavors are chocolate and strawberry, additional stabilizers and hydrocolloids added to prevent sedimentation of cocoa component and modify texture for creating milk shakes
Carbonated drinks	SMP base with flavors, sweeteners, and carbon dioxide
Fruit beverages and smoothies	Milk ingredients and fruit components (fruit juice, fruit pulp, or syrups) are combined together; additional sweeteners or emulsifiers can be added
Dry mixes containing milk powder as an ingredient	Dry ingredients may include: tea, coffee, or chocolate; milk powder and other ingredients, for example, flavors, sugar. Sold either as consumer packs or for use in vending and catering Dietary supplements and meal replacers containing SMP

use, while also providing whitening ability. Furthermore, milk powders also provide the important functional properties of nutritional fortification, solubility, emulsification, flavor, thickening, and body (viscosity modification). The nutritious and fresh/healthy aspect of dairy ingredients in combination with fruit flavors, carbonation, and textures is present in a wide range of beverages (Table 6). Dairy-based beverages have been reviewed by Ernest Mann [63,64].

Additional fortification of nutrients often includes vitamins A and D when SMP is used, as SMP contains only minute quantities of the fat-soluble vitamins.

Milk powders contain both casein and whey proteins. Low pH drinks (fruit/coffee/carbonated) are a challenge for casein as they precipitate out at pH 4.6. Acidic drinks require stabilizers to either maintain the pH above 4.6 during any thermal processing (UHT, sterilization), or to interact with the casein and prevent them from aggregating. Adjustment of the pH can be made by the addition of stabilizing salts, for example, di-potassium phosphate. Milk proteins can also be stabilized by pectin and carboxymethyl cellulose (CMC). Pectin prevents casein aggregation below pH 5.0 by adsorbing to the casein micelle, providing steric stabilization [65].

In contrast to casein, undenatured whey proteins have high solubility in acidic conditions, which is probably related to their biological function and their ability to resist digestion. Whey powders (whey protein concentrate, whey protein isolate) can be used in combination with SMP or instead of SMP. The high concentration of nutritious whey protein is desirable for sports and health drinks.

REFERENCES

1. Codex Alimentarius, FAO/WHO Food Standards. Codex standard for milk powders and cream powders. Codex Standard 207-1999. [Online] http://www.codexalimentarius.net/ 2002.
2. C.W. Holmes, I.M. Brookes, D.J. Garrick, D.D.S. MacKenzie, T.J. Parkinson, and G.F. Wilson. *Milk Production from Pasture*. New Zealand: Massey University, 2002, pp. 337–368.
3. P. Jelen. Standardization of fat and protein content. In: H. Roginski, J.W. Fuquay, and P.F. Fox, Eds., *Encyclopedia of Dairy Sciences*. London: Elsevier Science Ltd, 2003, pp. 2549–2553.
4. J.J. Kennelly. The fatty acid composition of milk fat as influenced by feeding oilseeds. In: IDF Bulletin 366. Brussels: International Dairy Federation, 2001, pp. 28–36.
5. C. Stanton, M. Coakley, J.J. Murphy, G.F. Fitzgerald, R. Devery, and P.P. Ross. Development of dairy-based functional foods. *Sciences des Aliments* 22: 439–447, 2002.
6. M. Pfeuffer and J. Schrezenmeir. Sphingolipids: metabolism and implications for health. In: IDF Bulletin No. 363. Brussels: International Dairy Federation, 2001, pp. 46–50.
7. M.A.J.S. van Boekel and P. Walstra. Effect of heat treatment on chemical and physical changes to milkfat globules. In: P.F. Fox *Heat-Induced Changes in Milk*, 2nd ed. Brussels: International Dairy Federation, 1995, pp. 51–65.
8. W.N. Eigel, J.E. Butler, C.A. Ernstrom, H.M. Farrell, V.R. Harwalker, R. Jenness, and R.M.C.L Whitney. Nomenclature of proteins of cow's milk: fifth revision. *J. Dairy Sci.* 67: 1599–1631, 1984.
9. D.M. Mulvihill and M. Donovan. Whey proteins and their thermal denaturation: a review. *Ir. J. Food Sci. Technol.* 11: 43–75, 1987.
10. H.E. Swaisgood. Chemistry of caseins. In: P.F. Fox and P.L.H. McSweeney, Eds., *Advanced Dairy Chemistry-1: Proteins*, 3rd ed. New York: Kluwer Academic/Plenum Publishers, 2003, pp. 139–201.
11. C.G. de Kruif and C. Holt. Casein micelle structure, functions and interactions. In: P.F. Fox and P.L.H. McSweeney, Eds., *Advanced Dairy Chemistry-1: Proteins*, 3rd ed. New York: Kluwer Academic/Plenum Publishers, 2003, pp. 233–276.

12. H. Singh and R.J. Bennett. Milk and milk processing. In: R.K. Robinson, Ed., *Dairy Microbiology Handbook*. 3rd ed. New York: John Wiley and Sons Inc., 2002, pp. 1–38.

13. T.C.A. McGann, W.J. Donnelly, R.D. Kearney, and W. Buchheim. Composition and size distribution of bovine casein micelles. *Biochim. Biophys. Acta* 630: 261–270, 1980.

14. T.A.J. Payens. Association of caseins and their possible relation to structure of the casein micelle. *J. Dairy Sci.* 49: 1317–1324, 1966.

15. D.G. Schmidt. Association of casein and casein micelle structure. In: P.F. Fox, Ed., *Advanced Dairy Chemistry-1: Proteins*. London: Elsevier Applied Science, 1982, pp. 61–86.

16. P. Walstra and R. Jenness. *Dairy Chemistry and Physics*. New York: John Wiley and Sons, 1984.

17. C. Holt. Structure and stability of bovine casein micelles. *Adv. Protein Chem.* 43: 63–151, 1992.

18. H. Singh and L.K. Creamer. Heat stability of milk. In: P.F. Fox, Ed., *Advanced Dairy Chemistry-1: Proteins*, 2nd ed. London: Elsevier Applied Science, 1992, pp. 621–656.

19. C. Schorsch, D.K. Wilkins, M.G. Jones, and I.T. Norton. Gelation of casein–whey mixtures: effects of heating whey proteins alone or in the presence of casein micelles. *J. Dairy Res.* 68: 471–481, 2001.

20. H. Meisel. Overview of milk protein-derived peptides. *Int. Dairy J.* 8: 363–373, 1998.

21. H. Korhonen and A. Pihlanto-Leppälä. Milk protein-derived peptides: novel opportunities for health promotion. In: IDF Bulletin 363. Brussels: International Dairy Federation, 2001, pp. 17–26.

22. J. Pisecky. *Handbook of Milk Powder Manufacture*. Copenhagen, Denmark: Niro N/S. 1997.

23. M.A. Augustin. Mineral salts and their effect on milk functionality. *Aust. J. Dairy Technol.* 55: 61–64, 2000.

24. P.F. Fox and P.L.H. McSweeney. *Dairy Chemistry and Biochemistry*. London: Blackie Academic and Professional, 1998.

25. J.J. Higgens, R.D. Lynn, J.F. Smith, and K.R. Marshall. Protein standardization of milk products. In: IDF Bulletin 304. Brussels: International Dairy Federation, 1995, pp. 26–40.

26. H. Singh and D.F. Newstead. Aspects of proteins in milk powder manufacture. In: P.F. Fox, Ed., *Advanced Dairy Chemistry-1: Proteins*, 2nd ed. London: Elsevier Applied Science, 1992, pp. 735–765.

27. D.F. Newstead. Effect of protein and salt concentration on the heat stability of evaporated milk. *NZ J. Dairy Sci. Technol.* 12: 171–175, 1977.

28. W. Rattray and P. Jelen. Thermal stability of skim milk with protein content standardized by the addition of ultrafiltration permeate. *Int. Dairy J.* 6: 157–170, 1996.

29. W. Rattray and P. Jelen. Protein standardization of milk and dairy products. *Trends Food Sci. Technol.* 7: 227–234, 1996.

30. A.J. Baldwin and J.D. Ackland. Effect of preheat treatment and storage on the properties of whole milk powder. Changes in physical and chemical properties. *Neth. Milk Dairy J.* 45: 169–181, 1991.

31. S.G. Anema. Effect of milk concentration on the irreversible thermal denaturation and disulfide aggregation of β-lactoglobulin. *J. Agric. Food Chem.* 48: 4168–4175, 2000.

32. A.J. Baldwin, A.G. Baucke, and W.B. Sanderson. The effect of concentrate viscosity on the properties of spray dried skim milk powder. *NZ J. Dairy Sci. Technol.* 15: 289–297, 1980.

33. D.J. Oldfield, C.M. Teehan, and P.M. Kelly. The effect of preheat treatment and other process parameters on the coffee stability of instant whole milk powder. *Int. Dairy J.* 10: 659–667, 2000.

34. F. Guyomarc'h, F. Warin, D.D. Muir, and J. Leaver. Lactosylation of milk proteins during the manufacture and storage of skim milk powders. *Int. Dairy J.* 10: 863–872, 2000.

35. M.A. Augustin. Dairy ingredients in chocolate: chemistry and ingredient interactions. *Food Aust.* 53: 389–391, 2001.

36. P.J.J.N. van Mil and J.A. Jans. Storage stability of whole milk powder: effects of process and storage conditions on product properties. *Neth. Milk Dairy J.* 45: 145–167, 1991.

37. H. Stapelfeldt, B.R. Nielsen, and L.H. Skibsted. Effect of heat treatment, water activity and storage temperature on the oxidative stability of whole milk powder. *Int. Dairy J.* 7: 331–339, 1997.

38. F.G. Kieseker and P.T. Clarke. The effect of storage on the properties of non-fat milk powders. *Aust. J. Dairy Technol.* 39: 74–77, 1984.

39. A.H. Jana and P.N. Thakar. Recombined milk cheese: a review. *Aust. J. Dairy Technol.* 51: 33–43, 1996.

40. C.H. McCrae and D.D. Muir. Heat stability of milk. In: P.F. Fox *Heat-Induced Changes in Milk*, 2nd ed. Brussels: International Dairy Federation, 1995, pp. 206–230.

41. H. Singh, L.K. Creamer, and D.F. Newstead. Heat stability of concentrated milk. In: *Heat-Induced Changes in Milk*, 2nd ed. Brussels: International Dairy Federation, 1995, pp. 256–278.

42. T.D. Davies and J.C.D. White. The stability of milk protein to heat: I. Subjective measurement of heat stability of milk. *J. Dairy Res.* 33: 67–81, 1966.

43. P.M. Kelly. Effect of some non-protein nitrogen components on the heat stability of skim-milk powders. *Ir. J. Food Sci. Technol.* 1: 129–135, 1977.

44. R.P.W. Williams. The relationship between the composition of milk and the properties of bulk milk products. *Aust. J. Dairy Technol.* 57: 30–44, 2002.

45. F.G. Kieseker and B. Aitken. An objective method for determination of heat stability of milk powders. *Aust. J. Dairy Technol.* 43: 26–31, 1988.

46. D.F. Newstead. Sweet-cream buttermilk powders: key functional ingredients for recombined milk products. In: *3rd International Symposium on Recombined Milk and Milk Products.* Brussels: International Dairy Federation, 1999, pp. 55–60.

47. D.D. Muir and A.W.M. Sweetsur. The influence of naturally occurring levels of urea on the heat stability of bulk milk. *J. Dairy Res.* 43: 495–499, 1976.

48. P.M. Kelly, A.M. O'Keefe, M.K. Keogh, and J.A. Phelan. Studies of milk composition and its relationship to some processing criteria: III. Seasonal variation in heat stability of milk. *Ir. J. Food Sci. Technol.* 6: 29–38, 1982.

49. H. Singh and L.K. Creamer. Denaturation, aggregation and the stability of milk protein during the manufacture of skim milk powder. *J. Dairy Res.* 58: 269–283, 1991.

50. M.A. Augustin, P.T. Clarke, and T. Greenwood. The heat stability of recombined evaporated milk made from skim milk powder produced in late autumn and winter. *Aust. J. Dairy Technol.* 45: 47–49, 1990.

51. H. Singh and R.P. Tokely. Effect of preheat treatments and buttermilk addition on the seasonal variations in the heat stability of recombined evaporated milk and reconstituted concentrated milk. *Aust. J. Dairy Technol.* 45: 10–16, 1990.

52. F.G. Kieseker. Recombined evaporated milk review paper. In: IDF Bulletin 142. Brussels: International Dairy Federation, 1982, pp. 79–90.

53. C.M. Teehan, P.M. Kelly, R. Devery, and A. O'Toole. Evaluation of test conditions during the measurement of coffee stability of instant whole milk powder. *Int. J. Dairy Technol.* 50: 113–121, 1997.

54. I. Haugaard Sorensen, J. Krag, J. Pisecky and V. Westergaard *Analytical Methods for Dry Milk Products*, 4th ed. Copenhagen, Denmark: A/S. Niro atomizer, 1978.

55. S. Geyer and H.G. Kessler. Effect of manufacturing methods on the stability to feathering of homogenized UHT coffee cream. *Milchwissenschaft* 44: 423–427, 1989.

56. T.J. Gruetzmacher and R.L. Bradley. Acid whey as a replacement for sodium caseinate in spray-dried coffee whiteners. *J. Dairy Sci.* 74: 2838–2849, 1991.

57. A.W.M. Sweetsur. The stability of instantized skimmed milk powder to hot coffee. *J. Soc. Dairy Technol.* 29: 157–160, 1976.

58. P.M. Kelly. Coffee stability of agglomerated wholemilk powder and other dairy creamer emulsions. [Online] http://www.ucc.ie/ucc/faculties/foodsci/Ncfrp/NCFRPfinalreport52.htm 2002.

59. I.R. McKinnon, E.K. Jackson, and L. Fitzpatrick. Instant whole milk powder micelles: stability in instant coffee. *Aust. J. Dairy Technol.* 55: 88, 2000.

60. A.B. McKenna, R.J. Lloyd, P.A. Munro, and H. Singh. Microstructure of whole milk powder and of insolubles detected by powder functional testing. *Scanning* 21: 305–315, 1999.

61. J.E. O'Connell, P.D. Fox, R. Tan-Kintia, and P.F. Fox. Effects of tea, coffee and cocoa extracts on the colloidal stability of milk and concentrated milk. *Int. Dairy J.* 8: 689–693, 1998.

62. J.E. O'Connell and P.F. Fox. Effects of phenolic compounds on the heat stability of milk and concentrated milk. *J. Dairy Res.* 66: 399–407, 1999.

63. E. Mann. Dairy beverages. *Dairy Ind. Int.* 63: 17–18, 1998.

64. E. Mann. Dairy beverages. *Dairy Ind. Int.* 66: 17–18, 2001.

65. R. Tuinier, C. Rolin, and C.G. de Kruif. Electrosorption of pectin onto casein micelles. *Biomacromolecules* 3: 632–638, 2002.

66. T.D. Davies and A.J.R. Law. The content and composition of protein in creamery milks in south-west Scotland. *J. Dairy Res.* 47: 83–90, 1980.

67. J.A. Phelan, A.M. O'Keefe, M.K. Keogh, and P.M. Kelly. Studies of milk composition and its relationship to some processing criteria: 1. Seasonal changes in the composition of Irish milk. *Ir. J. Food Sci. Technol.* 6: 1–11, 1982.

68. M.J. Auldist, B.J. Walsh, and N.A. Thomson. Seasonal and lactational influences on bovine milk composition in New Zealand. *J. Dairy Res.* 65: 401–411, 1998.

69. S. Tamminga. Effects of feed, feed composition and feed strategy on fat content and fatty acid composition in milk. In: IDF Bulletin 366. Brussels: International Dairy Federation, 2001, pp. 15–27.

70. H. Itabashi and L. Baevre. Trends in milk composition: results of a survey by IDF (Questionnaire 1993/A). In: IDF Bulletin 366. Brussels: International Dairy Federation, 2001, pp. 9–14.

15
Milk Powder

Alan Baldwin and David Pearce
Fonterra Research Centre
Palmerston North, New Zealand

CONTENTS

I. INTRODUCTION

The purpose of this chapter is to illustrate the wide variety of dairy powders manufactured, describe the principal manufacturing steps, comment on processing difficulties and show examples of how these powders are used. In addition, we list typical compositions and physical properties for these powders. The chapter is particularly directed to those who wish to utilize dairy powders in typical applications.

The information is drawn from the experience of those in the New Zealand dairy industry, supported by references to the literature where appropriate. There is a number of key texts on the science and technology of condensed products and milk powder [1–7]. Dairy powders are produced primarily on spray dryers and the reader is directed to the technology chapters in the earlier part of this book for more details on equipment design and operation.

A. Milk Composition

Milk contains proteins, fats, carbohydrates, and many of the vitamins and trace minerals essential for our daily nutrition. It has been consumed in fresh form since humans have been able to domesticate ungulates.[1] Milk for human consumption is sourced mainly from cows (in temperate climates) and buffaloes (in tropical climates, especially India). There is some use of sheep milk and goat milk, although this is largely for cheese manufacture.

Whole milk has several major components that exist either in solution or in colloidal suspension. The main milk carbohydrate, lactose, exists in milk in solution. Most of the minerals are in solution but a proportion of the calcium, magnesium, and phosphate is associated with the casein protein.

The Fonterra Research Centre was formerly known as the New Zealand Dairy Research Institute (NZDRI).
Notice of Disclaimer of Liability: This material is intended to provide accurate and professional information relating to the subject matters covered within. It has been compiled and written, and is made available, on the basis that the Fonterra Research Centre Limited and its employees are fully excluded from any liability to any person or entity for any damages whatsoever in respect of or arising out of any reliance in part or full, by such person or entity or by any other person or entity, upon any of the contents of this material for any purposes whatsoever.
[1] Hoofed mammals.

There are two major classes of protein, casein proteins and whey proteins, as well as a number of minor protein components (immunoglobulins, proteose peptones) and nonprotein nitrogenous compounds. The whey proteins exist in solution and, in some cases, may form dimers or higher associations. With heat treatment of the milk, the whey proteins change their structure and may also associate with casein proteins [8]. This heat treatment is given to the milk before evaporation and is very important in determining the performance of milk powders in applications such as recombined evaporated milk (REM) and recombined sweetened condensed milk (RSCM).

The caseins exist as a colloidal dispersion of aggregated casein molecules known as micelles. In the native state, the average micelle diameter[2] is of the order of 200 nm. The way in which the micelle is structured and held together is the subject of considerable debate — two views of micelle structure have been reviewed recently [9,10]. The fat also exists as a colloidal dispersion (an emulsion of fat globules). The average size of the milk fat globules in raw milk is about 3 μm (approximate range is from 0.2 to 6 μm). The fat globule size will be reduced by fluid handling in pumps and pipes or intentionally by homogenization. The milk fat globules have a membrane of surface-active lipoproteins, 50% of which are phospholipids. With the reduction in particle size, other materials, particularly the caseins and whey proteins — which are themselves effective surfactants — form part of the new surface formed by size reduction.

The composition and structure of milk is a complex subject and readers are referred to chemistry texts (e.g., [11,12]). Information on the important role of milk proteins in milk powder manufacture and product properties is also available [3].

II. SCIENTIFIC PRINCIPLES

The dry-milk-products industry has had more than a century of development. It is based on the following principles that have been gradually elucidated.

A. Principles

1. Moisture Removal

Spray drying is based on the formation of a cloud of small droplets by atomization (covered in detail in an earlier chapter of this book). This allows the relatively rapid transfer of heat into the particles and moisture out of the particles. In this way, drying to acceptable moisture content can be achieved in a time of the order of 10 sec.

2. Solubility

In making milk powders, the objective is to dry the material so that, when the powder is rehydrated, the structure prior to dehydration is reestablished. The casein proteins are sensitive to heat during drying and the drying process has evolved to minimize the degree to which the proteins are changed.

The key to a soluble milk powder is minimization of the temperature and the residence time in an intermediate moisture zone between concentrate and powder. The moisture regions at either end of the spectrum are much more stable and can withstand higher temperatures than when the product is between these relatively stable regions of moisture [13].

---

[2]The numerical value estimated depends on the definition of average diameter and the method of measuring the diameter, for example, transmission electron microscopy or light scattering.

When a wetted surface is in contact with unsaturated air flowing past, the surface is at the "wet bulb" temperature, which is much less than the temperature of the air with which it is in contact. Evaporative cooling greatly assists in reducing heat effects on the product during drying.

The steps taken to minimize heat effects are as follows:

1. The concentrate is formed into a thin sheet (roller drying) or atomized to droplets. This allows the product to dry in a relatively short time (some tens of seconds).
2. The product temperature is kept as low as possible (however, see the section on Stickiness).
3. In spray drying, the dryer is run cocurrent instead of countercurrent, even though the latter would result in greater thermal efficiency[3] [14].
4. The last stage of drying is carried out in a secondary dryer, resulting in lower air temperatures in the spray drying chamber.

3. Stickiness

Dairy products, when moist and/or warm, are sticky to a greater or lesser degree depending on the composition. Contributors to stickiness are lactose, fat and, to a lesser extent, protein.

There is a relationship between the relative humidity (RH) of the drying air in contact with the droplet, the moisture on the particle surface, and the resultant stickiness of the product. A consequence is that, given a certain composition, the temperature in the drying chamber must be increased to reduce the RH and hence the stickiness.

Spray dryers for milk products are designed to prevent the droplets hitting a surface until the material is relatively dry and not sticky. Effective atomization to reduce the drying time is a key parameter. In addition, the diameter of the chamber is selected to minimize contact between partially dried, sticky particles and the wall. Thus, dryers with disk atomization are usually of wide diameter and low height — squat-form dryers.

4. Particle Size Enlargement

The powder so dried may be too fine for some applications, particularly consumer reconstitution. The initial solution to this was to agglomerate powder in a separate operation. This can still be done, but it is more common to produce the agglomerated powder in the spray dryer, in a straight-through process. This is done by returning fines to the atomization zone in the dryer to agglomerate with the droplets. Careful design is required to ensure that the returned fines hit the droplets at a moisture content that ensures agglomeration. For reasons of fire and explosion safety, fines are returned in the same direction as the incoming hot air.

III.   DEVELOPMENT OF TECHNOLOGY/HISTORY

A.   Beginnings

Two different forms of drying were utilized in the early stages of development — roller drying and spray drying [15]. A number of different designs of both processes were developed; eventually, the spray dryer proved to give a better product and to be more readily scaled-up. Today, dairy powders are almost universally made by spray drying, but roller drying is still

---

[3]Countercurrent operation results in lower air temperatures when the product is vulnerable to heat.

commonly used for speciality products such as whole milk powder (WMP) destined for the manufacture of chocolate.

The first milk powders manufactured for consumers were those made by roller drying. A primary market provided milk solids for feeding infants. The history of an early international milk powder company, Glaxo,[4] that exported roller-dried baby foods from New Zealand to the United Kingdom in the early 1900s is well documented [16]. The following description links dryer developments to the quality of the milk powder and the development of the scale of the industry.

B. Milk Powder in the 1950s

Many countries have a shortfall between domestic milk production and consumer demand. In the 20th century, concentrated products such as evaporated milk and sweetened condensed milk were exported from Europe to other countries. In the 1950s, these products began to be made from milk powders and anhydrous milk fat. The imported solids were used "as received" or, in some cases, blended with local milk for further processing.

During this period, consumption of powders reconstituted by the consumer also expanded rapidly. Considerable research into milk powder, particularly into WMP, took place during World War II and soon after [1]. In the 1950s, the need to improve the dispersing properties of powders for consumer reconstitution led to the development of agglomerated powders by the "rewet" process [15].

C. Dryer Development

The major steps in the development of the milk drying industry were as given below.

1. Secondary Drying

This started in the early 1950s with pneumatic secondary drying. The adaptation of fluid beds to milk powder drying in the late 1960s took the industry in a new direction. The benefits of this approach were as follows:

- Improvement in powder quality — previously it was difficult to optimize moisture, bulk density, and solubility all together. The lower temperatures in the drying chamber gave more room to manoeuver.
- Increased thermal economy, arising from the opportunity to use higher inlet air temperatures, lower outlet air temperatures, and higher concentrate total solids (TS).
- The opportunity to use the fine particles removed from the fluid bed to produce an agglomerated powder.

The development of practical fines return-agglomeration processes was the key to the success of instant WMP (IWMP) as a consumer product. The "straight-through" process quickly displaced the rewet process because it eliminated double handling and because microbiological risks were managed more easily. IWMP has become a major food product in many countries that lack a dairy production infrastructure to supply large volumes of liquid milk.

---

[4]Glaxo developed into a food and pharmaceutical company and eventually exited the food industry.

2. Chamber Design

The next major dryer development was the incorporation of a fluidized bed into (integral fluid bed) or adjacent to (well-mixed bed) the spray dryer chamber. This allowed increases in the moisture content of the powder on to the bed, a reduction in the droplet drying time and hence a reduction in the size of the chamber. This development led to new dryer chamber designs. The most dramatic was the development of a chamber in which the air flowed down the center of the chamber, reversed flow, and exited from the periphery of the roof (designated the MSD, "multi-stage dryer"). A description of spray dryer configurations for milk powder drying is given elsewhere [6].

3. Membrane Processes

The 1970s saw the introduction of ultrafiltration as a unit operation into the dairy industry. The first applications were to the economical extraction of whey proteins from whey [17], leading to the development and marketing of a range of whey protein concentrates (WPCs). This was followed in the late 1980s and early 1990s by the use of ultrafiltration membranes to increase the protein content of skim milk [18,19]. Milk protein concentrates (MPCs) are now important products in international trade.

D. Scale of Manufacture

The development of road and rail transport of milk has led to amalgamation of dairy companies processing larger volumes of milk at a central site. Initially, multiple dryers in one building were installed to handle the milk, but, with increased confidence in the dryer design, larger capacity dryers were installed. Typical maximum capacities of milk powder dryers in different decades in New Zealand are given in Figure 1.

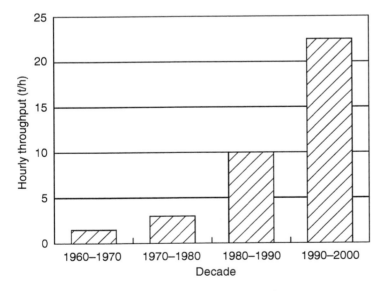

Figure 1 Maximum capacity of milk powder spray dryers.

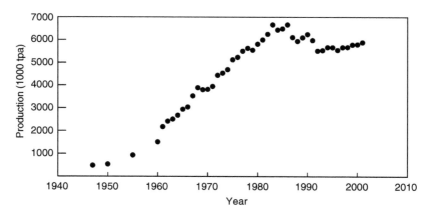

Figure 2   World production of milk powder over 54 years (1947–2001): 1947–1960 — (Extracted from Hall, C.W. and Hedrick, T.I. *Drying of Milk and Milk Products*, 2nd ed. Connecticut: AVI Publishing Co, Westport, 1971. pp. 6, 152–155, 179–181. With permission); 1961 – 2001—(From Food and Agriculture Organisation. 2002. FAOSTAT Agriculture database http: //apps.fao.org/cgi-bin/nph-db.pl?subset=agriculture. With permission.)

E.   Milk Powder in World Trade

The major markets in terms of milk volume are the European Community (EC), North America, and South America. The three principal regions of milk powder production for export trade, (based on 2001 figures) are the EC (25%), Australasia (25%), and South America (20%). Quantities of powder are consumed internally and the quantity traded on the international market is dependent on the production from Australasia and the surpluses, more or less, in the EC and North America. Entry of many dairy products into the EC and North America is subject to tariffs.

Figure 2 shows the dramatic growth in world milk powder production (combined skim milk powder (SMP) and WMP) from the 1960s through to the mid-1980s. Since this time, the production has steadied at around $6,000,000$ tonnes per annum.

IV.   DAIRY PROCESSING

Dairy processing is made up of a series of unit operations designed to produce a useable powder from raw milk. The set of unit operations used is drawn from the following list:

- Skim milk/cream separation.
- Pasteurization.
- Membrane processes.
- Evaporator preheating.[5]
- Enzymatic hydrolysis reactions.

---

[5]In the dairy industry, the term preheating, used in this context, means the heat treatment given to the milk *immediately prior to the first effect*. The first function of preheating is to raise the milk temperature above that of the first effect boiling temperature (thermal economy). The second function is to impart specific properties on the powder. It is defined by temperature (°C) and time (sec), for example, 95°C for 15 sec. See, for example, [5].

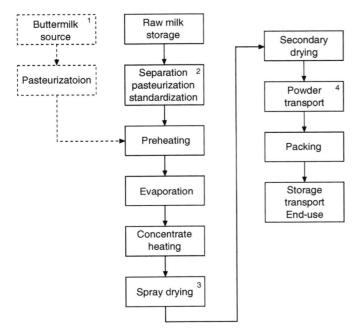

Figure 3 Unit operation sequence for a typical WMP or SMP process. (1) Buttermilk indicated (- - -); (2) the unit operation order may vary from plant to plant; (3) spray drying may or may not include agglomeration; (4) there may be a milling stage here, depending on the dryer/product combination.

- Vitamin and mineral addition
- Evaporation
- Concentrate heating
- Spray drying
- Fluid bed (or secondary) drying
- Packing
- Storage
- Transport

The order broadly follows the milk powder production process, so the actual unit operations and their order depends on the product being made and may be changed to accommodate particular requirements. The first steps in the WMP process exemplify the flexibility possible even for one type of product. Some factories separate, pasteurize, and then standardize.[6] Other factories separate, standardize, and then pasteurize.

The block diagrams (Figures 3–9) illustrate typical processes for seven powder types: milk powder, caseinate, MPC,[7] WPC, whey powder, formulated products (infant formulae, cream powder, cheese powder), and milk protein hydrolyzates. Details of the unit operations have been covered in earlier chapters of this book.

---

[6] Standardizing is the process of mixing several liquid streams together, prior to evaporation, to give a desired liquid composition. For example, fat standardizing is where cream and skim milk streams are mixed to give a whole milk stream with the required fat content.

[7] Numbers after MPC and WPC denote the protein content, for example, MPC56 means 56% protein in the powder. MPI means Milk Protein Isolate.

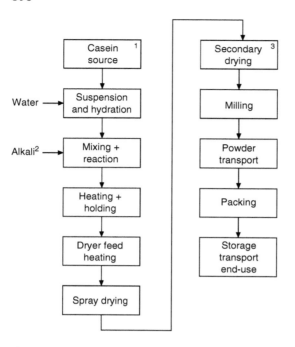

Figure 4   Unit operation sequence for a typical caseinate process. (1) casein may be in either pressed curd or dry powder (granular) form; (2) alkali source (sodium or calcium hydroxide, or any counter ion) will determine caseinate type (sodium or calcium); (3) depends on primary dryer type whether or not there is additional secondary drying.

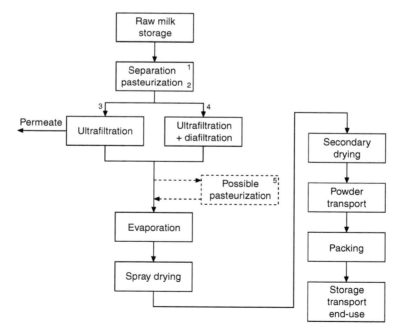

Figure 5   Unit operation sequence for a typical milk protein concentrate (or milk retentate) powder process. Applicable to MPC56, MPC70, MPC85, MPI. (1) An unit operation order away vary from plant to plant; (2) some plants use thermization (63–65°C/15 sec) in place of pasteurization, here; (3) MPC56 route; (4) MPC70, MPC85, MPI route; (5) pasteurization here depends on the required evaporator feed quality.

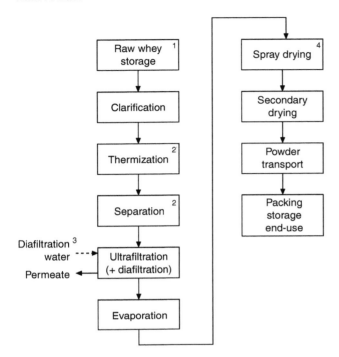

Figure 6 Unit operation sequence for a typical whey protein concentrate (WPC) powder process. (1) whey may be from either suspect or acid sources; (2) cheese whey only; (3) diafiltration for > 70% protein; (4) agglomeration may take place here.

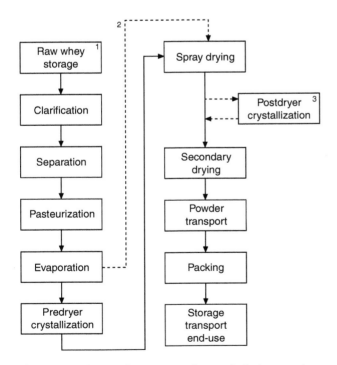

Figure 7 Unit operation sequence for a typical whey powder process. (1) Whey may be from either sweet or acid sources; (2) ordinary whey powder has no pre-crystallization stage; (3) belt process; (4) for a discussion of whey demineralization, see Ref. [21].

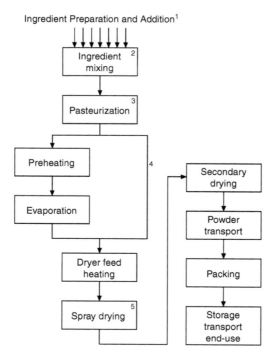

Figure 8 Unit operation sequence for a typical formulated powder process; includes infant formulae, cheese powders and cream powders. (1) ingredients used according to the recipe for the product; (2) missing is usually a batch process, although in-line liquid addition at various points in the process can occur. This will be product dependent; (3) depending on the product, the preheat and pasteurization steps may be combined; (4) process and evaporator feed solids dependent; (5) agglomeration may take place here.

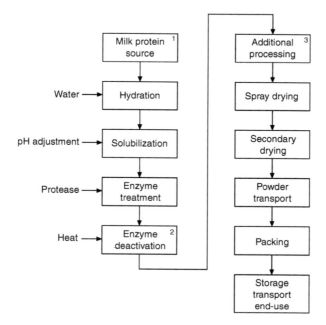

Figure 9 Unit operation sequence for a typical milk protein hydrolyzate process. (1) protein source will depend upon the desired product; (2) this step is not required for immobilized and membrane reactor systems; (3) this stage may include clarification, evaporation, filtration of flavor improvement.

V. TYPICAL PRODUCT COMPOSITIONS

Table 1 lists typical compositions of some major dairy powders. These compositions will vary from supplier to supplier.

VI. TYPICAL PRODUCT OPERATING CONDITIONS

Table 2 lists typical operating conditions for drying dairy powders. No attempt has been made to give a range of conditions for all the variables. The data are plant specific and many plants will be able to operate comfortably at conditions well outside those described here. These conditions should serve as a starting point for plant operation.

VII. PRODUCTION DIFFICULTIES IN DRYING

There are few, serious ongoing production difficulties in making WMP and SMP. The technology is mature, the engineering design is robust, and the equipment supply companies can recommend operating conditions to produce a good product with few problems. These are dealt with below. Product performance is not discussed here.

A. Fires and Explosions

Fires and explosions constitute the most serious threat to operator and plant safety. Fortunately, in a well-run and well-maintained plant, they are rare. For a dust explosion to occur, fuel (milk powder — it is combustible) and an oxidant (oxygen in the air) have to be present and there needs to be an ignition source. Most countries have a set of rules (e.g., [24]) that covers dryer design. All modern dryers have explosion protection as a standard feature.

In the dryer, there is a dust-laden atmosphere. For an explosion to occur, the dust:air ratio has to be in the right range: too dense and there is not enough oxygen to support the fire; too disperse and there is insufficient fuel. There can be spots in the dryer where there is a powder buildup. Some self-ignition data for dairy powders are available elsewhere [25]. Some of these data are presented in Table 3.

This table illustrates that a small powder layer on the surface of a dryer can be very dangerous, especially if it is near the inlet air duct. Layers can occur from several sources including:

- Drying a cohesive product with the wrong conditions.
- Wetting the chamber walls (with concentrate) during start-up.
- Poor atomization or spray/air mixing; this results in semidried particles hitting the walls.
- Leaking nozzles (a dag can form on the tip of a nozzle, which can drop off and pass through the chamber; this is particularly dangerous if the concentrate dries and then self-heats).
- Blockages in cyclones.
- Poor washing, leaving deposits.
- Injudicious placement or adjustment of the fines return lines.

Most of the prevention efforts should be in removing ignition sources from the vicinity of the drying chamber. Potential ignition sources can include a contaminated air supply (e.g., heated dust entering the chamber), friction (e.g., rubbing mechanical parts), static

Table 1 Typical Compositions of Some Dairy Powders

Product	WMP	SMP	BMP	Cream powder[a] CP55	CP70	Sodium caseinate	Cheese powder[b]	Cheese powder[c]	MPC56	MPC70	MPC85	Whey powder	WPC34	WPC80
Milkfat	26.8[d]	0.8	7.8	54.8	71.0	<1	32.0	42.0	1.3	1.4	1.0	0.9	2.7	5.7
Protein	25.0	34.0	31.0	15.6	12.0	92	19.6	19.1	57.1	70.0	85.0	13.0	35.0	80.8
Lactose	39.1	53.5	50.3	24.3	12.8	<0.3	35.1	28.3	30.1	17.0	3.0	74.0	51.3	6.5
Minerals (ash)	5.8	7.9	7.4	3.5	2.5	3.6	11.0	8.3	7.7	7.2	7.0	7.7	7.0	3.0
Moisture	3.3[e]	3.8	3.5	1.8	1.76	4.0	2.3	2.3	3.8	4.4	4.0	4.4	4.0	4.0

[a] Cream powder (CP) is a dairy powder with a fat content in the 35 to 80% range. The number denotes the fat content. WMP and CP70 contain 0.2% soy phospholipids.
[b] Snack powder.
[c] Powder for sauce applications.
[d] The fat content will vary depending on product requirements.
[e] A lower moisture content is recommended if the WMP is being stored in a hot climate.

Table 2 Typical Operating Conditions for Dairy Powders

Product	WMP	SMP	IWMP	ISMP	BMP	Whey powder	CP70	Cheese powder[a]	MPC70[b]	Na Cas[c]	WPC[d]
Feed solids (%)	12.5	9	12.5	12.5	9	13	—	—	20	—	10–20
Preheat T(°C)/t(sec)	80/15–120/180	80/15–120/180	80/15–95/30	80/15–95/30	80/15–120/180	75/0	—	—	—	—	—
Dryer feed solids (%)	49	48	49	48	47	60	50	—[a]	—	20–22	30–40
Concentrate heating (°C)	65	65	75	70	65	50	75	70–80	25	130	50
Dryer inlet air temperature (°C)	190	200	180	200	200	195	180	175	200	230	190
Secondary air drying temperature (°C)	100	100	100	100	100	90	50	90	—	100	65

[a] Highly formulation dependent.
[b] Sources include MPC 49–MPC 65 [18]; MPC 70 [22] and general overviews of MPC productions [19]. Most of these studies were performed on laboratory-scale plants and few of these references have all the conditions of interest. The above operating conditions were assembled from the following references: Whey powders [23]; other powders [6, 14].
[c] Sodium caseinate.
[d] Input solids are product dependent.

Table 3 Thickness of Layer Resulting in Self-Ignition at Two
Air Temperatures (200 and 100°C)

Product	Minimum thickness at 200°C air temperature (mm)	Minimum thickness at 100°C air temperature (mm)
SMP	17	170
WMP	10	170
BMP	8	100
Whey powder	13	320

discharge, electrical failure, impact sparks, and external heat. A sign that all is not well may be the appearance of scorched particles in the powder. (In one incident in New Zealand, scorched particles were seen in the powder shortly before an explosion occurred.)

Most of these potential ignition sources are removed from consideration by good manufacturing practice, which includes regular maintenance, plant checks etc., and some procedure for ensuring that the tasks are indeed carried out.

B. Stickiness

Most dairy powders are sticky, given the right conditions. This comes from the presence of lactose and fat in the powder, and the "right" combination of temperature, RH in the surrounding atmosphere, and water activity in the product. In spray dryers, it is the outlet air conditions (temperature and RH) that may give rise to powder stickiness. Unless otherwise stated, we are referring to spray dryer operation in this discussion.

The sticky point itself is hard to define as it means different things in different situations. For example, the stickiness observed during storage, leading to caking, is an equilibrium process, but the stickiness seen during spray dryer operation is not. In both examples, viscous flow is involved, but the timescales are vastly different and the definition of a sticky point will be different.

1. Stickiness in Drying

The glass transition curve defines the boundary between an immobile "glassy" state and a less viscous "rubbery" state of an amorphous material. The temperature region where this transition occurs is very important to food processing, particularly in relation to stickiness. The presence of a solvent (water in many food systems) acts as a plasticizer and dramatically influences the glass transition temperature (T_g) [26]. For example with lactose, when the RH increases from 20 to 30%, the lactose T_g decreases from approximately 47 to 34°C. There has been much investigation into the role of lactose in milk powder stickiness in both dryer operation and storage (see, e.g. [26–30]). Generally, carbohydrate T_g is recognized as a very useful parameter for predicting powder stickiness [26] and, the lower the T_g, the more sticky the powder will be. Preventing dairy powders from becoming sticky usually means running the dryer at a "safe" RH in the outlet air stream. As a first approximation, the sticky point of powders with 40 to 50% lactose (w/w) (e.g., SMP) will occur at about 25°C above the lactose T_g at the same RH. This point will vary depending on the exact composition and the dryer type and will need to be confirmed during plant operation. If the powder is sticky, the outlet RH needs to be lowered. Production demands can mean that this is usually achieved by increasing the dryer outlet air temperature. Dehumidification of the drying air is possible, but, except in the tropics, it is usually considered to be too expensive. A common first sign of a sticky problem can be a cyclone blockage.

Glass transition temperatures generally decrease with decreasing sugar molecular weight [31]. Hence, a milk powder in which the lactose has been hydrolyzed (to glucose and galactose) will be stickier when dried at a given set of outlet conditions than standard milk powders.

In some instances, it may be desirable to have powder sticking to the dryer walls. Such an example is the production of precrystallized whey powder. In the Stork (now known as CPS Powder) process [32], a dryer and a postcrystallization belt are used. A layer of powder is allowed to build up on the dryer walls. With the correct cone angle, the powder buildup is restricted and it sloughs off periodically, collecting on the belt for further development of crystalline lactose.

For fat-containing powders, the presence of liquid fat is thought to contribute to the stickiness during processing. Fat seems to preferentially appear on the surface of many powders [33,34]. If the air temperature is above the melting point (about 40°C), all the fat will be liquid. In the dryer, this causes smearing of the surfaces where there is a high velocity, dust-laden air stream. Examples include entry points to exhaust ducts from the main chamber, and particularly the cyclones. Introducing cool air to the air stream prior to the cyclones may help [14]. The trade-off between a high outlet air temperature to effect drying and a low outlet air temperature to avoid smearing is very difficult to achieve. In practice, some dryer fouling is accepted and short runs are standard. The storage of high fat powders is usually not a problem, except where the storage temperature is too high (above about 40°C) and/or if the pallets are stacked too high. In the latter case, the compressive load can cause caking.

For some very sticky products, a spray dryer will not be suitable and an integrated belt dryer may be the best choice [6].

2. Stickiness in Storage

Under normal conditions of water activity (A_w) and temperature, stickiness and the subsequent caking of milk powders are not usually a problem. However, if the powder is exposed to a warm, humid environment (e.g., repeated opening and closing of a consumer can in the tropics), or the original powder moisture is too high, the chances of caking are dramatically increased [28].

C. Losses

Losses from milk powder plants come from three major sources: start-up and shut-down, fouling, and dryer exhaust stack losses.

Start-up and shut-down losses come from low solids concentrate that is not processed further and from lower than optimal solids concentrate being atomized when the plant switches from water to product (producing very fine particles, which may be lost from the cyclones).

Fouling occurs when the product adheres to a surface. It is usually undesirable and needs to be removed periodically to facilitate good plant operation. In evaporation and drying plants, fouling is commonly associated with heating. The foulant is removed by chemical cleaning (clean-in-place or CIP) every 10 to 40 h from the evaporator (depending on the plant). Several workers [35–37] have examined milk fouling and present details about fouling mechanisms. Generally, one hopes to run the evaporator for as long as possible before cleaning. Dryers are much less prone to fouling and can run for many weeks on the same product before a clean is required. In fact, in some new plant installations, several matched evaporators feed a single dryer. In this way, production can be maintained even when one evaporator is being cleaned. This is economically most attractive on a large scale (>14 ton/h). Many mechanisms of heat-induced fouling have been hypothesized, and optimal cleaning regimes have been considered.[8]

Common parts of the plant where fouling occurs include the following:

1. Wet areas
   - Preheat area, especially direct steam injection (DSI) units
   - Tubes (often from a poorly designed or deformed distribution plate)
2. Dry areas
   - Dryer cone.
   - Cyclones.
   - Nozzles (leaks).
   - Dryer ceiling.
   - Internal beds (usually on poor start-up or shut-down).
   - Vibrating beds (normally OK, but, if lecithin is accidentally added as an aerosol, not a spray, fouling can occur).

Poor evaporator tube wetting and/or low milk pH (<6.4) can contribute to evaporator and preheat fouling.

Computational fluid dynamics (CFD) modeling of the dryer operation can suggest ways to change airflows (rotation, velocity etc.) to minimize dryer wall deposits [38].

Liquid atomization results in a wide particle size distribution (seen in the final powder). Stack losses of milk powder fines can be minimized by good equipment design. Cyclones and bag filters have particle size cut-off points (above which, the material is assumed to be collected). They are also designed to cope with a particular range of gas loading and flow [14]; when the conditions change, the gas cleaning equipment will not operate optimally and unacceptable losses may occur. To meet strict environmental requirements, it is common to have a baghouse (filter system) as well as cyclones. Indeed the newest dryer installations have washable baghouses in place of cyclones. Certainly, the addition of a baghouse all but eliminates stack emissions. The New Zealand guidelines are not more than 130 mg/m^3 exhaust gas, measured at the stack and corrected to 0°C [39]. It is also common to have a deposition guideline at the site boundary.

Stack losses usually increase as the concentration of the concentrate decreases (and the primary particle size decreases) [40]. In the case of MPCs and other protein products, the concentration to the dryer rarely exceeds 40%. This tends to give a powder with very small particles, and prone to dustiness (and losses). Raising the solids, where possible, will go some way to alleviating this problem. Also, higher fat products tend to have lower emissions, as can be observed with the differences between SMP and WMP stack losses.

D. Other Comments

1. Powder Transport

Transporting the powder from the dryer to the packing system is where fragile powders are often broken down. This is most often a problem with agglomerated or instant products, and in particular with low fat products. Difficulty arises from handling fragile agglomerates

---

[8] A four-yearly conference series has focused on this topic: Fouling, Cleaning and Disinfection in Food Processing, Cambridge, UK, 1994, 1998, and 2002.

in some of the less gentle conveying systems. As a rule, dense phase conveying, vibrating tubes, and bucket elevators are gentler on powder than traditional pneumatic systems. Care needs to be taken at the design phase to choose the correct conveyor.

2. Plant Economics

Generally the aim is to get as much product through the plant as possible, to repay the investment in a good time. The much lower relative cost of water removal in the evaporator compared with the spray dryer makes it obvious that it is much more beneficial to remove water in an evaporator than in a dryer. However, there are practical reasons why whole milk concentrate is not dried from 70% solids; 48% solids is usually recommended. Milk concentrate is a paste at 70% solids and completely unsuited to spray processing. The difficulties lie in the higher solids concentrate being more viscous and consequently more difficult to handle (pump etc.). In addition, the higher the solids, the more difficult it is to make powder with acceptable solubility. With some of the protein products, one quickly sees limiting behavior in the dryer feed. As a rule, the higher the protein, the lower are the solids possible. This can be seen with caseinate, for which the dryer feed is about 20% solids.

There is a fairly well-accepted envelope of operation for most powders (inlet and outlet air temperatures, evaporator preheat, concentrate heat temperatures, concentrate solids), generally determined by the composition. These variables have an influence on the powder properties and the economics.

VIII. POWDER PROPERTIES

A number of particle properties determine the physical characteristics of a spray-dried powder such as powder appearance, flowability, ability of a mass of powder to pack down (bulk density), and reconstitution in water. The reconstitution process is controlled by two main processing parameters: (1) formation of agglomerated particles, which are considerably larger, than standard, made by returning fine particles to the atomization zone, as described in the chapter on spray drying; (2) the heat treatment received by the particles during drying, that affects particle rehydration. The particle temperature is controlled initially by the wet bulb temperature and, later in the drying history of the particles, by the temperatures in the spray drying chamber. The wettability of the particle surface may also be a factor, depending on the temperature of reconstitution.

In this section we describe a number of milk powder properties. Definitions of some terms commonly used in powder technology are given in the Appendix.

A. Particle Properties

1. Particle Size Distribution

Powders are naturally made up of particles of different sizes. This distribution of sizes is often represented by a mean size that represents a size distribution in two (and only two) of its attributes. Information on the representation of particle size distributions is given in particle size measurement texts (e.g. [41]). A commonly used mean is d_{sv}, which has the same ratio of volume to surface as the distribution of particles.

In addition to mean size, the spread of a distribution (polydispersity) is usually as important as the mean size. Various measures of spread can be used. If the particle size data fits a normal (Gaussian) distribution, a standard deviation can be used to represent the

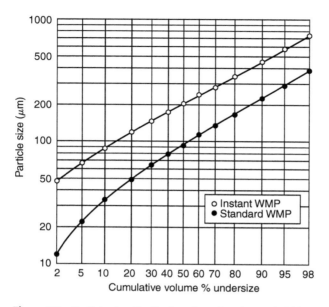

Figure 10   Particle size distributions (axes log size, probability) of regular and agglomerated (instant) powder, both from a two-stage dryer.

spread of the particle population.[9] However, in any troubleshooting, it is best to plot the distribution. A revealing way of plotting size analysis data is to plot with a logarithmic scale for the particle size axis and with a "probability" scale for the cumulative weight percent undersize axis [41].

Milk powder particles are often irregular in shape, even when dried from liquid droplets. Different methods of particle size analysis employ different principles of measurement and hence particle shape affects the size reported by the particle size technique.

Two main groups of milk powders can be distinguished according to particle size distribution:

1.   Regular powders
2.   Agglomerated powders

Typical size distributions of these powders (by a forward laser-light scattering instrument[10]) are shown in Figure 10.

Many of the physical properties of milk powders relate to the particle size distribution. The droplets made by the atomizer are fine, to ensure rapid drying. However, different dryers will produce somewhat different size distributions.

2.   Particle Density

Particle density is the mass per unit volume of a particle, excluding the open pores but including the closed pores.

---

[9]In troubleshooting single sizes may be reported — if it is known that the large size (e.g., 90% of the particles fall below) or the small sizes (e.g., 10% of the particles fall below) are particularly relevant.

[10]Malvern analyzer with a PS64 feeder.

3. Particle Shape

It is not easy to characterize particle shape quantitatively. Particle shape is assessed from viewing particles under a microscope, in particular a confocal microscope, or from scanning electron microscope micrographs. These techniques can show effects of the initial drying conditions, show the degree to which the individual particles are agglomerated and sometimes reveal internal voids.

4. Thermal Properties

Information on physical properties such as the thermal conductivity of milk powders has been documented [42].

B. Powder Characteristics

Many tests have been developed to measure milk powder properties and reconstitution properties. When results are reported, they are meaningful only when the test method is also reported. Typical milk powder physical characteristics are given in Table 4.

1. Bulk Density

Bulk density is the mass of powder that can be fitted into a given volume. The amount of material that can be fitted into a multiwall paper bag or bulk container, which is important in the economics of transporting powder, particularly when powder is transported to other countries for consumption, can be related back to bulk density. It is also very important in ensuring the correct fill volume in a consumer pack. The bulk density is very dependent on the vibration to which the powder has been subjected.[11]

   Bulk density is affected by a large number of variables such as the density of the components making up the product, the particle density, the shape of the particles, the characteristics (smoothness and stickiness) of the surface, and the width of the particle size distribution. Higher particle densities result from low inclusion of air in the droplet, high concentrate TS and shrinkage before the surface sets up to a rigid shape.

Table 4 Physical Characteristics of Typical Milk Powders

Property	Designation	Units	Range Minus	Range Plus	SMP	WMP	ISMP	IWMP	MPC70	MPC85
Bulk density[a]	100 taps	g/ml	0.05	0.05	0.75	0.6	0.45	0.48	0.58	0.38
Particle density[b]		g/ml			1.3	1.2	1.3	1.2	1.2	1.0
Particle size[c]	d_{50}	μm	40	40	90	90	150	125	80	100
	d_{90}	μm	70	100	220	220	450	400	200	250
Flowability[d]		g/sec			1.8	1	2.5	0.7	0.8	

[a] Machine tapping; Niro method No. A2a [43].
[b] Air pycnometer; Beckman air pycnometer manual.
[c] Laser-light scattering; Malvern Mastersizer manual.
[d] Niro drum (method A23a [43], result expressed as g/sec).

---

[11]This is also the case in the test method. A bulk density may be either loose poured or packed. To make a bulk density data point meaningful, the tapping conditions need to be quoted. Bulk density is often reported as bulk density at 100 taps by a standardized tapping machine.

Agglomerated powders have lower bulk densities than conventional powders; the irregular particles lock together and cannot pack together so closely. Two-stage drying usually results in some unintentional agglomeration, but these agglomerates are often broken down in powder transport. Multistage dryers (MSD) may require a deliberate size reduction step to match the bulk density achieved on other types of dryers.

2. Dustiness

The dustiness of a powder can be a problem — dust is objectionable to plant operators and increases housekeeping costs. It can be a problem, particularly in a food plant; if cleaning up is neglected, the milk powder can support the growth of microorganisms, which leads to microbiological hazards within the plant. Regular SMPs tend to be dusty. Tipping of powder should be kept to a minimum. In situations such as debagging where tipping is necessary, we recommend containing the dust at the source with small local extraction hoods.

3. Appearance/Flowability

Powder appearance is the characteristic of the powder first encountered by the end-user. Opposing characteristics are soft friable lumps and the cohesive/sticky nature of the powder (undesirable) versus granular and free-flowing properties (desirable). The latter can be achieved by fines-return agglomeration. A cohesive appearance can be an issue with a fine WMP. A subjective scale of appearance has been generated [44]. The cohesive nature of a powder can affect flow in hoppers and chutes. In addition, SMP exhibits slip stick behavior [45].

Friable or hard lumps indicate that the powder has been exposed to too high a temperature during storage or that it has been manufactured with too high a water activity.

4. Low Shear Reconstitution

The stages in reconstitution at low shear, for example, mixing by hand with a teaspoon, are illustrated in Figure 11. They are:

- Initial wetting
- Sinking of the powder mass through the surface of the water
- Breaking up of the powder mass
- Dispersion of the particles from the clusters of particles and agglomerates
- Dissolution of the individual particles

Very often, the process will not be complete. Some powder may remain as wet lumps, usually at the base of the container. In addition, some particles may take a considerable time to dissolve and will be unsightly if reconstitution is carried out in a transparent container.

When a mass of SMP is added to quiescent water, it will start to sink slowly. It must break through the meniscus layer at the surface of the water. If the powder mass disperses, the particles will start to dissolve. What often happens, however, is that some of the particles dissolve quickly, leading to a viscous material; this material can coat dry powder, leading to lumps that are hard to dissolve.

A solution to this problem is to increase the average particle size, particularly by removing the fine particles (fines). This has two effects: (1) the rate of dissolution is slowed down; (2) the particles individually have greater mass and can penetrate the water meniscus. This counterintuitive effect — larger particles, better dispersion — is the basis of the manufacture of "instant powders" for consumer reconstitution.

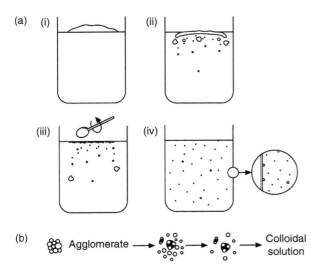

Figure 11   (a) Stages in the dissolution of milk powder: (i) wetting; (ii) sinking; (iii) dispersion of lumps and dissolution of agglomerates and particles; (iv) solution with remaining slow-to-dissolve particles in suspension. (b) Dissolution of an agglomerate.

With WMP, the particle size distribution is not the only factor that must be controlled. The fat at the surface of the WMP particles repels water. The particles prefer to associate with themselves rather than be wetted by the water, and the "balling-up" effect described above is accentuated. Thus, for reconstitution of WMP by the consumer, not only is the particle size distribution made coarser but also a naturally occurring surfactant (lecithin) is applied to the particle surface to overcome the water-repellent nature of the fat in cold water [46]. This process is the equivalent of an encapsulation process. The commercial processes for producing agglomerated and instant WMPs are described elsewhere [6].

5.   Moderate Shear Reconstitution

The dispersing effects described above are particularly relevant to consumer reconstitution. In using milk powder as an ingredient in recombining, the milk is generally mixed with an agitator. Getting the powder to disperse without forming lumps can still be difficult and the recommended methods are to introduce the powder into the inlet of a pump or the vortex created by an agitator. Systems to disperse powders are described in the section on Ingredient Milk Powders.

REM requires a concentration of solids-nonfat (SNF) in the water (at the initial stage of reconstitution) of approximately 21%. RSCM requires a concentration of SNF in the reconstitution water of 43%. The high total solids RSCM mixes inherently have high viscosity and dispersing the powder is correspondingly more difficult than when preparing REM. Problems can be experienced when a powder gives a mixture with a higher-than-normal viscosity. This may arise from factors such as high protein content, a greater than normal heat treatment in manufacture, or an excessively fine powder [47].

6.   Particle Solubility

Once the particles are dispersed, they generally take a short time to dissolve in the solution. If there has been damage to the proteins during drying, however, the particles may be very slow to dissolve. When the liquid is poured out of the receptacle, undissolved particles can be observed in the solution, on the walls of a glass or clear plastic receptacle, or on mesh screens (apertures say, 100 to 600 μm).

Increasing the water temperature above ambient will increase the rate of particle dissolution. However, at yet higher temperatures, say above 80°C, coagulation reactions will normally set in. The optimum temperatures for particle dissolution are thus intermediate temperatures of 50 to 60°C.

With high protein products, such as MPC, the effects of drying on solubility are accentuated. From our observations, such products may be very slow to hydrate at 20°C and may require homogenization or hydration for many hours before even a portion of the particles reach a colloidal state. In contrast, SMP and WMP particles dried with primary and secondary drying stages will dissolve rapidly in cold water to form a colloidal solution that has properties very close to those of the material from which it was spray dried.

C.   Effects of Storage

If milk powders are manufactured correctly and treated properly, they will generally be stable with time. However, some physical/chemical properties may change to a small degree with time. For instance, insolubility may increase (increase in the solubility index of WMP has been demonstrated [48], sediment formed in coffee will increase (unpublished FRC data) and possibly the particle solubility will increase.

The keys to stability are as follows:

- A suitable moisture content, giving a water activity of approximately 0.23. The moisture content should be related to the protein content — the lower the protein content, the lower should be the moisture setting (to ensure an appropriate water activity).
- A storage temperature that is ideally around 20°C. In practice, milk powders are often transported and stored in tropical countries. The deteriorative reactions will be faster, but detrimental effects should not be apparent if the milk powder is kept below 35°C. Such a temperature limit can usually be achieved for bulk powder if the product is warehoused and kept out of direct sunlight. Defects can sometimes arise when product in shipment, as deck cargo or in the top of holds, is exposed to extreme conditions even for short times.

A reaction that occurs during storage is the Maillard reaction, between lactose and protein. The most obvious defect is the formation of colored intermediates, a phenomenon known as "browning." The reaction proceeds by a number of steps [49] and the reaction pathway depends on the particular conditions [11]. At higher water activities and temperatures, the reaction proceeds at a significantly faster rate. It is essential that milk powders be kept below the glass transition temperature of lactose; otherwise the Maillard reaction will proceed rapidly. A relationship between water activity and the glass transition temperature of lactose can be found elsewhere [50].

An additional defect that may arise with high temperature is the formation of lumps, which may be either soft or hard (caking). Soft lumps are formed with fine WMP. If the water activity is too high, lactose crystal bridges may form between particles, giving rise to hard lumps.

It must be emphasized that attention to the correct water activity, reasonable storage temperatures, and packaging that provides an effective barrier to moisture uptake, will give a milk powder with good storage stability.

IX. PRODUCT TESTING

Early tests for milk powder quality promulgated by the American Dry Milk Institute (ADMI) (e.g. [51]) have been adopted worldwide. Niro Atomizer A/S has documented test methods for dried milk products [43]; another compilation is given by Pisecky [6].

Tests are, in the first instance, used to facilitate trade between the seller and the buyer. A secondary purpose is to ensure consistency of manufacture, which is particularly important for consumer products.

There are many variants on some of the milk powder tests and only the versions judged to be the most relevant, or commonly used, are mentioned.[12]

A. Standard Milk Powder Tests

1. Milk Quality

Development of acid arises from growth of lactic acid bacteria. This can be checked by titratable acidity [51]. A number of naturally occurring components of milk also contribute to titratable acidity [6].

Enzymes in milk, both naturally occurring and arising from microorganisms (e.g., psychrotrophs and thermophiles) can break down the milk fat (lipolysis). This can be measured by measuring free fatty acids [52,53].

2. Scorched Particles/Foreign Matter

The powder should be free of foreign matter and substantially free of scorched particles. There is a standard test for scorched particles [50]. The same pad assessed for scorched particles can be used to check for small particles of foreign matter.

3. Moisture

The traditional method for determining moisture uses oven drying [43]. Different times and temperatures may be used. Adequate circulation in the oven to give even temperature distribution is a necessity. Other methods that can be used are the Karl Fischer titration method [43] and the toluene method [51].

The components present in milk mean that moisture determination is not simple. Some moisture may be bound up in lactose crystals; the remainder is known as "free moisture" [43,54]. Moreover, the moisture test method itself may start reactions, particularly browning, that involve either uptake or release of water molecules [55].

4. Heat Treatment

The heat treatment of the milk, prior to evaporation, is a key element of milk processing. The heat treatment in the milk prior to evaporation affects mostly the whey proteins rather than the casein proteins. At temperatures above 65 to 70°C, the whey proteins start to change their conformation and solubility properties. The chemistry of some of the reactions involved has been investigated by a number of workers (e.g., refer to [8]).

The accepted milk powder test for heat treatment that measures the amount of undenatured whey protein (as indicated by solubility in saturated NaCl), is the whey protein nitrogen index (WPNI) test [51]. The classification system is low heat >6.0, medium heat from

---

[12]International methods would be the first choice, but some International Dairy Federation methods are more research orientated.

5.9 to 1.5 and high heat <1.5 mg/g powder [51]. For WMP, the results should be expressed on a SNF basis.

The WPNI value cannot distinguish between low temperature–long time and high temperature–short time heat treatments.[13] For recombining applications, end-use tests are preferred (see the Ingredient Powder Section).

5. Solubility

The standard test for ingredient products is one developed by the ADMI. It has been adopted by the IDF [56]. Subsequently, the ADMI has published a version based on a different mixer [57]. The ADMI/IDF test uses mixing into 24°C water by a high-speed mixer for 90 sec followed by centrifugation. A number of other solubility tests exist.

6. Microbiological Tests

The standard tests are for aerobic plate count (APC), yeasts and molds, *Escherichia coli*, and pathogens. Microbiological techniques are discussed elsewhere in this book.

B. Tests of Quality Following Storage

The main reaction that occurs in SMP during storage is the Maillard reaction — a reaction between lactose and protein, which proceeds much more rapidly at greater than normal[14] water activity and at elevated storage temperatures (above say 35°C). The reaction produces a large number of chemical species. The pigments formed can be measured by a colorimeter. Chemical tests can also be used to detect intermediate compounds such as hydroxymethyl furfural [58], or furosine [59,60].

The oxidation of milk fat can be followed by measuring the peroxide value (PV) [61,62]. The oxidation reaction produces a hydroperoxide (glycerol with OOH group). However, this compound is an intermediate in the oxidation reaction and may increase and then decrease[15] in concentration. Hence the PV should be followed from the time of manufacture if it is to provide meaningful results.

The changes in flavor with storage are best followed by sensory methods. There may be some degree of correlation with chemical species such as peroxide (e.g., in one study, [63], oxidized flavor correlated with PV, correlation coefficient $r = 0.74$). However, fatty acid breakdown compounds at concentrations of parts per 10^9 may contribute significantly to storage flavors of oxidized fat [64], meaning that individual flavor profiles develop depending on the composition of the individual fatty acids and the storage history.

C. Tests of Powder Properties

1. Sampling

Samples should be representative of the bulk. Ideally, powders should be divided from the bulk with a sample splitter.[16] In commercial plants, the recommended practice is to take a

---

[13]There are now various other methods for assessing the degree of heat treatment, all of which suffer from the same limitation.

[14]A suitable water activity for a milk powder of regular composition is approximately 0.23.

[15]In practice the reaction rarely proceeds this far in milk powder.

[16]This procedure step is often ignored. For free-flowing powders, for example, ISMP, sample dividing equipment, such as a spinning riffler, must be used to obtain a representative sample for laboratory analysis.

composite sample with an autosampler, which is then representative of the batch. Refer to Allen [41] for a comprehensive discussion of powder sampling procedures.

2. Flowability

Niro has developed a test that utilizes a rotating drum with a slit [6,43]. This method has been investigated by Chen [65] who advocated expressing the result as quantity of powder/elapsed time (g/sec). Some relative values for different products are given in Table 4.

3. Particle Size

Particle size is an important parameter in heat and mass transfer in drying, flowability, dispersing of powders, and many powder properties. The measurement of particle size and its interpretation is a specialist topic[17] and expert advice should be sought if particle size is being used to investigate a processing problem.

The two main methods used currently to measure the particle size of dairy products are sieve analysis and laser-light scattering. In sieve analysis, the powder, combined with a free-flow agent, is put on a stack of sieves and shaken on a machine with defined amplitude of vibration, for a defined time. The time is set according to the requirements of a sieve analysis standard [66].

The laser diffraction method uses light scattering of the particles as a dilute suspension in air. A computer fits a particle size distribution to the measured laser diffraction pattern [67].

For nonspherical particles such as milk powder agglomerates, sieve analysis measures the second largest dimension of the particles passing through the mesh. Laser-light scattering instruments calculate the particle diameter of a sphere with a volume equivalent to the volume of the particle.

A particular difficulty with agglomerated powders is the potential for the breakdown of agglomerates in the size analysis method. Our Fonterra Research Centre (FRC) laboratory has shown a reasonable degree of agreement between sieve analysis and laser-light scattering equipment (Sicrocco feeder and Malvern Mastersizer 2000) on some commercial agglomerated WMP.

4. Bulk Density

Bulk density is measured by placing powder in a container of defined volume and tapping with a defined force for a certain number of taps. The tapping action is usually mechanized [43].

5. Particle Density

It is easiest to determine the particle density of a milk powder by air pycnometer [43], because the protein and lactose dissolve in water and the fat dissolves in most organic solvents. For operation, refer to the instructions of the pycnometer manufacturer. SMP and protein powders[18] can be evaluated using an organic solvent, for example, isopropyl

---

[17]For instance, measuring the particles by volume, weight, surface, or number will give size distributions that appear to be different, for the same population of particles. In any size analysis method, a number of assumptions are made — the distribution obtained is likely to be different when one method is compared with another. It is best to compare results within one method and it is important to specify the method employed when reporting results.

[18]Also low free fat WMP. The concern with WMP is extraction of powder material into the solvent.

alcohol, in a density bottle (common laboratory method). The voidage of particles can be calculated from the particle density and the material density [43].

D.　Consumer End-Use Tests

1.　Wettability

Wettability tests usually measure the time to sink below the water surface of a mass of powder, rather than the tendency of the particle surface to be wetted. Powder is introduced in a standardized manner to a quiescent water surface. A number of tests have been devised but two tests are widely accepted [43,68]. These tests are briefly described by Pisecky [6].

2.　Dispersibility

The manner in which a powder mixes into water is one of the most difficult properties to quantify because the exact answer depends markedly on interactions between the test mixing conditions and the properties of the powder. In addition, the powder falls on to the water in a different way for each test (despite procedures to standardize this step). This can lead to a repeatability of about ±30% of the test value. We recommend two or three replicate analyzes to improve confidence in the results.

　　　We prefer a method developed at the FRC. This method is a simulation of what the consumer does — hand stirring — and a direct measurement of the wet undispersed material [6]. It can be applied for reconstitution in cold (e.g., 20 or 25°C), warm (e.g., 40 or 50°C) or hot (e.g., 75 or 80°C) water.

3.　Particle Solubility

The liquid obtained from a dispersibility test, at a defined time after commencement of reconstitution, can be assessed for the size of the undissolved particles. For routine plant control, this can be done either empirically by visual assessment of the film left after draining the liquid from a test tube, or by assessment against a photographic scale. For research work, a laser-light instrument may be employed.

4.　Reconstitution in Hot Drinks

Initially, powders made from ion-exchanged milk or specially formulated "non-dairy" coffee whiteners based on sodium caseinate and vegetable fat were supplied as milk powder for use in tea and coffee [15]. Later the widespread distribution of instant WMP led to its use in coffee. A coffee test that quantitatively measures the sediment formed in an instant coffee solution of defined concentration was developed by the NZDRI [6,69]. A test that evaluates performance in coffee by visual examination and counting the "floaters" and "flakes" in the solution and the sediment or "sinkers" that are present after settling has been advocated [6].

5.　End-Use Tests for Other Products

Some other tests purport to measure properties related to the end use. However, care must be taken to ensure that they bear a close enough relation to the conditions of the end use to be useful. Another difficulty with end-use testing is that it is labor intensive; the desired properties should be obtained by consistent manufacture rather than a large amount of testing.

E. Test Development

At the time of writing this chapter, the IDF is continuing the development of test methods that can be utilized internationally (Table 5). The tests that are under development currently are:

- Coffee test
- Scorched particles content — caseinates and casein
- Nitrogen solubility index
- Heat treatment intensity

Table 5   List of IDF Methods for Measuring Milk Powder Properties[a]

Property	Number	Title	Basis
Dispersibility	87:1979	Determination of dispersibility and wettability of instant dried milk	Hand stirring with spatula, filter through 150 μm screen, determining TS
Wettability	87:1979	Determination of dispersibility and wettability of instant dried milk	Powder sinking through quiescent surface
Titratable acidity	86:1981	Determination of titratable acidity	Titration with 0.1 N NaOH
Heat class	114:1982	Assessment of heat class. Heat-number reference method	Determination of undenatured whey protein by Kjeldahl nitrogen
Bulk density	134:1986	Determination of bulk density	Machine tapping of measuring cylinder
Insolubility	129A:1988	Determination of insolubility index	Mixing and centrifuging
Whey protein denaturation	162:1992	Assessment of heat treatment intensity	Determination of undenatured whey proteins by high performance liquid chromatography
Moisture	26A:1993	Determination of the water content of dried milk	Oven drying (102°C) to constant mass
Bulk density	134A:1995	Determination of bulk density	Machine tapping of measuring cylinder
Particle solubility — flecks	174:1995	Determination of white flecks number	Calculated from volume of reconstituted liquid passing through 63 μm aperture screen in 15 sec
Organoleptic attributes	99C:1997	Sensory evaluation of dairy products by scoring	Reconstituted powder appearance and flavor

[a]Excludes compositional methods.

X. INGREDIENT POWDERS

A. Introduction

Ingredient milk powder refers to milk powder that is purchased by a manufacturer and processed to provide a product for consumers. The processors mix the powder with water to the required concentration, add additional ingredients, give the product the necessary processing, pack it and sell it to the consumers. In some instances, consumer powders are sold to processors for repacking prior to sale.

Milk powders are used to make recombined single-strength milks (including pasteurized (PAST), extended shelf life[19] (ESL), ultrahigh temperature processed (UHT)), concentrated milks (REM, RSCM), and other milk products (directly acidified drinks, chocolate, ice cream and cultured products such as yoghurt, laban, yakult-type products, labneh). In fact, recombined analogs can be produced for all products made from fresh milk.

The type of milk product seen on the supermarket shelves is largely dependent on the dairy market development tradition and GDP per capita of the country in which the observation is made. As the consumers' incomes increase, so their demand for milk solids tends to move from "traditional" concentrated long-life products (REM, RSCM) to a product closer to fresh milk (UHT and pasteurized). They then demand a product that has less cooked flavor — more like fresh milk.

B. Processing

The utilization of ingredient milk powder comprises several processing steps (Figure 12). There is a wealth of information in the literature (e.g., [1,2,71–73]). An annotated bibliography on recombined milk products was recently published [74]. A general description of the process, applicable to both the single-strength and higher strength products, is given.

Reconstitution of dried milk has been practiced since before the time of Genghis Khan[20] (13th century). Then the dry matter solubility was dubious and the dissolution process haphazard, being occasioned by the motion of the horse and the rider. A degree of sophistication has been introduced since those days; attention being paid to several important factors, particularly the mixing equipment and water temperature. For milk powders, a water temperature of 40 to 60°C is recommended [74]. At temperatures above 60°C, heat-induced changes to the proteins may cause powder lumping in inefficient equipment. Below about 35°C, unmelted fat (even in SMP) appears to hinder dissolution, and hydration of the proteins is slower.

A good mixing system will ensure that there is no chance of powder lumps forming in the mix. There are numerous techniques for mixing powders into liquids. With an efficient dispersion system the temperature becomes less critical.

Mixing systems range from the traditional batch tank with an impeller (many impeller designs are possible) and (possibly) an external circulation loop, funnel (hopper) and pump,

---

[19]ESL milk can be considered a subset of pasteurized milk. The ESL heat treatment (e.g., 90°C/15 sec or 116°C/2 sec [71] is greater than the pasteurized treatment (72°C/15 sec) but unlike UHT milk, the packing is nonaseptic. Hence, ESL milk is still vulnerable to recontamination in the filling stage.

[20]J.M. Dent & Sons. The Travels of Marco Polo the Venetian, Everyman's ed., 1908, Chap. 49, pp. 129–130. The powder produced is similar to products like "*ägt*" found in the Middle East today.

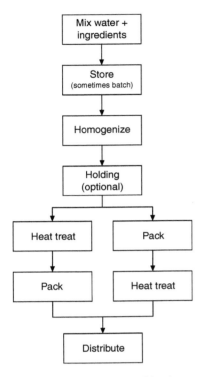

Figure 12 Generic recombination process.

and the venturi dispersion system to the more sophisticated, including various purpose-designed medium and high shear dispersion systems with and without applied vacuum to eliminate entrained air [73].

The main difference between the former and the latter is that the former units rely on shear in the pump (or impeller) and bulk liquid flow to break down any powder lumps and disperse the powder. The latter types use higher shear and resemble hammer mills [75]. To assist dispersion, the powder is introduced to the liquid in the eye of the pump impeller to prevent premature wetting. The ultimate choice of mixer will depend on scale, cost etc.

Recombined products can be made from either WMP or a mixture of SMP and milk fat. For some products, vegetable oils are used to create "fat-filled" products.

Most recombining systems tend to be batch or semibatch processes. Once mixed, it is common to allow some "hydration time" for complete dissolution of the powder. In most applications, 20 to 30 min is sufficient.[21]

The mixed product is usually subjected to homogenization at some point in the process. The main purpose of homogenization is to reduce the size of the fat globules so that the fat emulsion is stable during storage. The fat emulsion is stabilized by a coating of (milk) protein on the surface of the fat globules (this coating may be modified if a low molecular weight emulsifier is used) [2]. Homogenizer operating conditions tend to vary a little with

---

[21] David Newstead, FRC, 2002, personal communication.

the product and are dealt with below, although typically, homogenization is most effective if carried out at approximately 70°C.[22]

C. Recombined Evaporated Milk (REM)

Recombined evaporated milk is the recombined version of evaporated milk. The main processing difference is that the correct concentration is obtained by "diluting" powder rather than by removing water from the feed milk in an evaporator. It is unsweetened preserved milk in a concentrated form and as such, can be used diluted as a milk replacer or in the concentrated form as a coffee or tea whitener. REM is commonly used in countries in which there is a very small dairy industry, but a large demand for unsweetened milk solids. It can also be used as a milk replacer in cooking.

The process is relatively simple and follows the right hand path of Figure 12 (mix, store, homogenize, pack, heat, and distribute) but the ingredient and process requirements are more stringent.

Normal "standard trade" milk powders cannot be used reliably for the manufacture of REM — coagulation typically results. Special manufacturing conditions, that is, a high evaporator preheat treatment, are required [3]. The milk powder batch must then be tested for end-use (REM) suitability.

The powder and fat (if required) are mixed with water (at 50°C) to give the appropriate composition. There are two standard compositions. The British standard is 9% fat and 22% SNF, to give a liquid with 31% solids. This formulation is referred to as 9 : 22. The American standard is 8% fat and 18% SNF (8 : 18). In recent years, the American standard has assumed dominance. Other formulations also exist, for example, 6 : 18, 7 : 20, and 8 : 20.[23]

Following homogenization (two-stage, 160 bar first stage, 35 bar second stage), a sample of the batch is taken to test for heat stability. The heat sterilization step for REM is very severe (115–120°C for 20–13 min, plant specific) and the heat stability of the concentrated milk is very sensitive to pH variations. If precautions are not taken, the mixture can coagulate. In addition to selecting the correct powder, it is usual to add a stabilizing salt to prevent coagulation and control viscosity. As might be expected, viscosity and coagulation lie on a continuum. It is not an easy task to get a batch of REM both to be heat stable (i.e., not to coagulate) and to have the correct viscosity. The stabilizing salts that are used are monosodium dihydrogen orthophosphate (MSP) and disodium monohydrogen orthophosphate (DSP). The usual method of testing for stability is to fill cans with product from the new batch, add various amounts of the stabilizer, and seal and sterilize the cans with the previous batch. The test cans are checked for heat coagulation and the desired viscosity for that market. The correct amount of the correct stabilizer is added to the new batch and the batch is processed. Stabilizer addition rates range up to 0.06 mol of phosphate per kg of SNF. The stabilizer acts to change the pH of the milk. But, in addition, the phosphate ions play a significant role in stabilization [76]. BMP in the formulation will improve the heat stability [77], although the exact mechanism is unclear.

The REM manufacturing process leads to a slight caramelized flavor and color. This is accepted as traditional, indeed, regarded as a positive product attribute.

---

[22]There is scope to investigate the need for homogenization when WMP is the sole powder source in the mix.

[23]John Smith, Fonterra, 2002, personal communication.

D.   Recombined Sweetened Condensed Milk (RSCM)

Sweetened Condensed Milk (SCM) has been manufactured since 1857 and, in its recombined form, RSCM, since the late 1950s [78]. The process is more complex than that for REM, although the ingredient powder requirements are less stringent. It follows the left hand path of Figure 12 (mix, store, homogenize, heat, pack, and distribute). Additions to this generic process are the flash cooling stage (after the heating) and the addition of seed lactose to the final mix. The primary difference is that the sugar content of RSCM accounts for 43 to 45% (w/w) of the TS in the can. This has the effect of raising the osmotic pressure to prevent the growth of most bacteria. A heat treatment more severe than pasteurization (80°C for 3 min) destroys yeast and mold spores, meaning that the product can be kept for many months at ambient conditions.

In western countries, RSCM is used in dressings and confectionery. In South East Asia (SEA) and Latin America, it is used as a milk substitute and as a coffee whitener and sweetener. In parts of SEA, it is used as a spread on bread. In Latin America, it is caramelized, by boiling for 30 min, to make a dessert.

The American standard formulation is the one most commonly used and is 8% fat and 20% SNF (8:20). The British standard, like that for REM, is 9 : 22. This formulation is rare. The final solids of the mix is 73 to 75%.

The powder is mixed with water (at 50°C) until fully dissolved. The sucrose is then added and dissolved. If fat is required, it is melted (at about 50°C) and added at this point. From here, the mixture is pumped to a balance tank before being deaerated, homogenized (at 50 to 60°C, 70 bar) and pasteurized (80°C with holding for about 3 min (longer holding times are used) before being held in a second balance tank. Here, the mixture is pumped to a cooler (usually a flash cooler, the evaporation bringing the batch to the required TS) and cooled to about 32°C. At this point in the process, seed lactose is added either dry (as a dry powder) or wet (as a dispersed slurry in SCM). RSCM is supersaturated in lactose at these conditions [79] and crystals will grow. The seed lactose has a very small particle size (bulk less than 1 μm [80]) and is added to control the growth of lactose crystals in the RSCM. These seed crystals act as nucleation sites, allowing many small crystals to form, rather than a few large crystals. The crystal size limit in the final product is preferably <10 μm. Any larger and they give the product a sandy texture.[24]

Once crystallization has been completed (agitation in a controlled temperature silo for about 4 h), the RSCM is further cooled (usually a flash cooler) to about 20 to 25°C, packed in sterilized cans (usually 397 g) with little or no head space (<3 ml/can), so as to inhibit growth of surviving mold spores.

Difficulties with RSCM include problems with lactose crystallization and product viscosity. Lactose crystallization is usually controlled by seeding (as described above). However, excessive temperature fluctuations will cause growth of large crystals at the expense of small crystals — leading eventually to sandiness.

The RSCM viscosity can be a far more difficult problem to solve and has been the subject of work since at least 1965 [74]. In addition, the 1918 (2nd) edition of Hunziker's book mentions viscosity issues in SCM, so the problem is far from new and has still not been solved. Age thickening is the general term that applies to RSCM viscosity behavior. The nature of its composition and treatment means that the product tends to thicken over time [1]. In summary, if the initial viscosity (on canning) is about 1000 cP (10 P), then it is relatively easy to handle and will probably thicken to 10,000–12,000 cP

---

[24]David Mills, Fonterra, 2002, personal communication.

(100–120 P) after 6 m. Much higher is typical in some markets such as Bangladesh. If the initial viscosity is too low, then the RSCM thickens too slowly; if the initial viscosity is too high, then it tends to thicken too quickly.[25] Product composition (particularly the fat:protein ratio) and processing conditions (homogenization and pasteurization) are important. The effect of these variables will need to be determined on a plant-by-plant basis. Viscosity can be controlled, by selection of milk powder (evaporator preheat has a complex effect), homogenization and pasteurization conditions. In practice, the latter two parameters are seldom manipulated. The main viscosity control lies in powder selection and blending.

E. Single Strength Milks

1. Recombined UHT Milks

Ultra high temperature (UHT) milk is a single-strength product (approximately 12% solids). This product can be found throughout the world. There is a wealth of literature on all aspects of UHT milk (see, e.g. [2,81]). The majority of this literature relates to UHT milk made from fresh milk, not recombined milk. UHT milk is used in most domestic milk applications including drinking.

UHT recombined milk follows the left hand path of Figure 12. Some additional comments are given here. The heat treatment can be indirect, direct, or a combination of the two. Indirect heating can be a tubular (shell-and-tube or concentric tube) or a plate heat exchanger. With direct steam heating systems, either a milk-into-steam system (steam infusion) or a steam-into-milk system (direct steam injection [DSI]) can be used. In these systems, a flash vessel is used for cooling and to remove the steam condensed during heating. There are advantages and disadvantages with each system. Briefly, direct systems are capable of very high heating and cooling rates; they are also better suited to some viscous products than are simple indirect heating systems. Indirect systems have lower operating costs because maintenance costs are lower and more thermal regeneration is possible.

The choice of the heating system will largely determine the position of the homogenizer in the system. Direct systems tend to form aggregated material and hence it is usual to place the homogenizer after the heat treatment — to assist breaking down of any aggregates formed. This does, of course, mean that the homogenizer must be an aseptic device. Such devices have higher maintenance costs (particularly with regard to piston seals) than conventional homogenizers.

The powder, fat (if required), and water are mixed to the desired solids. Generally, the equipment used for the reconstitution/recombining can be simpler than that used for REM/RSCM because of the lower solids required in the final mix. The milk is given a heat treatment[26] (135–150°C, 8–2 sec, plant specific) and homogenized before temporary storage and aseptic packing. Fouling of UHT plants is a common problem. This is the subject of much research, but the bulk of the published work has been on fresh milk systems. Fouling of UHT plants by recombined and fresh milks has been examined [82,83]. They observed that the powder manufacturing conditions (particularly evaporator preheat treatment) can have a significant effect on the fouling of a plant, with higher powder manufacture preheat giving higher fouling rates in a UHT plant. The type of milk powder used is also important

---

[25]David Mills, Fonterra, 2002, personal communication.
[26]Some recombiners may use a separate pasteurization step as well.

with WMP tending to give lower fouling rates than SMP/AMF blends.[27] Preheating of recombined milk prior to UHT processing increases fouling drastically—contrary to what is generally believed to be the case with fresh milk [82].

Machines are available to fill many different aseptic package formats (blow-molded plastic bottles, plastic pouches, cartons, cans, form-fill-seal, glass bottles, bag-in-box [bulk only]). There is also a range of package-presterilizing methods such as hot, sterile air, UV light and warm hydrogen peroxide. Packaging suppliers will provide performance data for their equipment. The pack type used for a product is often determined by surveys of customer preferences. Packs now have a wealth of nutritional information and promotional material on them. There is also a move, especially in the United States of America, to have the product visible to the consumer. This can introduce some difficulties in finding a suitable packaging material. This is not to be encouraged as light quickly destroys riboflavin (vitamin B2). The performance of these systems (including the heat step) is usually described by the number of failures per number of filled packs (e.g., 1 in 5000) [81].

A detailed set of methods for assessing the quality of UHT milks has been published [83]. It covers practical issues such as gelation, fat separation, color and sediment in the milk and performance in coffee and other hot applications.

2. Pasteurized Milks

Pasteurized recombined milks are recombined in a similar process to recombined UHT milks. The heat treatment, however, is quite different. As the name suggests, a pasteurization treatment is given to the milk. As the milk is not sterilized, it must be refrigerated until it is consumed.

Typical pasteurizing conditions range from 72°C for 15 sec (sufficient to eliminate any pathogenic organism) to 85 to 90°C for 30 sec. In Taiwan near-UHT treatments are used because consumers prefer the cooked flavor. Recombined pasteurized milk is homogenized (sometimes) and deaerated prior to packing. It is usual to position the homogenizer upstream from the pasteurizer to avoid recontamination.

Recombined pasteurized milks are packed in either cartons (gable top packs) or plastic containers. The legal shelf life of these products is determined by individual country regulations and may vary from a few days to one to two weeks — under refrigeration.

F. Milk Protein Powders

Our definition of milk protein (or protein) powders, for this chapter, includes MPC56, MPC70, MPC85, MPI (milk protein isolate), WPC, WPI (whey protein isolate) and caseinate powders. They are added to a formulation for a particular property, either nutritional or functional, and are used in a vast range of food systems from health drinks and infant formulae to meat, cheese, and bakery products. Functional properties include water binding, gelation, whipping, and emulsification.

The protein powder is added (usually with other dry ingredients) to a liquid for further processing. Care must be taken with protein powders, especially in liquid systems, to ensure that the protein has had appropriate hydration conditions for the application (e.g., 30 min at 60°C or perhaps overnight at 4°C[28]) and that the solution containing the protein is not

---

[27] Susan Reelick, Fonterra, 2002, personal communication.
[28] Michelle Harnett, FRC, 2002, personal communication.

heated to above about 64°C. Above this temperature, the whey proteins can be denatured causing considerable modification to their properties. This can result in a destruction of desired functional properties and a loss of some biological activity [85].

The use of protein powders, especially in the nutritional business, is growing. In these uses, the selection between an MPC, caseinate, or WPC will depend upon the taste and nutritional profile required. For instance, MPC powders are considered to have a better flavor profile than caseinates. In these cases, the powder may be added in a dry blend with other ingredients to make, for example, sports formulations popular with bodybuilders.

Readers particularly interested in milk protein functional properties can refer elsewhere [86,87]. The latter reference is a book dedicated to the topic. There are several chapters that deal with milk protein manufacture and use. Tables 6 to 8 list some of the protein powder types and where they are used in specific nutritional and functional applications.

Table 6   Applications in which Nutritional Protein Products are Used

Application	Protein source				
	Caseinate	MPC	WPC	Hydrolyzates	TMP (total milk protein)
Liquid nutritional beverages	✓	✓	✓	✓	✓
Enteral products	✓	✓	✓	✓	✓
Nutritional bars	✓		✓	✓	✓
Instant nutritional powders	✓	✓	✓	✓	✓
Sports products (beverages, bars, supplements)	✓	✓	✓	✓	✓
Weight management	✓	✓	✓	✓	✓
Infant formulae	✓	✓	✓	✓	✓
Adult medical	✓	✓	✓	✓	✓
Low pH nutritional beverages			✓	✓	
Low allergenicity products				✓	
Protein-fortified foods	✓	✓	✓	✓	✓

Table 7   Functional Properties of Ingredient Protein Powders

Application	Protein source				
	Caseinate	MPC	WPC	Hydrolysates	TMP
Emulsion stability	✓	✓	✓		✓
Heat stability	✓	✓	(Some)	✓	✓
Opacity	✓	✓	✓		✓
Flavor		✓			✓
Water binding	✓		✓		✓
Protein fortification	✓	✓	✓	✓	✓
Heat gelation		✓			
Acid stability			✓	✓	
Whipping	✓		✓	✓	✓

Table 8 Technical Feasibility of Ingredient Protein Powders in Various Food Applications

Application	Caseinate	MPC	WPC	Hydrolysates	TMP
		Protein Source			
Bakery	✓	✓	✓		✓
Dressings	✓		✓		
Sauce/soup	✓	✓	✓		✓
Ham	✓		✓		✓
Reformed poultry	✓		✓		✓
Sausages	✓		✓		✓
Coffee whitener	✓				✓
Low fat spreads	✓				
Egg white replacer			✓		
Processed cheese	✓	✓	✓		✓
Cultured dairy	✓	✓	✓		✓
Cheese milk extension	✓	✓			✓
Nutritional bars	✓	✓	✓	✓	✓
Thermal processed beverages	✓	✓	✓	✓	✓
Powdered beverages	✓	✓	✓	✓	✓
Pasta/Noodles			✓		
Vegetable analogs			✓		
Surimi			✓		

Notes: While the table above shows technical feasibility, readers should refer to local food regulations to confirm the ingredient is permitted for use in that application. It is very important, especially with protein powders, to talk to your ingredient supplier to obtain the correct ingredient for your application, and to ensure the correct handling and processing conditions are used.

1. Whey Protein Concentrates

The primary use for whey-based powders is for specific nutritional properties. WPCs are used in a variety of applications including nutrition, dairy, meat, bakery, and convenience food products. Examples include custards, desserts, yoghurt, surimi, ham, cookies, quiche, pasta/noodles, dressings and sauces, and meat analogs. For these product types, whey proteins are used for combinations of gelation, water binding, foaming, acid stability, and emulsification properties.

2. Milk Protein Concentrates

MPCs are used in a range of applications such as the following:

- Cheese applications. The MPC can be added to milk or cheese formulations to enhance the protein content and/or the yield of the final product. The addition ratio and the ingredient choice are dependent on individual country regulations and trade practices.
- Nutritional drinks. To provide both casein and whey proteins in the same ratio as milk, but without the high lactose content.
- Yoghurt and cultured applications. The MPC can be used to enhance the content and/or provide textural benefits to the final product.

A more complete list of MPC properties is found elsewhere [88,89].

XI. CONSUMER MILK POWDERS

A. Powder Performance

Powders may be sold directly to the consumer for reconstitution in the home. An alternative to recombined milk products is the supply of powder direct to the consumer.[29] Dairy powders for direct reconstitution by consumers are made to general milk powder standards, but must also meet some extra requirements.

A key consideration for domestic use is reconstitution properties. On reconstitution, the powder should "wet" (water molecules must be attracted to, not repelled by, the particle surface), disperse, and reconstitute to leave an undetectable quantity of residual powder on the surface or at the base of the vessel. The particles must dissolve. This must be achieved by slow speed (hand) stirring.[30] It is also desirable for the powder to have a granular free-flowing appearance rather than a cohesive appearance. These properties, and how to control them, have been dealt with in detail under Powder Properties.

A wide range of conditions may be used in reconstituting milk powder. It is commonly assumed that instant milk powder will be reconstituted in cold or ambient-temperature water. However, in many Asian and Middle Eastern countries, the water used is warm or hot. This has arisen either as a practice to ensure microbiological control, by boiling the water, or as a cultural practice of drinking warm drinks. In addition, the powder may be added in the course of preparing a beverage. For example, the powder may be added to the infusion stage of making tea or added directly to coffee.

The direct addition of powder to coffee requires good stability of the powder in the hot and somewhat acidic conditions of the coffee. The main process control is a heat treatment of defined temperature and time at the concentrate stage of making the powder. Preheat treatment has a lesser effect, but high preheat treatments are detrimental to stability [6,90]. The mechanisms for stabilizing the powder are not well understood.

B. Nutritional Requirements

Another important consideration is nutritional benefits.[31] In the early stages of the development of the consumer-WMP market, the product gave basic nutrition, particularly for growing children. To enhance nutrition, vitamins were added, initially A and D; the product category was developed further to include additional vitamins. Since the early 1990s, consideration has been given to adding other nutrients to milk products [91,92]. Nutrients for fluid milk have been considered by the Fluid Milk Strategic Thinking Initiative [93]. The supplementation of food products has recently received considerable attention [94–98]. In Asia, the trend toward fortification of milk powders with specialist ingredients started in markets, such as Taiwan, where there was a much greater proportion of affluent consumers.

---

[29] There is often a progression in markets from concentrated shelf stable products (REM, RSCM) to powder to liquid products such as pasteurized milk. The latter products develop as consumers come to value convenience more than cost. It is also aided by the availability of a widespread cool chain distribution system. In a few markets, for example, Thailand, for historical reasons, the market has always been for liquid products.

[30] However, some households may use "milkshake" mixers or kitchen blenders for reconstitution.

[31] In this section, it is not intended to venture into detailed presentations of the role of ingredients in nutrition; rather, our aim is to outline the general trends in the development of markets for milk powders, particularly in the last 20 years.

Table 9 Examples of Fortifying Agents Claimed to Bring Nutritional Benefits to Consumers[a]

General application area	Fortifying agent	Use	Key consumer targets
Bones	Calcium	Build bones	Infants, children, and teenagers
	Calcium	Reduce bone loss	Postmenopausal women
	Inulin (prebiotic)	Enhance Ca absorption	Postmenopausal women
	Magnesium	Enhance bones	Generic
	Vitamin D	Enhance Ca absorption	Generic
	Vitamin C	Enhance Ca absorption	Generic
	Vitamin K	Enhance Ca absorption	Generic
	CPP	Dental	Children
Brain	Iron	Stimulate brain development	Infants
	Folate	Stimulate brain development	Infants
	Choline	Maintain brain activity	Seniors
Gut health	Probitic organisms	Promote healthy digestion	Children, adults
	Oligosaccharides	Promote healthy digestion	Children, adults
	Fibre	Promote healthy digestion	Children, adults
Immune system	Colostrum	Boost immune function	Children, adults
	Zinc	Boost immune function	Children, adults
	Selenium	Boost immune function	Children, adults
Heart health	Omega-3 fatty acids	Reduce artery deposits	Mid-years to seniors
Reduce cancer risk	Vitamin E	Anti-oxidant	Mid-years to seniors
	Green tea	Anti-oxidant	Adults

[a] The degree of substantiation of claims varies widely; not all fortifying agents may have been used commercially.

The drive to increase brand awareness by differentiation and the increasing knowledge of human nutritional requirements led to an increasingly wide range of additives. These have been targeted at a particular age range or segment of the population. Age brackets that can be considered are:

- Infant (<6 months)
- Follow-on (6–18 m)
- Growing up
- Teenage
- Young adult
- Middle years
- Senior

Examples of groups with particular nutritional needs are pregnant women, women breastfeeding, women in menopause, and men and women with high blood pressure.

Consideration needs to be given to whether a fortifying agent should be delivered in a food or as a dietary supplement. In some cases, a food constitutes a readily accepted delivery vehicle for a health ingredient. For example, milk is naturally rich in calcium; the calcium concentration can be further increased to deliver most of the daily requirements for calcium in a quantity of milk that is readily consumed on a daily basis. In addition, a number of other ingredients can be combined with calcium to increase absorption. Other speciality applications are also being advocated, for example, the use of casein phosphopeptide (CPP) for dental protection [99].

A list of some fortifying agents that have been suggested for milk powder are listed in Table 9. Detailed information on some of the current trends in the use of bioactives in milk products were given at an IDF conference [100]. The effective use of nutritional fortifying agents requires consideration of nutrition status, nutrient adsorption, product stability, and sensory acceptability [101].

XII. GENERAL HEALTH

A. Healthy Digestion

Some consumer products are aimed at enhancing health by promotion of healthy bacteria in the gut. One approach (probiotics) is the inclusion of microorganisms in products to help establish desirable flora in the gut. Such a feature is now available in milk powders, having been offered in yoghurt and cultured drinks for some years. Another approach (prebiotics) is the inclusion of compounds that stimulate healthy bacteria. The composition of milk can be altered by lactose hydrolysis to produce oligosaccharides that stimulate gut bacteria.

Another possibility is to reduce the lactose, to which some consumers are sensitive. This can be done by hydrolyzing the lactose to glucose and galactose. Lactose hydrolysis also has the effect of increasing slightly the sweetness of the reconstituted powder, without increasing the carbohydrate content.

B. Milk as a Source of Nutraceuticals

The minor ingredients of milk are being intensively studied [100,102]. In milk there are potentially many peptides with specific biological activity. Peptides from casein (CPP) enhance the solubility of minerals such as calcium and zinc (e.g. [99]). Others reduce the activity of the angiotension-converting enzyme that is involved in vasoconstriction and hence control of blood pressure [102]. Lactoferrin is a minor milk protein that is useful for nonspecific defense against bacteria, fungi, and viruses [102,103]. Oligosaccharides, glycolipids, and glycoproteins containing sialic acid residues have been suggested as aids to digestion and as antiinfectives [104].

C. Liquid Products

The features described above can be incorporated into liquid products made by the recombining processes described in this chapter. However, UHT products have an expected shelf life of 6 months; these pose particular challenges if suspension stability and maintenance of activity of the nutrients during the full shelf life are to be achieved.

XIII. DAIRY INGREDIENT SUPPLIERS

A literature search shows that there are: (1) more than 700 manufacturers and suppliers of milk powders; (2) 100 manufacturers and suppliers of whey powders; (3) 80 manufacturers and suppliers of milk proteins and caseinates; (4) 30 manufacturers and suppliers of cheese powders.

- Arla Foods, Sweden/Denmark
  www.arlafoods.com
- Campina, The Netherlands
  www.campina.com
- Dairy Farmers of America, USA (within the USA only)
  www2.dfamilk.com
- Euroserum, France
  www.euroserum.com
- Friesland Coberco Dairy Foods
  www.fcdf.nl
- Glanbia plc, Ireland/England
  www.glanbia.com
- Groupe Lactalis, France
  www.lactalis.com
- Hoogwegt Group, The Netherlands
  www.hoogwegt.com
- Ingredia Dairy Ingredients, France
  www.ingredia.com
- Kerry Group, Ireland
  www.kerrygroup.com
- Lactoprot Deutschland GmbH
  www.lactoprot.net
- Land O'Lakes Inc, USA
  www.landolakesinc.com
- Murray Goulburn Co-operative Co Limited, Australia
  http://www.rochsec.vic.edu.au/Pages/MG.Products.html
- NZMP, New Zealand
  www.nzmp.com
- Sancor Milkaut SA, Argentina
  www.sancor.com

ACKNOWLEDGMENTS

We would like to gratefully acknowledge the assistance of the Fonterra Research Centre for providing the resources for us to complete this chapter. We would also like to thank our friends and colleagues in the New Zealand dairy industry for their contributions by way of material or helpful comments on the completed text.

APPENDIX

DEFINITIONS OF SOME TERMS USED IN POWDER TECHNOLOGY

Adherence	The sticking of a powder to a wall or substrate
Bulk density	Mass per unit volume of powder under defined conditions of tapping or consolidation
Cohesiveness	The stickiness of a powder
Flowability	The ease with which a powder flows
Friability	The tendency of particulates to break down in size during handling or storage under the influence of light physical forces
Mean particle size	Dimension of a hypothetical particle that represents some characteristic of the whole distribution
Median diameter	The size above which 50% of the particles of a distribution lie
Particle density	The mass per unit volume of a (effective) particle, excluding the open pores but including the closed pores
Size distribution	The proportion of particles in different size ranges. Important to define whether number, volume, or weight distribution. The size analysis method also affects the size of the particle that is reported. May be percentage in size range or cumulative distribution. There are graphs with specialized axes, which can be informative

In any size analysis method, a number of assumptions are made. The distribution obtained is likely to be different when one method is compared with another. It is best to compare results within one method and it is important to note the method employed when reporting results. Also, note that measuring the particles by volume, weight, surface, or number will give size distributions that appear to be different for the same population of particles.

REFERENCES

1. Hunziker, O.F. *Condensed Milk and Milk Powder*, 6th ed. Illinois: Hunziker, La Grange, 1946.
2. Kessler, H.G. *Food Engineering and Dairy Technology*. Freising: Kessler Verlag, 1981.
3. Singh, H. and Newstead, D.F. Aspects of Proteins in Milk Powder Manufacture. *Advanced Dairy Chemistry Volume 1: Proteins*, Fox, P.F., Ed. London: Elsevier Applied Science, 1992, pp. 735–765.
4. Knipschildt, M.E. and Andersen, G.G. Drying of Milk and Milk Products. *Modern Dairy Technology*, 2nd ed. Robinson, R.K., Ed. London: Chapman & Hall, 1994, pp. 159–224.
5. Caric, M. *Concentrated and Dried Dairy Products*. New York: VCH Publishers, 1994, p. 66
6. Pisecky, J. *Handbook of Milk Powder Manufacture*. Niro A/S, Copenhagen, 1997, pp. 68–81, 140–142, 182, 199–204, 214–215.
7. Early, R. Milk Concentrates and Milk Powders. The *Technology of Dairy Products*, Early, R., Ed., 2nd ed. London: Blackie Academic & Professional, 1998, pp. 228–300.
8. Schorsch, C., Wilkins, D.K., Jones, M.G., and Norton, I.T. Gelation of casein-whey mixtures: effects of heating whey proteins alone or in the presence of casein micelles. *J. Dairy Res.*, 68: 471–481, 2001.
9. Walstra, P. Casein sub-micelles: Do they exist? *Int. Dairy J.*, 9: 189–192, 1999.

10. Horne, D.S. Casein interactions: Casting light on the black boxes, the structure in dairy products. *Int. Dairy J.*, 8: 171–177, 1998.

11. Walstra, P. and Jenness, R. *Dairy Chemistry and Physics*. New York: John Wiley & Sons Ltd, 1984, pp. 171–181.

12. Wong, N.P. Ed. *Fundamentals of Dairy Chemistry*, 3rd ed. New York: Van Nostrand Reinhold Company, 1988.

13. Parry, R.M. Milk coagulation and protein denaturation. *Fundamentals of Dairy Chemistry*, 2nd ed. Webb, B.H., Johnson, A.H., and Alford, J.A., Eds. Connecticut: AVI Publishing Co., Westport, 1974, pp. 603–661.

14. Masters, K. *Spray Drying Handbook*, 5th ed. Harlow: Longman Scientific and Technical, 1991, pp. 33, 444–488.

15. Hall, C.W. and Hedrick, T.I. *Drying of Milk and Milk Products*, 2nd ed. Connecticut: AVI Publishing Co, Westport, 1971, pp. 6, 152–155, 179–181.

16. Millen, J. *From Joseph Nathan to Glaxo Wellcome: The History of Glaxo in New Zealand*, 2nd ed. Glaxo New Zealand Ltd, Auckland: Glaxo Wellcome, 1997.

17. Hobman, P.G. Ultrafiltration and manufacture of whey protein concentrates. *Whey and Lactose Processing*, Zadow, J.G., Ed. London: Elsevier Applied Science, 1992, pp. 195–230.

18. Jiminez-Flores, R. and Kosikowski, F.V. Properties of ultrafiltered skim milk retentate powders. *J. Dairy Sci.*, 69: 329–339, 1986.

19. Novak, A. Milk Protein Concentrate. New Applications of Membrane Processes, IDF Special Issue No. 9201. Brussels: International Dairy Federation, 1991, pp. 51–66.

20. Food and Agriculture Organisation. 2002, FAOSTAT Agriculture database http://apps.fao.org/cgi-bin/nph-db.pl?subset=agriculture.

21. Hoppe, G.K. and Higgins, J.J. Demineralization. *Whey and Lactose Processing*, Zadow, J.G., Ed. London: Elsevier Applied Science, 1992, pp. 91–131.

22. de Castro, M. and Harper, W.J. Effect of drying on characteristics of 70% milk protein concentrate. *Milchwissenschaft*, 56: 269–272, 2001.

23. Kjærgaard Jensen,G. and Oxlund, J.K. Concentration and drying of whey and permeates. *Bull. Int. Dairy Fed.*, 233: 4–20, 1988.

24. New Zealand Department of Labour. Approved code of practice for the prevention, detection and control of fire and explosion in New Zealand dairy industry spray drying plant. New Zealand Department of Labour, Wellington, 1993.

25. Beever, P.F. Fire and explosion hazards in the spray drying of milk. *J. Food Technol.*, 20: 637–645, 1985.

26. Jouppila, K. and Roos, Y.H. Glass transition and crystallization in milk powders. *J. Dairy Sci.*, 77: 2907–2915, 1994.

27. Wallack, D.A. and King, C.J. Sticking and agglomeration of hygroscopic, amorphous carbohydrate and food powders. *Biotechnol. Prog.*, 4: 31–35, 1988.

28. Bronlund, J. The modelling of caking in bulk lactose. A thesis presented in partial fulfilment of the requirements for a PhD in Process and Environmental Technology, Massey University, NZ, 1997.

29. Adhikari, B., Howes, T., Bhandari, B.R., and Truong, V. Stickiness in foods. A review of mechanisms and test methods. *Int. J. Food Prop.*, 4: 1–33, 2001.

30. Hennigs, C., Kockel, T.K., and Langrish, T.A.G. New measurements of the sticky behavior of skim milk powder. *Drying Technol.*, 19: 471–484, 2001.

31. Roos, Y. Melting and glass transition of low molecular weight carbohydrates. *Carbohydr. Res.*, 238: 39–48, 1993.

32. Boersen, A. The production of non-caking whey permeate powder. Paper presented to the International Food Dehydration Conference: 2000 and Beyond. Melbourne, Australia, March 29–30, 2000.

33. Fäldt, P., Bergenståhl, B., and Carlsson, G. The surface coverage of fat on food powders analyzed by ESCA (electron spectroscopy for chemical analysis). *Food Struct.*, 12: 225–234, 1993.

34. Kim, E.H.-J., Chen, X.-D., and Pearce, D.L. Surface characterization of four industrial spray-dried dairy powders in relation to chemical composition, structure and wetting property. *Colloids Surf. B: Biointerfaces*, 26: 197–212.

35. de Jong, P. and Verdurmen, R.E.M. Concentrated and dried dairy products. In *Mechanisation and Automation in Dairy Technology*, Tamine, A.Y. and Law, B.A., Eds. Sheffield: Sheffield Academic Press, 2001, pp. 95–118.

36. de Jong, P. Modelling and optimization of thermal treatments in the dairy industry. A thesis presented in partial fulfilment of the requirements for a PhD. Department of Reactor and Catalysis Engineering, Technical University of Delft, Delft, The Netherlands, 1996.

37. Truong, H.T. Fouling of stainless steel surfaces by heated whole milk. A thesis presented in partial fulfilment of the requirements for a PhD. Department of Food Technology, Massey University, NZ, 2001.

38. Goldberg, J.E. Prediction of spray dryer performance. DPhil Thesis, Oxford University, Oxford, UK, 1987.

39. New Zealand Ministry for the Environment. Ambient Air Quality Guidelines July 1994. New Zealand Ministry for the Environment, Wellington, p. 26.

40. O'Donnell, C.P. Identification and minimization of product losses in a milk processing plant with particular reference to spray dryer stack losses. A thesis presented in partial fulfilment of the requirements for a PhD. Department of Agricultural and Food Engineering, University College Dublin, National University of Ireland, 1992.

41. Allen, T. *Particle Size Measurement*, 4th ed. London: Chapman and Hall, 1990, pp. 124–189.

42. Houska, M. (Ed.), Adam, M., Celba, J., Havlicek, Z., Jeschke, J., Kubesova, A., Neumannova, J., Pokorny, D., Sestak, J., and Sramek, P. Milk, Milk Products and Semi-products: Thermophysical and Rheological Properties of Foods. Food Research Institute, Prague, Institute of Agricultural and Food Information, Ministry of Agriculture, Prague, 1994.

43. Niro. *Analytical Methods for Dry Milk Products*, 4th ed. Niro Atomiser A/S, Copenhagen, Denmark, 1978, pp. 10–20, 26–27, 48–51, 83–85.

44. Baldwin, A.J. The appearance of whole milk powder as related to physical properties. *NZ J. Dairy Sci. Technol.*, 12: 201–202, 1977.

45. Bloore, C.G. and Baldwin, A.J. Flow properties of milk powder. 44th Annual Report. New Zealand Dairy Research Institute, Palmerston North, 1972, pp. 64–65.

46. Baldwin, A.J. and Sanderson, W.B. Factors affecting the reconstitution properties of whole milk powder. *NZ J. Dairy Sci. Technol.*, 8: 92–100, 1973.

47. Baldwin, A.J. and Woodhams, D.J. The dispersibility of skim milk powder at high total solids. *NZ J. Dairy Sci. Technol.*, 9: 140–151, 1974.

48. Baldwin, A.J. and Ackland, J.D. Effect of preheat treatment and storage on the properties of whole milk powder: changes in physical and chemical properties. *Neth. Milk Dairy J.*, 45: 169–181, 1991.

49. Hurrell, R.F. Reactions of food proteins during processing and storage and their nutritional consequences. *Dev. Food Proteins*, 3: 213–244, 1984. Hudson, B.J.F., Ed. London: Elsevier Applied Science.

50. Lloyd, R.J., Chen, X.D., and Hargreaves, J.B. Glass transition and caking of spray-dried lactose. *Int. J. Food Sci. Technol.*, 31: 305–311, 1996.

51. Standards for Grades of Dry Milks Including Methods of Analysis. American Dry Milk Institute, Bulletin 916, 2nd ed. Chicago, Illinois, 1971, pp. 9, 49–52.

52. International Dairy Federation. Determination of free fatty acids in milk and milk products. Bulletin 265, 1991. International Dairy Federation, Brussels.

53. Evers, J.M., Luckman, M.S., and Palfreyman, K.R. The BDI method — Part 1: determination of free fatty acids in cream and whole milk powder. *Aust. J. Dairy Technol.*, 55: 33–36, 2000.

54. Schuck, P. and Dolivet, A. Lactose crystallization: determination of α-lactose monohydrate in spray dried milk products. 1st International Symposium on Spray Drying of Milk Products. October 16–18, 2001, Rennes, France.

55. Boon, P.M. and Woodhams, D.J. The moisture content of milk powders. *NZ J. Dairy Sci. and Technol.*, 9: 151–155, 1974.

56. International Dairy Federation. Determination of insolubility index. Method 129A, 1988. International Dairy Federation, Brussels.

57. Standards for Grades of Dry Milks Including Methods of Analysis. American Dry Milk Institute, Bulletin 916. Chicago, Illinois, 1991.

58. Keeney, M. and Basette, R. Detection of intermediate compounds in the early stages of browning reaction in milk products. *J. Dairy Sci.*, 42: 945–960, 1959.

59. Henle, T., Zehetner, G., and Klostemeyer, H. Fast and sensitive determination of furosine. *Zeitschrift fur Lebensmittel Untersuchung und Forschung*, 200: 235–237, 1995.

60. International Dairy Federation. HPLC of Furosine for Evaluating Maillard Reaction Damage in Skimmilk Powders During Processing and Storage. IDF Bulletin No. 298, 1995. International Dairy Federation, Brussels.

61. Loftus-Hills, B.A. and Thiel, C.C. The ferric thiocyanate method of estimating peroxide in the fat of butter, milk and dried milk. *J. Dairy Res.*, 14: 340–353, 1945.

62. Newstead, D.F. and Headifen, J.M. A reappraisal of the method for estimation of the peroxide value of fat in whole milk powder. *NZ J. Dairy Sci. Technol.*, 16: 13–18, 1981.

63. Baldwin, A.J., Cooper, H.R., and Palmer, K.C. Effect of preheat treatment and storage on the properties of whole milk powder. Changes in physical and chemical properties. *Neth. Milk Dairy J.*, 45: 97–116, 1991.

64. Badings, H.T. Cold-storage defects in butter and their relation to the auto-oxidation of unsaturated fatty acids. *Neth. Milk Dairy J.*, 24: 144, 1970.

65. Chen, X.D. Mathematical analysis of powder diskharge through longitudinal slits in a slowly rotating drum: objective measurements of powder flowability. *J. Food Eng.*, 21: 421–437, 1992.

66. British Standards Institution. Methods for the use of fine mesh test sieves. British Standard 1796: part I. British Standards Institution, London, 1989.

67. International Standards Organisation. ISO 13320-1. Particle size analysis: Laser diffraction method. Part 1: General principles. International Standards Organisation, Geneva, 1999.

68. International Dairy Federation. Wettability. Method 87: 1979. International Dairy Federation, Brussels, 1979.

69. Baldwin, A.J., Kjægaard Jensen, G., and Nielsen, P. An examination of tests for the measurement of the solubility of instant whole milk powder. Report 251. Statens Forsogsmejeri, Hillerod, 1982.

70. International Dairy Federation. Inventory of IDF/ISO/AOAC International Adopted Methods of Analysis and Sampling for Milk and Milk Products, 6th ed, IDF Bulletin No. 350, 2000. International Dairy Federation, Brussels.

71. Lewis, M.J. and Heppell, N.J. *Continuous Thermal Processing of Foods: Pasteurization and Sterilization*. Gaithersburg, Maryland: Aspen Food Engineering, 2000.

72. International Dairy Federation. Proceedings of the IDF Seminar on Recombination of Milk and Milk Products, Singapore, October 7–10, 1982. IDF Bulletin No. 142. International Dairy Federation, Brussels.

73. Bylund, G. *Dairy Processing Handbook*. Tetra Pak: Lund, 1995, p. 382

74. Newstead, D.F. and Howell, J.S. Recombined milk products: The first 60 years. An annotated bibliography. A Fonterra Research Centre Handbook. Fonterra Research Centre, Palmerston North, 1997.

75. Snow, R.H., Kaye, B.H., Capes, C.E., and Srestry, G.C. Size reduction and size enlargement. *Perry's Chemical Engineers' Handbook*, 6th ed, Perry, R.H., Green, D.W., and Maloney, J.O., Eds. New York: McGraw Hill, 1984, pp. 8–38 ff.

76. Augustin, M.A. and Clarke, P.T. Effects of added salts on the heat stability of recombined concentrated milk. *J. Dairy Res.*, 57: 213–226, 1990.

77. Newstead, D.F. Sweet-cream buttermilk powders: key functional ingredients for recombined milk products. Proceedings of 3rd International Symposium on Recombined Milk and Milk Products, 1999. IDF Special Issue No. 9902. International Dairy Federation, Brussels, 1999.

78. Newstead, D.F. Recombined sweetened condensed milk. Proceedings of the IDF Seminar on Recombination of Milk and Milk Products, Singapore, October 7–10, 1982, pp. 59–62, IDF Bulletin No. 142. International Dairy Federation, Brussels.

79. Jenness, R. and Patton, S. *Principles of Dairy Chemistry*. London: Chapman and Hall, 1959.

80. Choat, T. Recombined sweetened condensed milk recombined filled sweetened condensed milk. In *Monograph on Recombination of Milk and Milk Products* (Technology and Engineering Aspects), International Dairy Federation, Brussels, p. 24, IDF Bulletin No. 116, 1979.

81. Burton, H. *Ultra-High-Temperature Processing of Milk and Milk Products*. London: Elsevier Applied Science, 1988.

82. Newstead, D.F., Groube, G.F., Smith, A.F., and Eiger, R.N. Fouling of UHT plants by recombined and fresh milk. Some effects of preheat treatment. Proceedings of a Conference Held at Jesus College, Cambridge, 6–8 April 1998; Fouling and Cleaning in Food Processing '98. Wilson, D.I., Fryer, P.J., and Hastings, A.P.M., Eds. Doc EUR 18804, European Commission (Brussels), 1999.

83. Fonterra Research Centre. Methods for Quality Assessment of UHT Milks. 2nd ed, Newstead, D.F., Ed. A Fonterra Research Centre Handbook. Fonterra Research Centre, Palmerston North, 2000.

84. Newstead, D.F. Observations on UHT plant fouling by recombined milk. 1994 Middle East Recombining Seminar "New Ideas for Quality and Innovation", Bahrain, 26–27 March, 1994. New Zealand Milk Products (Middle East) EC.

85. Mulvihill, D.M. and Fox, P.F. Properties of Milk Proteins. *Developments in Dairy Chemistry: Four Functional Milk Proteins*, Fox, P.F., Ed. London: Elsevier Applied Science, 1989, pp. 137–141.

86. Kinsella, J.E. and Whitehead, D.M. Proteins in whey: chemical, physical and functional properties. *Adv. Food Nutr. Res.*, 33: 343–438, 1989.

87. Fox, P.F. *Developments in Dairy Chemistry-4: Functional Milk Proteins*. Fox, P.F., Ed. 1989. London: Elsevier Applied Science.

88. Morfin, I. Hydration properties of 56% milk protein concentrate. A dissertation presented in partial fulfilment of the requirements for the postgraduate diploma in Dairy Science and Technology of Massey University, New Zealand, 1997.

89. Carr, A.J. The functional properties of milk protein concentrates. A thesis presented in partial fulfilment of the requirements for a PhD in Food Technology. Massey University, Palmerston North, New Zealand, 1999.

90. Oldfield, D.J., Teehan, C.M., and Kelly, P.M. The effect of preheat and other process parameters on the coffee stability of instant whole milk powder. *Int. Dairy J.*, 10: 659–667, 2000.

91. Berry, D. Making milk better. *Dairy Foods*, October, 102(10): 80–82, 2001.

92. Donnelly, P. Competitive integration of technology and marketing for bioactive dairy ingredients. Proceedings of International Dairy Federation Nutrition Conference, Paper 1.2, CD Rom. IDF Nutrition Conference, IDF World Dairy Summit, 2001, Auckland, New Zealand. International Dairy Federation, Brussels.

93. Fluid Milk Strategic Thinking Initiative. Fluid milk's role in the functional foods movement: milk's unique nutrient profile and functional ingredient opportunities. Official report to the industry, 2001. www.milkplan.org.

94. Angus, F. and Miller, C. Functional Foods 2000. Conference Proceedings, Leatherhead Publishing, Leatherhead, England, 2000.

95. Augustin, M.A. Functional foods: An adventure in food formulation. *Food Austr.*, 53: 428–432, 2001.

96. Heasman, M. and Mellentin, J. *The Functional Foods Revolution: Healthy People, Healthy Profits?* London: Earthscan, 2001.

97. Rowan, C. Fighting through the functional maze. *Food Eng. Ingredients*, October, 15: 68–81, 2001.

98. Young, J., Ed. *Guide to Functional Food Ingredients*. Leatherhead Publishing, 2001.

99. Reynolds, E. Calcium phosphopeptide complexes. Patent WO 98/40406. Held by University of Melbourne and Victorian Dairy Industry Authority, 1998.

100. International Dairy Federation. International Dairy Federation Nutrition Conference Proceedings, CD Rom. IDF Nutrition Conference, IDF World Dairy Summit, 2001, Auckland, New Zealand. International Dairy Federation, Brussels.

101. Haylock, S.J. Fortification of consumer milk products. Proceedings of International Dairy Federation Nutrition Conference, Paper 5.4, CD Rom. IDF Nutrition Conference, IDF World Dairy Summit, 2001, Auckland, New Zealand. International Dairy Federation, Brussels.

102. Steijns, J.M. Milk ingredients as nutraceuticals. *Int. J. Dairy Technol.*, 54: 81–88, 2001.

103. Baker, E.N., Baker, H.M., Koon, N., and Kidd, R.D. Lactoferrin: Bioactive properties and applications. Proceedings of International Dairy Federation Nutrition Conference, Paper 3.2. CD Rom. IDF Nutrition Conference, IDF World Dairy Summit, 2001, Auckland, New Zealand. International Dairy Federation, Brussels, 2001.

104. Kunz, C. and Rudlolf, S. Health benefits of milk-derived carbohydrates. Proceedings of International Dairy Federation Nutrition Conference, Paper 4.1. CD Rom. IDF Nutrition Conference, IDF World Dairy Summit, 2001, Auckland, New Zealand. International Dairy Federation, Brussels.

16

Properties of Culinary Powders: Salt and Sugar

C. I. Onwulata
U.S. Department of Agriculture
ARS, Eastern Regional Research Center
Wyndmoor, Pennsylvania

CONTENTS

I. INTRODUCTION

Salts and sugars are essential food powders used in great quantities, and both play major roles in food processing and technology. Salts and sugars have been used over the millennia for consumption and food preservation. They enhance the quality, texture, flavor, and add esthetic value to many foods. Salts and sugars function as preservatives inhibiting microbial growth, as flavor enhancers to improve taste, and as thickeners to enhance texture in foods (Marth, 1978). Salts harvested from deep underground mines or dried ocean water, and sugars derived from crops such as sugarcane or beets, are processed and formed into crystals or particles of various shapes. The crystal or particle size and shape allows for different functions, for example, large salt crystals are used on pretzels and finely ground sugar is used as dusting on puffed pastry such as on doughnuts. Salts (sodium chloride) and sugars (sucrose) have extremely low moisture and water activity when compared to other food powders, and are chemically pure compounds. The difference in functionality from one batch of salt or sugar to another (batch of salt or sugar particles) is (only from) the effect of the size and shape of the particles. Powders are composed of many crystals or particles.

Salts and sugars have complex morphological features such as pores, irregular shapes, and nonuniform particles that affect bulk properties, causing handling problems (Peleg, 1987). In general, the morphology of a given powder, a mass or batch of particles or crystals, affects physical properties such as bulk and particle density, solubility, and agglomeration tendency. Also, the bulk density of powder is closely related to other bulk properties such as flowability or compactibility force (Abdullah and Geldart, 1999). The size of the powder particle plays an important role in its compressibility or "packing," and there is greater reduction in volume with larger particles of most food powders when compressed. The form of particle whether filled or porous, the state, whether glassy or amorphous, makes all the difference. The properties of porous particles are difficult to characterize because they tend to be relatively very cohesive (Chang et al., 1998). So, (1) the morphology of the particle, (2) size, and (3) density, chiefly determines its characteristics. Particle size and shape are important forms that affect the physical characteristics of the substance and behavior such as adherence.

The single most important characteristic of powders is the crystal structure of the particle (Niman, 2000). Particle size is an established parameter easily characterized in the laboratory, but particle shapes are determined by sophisticated image analysis. Determining both shape and size accurately provides a good understanding of the functional characteristics of any particulate food powder, including salts and sugars. For example, the scanning electron microscope images of salts (see Figure 1) and sugars (see Figure 2), show a wide range of salt particle sizes and shapes. The sizes range from greater than 800 to less than 300 μm for pretzel (Figure 1[a]) and fine flour (Figure 1[f]), respectively. The other sizes fall in between (Figure 1[b] to 1[e]).

A practical use of size and shape in food application is the use of salts as toppings on pretzels as mentioned earlier in this section. Several grades of salts and sugars are available, ranging from large particulates to very fine powders. The grades are based on the following key attributes: (1) adherence, ability to attach to a surface; (2) blendability, ability to resist separation; (3) flowability, ease of dispensing; friability or grinding down, ability to

Figure 1 Scanning electron micrograph (100×) showing different sizes and shapes of salt particles. (a) Rock pretzel; (b) Top flake; (c) Lite mixture; (d) Culinox; (e) Star flake; and (f) Fine flour.

Figure 2 Scanning electron micrograph (100×) showing different sizes and shapes of sugar crystals. (a) Medium granulated; (b) Standard grain; (c) Fine grain; and (d) Sanding.

withstand handling; (4) liquid absorption, ability to absorb oil or fat; and (5) solubility, ease of dissolution (Niman, 2000). Other intrinsic qualities of salt and sugar powders such as moisture content, water activity (a_w), and chemical composition, exert great influence on bulk flow properties. For instance, the relative humidity of a powder influences bulk properties, can correlate flow difficulties, caking, and the packing of crystals into an aggregate by force (Peleg, 1985; Yan and Barbosa-Canovas, 1997). The different sizes and shapes of sugar crystals shown in Figure 2 are described in Table 2.

Other food products such as milk, corn and wheat flours, as well as salts and sugars, are at some point in the processing stages handled as powders. Major considerations in processing food powders is to have them in a state that minimizes flow difficulties when conveyed through pipes or chutes and in bulk handling equipment such as bins or conveyors. The ease with which powders flow depends on their particle size, density, shape, water activity, and moisture content (Teunou et al., 1999).

II. HISTORICAL NOTES

Salts, composed mostly of sodium chloride, have been mined from deep caverns or harvested from sun-dried ocean water and used in various applications since ancient times. Their most important application is in food processing and preservation. Salts perform many functions in food processing such as, helping to develop natural flavors, making foods more palatable, protecting foods and keeping them safe by retarding growth of spoilage microorganisms, enhancing the texture, color, aroma, and improving their appearance. Processed foods are the single most essential source of dietary salt in the mammalian diet. Salts are important for maintaining health and preserving life.

Sugar (sucrose) commonly used in food preparation come, to a large extent, from extracts of sugarcane (*Saccharum officianarum*) and sugar beets (*Beta vulgaris*), whereas liquid sugar is mostly derived from corn syrup. Processed sugar crystal used in food processing come in varying sizes ranging from large granules to finely ground crystals (see Figure 2).

Sugars perform a variety of functions in foods such as sweetening, bulking, preserving, texturing, and flavoring. Large amounts of sugars are used in bakery and confectionery products, and in beverages. The use of the sugar particles in food processing depends on their size. For example, bakers use coarse or fine powder, while confectioners use even finer powders. There is no standard definition of powder grades, so manufacturers' grades differ widely, and the names of the different grades vary. A general classification in order from large to small is: coarse, large, standard, and fine (Hui, 1992).

III. GENERAL KNOWLEDGE

Irani and Callis (1963) described a particle of a substance as: "that state of subdivision of matter whose shape depends on the process by which it was formed and on the intra-molecular cohesive forces present." A particle may be a single crystal or an amorphous mass of material. An aggregate is an assembly of particles held together by strong inter- or intramolecular or atomic cohesive forces. The size of a particle or aggregate is the representative dimension that best describes the degree of comminution of a particle formed from a larger size material, and for a spherically symmetric particle, is its diameter. As most solid particles are not spherical and symmetric, a description of their shape becomes essential. Naturally, particle shapes and sizes differ widely. Uniformity of crystal size prevents separation or settling of smaller crystals, an important property in dry blending of powders. For example, in confectionery, cocoa is blended with powdered sugar. Sphericity is a measure of the external surface area of a sphere to that of a particle of equivalent volume, representing the deviation of particle shape from spherical symmetry (Irani and Callis, 1963).

Particle size determines a particulate solid's flow ability and particle shape establishes its cohesive tendency, and both are directly related to its bulk property. Particulates having a very uniform size (± 20 μm), exhibiting low adhesive attraction properties such as fine salt and sugar powders, flow freely, and are the easiest to handle. Several kinds of particulate material require further treatment, such as size reduction to improve their flow characteristics. Next in flowability are irregular but nonadhesive particles such as cereal grains. But cereal grain flours and other food powders are generally more cohesive and less free flowing, and seldom exhibit particle segregation. The particles may be agglomerated and form dendritic appendages, having a branched, crystalline shape with the branches extending from the faces of the main body. These dendrites often exhibit electrostatic charge potential. These aggregates cohere due to both internal and external forces, and their nonuniform particle size and shape, nonuniform length to diameter ratios, and their interlocking tendencies. Typical examples are cereal grain products such as wheat flour, corn meal flour, and high-fat milk powders (Kolatac, 1996).

IV. PROPERTIES OF POWDERS

Powders can be characterized by different inherent properties such as by appearance, average size of particles, shape, bulk density, compressibility, or by the angle of slide and repose of a mass of the powder.

A. Particle Size Variations

Size is a very important particle property that affects behavior (See Section I). A particular size is achieved either by crystal growth or by size reduction through milling. If particle sizes are relatively uniform, powder handling characteristics and other properties are enhanced, and bulk flow problems are minimized. Irregular particle sizes inherently cause flow problems. The particle size distribution is derived from aggregates of a powder.

B. Particle Shape

The shape of a particle affects its handling and flow properties. Naturally, shapes vary, from spherical, typical of spray-dried food particles, to square or rhomboid, typical of crystal-grown particles of salts and sugars. Circularity, sometimes referred to as sphericity, is a measure of particle shape. For the measure of circularity, the circumference of the particle is divided by the perimeter, defined mathematically as: ($2\sqrt{\pi}$ (Area of the particle)/perimeter of the particle). Circularity of 1.0 describes a perfect sphere with the greatest ease of flow. As circularity declines below 0.9, the particles become coarser and flowability is reduced. Sphericity of an object expresses the shape character of the solid relative to that of a sphere of the same volume. It is a ratio of the diameter of a sphere of the same volume as the object to the diameter of the smallest circumscribing sphere. Sphericity is also calculated as the ratio of geometric mean diameter of the object to the major diameter. The largest intercept, "a," second largest intercept, "b," normal to "a" and the third intercept, "c," normal to both "a" and "b" are measured; and sphericity is computed as: $S = [(abc)\frac{1}{3}]/a$ (Mohsenin, 1980).

C. Bulk Density

Bulk density is a measure of a mass of particles' weight per unit volume. Bulk density of a mass of particles, measured without applied force or pressure is termed loose density, but with applied force or pressure, is packed density. Bulk density is a good indicator of flowability, the ability of the mass of particles to flow, and its handling characteristics. For example, when a large difference exists between loose- and packed-density values, powder handling problems often exist. Bulk density varies with changes in particle size distribution, particle shape, moisture content, headloads, flow velocities, material temperature, and friability. Therefore, as any of the preceding features change, the particle shape changes and bulk density changes too.

D. Compressibility

Compressibility of a powder is a measure of its particle strength, which depends on its form and shape. Compressibility also determines the powder's flow properties. When a mass of particles is compressed, the voids between particles are reduced or eliminated and the powder tends to become a solid mass with fixed shape. Free-flowing powders are less compressible while nonfree-flowing powders are more. Free-flowing powders are generally less than 20% compressible as a rule.

E. Angle of Repose and Angle of Slide

The angle a batch of particles forms when poured on a flat surface, from the top of the batch to its base, is the angle of repose. The slide angle is the angle at which the batch will begin to slide on a smooth, flat surface, by its own weight. Slide angle will vary depending on the moisture content, particle size and shape, and the smoothness of the flat surface. Angles of repose and slide are also indicators of powder flowability (See Section X).

F. Factors Affecting Properties of Powders

External factors such as moisture, water activity, temperature, and presence of flow agents or other powders affect the flow of powders. Moisture content affects flow considerably. Surface water increases cohesion. Attractive forces are created by liquid bridges mostly from free moisture on the surface of the particles, making particles more adhesive. These forces influence bulk properties, for example, raising the loose bulk density in powdered sugar and salt (Peleg, 1983). Water activity increases with increasing moisture content. Changes in temperature may affect the surface water and can induce a physical change or initiate a chemical reaction. Interparticle friction can generate heat, which can dissolve and soften the surfaces of the particles leading to clumping, agglomeration, or caking. Anticaking agents are added to powders to reduce interparticle forces and improve powder flow characteristics, for example, silica derivatives are used to aid the flow of powders.

V. PROCESSING AND MANUFACTURE

Sugars are carbon-based compounds with many alcohol groups (-OH) attached. Simple sugars, monosaccharides, include fructose, dextrose, xylose, arabinose, and galactose. The most common sugar is sucrose, composed of a molecule each of D-glucose and D-fructose. The historic source of sugars is sugarcane (*S. officinarum*), a tropical plant, containing approximately 12 to 14% sucrose. Sugar is also manufactured from sugar beet (*B. vulgaris*), containing 10 to 15% sucrose. Sugarcane processing involves several unit operations that separate the sugar from the pulp and refine it into crystals. Sugar processing involves shredding the pulp, pressing the juice out of the pulp, clarifying the juice with calcium carbonate, concentrating by vacuum evaporation and crystallization, yielding golden raw sugar containing approximately 96 to 98% sucrose.

Food-grade salt is 99.9% sodium chloride mined from salt wells or from dried salt seabeds. Salt is mined by pumping water into the salt mine to form a brine stream that is then pumped into a vacuum pan, evaporated, and refined to produce salt. The process affects the size of the crystals, which depends on the vacuum conditions, rate of growth, and drying conditions. Further processing, sorting, or milling of the crystals produces the different sizes and grades.

VI. SIZING SALTS AND SUGARS

The manufacturing process affects both shape and size of salt and sugar particles. Large crystals are often milled to reduce size, and sieved to classify into different grades for specific use. There are a variety of sizes and shapes. Different sizes and shapes are used by food processors because of their functional properties. These sizes range from very large particles or crystals greater than 2000 μm, to intermediate particles ranging from 400 to 600 μm, and specially fine particles less than 140 μm. These particles or crystals go by different names given by manufacturers (names are not standard), such as, extra coarse, granulated, medium coarse, fines, sanding or flour; dendritic and flakes, etc., characteristic of their basic morphology. Each size and shape provides unique functional characteristics that make the salt or sugar appropriate for the food processor's special need. For example, large salt crystals are used in the pickled foods and on pretzels; medium salt crystals are used in baking and confectionery, while sugar fines and flours are sprinkled on top of baked goods.

VII. EFFECT OF SHAPES AND SIZES

A. Materials

The sizes and shapes of several kinds of salt (Figure 1) and sugar (Figure 2) as described in Sections VI through XI are listed in Table 1 and Table 2. The salts measured are from Morton Salt (Morton International, Inc., Chicago, IL), with typical chemical compositions of sodium chloride (98.9%), calcium sulfate (0.92%), sodium sulfates (0.13%), and traces of magnesium chloride, moisture, and heavy metals (<1 ppm). The sugars characterized are from Domino Sugar Corporation (Baltimore, MD), with typical chemical compositions of sucrose (99.93%), moisture (0.03%), invert (0.02%), and ash (0.02%). Statistical analyses and means testing of the results were performed using the General Linear Methods (GLM)

Table 1 Physical Characteristics and Coding of the Salt Powders[a]

Product	Code	Shape	Median size (μm)	Size range (μm)	Class
Rock pretzel salt	Pretzel	Rhombohedral	1000	840–1200	Granulated[1]
Top flake salt	Topflake	Flakes	600	420–480	Flake[5]
Lite salt mixture	Lite	Cube	400	200–600	Cube[2]
Culinox fine salt	Culinox	Cube	300	200–420	Cube[1]
Star flake dendritic salt	Dendritic	Porous cube	250	150–420	Cube[4]
Fine flour salt	Flour	Cube	200	100–300	Cube[3]

[a] Based on manufacturers trade name and physical characterization.
[b] Class codes indicate: [1] granular free flowing; [2] Nonuniform powder; [3] Uniform powder; [4] Cohesive uniform powder; [5] Nonuniform interlocking fibrous flake.

Table 2 Physical Characteristics and Codings of the Sugar powders[a]

Product	Code	Shape	Median size (μm)	Size range (μm)	Class
Medium granulated	Medium	Cube	1400	1200–1700	Granulated[1]
Standard grain	Standard	Rectangular	1200	1000–1400	Granulated[1]
Medium fine grain	Fine	Cube	750	600–850	Cube[2]
Sanding #2	Sanding	Cube	350	250–500	Cube[3]

[a] Based on manufacturers trade name and physical characterization.
[b] Class codes indicate: [1] granular free flowing; [2] Nonuniform powder; [3] Uniform powder;

procedure (SAS, 1998). All evaluations were repeated four times. The salt sizes studied range from 100 to 1200 μm (Table 1), and the shapes vary from irregular to cubic and rhomboid. The size spread for salt is seen clearly in Figure 1, from the coarse rhomboidal particles (Figure 1[a]), to the cube-shaped particles (Figure 1[c] and Figure 1[d]), and the irregular fines (Figure 1[f]).

Sugar crystals studied ranged from 250 to 1400 μm (Table 2). The typical cube-shaped particles are shown in Figure 2.

B. Physical Properties

The physical properties, moisture, water activity (a_w), substance density, Hausner ratio, flow, angle of repose, and compression are described below.

1. Moisture

Moisture is a vital physical attribute of food powders. Though pure salts and sugars are provided as comparatively dry compounds, when blended with other foods, they pick up moisture, become cohesive, and form hard cakes. Moist powders are more deformable under pressure than dry or noncohesive ones and are more likely to cause flow problems.

Moisture content of the salt crystals was determined by drying in a vacuum oven using AOAC method 925.55 for salt, which calls for sieving the samples through a U.S. standard mesh 20, which is approximately 850 μm, but must pass through No. 80 mesh, which is approximately 180 μm. Approximately 10 g of salt was weighed into a 200 ml Erlenmeyer flask, spread evenly in the bottom. The flask with salt was heated for periods of 1 h at approximately 250°C until two consecutive weighings agreed within 5 mg (AOAC, 1995).

Moisture content of the sugar crystals was determined by drying in a vacuum oven using AOAC Method 925.175 (a) for sugar, which calls for weighing approximately 5 g of sugar in a flat cup (Ni, Pt, or Al with tight-fit cover), for 3 h at 100°C. The samples were removed and the process repeated until change in weight between successive dryings at 1 h intervals was ≤2 mg. For large grain sugars, Medium and Standard, the temperature was increased from 105 to 100°C in the final heating periods to expel the last traces of occluded water. Loss in weight was reported as a percent moisture (AOAC, 1995).

The moisture content of salt and sugar powders is presented in Table 3 and Table 4. The moisture values for the differently sized salt powders ranged from 0.05 to 0.18%; these values were statistically similar to each other ($p < .05$). Moisture values for sugar powders ranged only from 0.03 to 0.07%, and were also statistically similar to each other. These powders were all very dry, much less than 1% moisture; therefore, the size of the different

Table 3 Size, Moisture, and Water Activity of Salt
Powders

Product salt	Size (μm)	Moisture (%)	a_w
Pretzel	1000	0.06	0.68
Top flake	600	0.05	0.65
Lite	400	0.08	0.65
Culinox	300	0.07	0.66
Dendritic	250	0.18	0.66
Flour	200	0.10	0.62

Table 4 Size, Moisture, and Water Activity of Sugar
Powders

Product sugar	Size (μm)	Moisture (%)	a_w
Medium	1400	0.06	0.67
Standard	1200	0.04	0.64
Fine	750	0.07	0.65
Sanding	350	0.03	0.63

grades of salt and sugar is not related to moisture. Surface water content of food powder
particles is important because of the tendency for water films to form "liquid bridges" that
retard powder flow and increase compressibility, surface deformation, or fracture (Peleg,
1985). Bridging effects are unlikely for salt and sugar because of their extremely low overall
moisture content. Sugars and salts are unlikely to fracture or deform, except under great
stress, which is not typical of ordinary uses.

Cohesiveness of fine powdered foods increases with increasing water vapor pressure.
Temperature also affects cohesiveness (caking). Increasing temperature increases water
activity, therefore, higher temperature results in higher caking due to increased water activity
if water vapor is in contact with the food. Dry sugar is mostly in the glassy state and is usually
free flowing. When temperature rises, the water activity can increase and sugars sometimes
change into the sticky crystalline form. Crystalline sugar has increased cohesion and caking
tendency. Different sugars crystallize at different rates. For example, sucrose crystallizes
readily while fructose is resistant to crystallization. A typical absorption isotherm for sucrose
at 25°C showing the amorphous glassy state on the right and the crystalline state on the left
is depicted in Figure 3, with the critical transition zone of 35 to 45% relative humidity.

2. Water Activity (a_w)

A water activity meter is sometimes used for a_w determination. Water activity meters are
more suited for food systems. Water activity is an important food property related to the
previously mentioned intrinsic factors (see Section IV). Water activity is the ratio of actual
water vapor pressure (ρ) and the maximum possible water vapor pressure of pure water
(ρ_0) or saturation pressure at the same temperature [$a_w = \rho/\rho_0$]. The a_w of salt and sugar
powders are presented in Table 3 and Table 4. For the salt and sugar samples, an AquaLab
CX-2 Water Activity System (Decagon Devices., Pullman, WA) was used. Water activity
for both salt and sugar did not vary significantly, ranging only from 0.62 to 0.68 for salt
and from 0.63 to 0.67 for sugar at 25°C, though it has been thought that $a_w > 0.45$ has

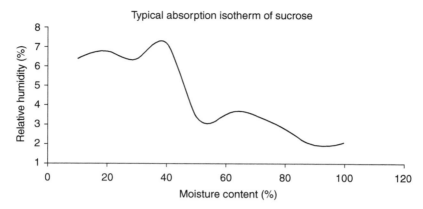

Figure 3 Typical absorption isotherm of sugar crystals, showing characteristic transition from the amorphous to the crystalline state around 40% moisture.

a propensity for caking (severe agglomeration). But this effect is solely dependent on the chemical composition of the powder (Moreyra and Peleg 1980; Peleg, 1977). When we compressed salt and sugar powders with a_w in the range 0.62 to 0.68, no cohesiveness, deformation, or caking was observed (see Table 5 and Table 6).

C. Sizes and Shapes

Particles with uniform shapes, such as spheres can be easily characterized by one dimension, such as diameter. Food powders are more complex particles and require more measurements to determine their shapes. The process of manufacturing particles give rise to a wide range of sizes and shapes in a given powder, and knowing particle size distribution is essential to understanding its relation to bulk properties. Dry sieving is a simple way of estimating the size range of a mixture of particles. The particle size distribution of salts and sugars was estimated by passing 500 g of each powder through a series of sieves with screen openings ranging from 150 μm to 2.00 mm. The distributions were determined using a mechanical sieve shaker, Retsch AS 200 Digit (Retsch, Rheinieche, Germany), set at amplitude of 60, for 30 min. The sieves were stacked top to bottom in order, 2.00 mm, 1.00 mm, 500 μm, 250 μm, 150 μm, and the fines >150 μm. The number reported for each powder is the average of three tests. The median sizes of salt and sugar particles ranged from 200 to 1000 μm for salt and from 350 to 1400 μm for sugar (see Table 1 and Table 2); the shapes varied, from cubes to rectangular to rhombohedral. The distribution of salt and sugar particles by size, are presented in Figure 4 and Figure 5. The distribution pattern of smaller salt particles is normal and skewed to the left, but the large particle Pretzel® salt had a right skewed distribution.

   The larger sugar crystals, Granulated and Standard, were skewed to the right, except for the Standard® particles, and the smaller particles, Fine and Sanding, had right-skewed distribution. The skew of a mass of crystals depend on the process of manufacture of the crystals, which gives them shape and hence distribution pattern. Peleg (1987) showed that the properties of individual particles, shape, size distribution, and skew affect flow properties. Small variations in surface properties (shape) result in significant differences in flowability. Particle size is one of the most important physical properties that affects the flowability of powders (Teunou et al., 1999). Also, particle size has a direct effect on compressibility (Barbosa-Canovas et al., 1987). Particle size distribution is often represented

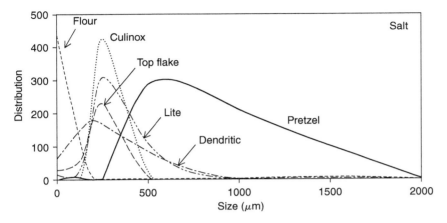

Figure 4 Particle size distribution of salt powders showing size-dependent patterns.

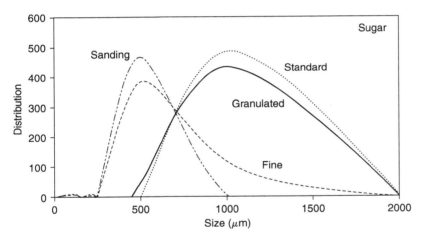

Figure 5 Particle size distribution of sugar powders showing size-dependent patterns.

as a distribution curve, either with the frequency as shown (Figure 4 and Figure 5) or as a cumulative distribution. Though size and shape are interrelated, different methods of measurement give different results. Particle size and shape affect bulk properties profoundly (Lewis, 1996). Accurately characterizing the particle sizes of food powders is now very important for manufacturers who have to meet international quality standards such as ISO 9000 and other performance standards such as accuracy and consistency while maintaining optimum quality (Phillips, 1997).

VIII. BULK PROPERTIES

Bulk properties such as density and compressibility depend on the behavior of a mass of particles or crystals with intrinsic individual properties. Bulk properties play a direct and major role in food packaging (Malave et al., 1985). They also provide indirect evidence of physical characteristics such as the extent of internal cohesion, which affects flowability and storage stability (Moreyra and Peleg, 1980).

Most particle density determinations are made with an air pycnometer. The gas pycnometer uses the ideal gas law to determine the volume of a given mass of sample within a chamber of known volume, after subtracting the gas volume of a given mass, which can be determined as a result of measuring change in pressure and gas volume. Bulk densities (g/cm^3) of salt and sugar powders are easily determined ordinary, direct measurements by dividing a given powder mass (g) contained in a polyethylene graduated cylinder, by its volume (cm^3).

The behavior of a compacted mass of particles, powder, as described by the "Hausner ratio," presents the relationship between tapped (packed) and loose bulk density (Hausner, 1967). Tapped density (ρ_T) is determined by measuring the density of the powders after "hand-tapping" the container 100 times at the rate of 50 taps/min. The loose-density value used to calculate the Hausner ratio, is the initial density of the filled cylinder before tapping.

A. Bulk Density

The bulk density of salt (Table 5) and sugar powders (Table 6) is significantly different ($p < .05$), owing to their chemical compositions. Salt powders are denser than sugar. The particle size of salt or sugar crystals makes a slight difference in the bulk density as seen in Table 5 and Table 6. Others have reported that particle size makes no difference (Moreyra and Peleg, 1980). The difference results from varying shapes that can interlock forming a rigid structure, and hence different densities. Salt and sugar particles are mostly noncohesive; therefore, the potential to form liquid bridges is minimal because of extremely low moisture content (see Table 5 and Table 6). It is expected that with differences in particle sizes the powders will pack differently resulting in size-dependent bulk densities. The effect of size on density is shown in Figure 1. Despite the wide range of sizes of salt and sugar crystals, the densities are similar, varying only (±0.05) for particles ranging from 200 to 1400 μm (see Table 5 and Table 6). Since density does not vary much with size, therefore, the size of the crystal is not a contributing factor when there is flow problem. Perhaps shape, rather than the particle size, affects bulk densities by bridging and consequently, results in flow problems.

The Hausner ratio, the ratio of tapped density and bulk density of salts and sugars (see Table 5 and Table 6) show that they are not very compressible. Salt crystals (0.84–0.90) are slightly more compressible than sugar (0.92–0.93) powders, but when compared to compressibility of other food powders such as milk, are hardly compressible at all (Konstance et al., 1995). For sugar powders, the Hausner ratio correlates moderately with

Table 5 Size, Bulk Density, and Hausner Ratio of Salt Powders

Product salt	Size (μm)	Bulk density (g/cm^3)	Hausner ratio
Pretzel	1000	2.18[a]	0.89[b]
TopFlake	600	2.20[a]	0.87[a]
Lite	400	2.10[b]	0.88[a]
Culinox	300	2.18[a]	0.90[b]
Dendritic	250	2.18[a]	0.87[a]
Flour	200	2.23	0.84[a]

$$\text{Hausner ratio} = \frac{(\text{Loose Density } [g/cm^3])}{(\text{Tapped Density } [g/cm^3])}.$$
Within a column, the same letter(s) are not significantly different.

Table 6 Size, Bulk density, and Hausner Ratio of Sugar
Powders

Product sugar	Size (μm)	Bulk density (g/cm^3)	Hausner Ratio
Medium	1400	1.60[a]	0.93
Standard	1200	1.59[b]	0.93
Fine	750	1.61[a]	0.92
Sanding	350	1.59[b]	0.92

Hausner Ratio $= \dfrac{\text{(Loose Density [g/cm}^3\text{])}}{\text{(Tapped Density [g/cm}^3\text{])}}$.

Within a column, the same letter(s) are not significantly different.

water activity, a_w, $R^2 = 0.77$, but for salt it does not ($R^2 = 0.22$). Hausner ratio values (<0.85) are known to result from an open bed structure formed by particles that are sometimes supported by liquid bridges and interparticle forces (Chang et al., 1998). Hausner ratio provides information on powder flow behavior as do other properties such as angle of repose, friction, and compaction (Grey and Beddow, 1969); it also predicts powder properties at different stages of processing including storage and transportation (Chang et al., 1998). The more the particle shape deviates from spherical, the lower the Hausner ratio, and the higher the compressibility (Hausner, 1967). Hence, irregularly shaped salt and sugar particles are harder to compress. That powder bulk properties depends more on shape than on any other physical attributes has been reported (Abdullah and Geldart, 1999).

IX. FLOW MEASUREMENTS

Flow of powder is also an important bulk handling property. For example, powders such as sugar dispersed in mixing tanks must possess satisfactory flowability to deliver the metered amount every time. Flow problems with powders occasionally lead to products made with insufficient material (Schubert, 1987). Powder flow can be described by two methods, dynamic (Section IX), that is moving powders, or static (Section X), that is powder at rest. The dynamic method is a measure of the free flow of a moving powder (Sjollema, 1963).

The mass flow of salt and sugar powders (g/sec) were measured by permitting 80 g of each powder to flow under gravity through 15 mm annular funnels. Flow rates of salt and sugar powders show that particle size is not a significant factor ($p < .05$) in flow (see Table 7 and Table 8).

Powder flow is characterized by two parameters, flow rate through an orifice (dynamic), and the angle (static) formed by a batch of powder on a flat surface. Dynamic flow varied among the salt and sugar powders measured. The small particle-sized powders, fine[®] salt (see Table 7) and Sanding[®] sugar (see Table 8), the smallest sized powders, were slightly more free flowing than the larger particles. In general, salt powders flowed more readily than sugar powders. Though sugar particles were larger than salt particles, the shape of sugar crystals, which were mostly square and cubic (Figure 2) retarded flow. Sugars being more hygroscopic picked up ambient moisture, and formed liquid bridges, which slowed flow. For sugar powders, the flow rates ranged only from 30 to 34 g/sec for particle sizes ranging from 350 to 1400 μm, showing that particle size was not as significant. Therefore the effect of particle size on flow depends on the type and shape of powder. In general, flowability increases as particles get larger and as the moisture decreases.

Table 7   Size, Angle of Repose, and Peak Flow of Salt Powders

Product salt	Size (μm)	Angle of repose (°)	Peak flow (g/sec)
Pretzel	1000	43[a]	44[a]
Topflake	600	48[b]	53[b]
Lite	400	49[c]	51[a]
Culinox	300	51[d]	61[a]
Dendritic	250	48[b]	48[b]
Flour	200	48[b]	105[c]

Peak flow 15 mm (g/sec).
Within a column, the same letter(s) are not significantly different.

Table 8   Size, Angle of Repose, and Peak Flow of Salt Powders

Product sugar	Size (μm)	Angle of repose (°)	Peak flow (g/sec)
Medium	1400	47[a]	30[b]
Standard	1200	44[b]	27[b]
Fine	750	47[a]	28[b]
Sanding	350	47[a]	34[a]

Peak flow 15 mm (g/sec).
Within a column, the same letter(s) are not significantly different.

X.   ANGLE OF REPOSE

Angle of repose is a static measure of flowability of a powder mass. It is the most commonly used method of measuring free flow. It measures the maximum angle a powder mass forms with the horizontal plane without sliding off. For determining the angle of repose, a batch of powder is formed in a standard way, and the angle between powder surface and the horizontal plane is measured (Sjollema, 1963).

The angle of repose (θ), of salt and sugar powders, was calculated by the method described by Sjollema (1963). A powder batch was formed with 80 g of each powder, which was allowed to flow through a funnel to form a batch; each angle of repose was calculated as [$\theta = \cotan h/r$] from the dimensions of the mass where h = height of the powder mass and r = radius of the base of the symmetrical conical pile. The angle of repose is the angle formed at the base of a flat surface.

Angle of repose, for all powders fell within a very narrow range, 43–51° for salt and 44–47° for sugar. There was no clear trend for the effect of particle size on angle of repose. For most food powders, powders with angle of repose greater than 50° are termed nonflowable.

Flowability of powders as measured by angle of repose depended on the powder particle values; values below 35° were classified as free flowing; values between 35 and 45° indicated some cohesiveness, while values greater than 55°, showed high cohesiveness with tendency to cause flow problems (Peleg, 1977; Chang et al., 1998). This classification has narrow application, and as found, encapsulated powders containing milk fat (particles ranging from 25 to 115 μm), with angles of repose between 44 and 46°, were free flowing

(Onwulata et al., 1995). From the results obtained for the angle of repose for both salt and sugar (43 to 51°), one would expect relatively free-flowing powders. However, only the pretzel salt which consisted of very large particles (840–1200 μm) and Standard sugar (1000–1400 μm) were free flowing. Large particle size in salts or sugars did not contribute to flow problems. White et al. (1967), reported that uniformity of size and shape improves powder flow. The powders that were free flowing (pretzel salt and Standard sugar), were of uniform sizes. Particle shape and uniformity of particles are greater contributors to powder flow and compressibility of salt and sugar powders, than factors such as moisture content, water activity, and size.

XI. SUMMARY

The flowability of salt and sucrose powders depends largely on the uniformity of particles within a given narrow range of sizes. Powders that exhibit normal Gaussian distribution tend to be more free flowing, than powders with skewed distributions. Salt powders were generally more free flowing than sucrose; size classification had only a slight influence. Powder shape is a critical factor that affects flowability. The larger particle mass in each group was the more free flowing. Therefore, uniformity of particle size as well as uniformity of shape is of greater import on flow than the absolute size. Smaller particle sized powders were not free flowing even in glassy noncohesive powders such as salt and sugar. Thus, water activity does not influence the flow properties of either salt or sugar powders.

REFERENCES

Abdullah, E.C. and Geldart, D. 1999. The use of bulk density measurements as flowability indicators. *Powder Technol.* 102: 151–165.

AOAC. 1995. Method No. 925-45. *Official Methods of Analysis*, 16th ed. Association of Official Analytical Chemists, Washington, DC.

Barbosa-Canovas, G.V., Malave-lopez, J., and Peleg, M. 1987. Density and compressibility of selected food powder mixtures. *J. Food Process Eng.* 10: 1–19.

Chang, K.S., Kim, D.W., Kim, S.S., and Jung, M.Y. 1998. Bulk flow properties of model food powder at different water activity. *Int. J. Food Prop.* 1: 45–55.

Grey, R.O. and Beddow, J.K. 1969. On the Hausner ratio and its relationship to some properties of metal powders. *Powder Technol.* 2: 323–326.

Hausner, H.H. 1967. Friction conditions in a mass of metal powder. *Int. J. Powder Metall.* 3: 7–13.

Hollenbach, A.M., Peleg, M., and Rufner, R. 1983. Interparticle surface affinity and the bulk properties of conditioned powders. *Powder Technol.* 33: 51–62.

Hui, Y.H. 1992. Sucrose. *Encyclopedia of Food Science and Technology*. John Wiley & Sons, Inc., New York.

Irani, R.R. and Callis, C.F. (Eds.). 1963. *Particle Size: Measurement, Interpretation, and Application*. John Wiley & Sons, Inc., New York.

Kolatac, R.P. 1996. Understanding particulate solids. Chemical Processing. Http: //www.nauticom.net/www/jhorst/paper1.htm

Konstance, R.P., Onwulata, C.I., and Holsinger, V.H. 1995. Evaluation of flow properties of spray dried encapsulated butteroil. *J. Food Sci.* 60: 841–844.

Lewis, M.J. 1996. Solids Separation Processes. In: *Solids Separation Processes in the Food and Biotechnology Industries. Principles and Applications*. Grandison, A.S. and Lewis, M.J. (Eds.). Woodhead Publishing Ltd., Cambridge, England.

Malave, J., Barbosa-Canovas, G.V., and Peleg, M. 1985. Comparison of the compaction characteristics of selected food powders by vibration, tapping and mechanical compression. *J. Food Sci.* 50: 1473–1476.

Marth, E.H. 1978. Preservatives, Chemical. In: *Encyclopedia of Food Science*. Peterson, M.S. and Johnson, A.H. (Eds.). The AVI Publishing, Inc., Westport, CT.

Mohsenin, N.N. 1980. *Physical Properties of Plant and Animal Materials*. Gordon and Breach Science Publishers, New York.

Moreyra, R. and Peleg, M. 1980. Compressive deformation patterns of selected food powders. *J. Food Sci*. 45: 864–868.

Niman, C.E. 2000. In search of the perfect salt for topping snack foods. *Cereal Foods World* 45: 466–469.

Onwulata, C.I., Smith, P.W., Craig, J.C. Jr., and Holsinger, V.H. 1994. Physical properties of encapsulated spray-dried milkfat. *J. Food Sci*. 59: 316–320.

Phillips, R. 1997. The benefits of particle sizing. *Food Process*. 66: 29–30.

Peleg, M. 1977. Flowability of food powders and methods for its evaluation — a review *J. Food Sci*. 60: 836–840.

Peleg, M. 1983. Physical Characteristics of Food Powders. In: *Physical Properties of Foods*. Peleg, M. and Bagley, E.B. (Eds.), pp. 293–323. AVI Publishing Co., Inc., New York.

Peleg, M. 1985. The Role of Water in Rheology of Hygroscopic Food Powders. In: *Properties of Water in Foods*. Simatos, D. and Multon, J.L. (Eds.). Martinus Nijhoff Publishers, Dordrecht, Netherlands.

Peleg, M. 1987. Particle size, distribution of food powders. *Manuf. Confectioner* 67: 95–96, 98.

SAS. 1998. *SAS/STAT Guide for Personal Computers*. Statistical Analysis Systems Institute. Cary, NC.

Schubert, H. 1987. Food particle technology. Part II. Some specific cases. *J. Food Eng*. 6: 83–102.

Sjollema, A. 1963. Some investigations on the free-flowing properties and porosity of milk powders. *Neth. Milk Dairy J*. 17: 245–259.

Teunou, E., Fitzpatrick, J.J., and Synnott, E.C. 1999. Characterization of food powder flowability. *J. Food Eng*. 39: 31–37.

White, G.W., Bell, A.V. and Berry, G.K. 1967. Measurement of the flow properties of powders. *J. Food Technology* 2: 45–52.

Yan, H. and Barbosa-Canovas, G.V. 1997. Compression characteristics of agglomerated food powders: effect of agglomerate size and water activity. *Food Sci. Technol. Int*. 3: 351–359.

17
Cocoa

Leanne de Muijnck
Product Service and Development Manager
ADM Cocoa Division
Archer Daniels Midland Company
Milwaukee, Wisconsin

CONTENTS

I. INTRODUCTION

Cocoa is one of the most appreciated food flavors in the Western hemisphere. It is available in a wide variety of colors and flavors, and used in numerous applications. It is appreciated by almost all ages, and consumption is not limited to any particular part of the day. Breakfast may include chocolate-flavored cereal, hot cocoa or chocolate milk could complement lunch; there are numerous chocolate-flavored products available for in-between meals, snacking, and desserts, and there are even ethnic dinner dishes calling for cocoa as an ingredient. But before the consumer enjoys any of these products many processing steps have to take place to produce the ingredient cocoa powder, and prepare it for use in the final application.

A good quality cocoa powder is relatively free flowing, stable and uniform in color and flavor, of good microbiological quality, and easy to handle by the user. A combination of bean choice and controlled processing parameters are important factors in manufacturing such a product.

This chapter will give an overview of the history of cocoa, cocoa bean characteristics, the manufacturing of cocoa, the variety of products that result from different processes, characteristics of these cocoas, and the application of cocoa powder.

II. A BRIEF HISTORY OF COCOA

Cocoa originated in Central and South America. Although the exact origins are not known, the closest estimates put the area of origin in and around the valleys of the Amazon and Orinoco rivers. The first time that people far from the areas of its origin were confronted with cocoa beans was thanks to Columbus. But it was only years later at the beginning of the 16th century that Cortez confirmed the remarkable value assigned to the cocoa beans. He found that the Aztecs used them both as a means of payment as well as a source of a beverage drunk at court and religious ceremonies [6].

Cortez took cocoa beans back to Spain, where they gained popularity at the royal court. They were mixed with sugar and hot water, and consumed as a drink. King Charles V of Spain quickly declared it a state secret, and the access to the product was limited to a small group of people. It took almost 140 years before the secret of cocoa filtered out of Spain and became well known in the rest of Europe. By the end of the 17th century, drinking cocoa had become so popular that it served as a source of tax revenue for governments. Cocoa was no longer a product just for the elite [6].

The Spanish started to spread the cocoa tree, and tried to cultivate it across all their colonized territories. These efforts were met with frustration. The cocoa plant, known as *Theobroma cacao L.,* required special growing conditions, and turned out to be susceptible to various diseases. Over time, the cocoa tree started to spread across the world. The tree first spread out in regions close to its origins, from Brazil and Mexico in the 15th century across Central America and the Caribbean islands in the 16th. By 1560, the Spaniards had introduced it to some of the Indonesian islands. They brought the bean to the West African island of Fernando Po, from where it was later transferred to the main land. The great growth in the cocoa trade in the 19th century saw its expansion across many other countries especially in West Africa and South East Asia [6].

Initially, the entire roasted bean, containing roughly 50% fat, was ground up, mixed with water and sugar and possibly some spices, and consumed. In the 18th century the Dutchman Conraad Johannes van Houten developed a mechanical pressing process to separate the ground chocolate liquor into cocoa butter and a partly defatted fraction. Another process developed by Van Houten was alkalization, or the "Dutch process," a procedure of treating cocoa with alkali. This was originally done to improve the solubility. It was found that at the same time taste and color were also changed [6].

Nowadays the alkalization process is widely used in the entire cocoa industry. It has given the manufacturers the possibility to create a wide variety of color and flavor profiles. The colors range from light yellowish brown to deep reddish brown to very dark brown. The flavor profile is determined by the choice of beans, the various processing conditions, and the degree of alkalization. A natural cocoa has a profile dominated by sour, bitter, astringent, and "chocolate" like notes. A red alkalized cocoa will give an intense cocoa impact, mild bitter notes, and no sour notes. A so-called black alkalized cocoa gives a very intense cocoa impact, combined with typical alkalization notes. The final application determines the preferred choice of cocoa powder.

So over a period of about 500 years cocoa changed from a being a product grown in a small region and consumed by a very limited group of individuals, to a product grown in multiple regions and known and enjoyed globally by many.

III. AGRICULTURE AND CULTIVATION

A. The Cocoa Tree

The cocoa tree belongs to the genus *Theobroma*, a group of about 20 species. The only species that is extensively cultivated is *Theobroma cacao* [11]. Within this group, two distinct subspecies were developed. Morris classified these as Criollo and Forestero. Later, a third group called Trinitario, a cross between Criollo and Forestero, was added [12]. Criollo beans are light-colored with a mild nutty character. Forestero cocoas are dark brown, strong-flavored, slightly bitter, and have a higher fat content. Nowadays the vast majority of the world crop falls within the Forestero group.

The original tree grew to a height of about 10 m at maturity, and preferred the shade of larger trees.

A successful cultivation of cocoa requires a climate mostly found within 10° North and South of the Equator. Modern breeding methods have now led to the development of trees of a standard around 3 m height, to allow for easy hand harvesting. They come to maturity in 7 years. When the evergreen cocoa tree reaches its bearing age, flowers and fruit begin to appear in modest amounts. These can be found on the tree at all seasons of the year, although typically two crops a year are harvested [6].

The fruits grow directly from the trunk of the tree and the thicker branches. While there may be several thousands of flowers on a mature tree, there are only a small number that mature into fruits or pods. These take some 6 months to grow from a fertilized flower, measure 10–15 cm at the center, and are 15–25 cm long. The pods contain some 40 beans or seeds. After fermentation and drying, one pod produces some 40 g of beans, one bean weighing typically around 1 g [6].

B. Harvest, Fermentation, and Drying

Mostly the pods are harvested by hand, using long-handled knives, and broken open to reveal the beans and the white pulp surrounding them. They are removed from the pods and subjected to fermentation. The traditional process in west Africa is simple; farmers place the pulp-covered beans on the ground, and cover them with layers of banana leaves. They allow the heap to remain for 4 to 7 days, depending on the variety of the bean. The fermentation that occurs is critical for the future development of color and flavor of the cocoa. During the first, mostly anaerobic, step of fermentation the pulp surrounding the beans liquefies and drains off. This allows fresh air, and oxygen, to enter the fermentation heap, and a more or less aerobic environment is created. The heat generated during this microbiological process kills the germinating potential. The acid produced during the process penetrates into the beans, and destroys the cellular walls of cells containing starch and protein, and cells containing various enzymes. As a result, enzymes and substrates come together, starch starts to break down into sugars, and protein is broken down into peptides of various chain lengths. The monosaccharides and monopeptides are the so-called flavor precursors, the building blocks for flavor development during roasting. An inadequate fermentation leads to a partial lack of flavor precursors, and will not generate a lot of typical chocolate flavor upon roasting [6].

Today there are a great variety of techniques that have been developed to simplify and perhaps accelerate the fermentation process. One such technique involves a process where beans are loaded into the first of a series of containers and then transferred. By the time the beans reach the last container they are considered fermented and ready for the next treatment. There is no definite answer as to how to improve current West African practices [6].

After fermentation the beans have to be dried. In West Africa the traditional method is usually to spread the beans out on mats or in trays in the open air under the sun. In Brazil the beans are typically laid out on broad mats on stilts above the ground level to dry. In the event of rain a roof can be slid across the mats and hot air is used to dry them. In Malaysia widespread use is made of mechanical rotary dryers. After drying the beans are bagged, and made ready for transport to buying stations and regional warehouses [6].

C. Major Growing Countries

The table shows the Output and Market Shares of the Major Growing Areas of the World [7].

	Major bean producing areas (×1000 MT and in %)								
	1980/1		1990/1		2000/1		2001/2 (e)		2002/3 (f)	
Africa	995	59	1440	57	1950	68	1940	68	2060	69
Central/South America	530	32	615	24	365	13	315	11	335	11
West Indies	50	3	50	2	55	2	55	2	552	
Asia and Oceania	100	6	420	17	490	17	540	19	550	18
Total World production	1655		2525		2850		2850		3000	

Note: e = estimate; f = forecast.

Cocoa is a natural product, and suffers all the risks inherent in nature. The flower is very susceptible to rain and temperature conditions during its development. The pod can be attacked by a variety of molds, insects, and rodents. The shell has an increased total plate count owing to fermentation. The quality of the crop may vary by year and by country of origin. Good cocoa beans should be well-fermented, dry, and free from abnormal odors and adulteration. The beans should be reasonably uniform in size, reasonably free from broken beans, fragments, and pieces of shell, and free of foreign matter. Any residues should be within legislation or international guidelines applicable at the time of shipment [6].

Cocoa powder and chocolate producers keep a close watch on the expected and actual volumes and quality of the crops from various countries. Cocoa beans are mostly traded by country of origin. Besides a natural annual variation that depends upon growing and weather conditions, beans from specific countries of origin are known to have some distinct coloring, flavor, and cocoa butter characteristics. This is a result of the type of beans, environmental conditions, traditional harvesting, fermentation, and drying practices in each country. Depending upon the final products, the producers of chocolate and cocoa will define the desired quantities and types of beans. A cocoa bean processor will often work with bean blends, in order to prevent being dependent upon one particular country of origin, and annual variations of various crops.

D. Major Consuming Countries

The world demand for cocoa beans, supported by relatively low cocoa prices, has steadily increased over recent decades as a direct result of increased world demand for chocolate and chocolate-flavored products. The cocoa bean grinding facilities are mostly close to the chocolate and cocoa powder market. The cocoa producing countries, especially in the west African region, process a modest part of the world crop.

A table showing the major cocoa bean consuming countries is given below [7]:

	Major cocoa consuming countries (based on bean grind) ($\times 1000$ MT and in %)									
	1980/1		1990/1		2000/1		2001/2(e)		2002/3(f)	
Netherlands	135	9	250	11	452	15	418	15	440	15
USA	180	12	271	12	445	15	398	14	410	14
Germany	20	1	290	13	227	7	195	7	190	6
Ivory Coast	60	4	112	5	285	9	290	10	285	10
United Kingdom	72	5	125	6	151	5	140	5	135	5
France	50	3	60	3	145	5	139	5	145	5
Malaysia	7	—	75	3	125	4	105	4	120	4
Former USSR	120	8	100	4	102	3	95	3	97	3
Indonesia	8	—	28	1	87	3	105	4	110	4
Others	858	57	929	41	1031	34	980	34	1048	35
Total world grindings	1510		2240		3050		2860		2980	

Note: e = estimate; f = forecast.

The bean grinding quantities do not indicate what is actually made from the cocoa beans. There are no specific data available about cocoa powder producing countries. Figure 1 shows that three intermediate products — cocoa liquor, cocoa butter, and cocoa powder — are initially made from the cocoa bean. It should be added that cocoa powder is ground cocoa cake, a product produced during the pressing of cocoa liquor. Some cocoa powder manufacturers process cocoa beans, and do the entire processing. Other suppliers start with cocoa cake, and process this into a variety of cocoa powders. The choice of raw material and process has an effect on the characteristics of the final cocoa powder. An overview of the world's cocoa products flow is presented in Figure 1.

IV. COCOA PROCESSING

A. Introduction

The principles of cocoa bean processing have not changed much over the last 150 years. Today cocoa beans are still cleaned, deshelled, roasted and optionally alkalized, ground into chocolate liquor, and liquor is subsequently pressed into cocoa cake and cocoa butter. The cocoa cake is pulverized into cocoa powder. The order of some of these steps may vary. Roasting may be carried out with either whole beans or cocoa nibs, and the starting material for alkalization could be nibs, cocoa cake, cocoa powder, or chocolate liquor.

In this section the purpose of the various processing steps, and how they relate to cocoa powder is discussed. A discussion on specific machinery, or comparison of manufacturing equipment of different suppliers is beyond the scope of this chapter.

B. The Raw Material

The condition of the starting material, the cocoa bean, is of paramount importance for the characteristics of the end product. Standard contracts for cocoa define a number of quality

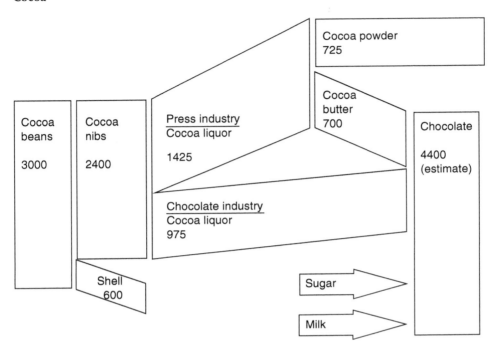

Figure 1 Diagram of the world's cocoa products flow [6].

standards. The product needs to be adequately fermented, should have less than 4% insect damage, less than 4% moldy beans, less than 6% combined insect damage and moldy beans, no foreign matter, less than 2% waste, less than 7.5% moisture, free of smoky or foreign flavors, and should have a reasonably uniform bean size. Many cocoa bean processors will separate the beans by country of origin, or will have an internal system to store beans. The definition of high quality depends upon the final product made from that specific lot. If the main purpose of the operation is to produce cocoa butter, the degree of fermentation of the cocoa beans is not very important. The processor will instead focus on the fat content of the bean, the absence of infestation, very few broken beans, and the quality of the cocoa butter. Cocoa cake will be considered to be a by-product, and flavor and color characteristics will vary. However, if the processors are aiming for a full-flavored chocolate liquor, or cocoa powder, the degree of fermentation becomes very important. Depending upon the finished product, the processor may look for specific bean origins or bean blends, and will adjust all process parameters [6].

C. The Production Process

A simplified diagram of cocoa bean processing is given in Figure 2. Some steps are optional, and the order of some of the processing steps depends upon the cocoa bean processor.

D. Blending Process

Like all products from nature, cocoa beans will vary in certain characteristics. The growing, harvesting, and drying conditions will vary from one year to another, and this may be

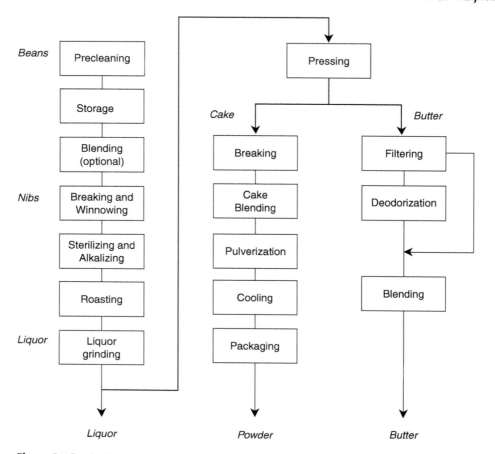

Figure 2 Production flow sheet [6].

reflected in bean size, flavor, and color potential. Prior to processing, individual cocoa bean lots will be analyzed and graded. These data will be used to determine optimal blends, and minimize fluctuating characteristics. Blending may occur at the very beginning of the process by blending beans, or further downstream by blending cocoa nibs or cocoa cake.

E. Cleaning, Breaking, Winnowing

The actual process starts with cleaning, breaking, and winnowing. The objective is to obtain clean broken deshelled kernels. The kernels must be as uniform in size as possible in order to realize an even roast.

First the beans are sieved and foreign matter such as bamboo, twigs, strings, stones, double, and clusters of beans are removed. The clean beans are broken to loosen the shell from the nib. An optional process between cleaning and breaking is micronizing. This short-time high-temperature heat treatment dries out the shell, and makes it more brittle. This helps in an effective separation of the cocoa kernels, also known as cocoa nibs, from the shell.

The breaking process takes place in multiple steps, to avoid an excess of fine particles. After breaking, the product is sieved into a number of fractions of even size, to reach an optimal separation during winnowing [6].

These fractions then go to the winnowing cabinets where the lighter, broken shell is removed by a stream of air. The breaking and winnowing processes separate the cocoa nibs from the shell. Legal requirements specify the shell content in the nibs have to be at a maximum of 1.75%. The residual shell is sold to agricultural mulch or fertilizer producers, and as cattle feed. The high fiber content of the shell has generated interest from health food producers, and the processed shell has been sold as a food ingredient [6].

F. Alkalization

Alkalization or dutching consists of a treatment of the cocoa with an alkali solution such as potassium carbonate or sodium hydroxide. It is applied primarily to develop a wide range of color and flavor profiles in cocoa powder, but also improves the microbiological condition. This step is optional in the overall process. Without alkalization a "natural" cocoa powder is the end product [6].

There are several methods and several types of equipment to carry out an alkalization process. Examples are pressured vessels, vessels under atmospheric pressure, tanks, and rotating drums, all with varying capacities. The processing parameters that determine the color and flavor of the end product are processing time, temperature, type and amount of alkali, amount of moisture, amount of air, pressure, and the design of vessels or drums [14]. Each processor has a unique combination of processing equipment and recipes. While it is possible to group powders from various suppliers into similar groups, each powder may have a unique combination of color, flavor, and some other characteristic.

Alkalization can be introduced in various steps of the production process. Besides nibs, cocoa cake, cocoa powder, or chocolate liquor may be alkalized [9]. The choice of starter material depends upon both technical and economical factors. A producer of both cocoa powder and cocoa butter will process cocoa beans, and most likely include a nib alkalization process. A producer of solely cocoa powder, with no interest in cocoa butter, will apply a cake or powder alkalization step. Depending upon the stage at which alkalization takes place, different results will be obtained. Nib alkalization is often preferred as it combines optimal flavor and color development with minimal alkali usage. However, for the production of very dark cocoa powders a cake alkalization step is also quite common.

Important factors in the choice of cocoa powders for specific applications are color and flavor, both heavily affected by the degree of alkalization.

G. Roasting

The roasting process has the objective of developing the flavor, reducing the water content, as well as improving the microbiological condition. During roasting, the flavor precursors, formed during fermentation and drying, are responsible for the development of the typical chocolate notes. The starting material, roasting conditions, the degree of roasting, as well as the type of equipment further determine the flavor profile of the final product [6]. While most cocoa powder producers prefer a nib roasting, a more traditional process, applied by some chocolate producers, is a whole bean roast. The moisture content, the origin of beans or nibs, and the desired flavor profile dictate the processing conditions. For a natural cocoa or chocolate liquor made from light-colored Java beans or delicate Arriba AAA beans, a mild roast will bring the best out of the beans. On the other hand, strongly alkalized nibs with a high moisture content will require much longer processing times and most likely a higher roasting temperature. Each processor will develop its own roasting procedures, given the product characteristics desired.

H. Nib Grinding

The roasted nib is ground typically in a multistage process. The nibs have a cellular structure, with the fat in solid form locked within the cells. During grinding the cell walls are ruptured, and the broken kernels change from a solid into a liquid mass of cocoa particles suspended in cocoa butter [6]. Examples of machines used for grinding are stone mills, pin or hammer mills, three roller refiners, and vertical ball mills [9,14]. The fineness of the final cocoa powder is determined by the fineness of the liquor, and grinding or pulverizing of the cocoa cake. A fine grind of the liquor phase will minimize grinding requirements in the cake phase, and avoid overheating of the solids in this dry phase. Some cocoa powder manufacturers argue that, in order to avoid flavor defects arising during cake grinding, the liquor grinding phase should determine the fineness of the final cocoa powder.

I. Pressing

Cocoa nibs and consequently chocolate liquor contain 50 to 55% fat or cocoa butter. The fat is partially removed from the cocoa liquor by means of hydraulic presses, applying pressures as high as 450 kg/cm^2, during a certain amount of time. The resulting cakes have a fat content between 10 and 24% [4]. After pressing, the cakes are broken into kibbled cake. Kibbled cake is typically stored by fat content and degree of alkalization, and may be blended before pulverization.

Cocoa cake is traded between companies, and may be the end product for one and the starter material for the other. Cocoa powder producers who start with cocoa beans as their raw material have full control over the choice of beans and all processing parameters. Besides that, the entire process is carried out in an enclosed system. This will typically result in an end product of better microbiological quality than powder made from traded cocoa cake. The cocoa cake user has no control over initial processing conditions, and more handling and processing steps are involved. The availability of certain cakes may vary over time, and as a result the range of characteristics may be wider than products resulting from a fully controlled cocoa bean processing plant. Both processes are well accepted and practiced globally, and GMP procedures in either case will result in a safe end product. The specifications of the finished product, though, may reflect the processing history of the product.

J. Cocoa Powder

The final steps in cocoa powder production are grinding or pulverizing cake, cooling, and packaging. Grinding is typically carried out with hammer and disk or pin mills. The fineness of the cocoa powder is determined by a combination of factors, such as the fineness of the liquor, type of alkalization, and the final grinding step. Often grinding lines are provided with cooling equipment to ensure that the cocoa butter present in the powder crystallizes and settles in its stable form [4]. This helps prevent any discoloration (fat bloom) and lump formation later in the bags. Lump formation can be caused by insufficient crystallization of the fat present at the moment of filling, or by melting and recrystallization of fat during storage. The latter case is likely to happen at a storage temperature higher than 25°C, and especially in combination with temperature variations caused by a day–night cycle [13].

After cooling, the cocoa powder typically passes sieves and magnets to remove any coarse cocoa and metal particles, and the powder is packaged. Packaging size varies, but most common sizes are 25 kg and 50 lb bags, and the so-called Big Bags, or Flexible

Intermediate Bulk Containers, typically holding between 1600 and 2200 lb, or between 700 and 1000 kg.

K. Processing Variations

As mentioned before, there are a couple of well-known variations to the process. The choice of process and equipment all depends upon the desired end characteristics of the cocoa.

Roasting may be carried out with either whole beans or cocoa nibs. In the case of whole bean roasting, practiced mostly by chocolate manufacturers, a modest alkalization step may be included in the liquor phase.

Alkalization can be executed with nibs, cake, powder, and liquor. Theoretically, whole bean alkalization is possible, but it is quite an inefficient process, and very rarely applied. In the case of cake or powder alkalization an additional drying step is included in the process. The resulting product has a tough texture, and during grinding more energy is needed to achieve the fineness of a similar nib alkalized product.

Besides a traditional hydraulic pressing system, cocoa butter may also be separated from cocoa solids by means of an expeller process. If alkalization is included in the process, it will be carried out after the expeller step.

V. COCOA CHARACTERISTICS

A. Introduction

The two most obvious attributes of cocoa are its ability to give color and flavor to a wide variety of food products. Often the consumer will directly associate a brown color with a chocolate flavor, and the darker the color the stronger the flavor expectation will be.

Besides color and flavor a range of other characteristics define the powder, and have an important functional impact on the end product in which the cocoa is used. Examples of these characteristics are pH, fineness, fat content, alkalinity, wettability, moisture absorption, solubility, density, and microbiological qualities. Manufacturing parameters and other ingredients in the formula of a certain application may distinctively influence the overall performance of cocoa powder in the final product as well.

When choosing a cocoa powder for a certain application the desired characteristics of the end product, as well the as overall recipe and the processing parameters all need to be taken into consideration. On top of these technical characteristics, economical motives and availability may affect the choices.

B. Standards of Identity

Many countries have defined cocoa in their food law. Depending on the time that these food laws were initiated and the then prevailing chemical and physical analytical capabilities as well as the process-technical advancements, these laws may deviate on essential elements. In many instances, a difference exists in the product definition and the legal specification of the product [6].

The following regulations give the following descriptions:

	Low fat c.p. (% fat)	Fat-reduced c.p.(% fat)	Cocoa c.p. (% fat)	Breakfast cocoa (% fat)
USA: 21 CFR 163	<10		$10 \leq 22$	22 or more
EU: Directive 2000/36/EC		<20	20 or more	
CodexStan105-1981,Rev.1-2001	<10	$10 \leq 20$	20 or more	

In many countries, the fat content determines the name of the product. Many more rules, on a country-by-country basis, concern the permitted production processes, raw materials used, product specifications, labeling requirements, extraneous matter, and even packaging [6].

In this section an overview of the different characteristics of cocoa, and the factors affecting each attribute, are discussed. Cocoa will be used as a general term, and does not refer to the standards of identity of any specific country. If relevant, the fat content of the powder will be taken into account.

C.   Composition

The composition of cocoa powder depends upon the starter material as well as various processing parameters like pressing time, and type and degree of alkalization. Further, the methods of analysis and the methods of calculating macronutrients determines the reported caloric value and composition of individual cocoas [10]. Even though there are several generic databases available, a cocoa user will need to contact his supplier for the exact nutritional data of that particular product.

The fat content of the powder depends directly on the pressing parameters. In general the content is somewhere between 10 and 24%. The protein content is roughly between 18 and 21%. The total carbohydrate content has been reported in between 45 and 60%. The sugar content is very low, less than 2.5%, while the dietary fiber content has been reported between 25 and 35%. Cocoa contains a number of minerals and a small amount of vitamins. The amount of sodium and potassium is highly affected by the degree of alkalization [10]. Recent research into nutritional benefits of cocoa show that cocoa can be a valuable source of copper.

From a nutritional point of view, one of the most interesting set of components in cocoa is the group of phenolics, including catechin, epicatechin, quercetin, and clovamide. The components are known antioxidants, and have raised a lot of interest. The purified components have been studied extensively, and have shown various potential beneficial effects like antigenotoxicity, anticarcinogenic activity, and antioxidant activity. If these antioxidants are active in the human body, they could help slow down the buildup of artherosclerotic plaque [12]. Testing components and extracts *in vitro* shows very promising results. This has generated a lot of interest in evaluation of beneficial effects of chocolate with regards to especially coronary heart diseases, and the results are encouraging.

D.   Color

1.   *Color Formation*

The color of a food item is of critical importance to consumers as well as to food producers. The color is part of the overall attractiveness of an item, and creates expectations with regard to flavor. The consistency of the color reinforces the image of a constant product quality. Cocoa is available in a wide variety of colors, ranging from light yellowish brown to deep red brown to very dark blackish brown [6]. A combination of the choice of raw material as well as processing parameters determines the color of the finished cocoa. Unfortunately, there are no standards available that define what, for example, a "red," "brown," "lightly alkalized," or "black" cocoa is. Consequently, every manufacturer of cocoa powder uses his own terminology, and care must be take when comparing a "red/brown" cocoa from one supplier to a powder with the same description from another source. Besides an inconsistency in the description of the color, every supplier may have his own unique way of measuring the color of cocoa. It may be measured "dry," in water, in milk, or in a fat/sugar dispersion, while ultimately the color of the final application is the only thing that matters to the consumer.

The table below shows a general description of the degree of alkalization in combination with often used color descriptions. While the ranges vary per supplier, each supplier will use similar terminology to identify the degree of alkalization within his own product portfolio.

Degree of alkalization	Color indication
None: natural cocoa	Yellowish/brown
Lightly alkalized	Light to medium brown
Medium alkalized	Medium to deep brown
Red alkalized	Red/brown
Deep red alkalized	Deep red
Black alkalized	Very dark brown to black

The components responsible for the color formation in cocoa are the flavonoids, a subgroup of the polyphenols. They are present in fresh unfermented beans at a level of about 15%. The flavonoids of particular interest as flavor precursors are anthocyanidines and procyanidines. Anthocyanidines are esters of anthocyanines and sugars, and are responsible for the purple color of un- and underfermented beans. The colorless procyanidines are mostly present as mono-, di-, and trimers of epicatechin. During fermentation, the sugar esters are hydrolyzed by enzymes. The sugar molecules are of interest for flavor formation, while the free antho- and procyanidine molecules are oxidized by enzymes to quinones. Quinones react with amino acids and proteins, forming covalently bonded complexes. In this way they form very strongly colored pigments. They also react with other flavonoids, forming high molecular weight condensed tannins. If the molecular weight of the tannin is above 3000 it forms complexes with proteins by hydrogen bonding. As oxidation is involved, these reactions take place during the second, oxidative stage of fermentation and during sun-drying of the beans. The result is a brown pigment, which is stable and insoluble in water [6]. The degree of fermentation as well as the type of beans determines the color of the beans after drying. In certain *Theobroma cacao* species, such as the Criollo, the beans do not contain this purple pigment, and after fermentation the beans are still very lightly colored. Part of the evaluation of a lot of beans for quality purposes is determined by the cut test. By assessing the color of the interior of a certain amount of beans the degree of fermentation of the lot, and hence color and flavor potential, is checked.

During alkalization, it is possible that the polymerization and condensation reactions will continue. Numerous large dark pigments are produced, but very little details about the exact chemical reactions are known. The direction of the color formation depends upon the kind of bean, degree of fermentation, type and amount of alkali, time, temperature, moisture, amount of oxygen, as well as the type of equipment. The processor of alkalized cocoa powders will use his experience to determine which unique combination will result in the desired color profiles.

2. External Color

The color of cocoa powder as such is the so-called external or dry color. This color is important if cocoa powder is to be sold as such to the end consumer, or if used in a dry mix. The dry color is strongly influenced by the optical effect in which the fat on the solid particles affects the light absorption. The higher the fat content, the darker the color will appear to be. The crystallization form of the cocoa butter in the solid particles determines the strength of this optical effect. The crystals should be small and in the stable form. When cocoa powder is subjected to temperature fluctuations, discoloration will occur owing to the change in crystalline size or form of the cocoa butter. Too slow cooling or too rapid cooling without tempering, will result in larger particles and a grayish hue of the cocoa powder [6]. The fineness of the powder affects the dry color as well. A finer grind results in a lighter, dry color.

3. Intrinsic Color

The intrinsic color of cocoa powder is the color that the product made with the powder will have. For most applications this is far more important than the dry color. The choice of cocoa powder should be based on the performance of the powder in the finished application. While the fat content in cocoa creates an optical effect, the solids in cocoa are responsible for the true color. A 22–24% fat cocoa may appear darker than a 10–12% fat cocoa of the same degree of alkalization. However, in a wet or fat-based application the 10–12% cocoa will result in a darker end product, because it has a higher amount of cocoa solids. The fineness of cocoa has adverse effects on the dry as it has on the intrinsic color. A finer grind will result in a lighter, dry color, but a darker, intrinsic color. The degree of the effect depends upon the application.

The color of the finished food product depends upon more than just the type of cocoa powder. The amount of cocoa powder obviously affects the finished color. Other ingredients, like milk fat, caseinates, amount of sugar, amount of fat, and moisture to name a few, clearly affect the appearance of the finished good as well. Coloring agents used in combination with cocoa will have a huge effect. In addition, the manufacturing process of the food item has a clear influence. If air is whipped into the product the overall color will lighten. During a baking process, at a high pH and high moist conditions part of the alkalization process may continue, and dark red tones could be created. All these aspects need to be taken into consideration when choosing a cocoa.

E. Flavor

Just like color, the flavor of cocoa has a very important effect on the attractiveness of the finished good. It is quite easy to demonstrate the range of flavors available in a simple medium like hot water and sugar. However, one needs to take the final formula and processing parameters into account, and must carry out testing as such, when screening the flavor profiles that can be achieved by applying different cocoas.

1. Flavor Formation of Cocoa

The most important factors with regard to the formation of cocoa flavor are cocoa bean variety, fermentation and drying, alkalization, and roasting. The actual typical chocolate or cocoa flavor is formed during alkalization and roasting, while bean variety, fermentation, and drying determine the flavor potential.

The variety of cocoa, whether it is a Criollo, Forestero, or a hybrid, greatly affects the flavor potential of the bean. The growing conditions have an effect as well. Roasted Criollo beans may have a delicate, nutty, and sometimes floral/fruity flavor profile. When roasted, Forestero beans contain a more robust, basic, and less complicated bitter chocolate profile. Typically beans are identified by their country of origin, and cocoa bean processors have generic expectations of the flavor profile by country of origin. Sometimes chocolate manufactures will buy from certain farms or haciendas if they are looking for a very specific flavor profile.

The degree of fermentation determines the amount of flavor precursors. During the second phase of fermentation, enzymatic activity results in the formation of amino acids and peptides from protein, and the presence of glucose and fructose because of the break down of sugar, starch, and the anthocyanidines. Peptides and amino acids on the one hand, and reducing sugars on the other are the precursors for the formation of volatile flavor components formed by Maillard reactions during the later stages of processing of the cocoa beans. The enzymatic activity also results in the formation of tannins from flavonoids, thereby reducing some astringent and bitter notes. If the applied fermentation time is too short for the bean variety in question, there will be a limited formation of flavor precursors, and as a result a decreased flavor potential [6].

During alkalization the pH is raised, resulting in a sharp reduction of acidic notes. The overall cocoa notes intensify, being at a maximum with medium to red alkalized cocoas. Strongly alkalized products start to show very harsh, acrid, and strong bitter notes. In the next paragraph an example of the flavor profiles of cocoa powders going up in the degree of alkalization in a hot water and sugar solution is given.

During roasting, the ultimate flavor profile is achieved. The chemical processes are complex and not completely understood, but it is well known that Maillard reactions started during fermentation and drying, continue at this stage of processing. This results in a significant amount of aldehydes, some of which are directly correlated to a typical chocolate note. Other components created during this process contribute to the roasted character of the flavor profile. The overall flavor profile of roasted cocoa beans is complex, and more than 500 volatile components have been identified in the headspace of roasted cocoa. The intensity of the roasting step determines the amount of components formed, as well as the removal of volatile components like esters and low molecular acids. The variety of beans will affect the choice of roasting conditions. Certain beans from Ecuador and Venezuela contain a relatively high amount of volatile esters, responsible for fruity wine like top notes, and only under mild roasting conditions will these notes be retained.

Studies aimed at correlating a so-called headspace analysis to certain flavor characteristics are ongoing, and could be very valuable as a quality parameter. While no standard analysis is known yet, a lot of progress is being made by certain private organizations.

2. Evaluation of Cocoa Flavor

There are many factors that influence the overall flavor profile of any product that contains cocoa powder. When choosing or comparing different cocoa powders one needs to keep recipes, preparation, and testing conditions consistent, in order to get reliable results. Figure 3 shows a chart of the flavor of four cocoa powders with an increasing degree of

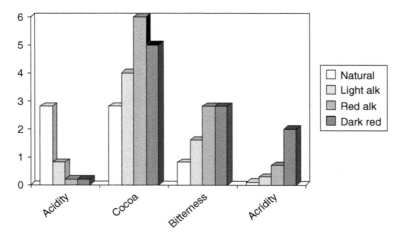

Figure 3   Flavor of cocoa in a sugar/hot water solution [6].

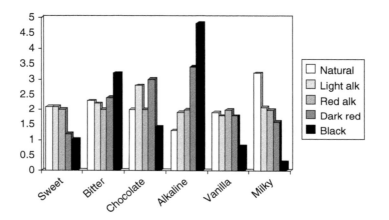

Figure 4   Flavor pudding with cocoa of various degrees of alkalization.

alkalization. For every powder, the panelists were asked to rate each descriptor on a scale from 0 to 6. The solution contained 4% cocoa and 5% sugar, in hot water.

The following should always be taken into consideration when testing cocoas for flavor:

- Ingredients like sugar, fat, flour, water, milk, and flavorings all contribute to the flavor profile, and interact with cocoa. As a result the same cocoa powders react differently in different recipes. Figure 4 and Figure 5 illustrate this concept. Five cocoa powders with an increasing degree of alkalization were tested in Devil's food cake and in a chocolate-flavored pudding. A panel was given a set of descriptors typical for each application, and asked to rate the products against the reference, which was the product with the red alkalized cocoa. In Figure 4 and Figure 5 those descriptors were selected that were affected by the type of cocoa used. It is obvious that the effect very much depends upon the application. Sweetness, cocoa intensity, bitterness, and typical alkalization notes decrease or increase at different rates in cake or pudding.

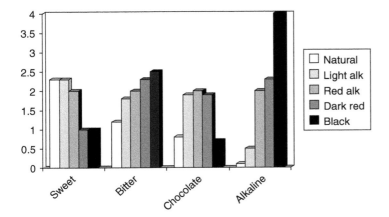

Figure 5 Flavor of cake with cocoa of various degrees of alkalization.

- The temperature of the food item highly affects the flavor perception. Therefore the choice of, for example, a cocoa for a cold chocolate milk should not be based on the evaluation in a hot water and sugar solution.
- Typical matrix characteristics like viscosity, viscoelasticity, mouthfeel, fineness, and texture play an important role in flavor release and flavor perception.

F. Fat Content

Every cocoa supplied by cocoa powder manufacturers has a defined fat content, typically within a range of 2%. The most common powders have a fat content between 10 and 12% or between 22 and 24%. The fat content has an effect on the appearance or dry color, flavor, and functionality in certain applications. Cocoa butter by itself has a mild flavor, while the solids are responsible for the typical cocoa impact [6]. Therefore, in most applications the flavor of products made with equal amounts of 10–12% fat powders will be more intense than the same product made with 22–24% fat cocoa. However, in cold products like ice cream the quick melting cocoa butter increases the flavor release, and also contributes to a creamy texture [4]. High-fat cocoas are not recommended for use in compound coating, because the high amount of cocoa butter has an adverse effect on gloss when lauric fats are used. Powders with a 10–12% fat content are less susceptible to lumping and are more free flowing. Therefore, these powders are better suited for products like vending mixes [3].

G. pH

The pH of cocoa powders varies between 5.3 and 6.0 for natural cocoas and can be as high as 8.7 for black cocoa powders. The pH of natural cocoa may vary somewhat over time and by the type of bean, but is very consistent between the given ranges. The pH of alkalized cocoa is a direct result of the intensity of the alkalization treatment. Type and amount of alkali are key factors [6]. Since the grouping of cocoa powders is arbitrary, there is no specific pH range for each group. The table below is indicative of the above concept and is by no means binding for any given cocoa processor. One needs to also keep in mind that by varying the processing conditions it is possible to create, for example, red powders with the same color, but with a difference of 1.5 in pH.

Grouping	Indicative pH range
Natural cocoa	5.3–6.0
Lightly alkalized	6.0–7.5
Medium alkalized	6.5–7.8
Red alkalized	7.2–8.4
Deep red alkalized	7.5–8.4
Black alkalized	7.5–8.7

The pH of cocoa usually has a limited influence on the pH of the final product because of the amount used and the buffering capacity of most products. Alkalinity and ash content of cocoa powder is related to the amount of acids and ash present in the cocoa beans or nibs prior to alkalization.

H. Fineness

The fineness of cocoa powder is usually determined by settings during the liquor and/or cake grinding phase of the production process. Proper tempering is important to avoid lump formation during storage. The fineness is quite important in most applications. A finer grind results in a greater surface area of the product, and a positive effect of the color intensity of the end product. It may affect flavor release and mouthfeel, and plays an important role in viscosity of syrups and settling out in liquid products [6]. In bakery products, the fineness is less important with regard to flavor than in most dairy products, but affects water absorption and functionality of the dough. The finer the particles, the more moisture can be absorbed.

Most cocoa processors express the fineness as the minimum amount of product that passes through a defined sieve in a liquid medium. The particle size distribution, on the other hand, may be a useful tool if high amounts of fine or coarse particles influence functionality during processing.

I. Rheology and Water Absorption

Cocoa powder has an important effect on the rheology and viscoelastic properties of many products that it is used in. Food systems where water is the continuous phase react differently to the presence of cocoa than products with fat as the continuous phase. Cocoa is able to absorb moisture up to 100% of its own weight. In comparison, flour can absorb up to 60% moisture of its own weight. As a consequence of this strong water absorption of cocoa powder, a stiffer dough and dryer baked product with more breakage will occur if no moisture correction is made when cocoa replaces part of the flour [5,6].

In high moisture products like chocolate milk, the usage level of cocoa by itself does not create a significant effect on the viscosity. However, in order to keep the insoluble cocoa powder particles from settling to the bottom, and create an unattractive looking product, a stabilizer is added to the product. A stabilizer like carrageenan forms a network together with cocoa particles, casein, and protein-covered fat globules, and significantly increases viscosity of the finished product [2].

In high viscose products that are rich in sugar, like syrups and toppings, cocoa interacts with sugar particles, which may lead to an undesirable after-thickening effect. Syrup processing parameters, usage of high fructose corn syrups, amount of sugar, as well as the amount of cocoa are key factors in this phenomenon [6].

In fat continuous, low-moisture products like chocolate or compound coating cocoa, particles affect viscosity by fat absorption and by introducing some moisture to the food system.

J. Wettability and Dispersability

Under typical food conditions, cocoa powder is about 30% soluble in water. The remaining part does not disperse well in water, owing to the presence of hydrophobic fat crystals on the outer side of the cocoa particles. Cocoa particles tend to float on the surface when added to cold milk or water. The wettability is poor, and the product does not disperse well. The wettability and the dispersability can be improved by mixing cocoa with lecithin, mixing with sugar, adding an agglomeration step, or a combination of these steps [6].

K. Product Specifications

While discussing specific cocoa powders falls outside the scope of this chapter, suppliers capture a number of the above-mentioned characteristics in the specification of their products. Color and flavor are typically checked against internal standards, and described in specifications as "up to standard." In some cases color values are given, but the method used should always be defined. For pH, fineness, and moisture content a range or a minimum/maximum value is given. The actual number and the width of the range depend upon the processes, and the control a processor has. Each specification includes microbiological and pathogen numbers. While the standards for pathogens are fairly consistent across different suppliers, the total plate count (TPC) is an example of a process dependent parameter. A cocoa bean processor with a totally enclosed system is able to achieve lower TPC numbers than a processor of cocoa cake, purchased in the open market. The final application, including recipe and processing steps, determine the toleration for micro standards.

VI. COCOA APPLICATION

A. Introduction

The wide variety of cocoas available for food products makes it a challenge to find the "best" one for different applications. A supplier is often asked the seemingly simple question about which cocoa to use for which application. While it is possible to give some general directions and examples of their use in specific products, there is no one specific cocoa per specific application. The choice of cocoa depends on the desired color and flavor profile, the functionality in a specific formula, other ingredients, and the preparation of the finished product. Besides technical considerations, the price may have an effect on the choice of cocoa as well as the usage level.

In this section the main cocoa application groups will be discussed, with focus on the functionality of cocoa, and the most common choices of cocoa types.

B. Dairy Products

The products considered to be part of this group are chocolate milk, milk shake, custard, mousse, pudding, whipped toppings, and fermented dairy products. All these products are made, to a great extent, from liquid milk. In general, 10 to 12% fat light, medium and red alkalized cocoa powders are used in these applications. Natural cocoa powders generate a weak flavor in these dairy products, and anything of a higher degree in alkalization than red cocoas may result in an unattractive color and flavor profile. While there are exceptions to this rule, it holds true for the vast majority of the cocoa used. The different dairy applications and food systems create a couple of challenges.

In chocolate milk, stabilization is a concern. Cocoa powder is roughly 30% water soluble. The insoluble particle will, over time, settle out as sediment. Without any type of stabilizer the sediment will become a compact layer, and this layer will be difficult to disperse back into the product by simply shaking, for example, the bottle, jar, or carton package. With the addition of a stabilizer like kappa-carrageenan, a network with milk protein and cocoa particles will be formed, holding the cocoa particles in suspension. The optimum amount of carrageenan depends upon the amount of fat and protein in the milk, the pH and degree of alkalization of the cocoa, and the heat treatment of the chocolate milk. Other stabilizers like xanthan gum and modified starches are known to be used as well in chocolate-flavored drinks. Another consideration is the fineness of cocoa. Larger particles will settle faster than finer particles, and give an unpleasant mouthfeel. Finally, a cocoa powder of excellent microbiological quality is important for all dairy applications [2].

In puddings, mousses, and custards, stabilizers and/or thickening agents are a standard ingredient. The viscosity of the dairy products is much higher than for milk, and settling of cocoa is not an issue. However, fineness in relation to mouthfeel, a smooth appearance, and the absence of dark specks is of utmost importance. The color of pudding-type products depends not only on the type of cocoa powder, but also on the fat content and the type of dairy component. A higher percentage of fat will give the product a lighter appearance, and greater variation in color hues derived from cocoa are dependent on the percentage of fat in the cocoa. Furthermore, a product based on a high percentage of whey ingredients will result in a different color than a milk-based product with a similar fat content [1].

C. Ice Cream and Frozen Desserts

Chocolate and cocoa rank second only to vanilla as flavoring for ice cream. Cocoa is easily incorporated as an ingredient during ice cream manufacturing. And while it changes color, flavor, and handling characteristics, it does not pose any complicated technical problems. In general light, medium and red alkalized cocoas are used in ice cream and frozen dessert. As for dairy products, natural cocoas generate a relatively weak flavor, and higher alkalized cocoas give too much of a pungent alkalized note. While 10 to 12% fat cocoas are more economical to use, high-fat cocoas are often incorporated in indulgent and high quality ice creams.

During the manufacturing of ice cream, cocoa is typically mixed with other dry ingredients, before adding milk or other liquid ingredients. The various cocoa powders and their usage level offer a wide variety of possible colors. The obvious flavor impact of cocoa requires an increase of the overall sweetener content, in order to compensate for bitter notes originating from cocoas. Addition of cocoa and increase of sugar content leads to a higher total solids content of the base mix. This in turn affects the freezing point and the overrun, being the increase in volume due to the incorporation of air during freezing of the base mix. The manufacturer will need to adjust processing parameters to arrive at the desired overrun in the case of chocolate-flavored ice cream [4].

As in dairy applications, the cocoa used in ice cream and frozen desserts needs to be of excellent fineness and microbiological quality. A coarse cocoa results in dark spots and a grainy texture. Furthermore, the color and flavor should be consistent, in order to prevent variation in the finished product.

D. Bakery Products

There are a tremendous variety of chocolate-flavored baked goods available worldwide. Examples are Devil's food cake, sponge cake, fudge cake, brownies, pounds-pounds cake, sandwich cookies, extruded cookies, cereal, and chocolate-flavored meringue. There are many region-specific products and recipes, and regional preferences for certain products. The type of powders used are mostly 10 to 12% fat, while natural as well as any degree of alkalized cocoas may also be used. Natural cocoas are used in light colored and flavored cookies, light to red alkalized cocoas are used in all sorts of cakes, and dark brown to black alkalized cocoas are used in sandwich cookies. The main functionality is color and flavor, but cocoa also has an impact on dough properties and the texture of the finished product. Ingredients such as flour, shortening, and sugar require a higher cocoa usage level than a dosage typical for chocolate milk or ice cream.

During alkalization cocoa becomes darker and more reddish. In a high, moist, baking environment, with alkali present as baking salts, this process may continue, and result in extra color development.

In a moist environment, cocoa is able to absorb 100% of its own weight. This means that in dough and batters, cocoa compete with other ingredients like flour and sugar for the available moisture. When cocoa is incorporated in a dough or batter, one needs to make an adjustment for the moisture content, in order to avoid a too stiff dough structure, resulting in a reduced spread during baking, and too dry spread of an end product. Alkalized cocoas tend to bind a bit more moisture than natural cocoas. While fineness does not have much of on an impact on the appearance of the finished good, it does affect the dough functionality. A finer cocoa will absorb more moisture than a coarser grind product [5].

E. Confectionery Coating, Fat-Based Fillings

Confectionery coatings and fat-based fillings are fat–sugar based systems. The amount and type of fat determines the appearance, melting, and flowing characteristics. Very often 10 to 12% fat natural cocoas are used for this application, but alkalized cocoas may be included to darken the color and increase the flavor impact. Especially in ice cream coatings, higher alkalized powders create a more intense color and flavor profile that is not dulled by the low temperature of the ice cream [6].

Cocoa is added for color and flavor reasons. Natural cocoa is closest in color and flavor to the solids mostly used in regular chocolate products. It is also typically the most economical cocoa to be used in cost-sensitive items. The fineness of cocoa is of lesser importance since the production of coatings and fillings includes a grinding step. Microbiological numbers are also less critical than for water-based applications, because the water activity is far below a level that could promote microbiological growth in the finished good. When lauric fats are used, there is a risk of saponification if lipase is present in cocoa or other ingredients [6].

F. Syrups, Water-Based Fillings

This group of products contains a certain amount of moisture, and typically undergoes a cooking step. Mostly light, medium, and red alkalized cocoas are used. Cocoa is used for flavor and color impact, but also has a significant effect on the texture of the product. In a high sugar–low moisture environment, cocoa and sugar particles will form a network over time. This will increase the viscosity, and may lead to an undesirable so-called after-thickening effect. In order to reduce the risk of this effect, it is advised to use no more than

10% cocoa. Natural cocoas tend to promote after thickening faster than alkalized cocoas, and a finer ground cocoa, with more particle per weight, may promote after thickening faster than a coarse ground cocoa. Even though water is present the water activity is typically too low to create microbiological problems.

G. Instant Mixes, Dry Mixes

Instant cocoa products are generally added to cold milk or water. Just adding a regular cocoa powder to cold water or milk and stirring will not give an attractive looking product. Cocoa powder does not disperse easily because of the hydrophobic character of part of the cocoa solids. Cocoa tends to form lumps when dispersed in a cold liquid. One way to improve dispersability is to mix cocoa powder with other solids like sugar and/or milk powders. Another way is to mix cocoa powder with lecithin. If processed well, lecithin will form a thin layer around cocoa powders. A lecithin molecule is made up of hydrophylic and hydrophobic parts. The hydrophobic part anchors itself to the cocoa butter on the cocoa solid particle. The hydrophylic part of the lecithin molecule is directed to the outside of the cocoa particle. In this way a cocoa particle is created whose outer surface has a hydrophylic character, and is easily dispersed. A third way to improve dispersibility is to agglomerate cocoa, sugar, and other solid particles. During this process, sugar crystals are moistened with water or steam, (lecithinated) cocoa and other solid particles adhere to the wet sugar crystals, and the agglomerated particles are dried. In this way sugar–cocoa agglomerates are created that are easily dispersed in cold milk or water [3,6].

In general light, medium, and red alkalized cocoa are used in instant products. In addition to instant mixes a lot of premixes are sold to smaller dairies and bakeries. The composition of these blends, and the cocoas to be used, are determined by the end product.

H. Regional Variation

The total powder usage as well as the types of powders used varies in the different global regions. In developed markets like western Europe, the United States, and Canada mostly alkalized higher quality cocoa powders are used. In developing markets like Mexico and eastern Europe, the demand for economical 10 to 12% fat natural cocoas is high. The demands of specific types depend upon the size of the bakery, dairy, and confectionery market, as well as the development of economical versus high quality and highly indulgent products. A cocoa supplier needs a good understanding of the regional markets in order to operate in an efficient and competitive way.

VII. CONCLUSION

Over a period of hundreds of years cocoa has made a transformation from a sacred item for religious ceremonies, to a luxury item for the Spanish court, to an enjoyable product available to people around the globe. Driven by technical developments, cocoa has evolved from a plain yellowish/brown powder to a product with a wide variety of attractive colors and flavors, suitable for numerous different applications. The presence of phenolic components has generated a lot of interest from a nutritional point of view. Therefore, it is no surprise that cocoa is still used in many product development projects, and the annual global cocoa powder demand may very well grow in the coming years. *Theobroma cacao L.*, the food of the gods, still has a growing popularity, and is looking at a bright future.

REFERENCES

1. ADM Cocoa, 1997, Chocolate Flavored Desserts, Technical Information Bulletin.
2. ADM Cocoa, 1998, Chocolate Milk a Complicated Product, Technical Information Bulletin.
3. ADM Cocoa, 1998, Chocolate Powder and Dry Mixes, Technical Information Bulletin.
4. ADM Cocoa, 1998, Cocoa Powders and Ice Cream, Technical Information Bulletin.
5. ADM Cocoa, 1998, Cocoa Powders in Bakery Application, Technical Information Bulletin.
6. Anon., 1999. The De Zaan Cocoa Products Manual, ADM Cocoa.
7. Anon., 2003, Quarterly Bulletin of Cocoa Statistics, ICCO.
8. Beckett, S.T. (Ed.), 1994, *Industrial Chocolate Manufacture and Use*, 2nd ed. Blackie Academic & Professional, Glasgow.
9. Bixler, R.B. and Morgan, J.N., 1999, Cocoa Bean and Chocolate Processing. In *Chocolate & Cocoa Health and Nutrition*, Ed., Knight, I. pp. 43–60, Blackwell Science Ltd, Oxford.
10. Cheney, S., 1999, Analysis and Nutrient Databases. In *Chocolate & Cocoa Health and Nutrition*, Ed., Knight, I. pp. 63–75, Blackwell Science Ltd, Oxford.
11. Cook, L.R. and revised by Meursing, E.H., 1982, *Chocolate Production and Use*, Harcourt Brace Janovich, New York.
12. Jardine, J.J., 1999, Phytochemicals and Phenolics. In *Chocolate & Cocoa Health and Nutrition*, Ed., Knight, I. pp. 119–142, Blackwell Science Ltd, Oxford.
13. Kattenberg, H.R. and Muijnck, L. de, 1993, The shelf life of cocoa products as ingredients for the food industry. In *Shelf Life Studies of Foods and Beverages: Chemical, Biological, Physical, and Nutritional Aspects*, Ed., G. Charalambous, Elsevier Science Publishers B.V., Amsterdam.
14. Minifie, B.M., 1989, *Chocolate, Cocoa, and Confectionery: Science and Technology*, 3rd ed. Van Nostrand Reinhold, New York.

Encapsulated Food Powders

18

Spray-Dried Microencapsulated Fat Powders

M. K. Keogh
Teagasc
Dairy Products Research Centre
Moorepark, Fermoy,
County Cork, Ireland.

CONTENTS

I. MICROENCAPSULATION: DEFINITION AND TERMINOLOGY

Emulsions, which have been stabilized by homogenization in the presence of an emulsifier, can be spray-dried to produce microencapsulated fat powders. Proteins, modified starches, or a hydrocolloid can be used as emulsifying agents. A non-emulsifying water-soluble material such as sugar or hydrolyzed starch is also used as filler. Combined emulsifying-filler agents such as modified starches or Acacia gum can also be used [1]. A comparison between

477

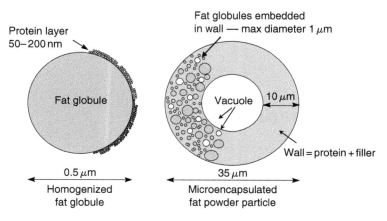

Figure 1 Comparison of homogenized and microencapsulated fat.

a homogenized fat globule and a microencapsulated fat powder particle is shown in Figure 1. Fat globules, which have been homogenized in the presence of milk proteins, have a median particle size of 0.5 μm or less depending on the homogenization pressure and number of passes and an adsorbed protein layer that is 50 to 200 nm thick [2]. The prefix micro implies that all or at least 90% of the globules should be <1 μm. By contrast, spray-dried microencapsulated fat powder particles have a median particle size of 35 to 75 μm or greater. The oil globules in the powder particles retain the layer(s) of adsorbed emulsifier during drying, and in turn are embedded in a continuous phase of inert filler. In the U.S. literature, especially, fat is referred to as the core material and the emulsifier and inert filler as the wall materials. Any fat can be used as core, which implies that one of the wall materials must have emulsifying properties. Spray-dried powder particles also contain a large central vacuole up to 15 μm or more in diameter with small to medium-sized vacuoles in the wall [3]. One advantage of microencapsulation is that any fat or oil can be converted to powder form, using any emulsifying or filler materials of choice. Low levels of fat on the surface of the spray-dried particles should result in better powder flow properties or the production of higher fat powders using conventional dryers. Avoiding contact between the fat core and prooxidant materials present such as atmospheric oxygen should extend the shelf life of the powder product.

II. HYPOTHESIS

The hypothesis underlying microencapsulation, as a technique, is that the smaller the emulsion fat globule size, the lower the rate of aggregation and coalescence of the globules [4]. In the spray-drying stage, the lower the extent of coalescence, the lower the rate of diffusion of oil on to the surface of the powder particles [5], the lower the surface fat [6] and the lower the rate of oxidation [7]. The surface composition (expressed as percent fat, protein, and filler) of powder particles can be measured by electron spectroscopy for chemical analysis (ESCA) [8]. While cholesterol oxidation was shown to be related to the level of surface fat in cholesterol-spiked tristearins and triolein [7], fish oil off-flavor was shown to be related to the solvent extractable or free fat content of fish oil powders [9]. However, the oxidative stability of milk fat powders was shown not to be related to free fat [10]. Surface fat does not correlate well to free fat, since the latter also includes some near-surface fat [11]. Free fat comprises of surface fat, subsurface fat, pore fat from the pores of the powder particle, and dissolution fat that is dissolved by the solvent [12]. Emulsion fat globule diameter, fat globule stability over time, and poorly encapsulated fat are therefore of fundamental importance for the shelf life of microencapsulated fat powders.

III.   ROLE OF EMULSION COMPOSITION

A.   General

Two research groups have extensively studied microencapsulation by spray drying using dairy ingredients. One group [2,13–18] used anhydrous milk fat at levels of 10–75%. Whey protein concentrates (WPC) with 50–75% protein and whey protein isolate (WPI, 95% whey protein) were employed as encapsulating proteins with lactose and other carbohydrates in various ratios as fillers. The emulsions (10–30% solids) were homogenized at 50 MPa by 4 passes. They used the term "*MicroEncapsulation Yield*" (MEY) of powders, defined as the powder fat content (%)/emulsion fat content (%), which varied from 91 to 99%. They also employed the term "*MicroEncapsulation Efficiency*" (MEE), which was equivalent to 100 — free fat content (g/100 g fat).

The other group [6–8,19–21] focused mainly on sodium caseinate, but compared whey proteins to sodium caseinate as emulsifiers, with lactose in various ratios as filler. Valuable insights into the mechanisms of free and surface fat formation were obtained by using fats with different melting points [6]. They used a microfluidizer to homogenize the emulsions at 100 MPa × 8 passes. These homogenization conditions were twice those used by the other group. The emulsions contained 10% solids only. This group developed the technique of ESCA to measure the composition of the surface (sampling depth ≈ 60 to 120 Å) of food powder particles. In this way, they were able to compare surface fat (percentage of total particle surface area covered by fat) to free fat (g/g powder) values. Figure 2 [20] shows the surface coverage by sodium caseinate (y_1-axis), and the surface tension (y_2-axis) versus sodium caseinate concentration (x-axis). It can be seen that at very small concentrations of sodium caseinate, there is a rapid decrease in the surface tension and simultaneously an increase in the surface coverage of the particle by sodium caseinate. They also demonstrated the behavior of a nonsurface-active material such as glycine in an aqueous system. As glycine does not reduce surface tension significantly, there was negligible coverage of the surface by the amino acid. The behavior of bovine albumin was intermediate. In other words, the ability of a protein to coat the surface of a powder particle is largely related to the surface tension-reducing properties of the protein. Since sodium caseinate is the most surface tension-reducing food protein known [22] it should also be the best microencapsulating protein. This ability of sodium caseinate to adsorb to a fat globule and coat a powder particle implies that the higher the surface coverage of the protein, the lower will be the surface fat.

B.   Fat: Globule Size, Content, and Melting Point

It has been shown by various workers [17,21], that as the emulsion fat content increases, the median globule diameter of the emulsion increases. Emulsions with a high fat load of milk fat prepared using whey proteins under high homogenization pressure exhibited a bimodal particle size distribution due to clustering of the protein-coated droplets. This will translate into an increase in diameter of the largest globules, as indicated by the D(v, 0.9) value [23].

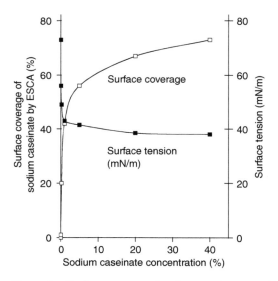

Figure 2   Surface coverage of sodium caseinate and the surface tension of the spray-dried solutions. (From Fäldt, P. and Bergenståhl, B. *Colloids Surf.* 90: 183–190, 1994. With permission.)

Thus, the latter value is a good indicator of emulsion fat globule coalescence before drying. The free fat content and surface fat coverage of powder particles will also increase.

The extent of surface fat coverage was also affected by the melting point of the fat used [7]. A low melting point oil gave rise to low levels of surface fat due to efficient encapsulation of the oil. A fat with a melting point higher than the temperature of the powder in the drying chamber gave powders with negligible levels of surface fat. This confirmed the view that, since the fat remained solid during drying, diffusion of the fat to the surface of the particle did not take place. When an intermediate melting fat was used, powders with moderate levels of surface fat were produced, because the fat was in a semicrystallized state during processing and storage, leading to more fat globule instability, coalescence, and diffusion and thereby leading to higher levels of free and surface fat.

C.   Emulsifier: Milk Protein Content, Type (Sodium Caseinate, Whey Protein Powders, Skim Milk Powder)

There have been a number of reports on the use of sodium caseinate and whey proteins as encapsulants and lactose as filler [8,14,17,23]. Skim milk powder (SMP, 36% protein) contains the parent micellar form of casein with whey proteins in an approximately 4:1 ratio. Sodium caseinate has a relatively uniform composition (90% protein), but commercial whey protein concentrates contain 35 to 75% or more protein and whey protein isolates have at least 90% protein. Sodium caseinate is a much better emulsifier than whey proteins, because it is a more hydrophobic and flexible molecule that reduces surface tension more quickly and to a greater extent than the larger globular bovine serum albumin or smaller less hydrophobic amino acids [20]. The superiority of sodium caseinate over whey proteins as a microencapsulant in terms of reduction of surface tension, free fat, and surface fat, especially in combination with lactose was also demonstrated [8]. However, greater stability to oxidation was shown for fish oil powders microencapsulated using skim milk powder (SMP) [9]. In that work, the use of SMP as microencapsulant proved superior to sodium caseinate in reducing off-flavor development in fish oil powders during storage. Since the form of casein (micellar SMP versus nonmicellar caseinate) was confounded with the vacuole volume of the powder used (7 versus 21 ml/100 g powder), the increased shelf life using SMP may have been due to either factor.

As the whey protein:lactose ratio increases, the reconstituted emulsion globule diameter becomes larger, because increasing protein levels increases globule cluster formation during drying [23]. It was also shown [24] that whey protein emulsions will only display globule cluster formation if the whey proteins are denatured prior to emulsification and if calcium or other divalent salts are present, but not if sodium or monovalent salt levels are high. The free fat level increases because fat is more easily extracted through the hydrophobic protein material. Increased coalescence and powder porosity lead to higher surface fat levels.

D. Filler: Lactose

The hydrophilic filler forms the continuous phase in spray-dried powder particles. Such a hydrophilic material reduces protein–protein interactions and thus fat globule aggregation/coalescence in the emulsion formation and drying steps. The filler also repels or prevents the polar solvent from reaching the fat in the powder particles for extraction as free fat. Increased lactose always reduces surface fat in combination with sodium caseinate but not with whey proteins [21]. Lactose has a high glass transition temperature, but has the disadvantage of being less soluble than sucrose, maltose, or hydrolyzed starch during emulsion preparation.

E. Milk Salts

Higher whey protein:lactose ratios at low salt levels in an emulsion reduced fat globule aggregation as measured by the D(v, 0.9) value [23]. However, in the presence of salts and especially at lower homogenization pressures, fat globule repulsion was reduced and aggregation increased. Even in the absence of salts, higher whey protein:lactose ratios increased aggregation of fat globules in the emulsion during drying, which resulted in higher D(v, 0.9) values in reconstituted emulsions and in turn higher free fat in the powder.

F. Crystallization of Lactose

Lactose is in the amorphous state in milk powders. However, when milk powder is exposed to atmospheres of high water activity, amorphous lactose is transformed to the α-crystalline state [25], new pores are generated, particle size increases, and the powder porosity changes [26]. Coalescence occurs resulting in fat pools, increased free fat and fat release on to the surface. Addition of only a small amount of lactose reduces fat globule coalescence during drying. Under these conditions, an increase in free fat and surface fat does not occur even under humid conditions [27].

G. Alternative Emulsifier-Fillers: Gum Acacia, Lipophilized Starches

Alternative wall materials could be used, for instance, Acacia gum (also known as gum Arabic or gum Senegal), a low viscosity hydrocolloid, which uniquely has a protein moiety or fraction and therefore can serve as a combined emulsifier-filler. Earlier work on microencapsulation, especially of orange oil, successfully used this material as microencapsulant [1].

Modified starches can provide a more reliable supply of combined emulsifier-filler material. Starch is hydrolyzed or pyrolyzed to reduce viscosity and increase solubility, and repolymerized (dextrinized) to form highly branched small polymers, which trap oil globules and provide steric hindrance. Hydrophobic sidechains are added to the depolymerized starches by reaction with sodium octenyl succinate. The degree of hydrophobicity, determined by the number of sidechains and the size distribution of the starch molecules can both be controlled for optimum performance [28].

H. Air

The effect of air in powders had not been investigated [1]. In a study of the stability to oxidation of a microencapsulated fish oil powder, the free fat, surface fat, and vacuole volume of the powder particles were measured and related to the shelf life as monitored sensorially [9]. The shelf life increased as the vacuole volume of the microencapsulated powders decreased. However, as the vacuole volume and form of casein used as emulsifier were confounded, the increased shelf life may have been due to either factor.

IV. ROLE OF PROCESSING

A. Homogenization: Pressure, Number of Passes

Reducing fat globule size depends on the homogenization pressure and the number of passes in addition to the composition of the emulsion already outlined. A median fat globule size of 0.34 to 0.54 μm was obtained using anhydrous milk fat (AMF) in WPI alone or in combination with carbohydrate fillers at a core load of 75 g AMF/100 g emulsifier + filler [17] using a pressure of 50 MPa × 4 passes. Similar values were obtained [27] using a microfluidizer at 100 MPa × 8 passes. The D(v. 0.9) value was also used [23] as an indicator of fat globule aggregation. The D(v. 0.9) value ranged from 0.76 μm after 2 passes to 0.54 μm after 4 passes at a pressure of 45 MPa in an emulsion containing 62.5 g AMF/100 g WPI + lactose. In subsequent work using fish oil, the lowest D(v. 0.9) value obtained was 0.54 μm after 5 passes at 50 MPa and a D(v. 0.5) value of 0.42 μm. The results indicated that further reductions in the globule size of fish oil emulsions, microencapsulated using SMP would increase shelf life with associated decreases in free fat but not in the surface fat of the powders.

B. Spray Drying Conditions

Spray drying conditions such as air inlet temperature, air outlet temperature and nozzle size affect the microencapsulated powder properties. Drying conditions should be chosen to minimize free and surface fat. These are low inlet and outlet temperatures to reduce the viscosity and diffusivity of the fat during drying. Larger nozzles and higher emulsion solids levels give larger particles of lower surface area with lower free and surface fat [29].

REFERENCES

1. Reineccius, G.A. Spray drying of food flavours. Symposium Series No. 370 on Flavour Encapsulation, Washington, D.C., 1988, pp. 63–72.
2. Rosenberg, M. and Lee, S.L. Microstructure of whey protein/anhydrous milkfat emulsions. *Food Struct.* 12: 267–274, 1993.

3. Verhey, J.G.P. Vacuole formation in spray powders particles. 1. Air incorporation and bubble expansion. *Neth. Milk Dairy J.* 26: 186–202, 1972.

4. Walstra, P. Overview of emulsion and foam stability. In: Dickinson, E. Ed. *Food Emulsions and Foams*. London, UK: Royal Society of Chemistry, 1986, pp. 242–257.

5. van Boekel, M.A.J.S. and Walstra, P. *Colloids Surf.* 3: 109, 1991.

6. Fäldt, P. and Bergenståhl, B. Fat encapsulation in spray-dried food powders. *J. Am. Oil Chem. Soc.* 72: 171–176, 1995.

7. Granelli, K., Fäldt, P., Appelqvist, L.A., and Bergenståhl, B. Influence of surface structure on cholesterol oxidation in model food powders. *J. Sci. Food Agric.* 17: 75–82, 1996.

8. Fäldt, P., Bergenståhl, B., and Carlsson, G. The surface coverage of fat on food powders analyzed by ESCA (electron spectroscopy for chemical analysis). *Food Struct.* 12: 225–234, 1993.

9. Keogh, M.K., O'Kennedy, B.T., Kelly, J., Auty, M.A., Kelly, P.M., Fureby, A., and Haahr, A.M. Stability to oxidation of spray-dried fish oil powder microencapsulated using milk ingredients. *J. Food Sci.* 66: 217–224, 2001.

10. Buma, T.J. Free fat in spray-dried whole milk. 4. Significance of free fat for other properties of practical importance. *Neth. Milk Dairy J.* 25: 88–106, 1971.

11. Buchheim, W. Electron microscopic localization of solvent-extractable fat in agglomerated spray-dried whole milk powder particles. *Food Microstruct.* 1: 233–238, 1982.

12. Buma, T.J. Free fat in spray-dried whole milk. 10. A final report with a physical model for free fat in spray-dried milk. *Neth. Milk Dairy J.* 25: 159–174, 1971.

13. Rosenberg, M. and Young, S.L. Whey proteins as microencapsulating agents. Microencapsulation of anhydrous milkfat — structure evaluation. *Food Struct.* 12: 31–41, 1993.

14. Moreau, D.L. and Rosenberg, M. Microstructure and fat extractability in microcapsules based on whey proteins or mixtures of whey proteins and lactose. *Food Struct.* 12: 457–468, 1993.

15. Moreau, D.L. and Rosenberg, M. Oxidative stability of anhydrous milkfat micro-encapsulated in whey proteins. *J. Food Sci.* 61: 39–43, 1996.

16. Moreau, D.L. and Rosenberg, M. Porosity of whey protein-based microcapsules containing anhydrous milkfat measured by gas displacement pycnometry. *J. Food Sci.* 63: 819–823, 1998.

17. Young, S.L., Sarda, X., and Rosenberg, M. Microencapsulating properties of whey proteins. 1. Microencapsulation of anhydrous milk fat. *J. Dairy Sci.* 76: 2868–2877, 1993.

18. Young, S.L., Sarda, X., and Rosenberg, M. Microencapsulating properties of whey proteins. 2. Combination of whey proteins with carbohydrates. *J. Dairy Sci.* 76: 2878–2885, 1993.

19. Fäldt, P. Surface composition of spray-dried emulsions. Ph.D. dissertation. Lund University and Institute for Surface Chemistry, Stockholm, Sweden, 1995.

20. Fäldt, P. and Bergenståhl, B. The surface composition of spray-dried protein-lactose powders. A — physicochemical and engineering aspects. *Colloids Surf.* 90: 183–190, 1994.

21. Fäldt, P. and Bergenståhl, B. Spray-dried whey protein/lactose/soy-bean oil emulsions 1. Surface composition and particle structure. Paper 6 in dissertation, ref. 19, 1995.

22. Dalgleish, D.G. Protein-stabilized emulsions and their properties. In: Hardman, T.M. Ed. *Water and Food Quality*. London, UK: Elsevier Applied Science, 1989, pp. 211–250.

23. Keogh, M.K. and O'Kennedy, B.T. Milk fat microencapsulation using whey proteins. *Int. Dairy J.* 9: 657–663, 1999.

24. Rientjes, G.J. and Walstra, P. Factors affecting the stability of whey-based emulsions. *Milchwiss.* 48: 63–67, 1993.

25. Saito, Z. Particle structure in spray-dried whole milk and in instant skim milk powder as related to lactose crystallization. *Food Microstruct.* 4: 333–340, 1985.

26. Aguilar, C.A. and Ziegler, G.R. Physical and microscopic characterization of dry whole milk with altered lactose content. 2. Effect of lactose crystallization. *J. Dairy Sci.* 77: 1198–1204, 1994.

27. Fäldt, P. and Bergenståhl, B. Spray-dried whey protein/lactose/soy-bean oil emulsions 2. Redispersability, wettability and particle structure. Paper 7 in dissertation, ref. 19. 1995.

28. McGlinchy, N. National Starch & Chemical Ltd., U.K., personal communication, 1993.

29. Pisecky, J. *Handbook of Milk Powder Manufacture*. A/S Niro Atomizer Ltd., Soeborg, Copenhagen, 1997, pp. 149–189.

19

Single- and Double-Encapsulated Butter Oil Powders

C. I. Onwulata
U.S. Department of Agriculture
ARS, Eastern Regional Research Center
Wyndmoor, Pennsylvania

CONTENTS

I. INTRODUCTION

Encapsulation of powders with different components plays a very important role in foods especially for the delivery of bioactive additives (Jackson and Lee, 1991; Duxbury and Swientek, 1992). Compound powders with "soft" cores provide means for protecting sensitive components with pharmaceutical benefits. Protecting loss of active nutrients, such as carotenoid activity during processing, was the essential role reported for the encapsulation

---

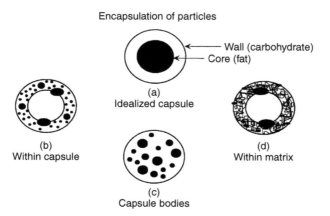

Encapsulation of particles

(a)
Idealized capsule

Wall (carbohydrate)
Core (fat)

(b)
Within capsule

(c)
Capsule bodies

(d)
Within matrix

Figure 1 Illustration of mode of encapsulation of butter oil in various carbohydrate matrices. (a) idealized capsule; (b) Modified corn starch capsule; (c) sucrose capsule, and (d) all purpose flour capsule. Dark spots or bodies represent fat capsules.

of beta-carotene (Wagner and Warthesen, 1995). Marine omega-3 fatty acids have been encapsulated in a starch/oil emulsion and spray-dried to provide this protection in food powders (Andersen, 1995). The entrapment of sensitive ingredients within a continuous film or coating protects them from environmental factors such as moisture, air, or light. Encapsulation, however, leads to demonstrable difficulties in handling, owing to changes in bulk properties of the powders. This was particularly true for encapsulated butter oil, where the encapsulated powders showed a propensity for stickiness and lumping (Konstance et al., 1995; Onwulata et al., 1995, 1996). The amount of the core material retained after encapsulation depends on the coating material used (Onwulata et al., 1994). The main purpose of preparing spray-dried fats is to enhance their handling properties and stability during storage, transport, and blending with nonfat ingredients. Butter oil, to be made into a powder, requires a carrier because it contains appreciable amounts of low melting triglycerides. As the amount of fat to be encapsulated increases, the choice of carrier constituents becomes more and more important, as fats on the surface of powders retard flow (Onwulata et al., 1995).

Volatiles have been entrapped and retained in sucrose, maltose, or lactose by freeze drying the amorphous sugar to the crystalline state (Flink and Karel, 1970). There has been remarkably high volatile retention in such powders. The surface membranes are impermeable with high resistance to diffusion at low water content or to the formation of inclusion complexes (Menting and Hoagstad, 1967; Flink and Karel, 1970).

A second encapsulating coat is sometimes provided to give a better moisture barrier. Such a barrier could be created through a combination of waxes coating the surfaces of the single powder (Figure 2) to form the double coat. Minimizing moisture uptake is extremely important for encapsulated powder stability and effectiveness (Onwulata et al., 1995). The proper choice of material for the second coat enhances delivery of the functional content (Pothakamury and Barbosa-Canovas, 1995). This is essential for controlled release of components in foods, an aspect that is proving to be increasingly useful for functional (nutraceutical) powders. The physical properties and structures of double-encapsulated powders containing butter oil is here compared to the characteristics of single-encapsulated powders such as bulk density, powder flowability, cohesiveness, and compressiblity (Figure 2).

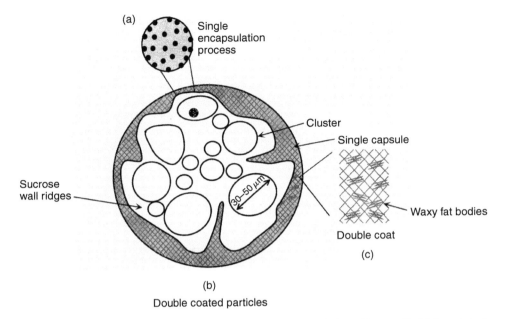

Figure 2 Schematic depiction of the mode of double encapsulation of butter oil. (a) Intact single-encapsulated particle, (b) single-encapsulated particles encapsulated with the double coat, (c) matrix of the surface of the double coat.

II. DESCRIPTIONS

Butter oil was obtained from a commercial manufacturer (Land O'Lakes, Inc., Arden Hills, MN). The encapsulating agent was granulated sucrose. Spray-dried powders containing 50% butter oil were prepared in our pilot plant as previously described (Onwulata et al., 1994). Encapsulated powders were formulated to have 50% butter oil, 5% emulsifier, and 5% skim milk powder, with the remainder encapsulant. The emulsifiers used were mono- and di-acylglycerides (American Ingredients Co., Kansas City, KS). The processing sequence for the single coat capsule was as follows: sucrose was dry-blended with nonfat dry milk solids and the mixture dispersed in water to form a pasty slurry of ≈25% total solids. The anhydrous butter oil and the emulsifier were heated to 45°C with stirring. The two blends were combined (40% total solids) and mixed for 5 min with a milk stirrer, after which the slurry temperature was slowly raised to 62.8°C with constant stirring. The mixture was homogenized at 17.2 MPa with a Manton–Gaulin Triplex homogenizer (Model 100 DJF3855, APV Gaulin, Inc., Everett, MA), followed by spray drying in a compact dryer (APV Crepaco, Inc., Attleboro Falls, MA) at an inlet temperature of 193.3–196.1°C and an outlet temperature of 82.2–87.8°C. Powders were produced batch-wise, collected from the dryer after 30 min and stored at 4°C.

III. METHOD OF MANUFACTURE

The process for single encapsulation is mentioned in the chapter describing butter oil powders encapsulated in sugars and starches. Double encapsulation was done with the application of a mixture of vegetable waxes (hydrogenated stearines), Sterotex NF® (Abitec

Corp., Columbus, OH) and food grade linear alcohol polymers, Unilin 350® (Petrolite, Tulsa, OK) at a 99:1 Sterotex to Unilin ratio. The waxes were melted at 105°C. The molten waxes, which are normally in the alpha form, were converted to the more stable beta form using the Beta® process device (Encapsulation Systems Inc., Sharon Hills, PA). The molten wax was sprayed onto the preformed single-encapsulated particles. The moist particles were subjected to another pressure shock wave at 414 kPa, to cause the wax to form a second coat on the particles. The waxed powders were subsequently cooled, sifted through a 20 mesh sieve and stored. The process and the M-CAP® device used for double encapsulation have been reported (Redding, 1990, 1993). The single-encapsulated powders passed through a 500 μm sieve, and the double-encapsulated powders passed through a 1000 μm sieve.

A. Moisture and Moisture Uptake

The moisture contents of the powders were determined by drying in a vacuum oven at 102°C for 4 h (AOAC, 1984). Moisture determinations were made in triplicate. Moisture uptake by the powders was determined at 25°C by equilibrating 10 g powder samples at 52% relative humidity. Vapor pressure was maintained with a saturated salt solution of $Mg(NO_3)_2$ (Rockland and Nishi, 1980). Measurements were made in duplicate.

Measured physical properties of single- and double-encapsulated powders were compared (Table 1). The moisture content was characteristic of powders encapsulated with sucrose (Onwulata et al., 1995). Double-encapsulated powders had higher moisture, 36% more than single-encapsulated powders. The moisture content of the particles was below the critical water activity (0.4 a_w) level and did not lead to crystallization of sucrose, which would have been detrimental to flow properties.

The moisture uptake rates of the different powders (Figure 3) fit an exponential function, the rate of uptake being dependent on the nature of the powder and exposure conditions. Sucrose is highly hygroscopic and its moisture uptake was higher than that of the single-encapsulated powder. Further reduction in moisture was shown with the double-encapsulated powder. In terms of moisture sorption or as a moisture barrier, the double-encapsulated powder was more efficient. This functionality enhances the benefits of encapsulation, and may ameliorate or eliminate the need for special packaging to prevent moisture uptake during storage (Onwulata and Holsinger, 1995).

Table 1 Physical Properties of Single- and Double-Encapsulated Powders

	Moisture %	Loose density (g/cm^3)	Bulk density (g/cm^3)	Total fat %
Sucrose (granulated)	0.12	0.84	1.58	—
	±0.01	±0.01	±0.02	
Single-encapsulated	1.13	0.24	1.10	50.9
	±0.05	±0.01	±0.01	±0.91
Double-encapsulated	1.53	0.26	1.09	62.9
	±0.13	±0.01	±0.01	±1.01

Note: "—": No fats present; ±: standard deviations.

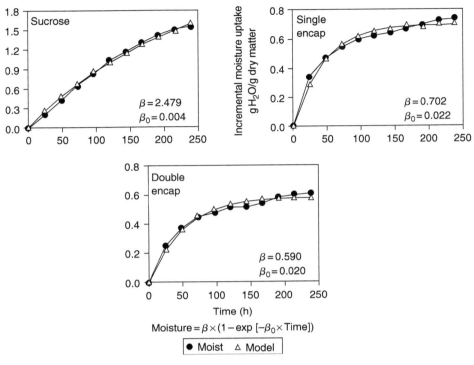

Figure 3 Moisture uptake of sucrose and single- and double-encapsulated powders.

B. Mass Flow Rate

The mass flow of the powder (g/sec) was measured by permitting 80 g to flow through funnels of outlet diameter 30 mm with gentle shaking using a Synthron Shaker (FMC/Synthron, Homer, PA) at the setting of 40, for those powders that would not flow without mechanical agitation (double-encapsulated powders). Time of flow was recorded and relative flow rate was calculated as powder weight (g) divided by time (sec). Flow rates were measured in triplicate. Flow properties of the powders (Table 2) showed that crystalline sucrose was free flowing (P < 0.1), with flow increasing under mechanical shaking. Single-encapsulated powders were a log order of magnitude less free flowing (Konstance et al., 1995). Double-encapsulated powders were 50% less flowable due to their size and the waxy top coat. With encapsulated powders, as fat content of the powder increased, flowability decreased sharply. Flowability was also influenced by the ratio of encapsulated to extractable fat; powders with higher levels of unencapsulated fat on the surface (therefore, a greater amount of extractable fat) tended to stick together and form lumps, which impeded flow (Onwulata et al., 1994). Flowability of a powder is determined by both the physical properties of the powder and the geometry of the particles. Several experimental methods measure relative flow characteristics. Flow dynamics are influenced by particle density, bulk density, particle shape and size as well as composition (White et al., 1967).

C. Density

Particle density was determined in triplicate with an air pycnometer (Horiba Instruments Inc., Irvine, CA). Bulk densities of all powders were determined by dividing the

Table 2 FLow and Compressibility Properties of Single- and Double-Encapsulated Powders

	G_flow (g/sec)	M_flow (g/sec)	% Compressibility[a](N)		
			100	200	300
Sucrose	190.0 ±3.2	228.6 ±1.6	4.7	6.9	7.4
Single-encap	71.5 ±6.0	72.2 ±6.0	25.5	32.5	35.0
Double-encap	—[b]	36.6 ±3.2	36.3	43.6	50.9

Note: G_Flow: gravitation flow: M_Flow: mechanical agitation to enhance flow.

[a] Compressibility: preset compression force (N). ±: standard deviations.
[b] No flow.

powder mass (g) contained in a 200 ml stainless steel cylinder (A/S Niro Atomizer, Copenhagen, Denmark) by its volume (cm^3). Packed bulk densities were calculated from the weight of powder contained in the cylinder after being tapped 100 times. Density measurements (g/cm^3) were done in triplicate.

The densities, both loose and bulk, were typical of encapsulated powders, and were lower than those of the sucrose powders used. Single- and double-encapsulated powders had similar densities, but the total fat content differed. There was a 25% increase in total fat content in the double-encapsulated powder due to the added vegetable waxes. There was also a 14% increase in extractable fat in the double-encapsulated powder. However, percentages of total fat extracted were similar: 25.9% for the single- and 24.3% for the double-encapsulated powders. Physical properties such as fat content, moisture, and product density affect powder flow (Peleg, 1977). The encapsulated material was affected less by these properties when the outer coat was composed of amorphous sucrose or other sugars. Double-coated powders were less free flowing when the outer coat was a waxy vegetable matrix, because fat on the surface of powders has a tendency to cause the particles to adhere to one another or agglomerate (Onwulata et al., 1995).

D. Compressibility

Compressibility was determined as described by Moreyra and Peleg (1980). The powders were carefully poured into a sample cell (30 mm high/45 mm diameter) and the loose density (ρ) was determined from the weight and known volume. The sample cell was mounted on the base plate of a Model 4200 INSTRON Universal Testing Machine (Instron, Canton, MA). The powders were compressed at a crosshead speed of 10 mm/min using a 50 kg load cell to a preselected force of $\approx$40 kg. Powder compressibility was determined by evaluating the slope of the plot of bulk density and the logarithm of compressive stress $(1 < \log \sigma < 4)$ using $\rho_D = a + b \log \sigma$; (Sone, 1972); where ρ_D = bulk density (g/cm^3) at corresponding σ, σ = compressive stress (g/cm^2) and a, b = empirical constants with "b" representing compressibility. Three data points were used.

Compressibility as a measure of internal cohesion and mechanical strength at various loads (Table 2) increased ($p < .05$) for single-encapsulated powders as well as ($p < .01$) for the double-encapsulated powders (Figure 4). Compressibility increased with increased force depending on type of encapsulation and there was a concurrent increase in cohesiveness.

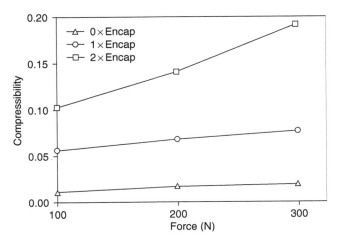

Figure 4 Relative compressibility of sucrose (0-Encap), single-encapsulated (1×-Encap) and double-encapsulated powders (2×-Encap).

Increased compression did not lead to powder caking or expulsion of encapsulated fat either in the single- or the double-encapsulated powder (data not shown). The sucrose powder was less compressible than its single- and double-encapsulated counterparts (Konstance et al., 1995). Carr (1976) had shown that compressibility, under relatively small loads, was a sensitive index of cohesiveness and could be used to predict potential flow problems.

E. Particle Sizes

Particle sizes were determined optically with an Accusizer® 770 (Particle Sizing Systems, Inc., Santa Barbara, CA). The particles were counted individually with a dry powder attachment and the population distribution was calculated from accumulated counts. The particle size distribution (Figure 5) showed that the sucrose and single-encapsulated powders ranged from 20 to 800 μm in size, while the double-encapsulated powders ranged from 50 to 1300 μm. The mean distribution showed that double-encapsulated powders (750 μm) were two times larger than the sucrose and single-encapsulated powders (350 μm). Larger sizes increase the potential for flow problems. The double-encapsulated powder had a normal Gaussian distribution. Single-encapsulated particles fell within a narrower particle size range, with relatively uniform distribution, but the double-encapsulated powders had a broader size range. Uniformity in particle size and shape improves powder flow (White et al., 1967). Previous work (Onwulata et al., 1994) with milk fat encapsulated in sucrose showed little or no milk fat present on the particle surface to interfere with flow. Therefore, flow difficulties with double-encapsulated powders were probably attributable to particle size and the waxy second coat.

F. Scanning Electron Microscopy

The topography of intact and fractured particles was examined by secondary electron imaging using a scanning electron microscope (Model 840A, JEOL, USA, Peabody, MA). Three methods of preparation were used. For superficial structure of intact particles, small samples of powder (<1 cc) were applied to an adhesive layer (Spot-0-Glue, Avery, AZ) affixed to 12-mm-dia A1 stubs. Excess particles were removed by directing a jet of dry

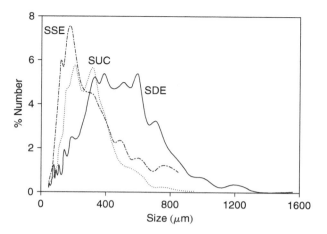

Figure 5 Particle size distribution of sucrose (SUC), single-encapsulated (SSE), and double-encapsulated powders (SDE).

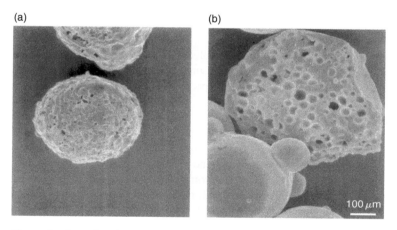

Figure 6 Scanning electron photomicrographs of (a) intact and (b) dry fractured particles containing 50% butter oil encapsulated in sucrose.

nitrogen gas (at 276 kPa) at the surface of the stub. For superficial structure of dehydrated particles, small samples of powder were immersed in ice-cold ethanol, followed by critical point-drying from liquid carbon dioxide. The dried particles were applied to A1 stubs as described above. For topographical study of internal particle structure, single selected powder particles were immersed in liquid nitrogen and manually cut with the edge of a cold stainless steel razor blade, or broken with cold tips of Dumont #5 tweezers (Electron Microscopy Sciences, Ft. Washington, PA). After "fracturing," the pieces of particles were placed in an A1 foil envelope under liquid nitrogen and transferred to a freeze drying apparatus. Dried pieces of particles were glued to A1 stubs with internal faces oriented upward. All samples were coated with a thin layer of Au in a sputter coater (Model LVC-76, Plasma Sciences, Inc., Lorton, VA).

SEM images of single-encapsulated powders (Figure 6) showed the surfaces had irregular ridges with ovoid dimpling, and the particles were somewhat agglomerated

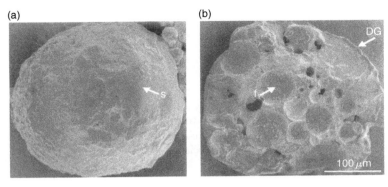

Figure 7 Scanning electron photomicrographs of (a) intact and (b) dry fractured particles containing 50% butter oil encapsulated in sucrose and double-coated with vegetable waxes (s: uneven, indented surface ridges, DC: double coat, f: fat locations).

(Figure 6[a]). The internal structures (Figure 6[b]) were consistent with our earlier results that showed unique microcavities uniformly distributed within the particle, without central voids (Onwulata et al., 1996). Surfaces resembled those described by Buma and Henstra (1971) for some spray-dried milk products and by Rosenberg and Young (1993) for milk fat encapsulated in whey proteins.

In contrast, the surface of the double-encapsulated powder, (Figure 7[a]) was very smooth, indicating the effect of the double coat. The fractured particle shows the enclosure of the single- encapsulated particle within the double capsule (Figure 7[b]). The fine particle structure of the double-encapsulated surface is a matrix formed with embedded hydrogenated stearine materials, with a mesh of linear polymeric alcoholic materials.

IV. DISCUSSION

The mode of double encapsulation (Figure 2) is that either a single large particle (Figure 2[a]) or several smaller particles (Figure 2[b]) are encased within the matrix of the double-encapsulated powder (Figure 2[c]). Therefore, a double-coated powder could be a simple or compound powder with more than one particle encapsulated. There are many possibilities for double-encapsulated powders, especially, considering the large number of compounds that could be used as the second coat. Double-encapsulated powders had properties that differed notably from those of the single-encapsulated powders or crystalline sucrose. Large particle size, lower density, and waxy surfaces impeded powder flow. However, the double-encapsulated powders had higher resistance to moisture sorption. Through choice of wall material (such as sucrose) and the powder preparation methods described here, an effective carrier for sensitive ingredients may be realized. Spray-dried powders with 50% butter oil encapsulated in sucrose were double-encapsulated by dispersion in a molten matrix of vegetable waxes (high-pressure-treated by a patented process) followed by pressure treating at 414 kPa and sieving. A 20 to 40% increase in particle size resulted from double encapsulation and powders were less flowable ($p < .01$). Scanning Electron Microscopy of double-encapsulated powders showed one or more sucrose-encapsulated particles embedded in a matrix of solid vegetable wax. Double coating reduced moisture uptake by 20%, possibly ameliorating the need for special packaging during storage. The flow and mechanical behavior of the encapsulated powders were different ($p < .05$) from the other powders studied.

REFERENCES

Andersen, S. 1995. Microencapsulated marine omega-3 fatty acids for use in the food industry. *Food Technol. Europe* December 1994/January 1995: 104–106.

Anon. 1978. Analytical Methods for Dry Milk Products. Niro Atomizer Co., Copenhagen, Denmark.

AOAC. 1984. *Official Methods of Analysis*, 14th ed. Association of Official Analytical Chemists, Washington, DC.

Buma, T.J. and Henstra, S. 1971. Particle structure of spray-dried milk products as observed by a scanning electron microscope. *Neth. Milk Dairy J.* 25: 75–80.

Carr, R.L. 1976. Powder and granule properties and mechanics. In: *Gas-Solids Handling in the Processing Industries*, Marchello, J.M. and Gomezplata, A. (Eds.). Marcel Dekker, Inc., New York.

Duxbury, D.D. and Swientek, R.J. 1992. Encapsulated ingredients face healthy future. *Food Process.* February: 38–46.

Flink, J. and Karel, M. 1970. Effects of process variables on retention of volatiles in freeze-drying. *J. Food Sci.* 35: 444–447.

Jackson, L.S. and Lee, K. 1991. Microencapsulation and the food industry. *Lebensm.-Wiss. u.-Technol.* 24: 289–297.

Konstance, R.P., Onwulata, C.I., and Holsinger, V.H. 1995. Evaluation of flow properties of spray-dried encapsulated butter oil. *J. Food Sci.* 60: 841–844.

Menting, L.C. and Hoagstad, B.J. 1967. Volatiles retention during the drying of aqueous carbohydrate solutions. *J. Food Sci.* 32: 87–90.

Moreyra, R. and Peleg, M. 1980. Compressive deformation patterns of selected food powders. *J. Food Sci.* 45: 864–868.

Onwulata, C.I., Smith, P.W., Craig J.C. Jr., and Holsinger, V.H. 1994. Physical properties of encapsulated spray-dried milkfat. *J. Food Sci.* 59: 316–320.

Onwulata, C.I. and Holsinger, V.H. 1995. Thermal properties and moisture sorption isotherms of spray-dried encapsulated milkfat. *J. Food Process. Pres.* 19: 33–51.

Onwulata, C.I., Smith, P.W., and Holsinger, V.H. 1995. Flow and compaction of spray-dried powders of anhydrous butter oil and high melting milkfat encapsulated in disaccharides. *J. Food Sci.* 60: 836–840.

Onwulata, C.I., Smith, P.W., Cooke, P., and Holsinger, V.H. 1996. Particle structures of encapsulated milkfat powders. *Lebensm.-Wiss. u.-Technol.* 29: 163–172.

Peleg, M. 1977. Flowability of food powders and methods for its evaluation — a review. *J. Food Process. Eng.* 1: 303–328.

Pothakamury, U.R. and Barbosa-Canovas, G.V. 1995. Fundamental aspects of controlled release in foods. *Trends Food Sci. Technol.* 6: 397–406.

Redding, B.K. 1990. Apparatus and method for making microcapsules. U.S. Patent 4,978,483.

Redding, B.K. 1993. Method for inducing transformations in waxes. U.S. Patent 5,209,879.

Rockland, L.B. and Nishi, S.K. 1980. Influence of water activity on food product quality and stability. *Food Technol.* 34: 42–59.

Rosenberg, M. and Young, S.L. 1993. Whey proteins as microencapsulating agents. Microencapsulation of anhydrous milkfat-structure evaluation. *Food Microstruct.* 12: 31–41.

Sone, T. 1972. *Consistency of Foodstuffs*. D. Reidel Publishing Co., Dordrecht, The Netherlands, pp. 133–142.

White, G.W., Bell, A.V., and Berry, G.K. 1967. Measurement of the flow properties of powders. *J. Food. Technol.* 2: 45–52.

Wagner, L.A. and Warthesen, J.J. 1995. Stability of spray-dried encapsulated carrot carotenes. *J. Food Sci.* 60: 1048–1053.

20

Milk Fat Powders Encapsulated in Sugars and Starches

C. I. Onwulata
U.S. Department of Agriculture
ARS, Eastern Regional Research Center
Wyndmoor, Pennsylvania

Contents

I. INTRODUCTION

Encapsulation, the process of coating or enclosing any matter in a capsule with other materials, is used widely in drug delivery, but is gaining wider application in many industries

495

including the food industry. The food industry uses encapsulated powders to preserve delicate ingredients in the form of powders. One such example is milk fat. The food capsules can be engineered for targeted delivery or release under specific conditions of temperature, shear, or moisture, or under acidic or alkaline environment. In this chapter, the process used to protect milk fat or butter oil is described, an encapsulation process that resulted in shelf-stable powders.

A.   Definitions

For encapsulating or coating one material over another, there are various means employed by manufacturers, depending on the material, the cost, the final size of particles desired, and the form, either powder or gel. Some of these processes are: spray drying, fluidized-bed coating, liquid coating or dry powder coating, coacervation, and others. While most of these processes are suitable for pharmaceuticals and other industries because of their high costs, food powder encapsulation is done mostly by spray drying because it is cheaper (Clark, 2002).

Spray drying is a process of reducing small emulsion droplets into powders. Biphasic emulsions, for instance, starch and butter oil, are homogenized and atomized (sprayed) into a chamber of heated air. In the chamber, the atomized small droplets are dried in the hot air current; dry powders are separated from the hot air through a cyclone. In the atomizing process, small spheres are formed, and these dry as fine rounded particles. If the encapsulated powder is emulsified, depending on the core loading, in the atomizer, the hydrophobic fat core folds inside the droplet while the more hydrophilic starch or protein coating forms the shell.

Fluid bed encapsulation of particles can be with liquid or very fine particles of coating material. This is accomplished by creating a fluid bed of the core particles by forcing air currents into a closed chamber, or in a drum rotating at high speed, then spraying the new coat material on the fluidized particles. Enough time is allowed for each particle in the fluid bed to be coated and dried in the chamber.

Coacervation is the process of inducing a more soluble polymer to come out of solution as two polymers by manipulating the ionic concentration of the desired coat material. Beads formed by the coat material contain the less soluble polymer as the core. The beads are then harvested and dried. Typical use of these powders in food are in protecting vitamins and minerals, masking off-flavors and tastes, delayed release of the core, uniform distribution of minute components, and preventing heat damage during cooking. The spray drying process was used for the manufacture of encapsulated butter oil powders described in this chapter.

B.   Properties of Encapsulated Butter Oil

Encapsulated powders are not always free-flowing dried powders. Some encapsulated powders are difficult to handle because of changes in bulk properties of the powders, that is, they are less dense. This is particularly true of encapsulated butter oil where the encapsulated powders show a propensity for stickiness and lumping. The flow problems result from inter/intraparticular forces (hydrophobic/hydrophilic), influence of powder particle size and shape, and from the effect of the moisture. Conditioners or anticaking agents are used to enhance powder flow. They work by reducing interparticle forces and minimizing cohesiveness (Peleg et al., 1973). Adding flow conditioners changes the bulk properties of encapsulated powders, making them less dense and more flowable. However, with some encapsulated powders, adding flow conditioners causes the smaller particles to adhere to the surfaces of larger particles thereby forming lumps. The behavior of mixed particles, density, and compressibility, depend on particle sizes (Barbosa-Canovas *et al.*, 1987).

C.   Manufacture of Encapsulated Butter Oil Powders

Spray drying of butter with functional coatings such as starch, maltodextrin, or gums enhances stability of the dry powders as microcapsules are formed to protect the milk fat from oxidative deterioration during storage. Such powders are easily recombined or incorporated as ingredients into many food systems (de Man, 1984; Frede and Ehlers, 1991). The choice of coating is critical as the material will influence emulsion stability before drying, flowability through the atomizer, and mechanical stability after drying. For example, flavorings and citrus oils have been encapsulated in food gums and modified starches that behave like gums (Tripp *et al.*, 1971; Bangs and Reineccius, 1990). For the production of butter powders, the solids-no-fat matrix may consist of milk protein products such as nonfat dry milk, sodium caseinate or whey proteins, various sugars, starches, gums, emulsifying agents, and sodium citrate (Frede and Ehlers, 1991). Powder stickiness and lumpiness are directly related to emulsion stability; the more stable the emulsion the more free flowing is the encapsulated powder.

D.   Efficiency of Encapsulated Butter Oil

The shelf life of encapsulated butter oil is limited. Anhydrous milk fat develops oxidative off-flavor when stored at 25°C or higher within two to three months when exposed to air (Hamm *et al.*, 1968; Kehagis and Radema, 1973; Keogh and Higgins, 1986). When milk is dried and powdered, the milk fat oxidation rate is slower. Milk powder is stable for 9 to 12 months when stored in air at 30 and 35°C, respectively (Baldwin *et al.*, 1991; van Mil and Jans, 1991). It is believed that the antioxidant properties of the sulfhydryl groups and the presence of Maillard reaction products in spray-dried milk is the cause of improved stability. Encapsulation is known to improve the stability of oil products by inhibiting oxidation (Beatus *et al.*, 1985; Anandaraman and Reineccius, 1986), and the proposed mechanism for this improved stability is the exclusion of oxygen by the coating.

Recently, it was reported that oxygen uptake of butter oil encapsulated in whey protein isolate and whey protein isolate: lactose was negligible (Moreau and Rosenberg, 1996). The mechanisms for the stability of the milk fat were the oxygen barrier and antioxidant properties of whey proteins. In an effort to improve the functionality of butter oil and to increase its storage life, anhydrous butter oil was encapsulated in sucrose, flour, or modified starch (Onwulata *et al.*, 1994).

II.   MATERIALS

Anhydrous butter oil was purchased from a commercial source (Land o' Lakes, Minneapolis, MN). Coatings selected were sucrose (Domino's, Domino Sugar Corp., New York, NY); N-starch, all-purpose flour (ADM Milling Co., Kansas City, MO); and M-starch, modified starch (Capsul, National Starch and Chemical Co., Bridgewater, NJ). An emulsifying agent, comprising mono- and di-glycerides (American Ingredients Co., Kansas City, MO), was also used. Nonfat dry milk served as the protein source (Maryland & Virginia Milk Producers Association, Inc., Laurel, MD). Sample preparation was carried out by following a full $2 \times 3 \times 3$ factorial design, completely randomized and replicated. Butter oil was emulsified at three fat levels, 40, 50, and 60%, with three coating materials, sucrose, unmodified starch (all-purpose flour) or modified starch, added emulsifier, and nonfat dry milk.

A. Manufacture of Encapsulated Butter Oil Powders

Encapsulated powders were formulated to have 40 to 60% milk fat, 5% emulsifier, and 5% skim milk powder. The processing sequence was as follows: the chosen coating was dry-blended with nonfat dry milk solids. The blended mixture was dispersed in water to form pasty slurry of approximately 25% total solids. The anhydrous butter oil and the emulsifier were heated to 23.9°C. The two blends were combined (40% total solids) and mixed for 5 min with a milk stirrer, after which the slurry temperature was slowly raised to 62.8°C with constant stirring, and homogenized at 17.2 MPa with a Manton–Gaulin Triplex homogenizer (Model 100 DJF3855, APV Gaulin, Inc., Everett, MA), followed by spray drying in a compact dryer (APV Crepaco, Inc., Attleboro Falls, MA) at an inlet temperature of 193.3–196.1°C and an outlet temperature of 82.2–87.8°C. Powders were produced batch-wise, removed from the dryer after 30 min and stored at 4°C. When unmodified starch (all-purpose flour) was used as the coating, it was necessary to homogenize at 10.3 MPa and 54.4°C to accommodate its pasting properties. The milk protein content ranged from 2% for powders made with butter oil to 4% for powders made with cream as the fat source. Spray-dried encapsulated milk fat powders were prepared from stable emulsions containing 40 to 60% milk fat and carbohydrate matrices. Moisture content of the spray-dried powders was from 1 to 4%. Lowest free fat content (<10%) was found in powders with 40% fat, encapsulated in sucrose. Angles of repose ranged from 37 to 46°, and could be correlated with powder flow ($p = 0.01$). Bulk density was dependent on the coating and declined with increasing fat content. Particle size distribution ranged from 20 to 120 μm with 80% of the particles below 100 μm. Powders with best physical properties were made with 40 to 50% butter oil encapsulated in sucrose.

III. CHARACTERIZATION

A. Particle Structure

Particle structure of the microcapsules were evaluated with an optical microscope equipped with optics for phase contrast (Olympus microscope, model BH2; Olympus Corp., Lake Success, NY). The particle structure of fractured microcapsules containing butter oil is shown in Figure 1. The structures were distinct and dependent on the carbohydrate matrix. Mean particle size as measured by image analysis, increased as the amount of encapsulated milk fat increased from 40 to 60% (Onwulata et al., 1994). Clustering of particles was the result of surface milk fat on the particles, which was influenced by the loading of milk fat encapsulated to coating.

B. Moisture

Moisture was determined by an AOAC (1984) method by drying under vacuum for 4 h at 102°C. Moisture content of the spray-dried powders varied from 1 to 4% (Table 1), with highest moisture (3.76%) in the sample prepared with unmodified starch and anhydrous butter oil. The amount of fat and the type of coating in the emulsion significantly affected the moisture content. The affinity for water was largely dependent on the coating used, but no powder had moisture content greater than 4%. Nonfat dry milk and whole milk powders

(a)   (b)   (c)

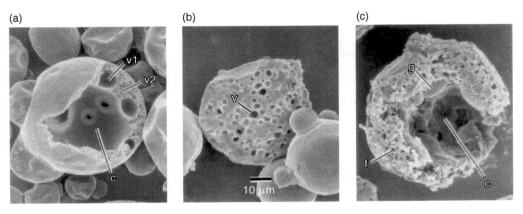

Figure 1   Scanning electron micrographs of microcapsules of butter oil in carbohydrate matrices-internal matrices. (a) Sucrose with 40% (b) M-Starch with 40% and (c) N-Starch with 40%

Table 1   Moisture Content of Spray-Dried Powders[a]

	BO (% moisture) for wall materials		
Milk fat	Sucrose	M-Starch	N-Starch
40	1.97 (0.7)	1.44 (0.2)	3.10 (0.6)
50	1.61 (0.2)	1.21 (0.4)	3.76 (0.5)
60	1.19 (0.6)	1.00 (0.3)	1.22 (0.1)

Note: M-Starch = Modified starch. N-Starch = All-purpose flour. BO = Butter oil. Standard deviation values is given in parenthis.

[a] Moisture on wet basis.

have moisture contents ranging from 2 to 4%. The moisture content is critical in dehydrated products, because it has long been known that a small residue of water appears to be a major factor in inhibiting fat oxidation (Koch, 1962). No optimum moisture content for butter powders is known.

C.   Bulk Density

Bulk densities of all powders were determined by dividing the weight of powder (g) contained in a 200-ml stainless steel cylinder (A/S Niro Atomizer, Copenhagen, Denmark) by its volume in cm³. Packed bulk densities were calculated from the weight of powder contained in the cylinder after being tapped 100 times. Density measurements (g/cm³) were done in triplicate.

Bulk densities of the spray-dried powders were dependent on the coating used. The highest bulk densities were found in powders made with N-starch, followed by sucrose and modified starch. A decrease in bulk density was observed for most powders when fat content increased to 50%. Occlusion of air within the microcapsules was determined by microscopic examination. Occlusion of air bubbles in the particles lowered bulk density. Bulk density of

Table 2 Flow Properties of Spray-dried
Powders[a]

| Milk fat | BO (g/sec) for wall materials | | |
	Sucrose	M-Starch	N-Starch
40	5.71 (0.01)	0.18 (0.08)	0.27 (0.03)
50	4.21 (0.40)	0.20 (0.04)	0.45 (0.08)
60	1.15 (0.28)	0.29 (0.09)	0.99 (0.22)

Note: M-Starch = Modified starch. N-Starch = All-purpose flour. BO = Butter oil. Standard deviation is given in parenthesis. Moisture content of powders was not adjusted.

[a] Flow Rates.

encapsulated powders is an important transport and handling property because the particles can become compacted and form large lumps, causing flow problems.

D. Flow Properties

Flow characteristics were evaluated by permitting 80 g of powder to flow through a funnel to form a heap; angle of repose was calculated as $[\Theta = \cotan\ h/r]$ from the dimensions of the pile where h = height of the powder pile and r = radius of the base of the pile. The angle of repose, defined as the base angle formed when a given weight of powder flows through a funnel of known dimensions to form a pile, was measured by permitting 80 g to flow through funnels of outlet diameters 1.27 to 2.54 cm with gentle shaking (FMC/Synthron, Homer, PA); time of flow was recorded as flow rate (Sjollema, 1963). Relative flow rate was calculated as powder weight (g) divided by time (sec). Measurement of powder flow, compared to the flow of nonfat dry milk at 3.0 g/sec, showed that butter oil powders encapsulated with sucrose were more free flowing than the others (Table 2); as fat content of the powder increased, flowability decreased sharply. Flowability was also influenced by the fat loading; powders with higher levels of unencapsulated fat on the surface tended to stick together and form lumps, which impeded flow.

An average angle of repose has been reported for nonfat dry milk (43°) (Sjollema, 1963). Angles of repose measured for the powders ranged from 37 to 46° with a mean angle of 40° (Table 3). It has been suggested that powders with angles of repose less than 35° should be considered to be free flowing, those with angles between 35 and 45° as cohesive, while powders with angles of repose greater than 55° as having little or no flow (Sjollema, 1963). Encapsulated butter oil powders with sucrose as coating were free flowing. Sources of stickiness were from uncoated fat on the surface of the powder particles (Peleg et al., 1973; Peleg, 1983). Encapsulated powders with high-fat levels have flow properties different from those of other food powders. These powders, with angles of repose between 37 and 42° were relatively free flowing in spite of particle aggregation due to uncoated fat.

E. Particle Size Distribution

Particle size distribution was estimated by passing 100 g of each powder through a series of sieves with screen openings ranging from 100 to 500 μm. The stack was shaken and tapped

Table 3   Angle of Repose of Spray Dried
Powders[a]

	BO (degrees) for wall materials		
Milk fat	Sucrose	M-Starch	N-Starch
40	45.6 (1.2)	39.7 (2.3)	38.4 (1.2)
50	38.0 (1.4)	40.9 (1.4)	38.5 (1.7)
60	40.8 (4.9)	42.0 (1.9)	40.5 (2.6)

Note: M-Starch = Modified starch. N-Starch = All-
purpose flour. BO = Butter oil. Standard
deviation given in paranthesis.

[a] Angle of repose.

with a Rotap shaker (Tyler Co., Cleveland, OH) for 5 min. Powder distribution by weight was recorded and cohesiveness was estimated by the percentage of powder that aggregated or did not pass through the 500 μm sieve. The particle distribution for powders with butter oil or cream showed a distribution range of 20 to 120 μm, with 80% of the particles less than 100 μm in diameter (Onwulata *et al.*, 1993). The powders were cohesive, forming large aggregates, their number increasing with increasing fat content. Sucrose containing powders were relatively free-flowing and nonlumpy, compared to the other powders. The particle clusters formed were caused by large amounts of uncoated fat that acted as a binder, increasing the cohesiveness. The sieve size distribution was directly related to the fat content of the powder; as the percent fat increased, the uncoated fat increased, and cohesiveness (stickiness) increased, producing lumpy chunks.

F.   Storage of Encapsulated Butter Oil

Twenty-one samples (approximately 140 g each) of each powder batch were packaged on the day after encapsulation was completed in ZipLoc® heavy duty freezer bags (DowBrands L.P. Chemical Co., Indianapolis, IN) which were made of high-density polyethylene that has steady state transmission rates for water and oxygen of 1.0 to 1.5 g/100 in^2/24 h and 420 ml/100 in^2/24 h, respectively, and used for determination of peroxide value and 2-thiobarbituric acid (TBA) value. In addition, 21 (1 g) samples of each encapsulated butter powder batch were sealed in crimp top glass vials (22 × 38 mm, 9 ml vol) with polytetrafluoroethylene butyl septa and aluminum caps. These 126 samples were used to determine the oxygen uptake of the butter powders during storage. Each vial was tested once. Samples were stored in the dark at room temperature (22°C), in a freezer (−20°C), or in an incubator (40°C) for up to 6 months and were removed at one month intervals for testing. Samples of each batch were analyzed immediately for oxygen levels after spray drying to establish initial value for oxygen. Oxygen levels over a six month period is presented in Figure 2.

1.   Oxygen Assay

The oxygen content of each of the vials was analyzed by withdrawing a 1 ml sample of the air in the vial with a 2.5 ml gas-tight syringe (Hamilton Series 1000, Alltech Associates, Inc. Deerfield, IL) and injecting it into a Gow-mac Series 580 gas chromatograph (Gow-mac Inc., Bridgewater, NJ) equipped with a CTR 1 column (Alltech, Alltech

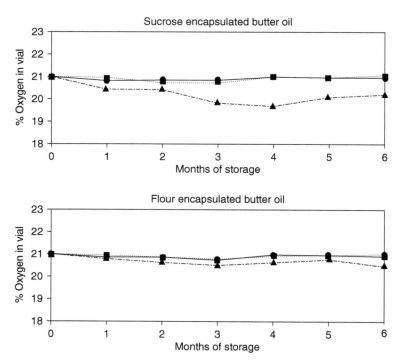

Figure 2 •–Butter powder stored in freezers (−20°C), ■- - - butter powder stored at room temperature (22°C), ▲ - - - butter powder stored in cubators (±40°).

Associates, Deerfield, IL), 6 ft × $\frac{1}{4}$ in outer column packed with activated molecular sieve and a 6 ft × $\frac{1}{8}$ in inner column packed with porous polymer mixture. This column configuration allows for the simultaneous analysis of O_2, N_2, and CO_2. The helium flow rate was 120 ml/min and the equipment was run at ambient temperature. Chromatograms were integrated and displayed on an HP 3396A Integrator (Hewlett-Packard Co., Avondale, PA). With this system, only four peaks appeared after injection: an injection peak with a retention time (RT) of 0.14–0.2 min, a CO_2 peak with a RT of 0.45–0.6 min, an O_2 peak with a RT of 1.6–2.0 min, and a N_2 peak with a RT of 2.5–3.4 min. No other large peaks were detected throughout the study and only one injection/vial was made. Oxygen content is reported as percent of detected peaks excluding the injection peak. The standard curves used for sucrose butter oil were constructed by adding 1.5 ml of 5% TCA containing 208 mg sucrose/ml to 1 ml of MDA solutions of varied concentrations; for flour butter oil, by adding 1.5 ml of 5% TCA. Individual curves were constructed for each type of sample because each type of sample formed complexes and had very different sugar contents.

2. Oxidative Stability of Powders

The oxygen content of butter oil powder over the entire storage time is shown in Figure 1. Analysis of Variance (ANOVA) of the data showed that the temperature of storage had significant effects on the O_2 content. The Bonferroni T test for type of powder confirms that the vials containing sucrose encapsulated butter oil had significantly ($P < 0.05$) lower O_2 content than did those with flour encapsulated butter oil. When only temperature of storage was considered, vials stored at 40°C also had significantly ($P < 0.05$) lower O_2 content.

Table 4 Peroxide Values for Encapsulated Powder in Milliequivalents of Peroxide/ Kg of Powder

Months of storage	Flour encapsulated butter oil temperature of storage			Sucrose encapsulated butter oil temperature of storage		
	−20°C	22°C	40°C	−20°C	22°C	40°C
0	0.131[ab]				0.115[abcdef]	
1	0.128[abc]	0.113[abcdef]	0.115[abcdef]	0.108[abcdef]	0.112[abcdef]	0.121[abcd]
2	0.148[a]	0.125[abcd]	0.125[abcd]	0.121[abcd]	0.135[ab]	0.115[abcdef]
3	0.060[ghijk]	0.058[ghijk]	0.056[hijk]	0.052[kj]	0.054[ijk]	0.047[k]
4	0.097[bcdefghij]	0.082[defghijk]	0.072[fghijk]	0.102[bcdefg]	0.110[abcdef]	0.124[abcd]
5	0.101[bcdefgh]	0.117[abcdef]	0.080[defghijk]	0.083[cdefghijk]	0.093[bcdefghij]	0.118[abcde]
6	0.080[defghijk]	0.081[defghijk]	0.072[efghijk]	0.074[efghijk]	0.099[bcdefghi]	0.097[bcdefghij]

Note: Peroxide values with superscripts in common are not significantly different by the Bonferroni (Dunn) T test.

While the differences in percentage of O_2 in headspace were statistically significant, they were actually relatively small compared to those reported for unencapsulated butter oil. Moreau and Rosenberg (1996) reported that unencapsulated butter oil had only 7% O_2 in headspace after three months at 40°C without light and that their encapsulated butter oil showed no significant decreases in O_2 content during storage. The other indicator of oxidation, peroxide values, of the butter oil powders are shown in Table 4. ANOVA for peroxide content showed that the type of coating and the temperature of storage alone did not significantly affect the peroxide content; however, storage time did show a significant effect with temperature of storage and with storage time on the peroxide content. Generally, as the storage time increased, the peroxide content decreased especially for flour encapsulated butter oil. As the storage temperature increased, the peroxide content decreased for flour encapsulated butter oil, while the opposite was generally true for sucrose encapsulated butter oil. A decrease in peroxide values was noted by Kehagis and Radema (1973) for non-encapsulated butter stored at −10 and 0°C with or without the addition of antioxidants, and these peroxide values were in the range that we report. At the higher storage temperatures, they noted rapid increases in peroxide value within four months even in the presence of antioxidants or when air was excluded by nitrogen. Overall, the peroxide content of our samples never reached a level that would indicate any active oxidation. The mean peroxide values throughout our storage study were 0.098 and 0.095 meq O_2/kg powder for sucrose encapsulated butter oil and flour encapsulated butter oil, respectively, somewhat less than the International Dairy Federation (1977) standards for anhydrous butter oil of 0.2 meq O_2/kg butter (Keogh and Higgins 1986). van Mil and Jans (1991) reported that the best quality dried whole milk had peroxide values less than 0.3 meq O_2/kg fat, about the range of peroxide values we have for our encapsulated butter oil and recently, Christensen and Holmer (1996) reported as "low," peroxide values of up to 1 meq/kg for butter stored at 20°C in the dark.

IV. SUMMARY

The main purpose of preparing spray-dried fats is to enhance their handling properties, for example, storage, transport, and blending with nonfat ingredients. Milk fat must be powdered by using a carrier because it contains appreciable amounts of low melting triglycerides. As the fat content increases, the choice of carrier constituents becomes more critical. Working in the medium fat range, allowed for greater ease in processing, and also reduced fat products are more desirable as ingredients in "light" processed foods of reduced calorie content. Relatively free-flowing powders containing 40 to 60% milk fat were successfully spray-dried, with a variety of coatings. The best powders contained sucrose as the coating. Such products can be used readily as food ingredients in processed foods where sweetness is required. Their long-term storage stability has been reported for up to 12 months (Moreau and Rosenberg, 1996).

REFERENCES

Anandaraman, S. and Reineccius, G.A. 1986. Stability of encapsulated orange peel oil. *Food Technol.* 40(11): 88–91.

Anonymous 1978. Analytical Methods for Dry Milk Products. Niro Atomizer Co., Copenhagen, Denmark.

Anonymous 1989. Butter: The World Market. International Dairy Federation, Brussels, Belgium.

AOAC. 1984. Official Methods of Analysis, 14th ed. Association of Official Analytical Chemists, Washington, DC.

Baldwin, A.J., Cooper, H.R. and Palmer, K.C. 1991. Effect of Pre heat treatment and storage on the properties of whole milk powder: Changes in sensory properties. *Neth. Milk Dairy J.* 45: 97–116.

Bangs, W.E. and Reineccius, G.A. 1990. Characterization of selected materials for lemon oil encapsulation by spray-drying. *J. Food Sci.* 55: 1356–1358.

Barbosa-Canovas, G.V., Malave-Lopez, J. and Peleg, M. 1987. Density and Compressibility of selected food powder mixtures. *J. Food Processing Eng.* 10: 1–19.

Beatus, Y., Raziel, A., Rosenberg, M., and Kopelman, I.J. 1985. Spray-drying microencapsulation of paprika oleoresin. *Lebensm.-Wiss. U.-Technol.* 18: 28–34.

Christensen, T.C. and Holmer, G. 1996. Lipid oxidation determination in butter and dairy spread by HPLC. *J. Food Sci.* 61: 486–489.

Clark, J.P. 2002. Food encapsulation: capturing one substance by another. *Food Technol.* 56(11): 63–65.

deMan, J.M. 1984. Butter as an ingredient for processed cereal products. In *Dairy Products for the Cereal Processing Industry*, Vetter, J.L. (Ed.). American Association of Cereal Chemists, St. Paul, MN, pp. 155–165.

Frede, E. and Ehlers, F. 1991. High-fat powdered products. In Bulletin No. 290. International Dairy Federation, Brussels, Belgium, pp. 26–27.

Hamm, D.L., Hammond, E.G., and Hotchkisss, D.K. 1968. Effect of temperature on rate of autoxidation of milk fat. *J. Dairy Sci.* 51: 483–491.

Kehagis, C. and Radema, L. 1973. Storage of butteroil under various conditions. *Neth. Milk Dairy J.* 27: 379–398.

Keogh, M.K. and Higgins, A.C. 1986. Anhydrous milk fat 1. Oxidative stability aspects. *Ir. J. Food Sci. Technol.* 10: 11–22.

Koch, R.B. 1962. Dehydrated foods and model systems. Ch. 13, In *Symposium on Foods: Lipids and their Oxidation*, Schultz, H.W., Day, E.A., and Sinnhuber, R.O. (Eds.). The Avi Publishing Co., Inc., Westport, CT, pp. 230–254.

Moreau, D.L. and Rosenberg, M. 1996. Oxidative stability of anhydrous milkfat microencapsulated in whey proteins. *J. Food Sci.* 61: 39–43.

Onwulata, C.I., Smith, P.W., Craig, J.C., Jr., and Holsinger, V.H. 1994. Physical properties of encapsulated spray-dried milkfat. *J. Food Sci.* 59: 316–320.

Onwulata, C.I., Smith, P.W., Cooke, P.H., and Holsinger, V.H. 1996. Particle structures of encapsulated milkfat powders. *Lebensm.-Wiss. U.-Technol.* 29: 163–172.

Patel, A.A., Frede, E., and Buchheim, W. 1987. Physical and technological aspects of the manufacture of butter powder. I. Effects of proteins, glycerol monostearate and tri-sodium citrate on the structural stability. *Kieler Milchw. Forsch.* 39: 191–200.

Peleg, M. 1983. Physical characteristics of food powders. Ch. 10, In *Physical Properties of Foods*, Peleg, M. and Bagley, E.B. (Eds.). AVI Publishing Co., Inc., New York, pp. 293–321.

Peleg, M., Mannheim, C.H., and Passy, N. 1973. Flow properties of some food powders. *J. Food Sci.* 38: 959–964.

SAS Institute, Inc. 1991. SASJ User's Guide: Basics, Version 6.03 Edition, SAS Institute, Inc., Cary, NC.

Sjollema, A. 1963. Some investigations on the free-flowing properties and porosity of milk powders. *Neth. Milk Dairy J.* 17: 245–258.

Tripp, R.C., Amundson, C.H., and Richardson, T. 1971. Spray-dried high-fat powders. *J. Dairy Sci.* 49: 694–695.

van Mil, P.J.J.M. and Jans, J.A. 1991. Storage stability of whole milk powder: effects of process and storage conditions on product properties. *Neth. Milk Dairy J.* 45: 145–167.

Index